Bo Guang Zhao • Kazuyoshi Futai • Jack R. Sutherland • Yuko Takeuchi
Editors

Pine Wilt Disease

Bo Guang Zhao • Kazuyoshi Futai •
Jack R. Sutherland • Yuko Takeuchi
Editors

Pine Wilt Disease

With 139 Figures, Including 26 in Color

Springer

Bo Guang Zhao, Ph.D.
Professor
College of Forest Resources and Environment
Nanjing Forestry University
159 Longpan Road
Nanjing 2110037, People's Republic of China

Kazuyoshi Futai, Ph.D.
Professor
Laboratory of Environmental Mycoscience
Graduate School of Agriculture, Kyoto University
Kitashirakawa Oiwake-cho, Sakyo-ku
Kyoto 606-8502, Japan

Jack R. Sutherland, Ph.D.
Former Principal Research Scientist
Pacific Forestry Centre
506 West Burnside Road, Victoria, British Columbia
V8Z 1M5, Canada

Yuko Takeuchi, Ph.D.
Assistant Professor
Laboratory of Environmental Mycoscience
Graduate School of Agriculture, Kyoto University
Kitashirakawa Oiwake-cho, Sakyo-ku
Kyoto 606-8502, Japan

Library of Congress Control Number: 2008926209

ISBN: 978-4-431-75654-5 Springer Tokyo Berlin Heidelberg NewYork
e-ISBN: 978-4-431-75655-2

Springer is a part of Springer Science+Business Media
springer.com

Printed in Japan

Typesetting: SNP Best-set Typesetter Ltd., Hong Kong
Printing and binding: Kato Bunmeisha, Japan

Printed on acid-free paper

We dedicate this book to Dr. Y. Tokushige, Dr. T. Kiyohara, Dr. W.R. Nickle, Dr. Y. Mamiya, and all other pioneers in this field.

Foreword

At the turn of the 20th Century, Japanese foresters began to notice the widespread mortality of pines around the port city of Nagasaki. For several decades the mortality spread northward through the main island of Honshu. At the time was thought that wood boring beetles, prevalent in the dead trees, were the cause of the mortality. The epidemic of pine mortality continues to this day; however, the nature of the investigation into the epidemiology of the disease changed with the determination by Kiyohara and Tokushige in 1971, that the causal agent to the rapid wilting of Japanese red and Japanese black pines was not the beetle but rather a species of nematode vectored by the beetle.

This discovery was not anticipated as nematodes were not considered the casual agents of tree disease. Indeed, a quick review of forest pathology texts from that time would provide a paucity of significant information about nematodes. The only nematode-caused tree disease to be found in the literature of the time was red ring disease of coconut palm caused by the weevil-vectored nematode, *Bursaphelenchus cocophilus.*

With the identification of the pine wood nematode, serious research on the biology and ecology of pine wilt began in Japan. A few years later, at the suggestion of visiting researchers from Japan, pine wood nematodes were recovered from the wood of a dead Austrian pine in Columbia, Missouri, USA. A first thought to be an introduction of the nematode into North America it was soon realized that the nematode was native to North America and was likely introduced into Japan in the early 1900s.

Human activity has resulted in the intentional and accidently distribution of plant and animal species beyond their historical geographical distributions. Most unintentional introductions fail to establish in their new environment; however, there have been many striking examples of ecosystems being threaten by the establishment of an exotic organism. Among the dramatic examples in forest ecosystems are chestnut blight, Dutch elm disease, gypsy moth and the hemlock woolly adelgid. The introduction of the pine wood nematode into Japan and its subsequent spread to China, Korea and more recently Portugal is yet another example.

The research community continues to make impressive strides in understanding the complex relationships that govern the interactions between the nematode, the beetle, the tree and associated microorganisms. The system is a beautifully synchronized series of interactions that involves numerous cross species signals that act to synchronize tree decline, beetle colonization, nematode development and nematode attraction to newly-formed adult beetles.

This volume represents the state of knowledge of the pine wilt system in eastern Asia and the efforts to control its spread. As such it will become a valuable resource for researchers, resource managers and students.

Marc J. Linit
Professor of Entomology and Associate Dean for Research and Extension
University of Missouri, Columbia, Missouri, USA
IUFRO Coordinator
Working Party 7.02.10—Pine Wilt Disease

Preface

The year 1905 witnessed the first description of pine wilt disease at Nagasaki on the island of Kyushu in Japan, but the pine wood nematode, *Bursaphelenchus xylophilus*, was not identified as the causal agent of the disease until 1971, when Kiyohara and Tokushige demonstrated the pathogenicity of the nematode by inoculating 25-year-old *Pinus densiflora*. The analysis of DNA from several studies indicates that the pine wood nematode was introduced to Japan from the United States. Since the first outbreak on the island of Honshu at Aioi in 1914, the disease has been found in North America (Canada, the United States, and Mexico), in East Asia (Japan, Korea, and China), and in Europe (Portugal). It has now become a worldwide threat to forest ecology and international trade.

Thanks to vigorous work and unremitting efforts devoted to research into pine wilt disease by scientists around the world, we now have a better understanding of the disease, e.g., systematics, diagnostics of the nematode, and the biology, ecology and modeling of the insect vectors, physiology and resistance of the host tree, and interactions between the nematode and its associated microbe. However, more mysteries behind the symptoms remain to be explained. For example, many scientists agree that toxins are involved in the disease, but the origin or origins of certain toxins are not entirely clear. Reports have shown that axenic *B. xylophilus* lost its pathogenicity and did not result in wilting symptoms when host seedlings were inoculated with it. Meanwhile, inoculation with a mixture of pathogenic bacteria carried by the nematode and the axenic nematode resulted in wilt and death of seedlings in *Pinus thunbergii*. Moreover, mutualistic symbiosis between the nematode and the bacteria carrying it was discovered. A hypothesis was then proposed, namely, that pine wilt disease was a complex one caused by both the nematode and its associated pathogenic bacteria, which sheds light on a new direction for research. I think better understanding of the etiology of the disease will pave the way to stopping its spread. Therefore, further basic research is needed in interactions between *B. xylophilus*, the bacterial pathogens, and the molecular biology of pathogenicity in the disease. However, some scientists, on the basis of their own experiments, hold opposing opinions, contrary to the above-mentioned hypothesis. I think that is quite normal. Debate among those with differing points of view will accelerate the development of a better understanding of certain scientific issues. We welcome critical reviews from readers of all the contents of this book.

When Dr. K. Futai paid a visit to China in 2006, we discussed the possibility of compiling a book on pine wilt disease for the purpose of providing useful assistance for researchers, university graduates, and foresters, and both of us believed that it was the right time to compile a book integrating recent research progress and critical assessment of the state of our understanding of the disease. Since then, many prestigious scientists knowledgeable about the disease and the research have been invited to join our efforts by contributing their intelligence and wisdom to the compilation of the book.

I would like to thank all the contributors to the book whether or not they are listed as authors, and I appreciate those who conducted most of the field work and scientific instrument analyses. Special acknowledgments are also given to the sponsors and providers of funds for their support for the research resulting in publication of this book. My project team thanks the Natural Science Foundation of China (Key Projects Nos. 30430580 and 30030110), the State Forestry Administration, the People's Republic of China (Project No. 20070430), and the Nanjing Forestry University for their financial support of our research.

Nanjing, May 2008

Professor Bo Guang Zhao
Nanjing Forestry University
People's Republic of China

Contents

Contributors

Takuya Aikawa (Chapter 13)
Tohoku Research Center, Forestry and Forest Products Research Institute, 92-25 Nabeyashiki, Shimo-Kuriyagawa, Morioka 020-0123, Japan

Kazuyoshi Futai (Chapters 1, 2, 7)
Laboratory of Environmental Mycoscience, Graduate School of Agriculture, Kyoto University, Kitashirakawa Oiwake-cho, Sakyo-ku, Kyoto 606-8502, Japan

Koichi Hasegawa (Chapter 11)
Laboratory of Environmental Mycoscience, Graduate School of Agriculture, Kyoto University, Kitashirakawa Oiwake-cho, Sakyo-ku, Kyoto 606-8502, Japan
Institute for Biological Function, Chubu University, 1200 Matsumoto, Kasugai 487-8501, Japan

Naoto Kamata (Chapter 32)
University Forest in Chichibu, University Forests, Graduate School of Agricultural and Life Sciences, The University of Tokyo. 1-1-49 Hinodamachi, Chichibu 368-0034, Japan

Natsumi Kanzaki (Chapters 8, 9, 14)
Forest Pathology Laboratory, Forestry and Forest Products Research Institute, 1 Matsunosato, Tsukuba 305-8687, Japan

Taisei Kikuchi (Chapter 10)
Forest Pathology Laboratory, Forestry and Forest Products Research Institute, 1 Matsunosato, Tsukuba 305-8687, Japan

Keiko Kuroda (Chapters 20, 21)
Kansai Research Center, Forestry and Forest Products Research Institute, Momoyama, Fushimi, Kyoto 612-0855, Japan

Rong Gui Li (Chapter 24)
Department of Biology, Qingdao University, Qingdao 266071, People's Republic of China

Noritoshi Maehara (Chapters 26, 29, 30)
Tohoku Research Center, Forestry and Forest Products Research Institute, 92-25 Nabeyashiki, Shimo-Kuriyagawa, Morioka 020-0123, Japan

Johji Miwa (Chapter 11)
Graduate School of Bioscience and Biotechnology, Chubu University, 1200 Matsumoto, Kasugai 487-8501, Japan

Manuel M. Mota (Chapter 6)
NemaLab-ICAM, Department of Biology, University of Évora, 7002-554 Évora, Portugal

Katsunori Nakamura-Matori (Chapter 16)
Tohoku Research Center, Forestry and Forest Products Research Institute, 92-25 Nabeyashiki, Shimo-Kuriyagawa, Morioka 020-0123, Japan

Mine Nose (Chapter 34)
Laboratory of Silviculture, Graduate School of Bioresource and Bioenvironmental Science, Kyushu University, 6-10-1 Hakozaki, Higashi-ku, Fukuoka 812-8581, Japan

Long Ke Phan (Chapter 36)
Institute of Ecology and Biological Resources, Vietnamese Academy of Science and Technology, 18 Hoang Quoc Viet, Nghiado, Caugiay, Hanoi, Vietnam

Mitsuaki Shimazu (Chapters 31, 35, 37)
Insect Management Laboratory, Forestry and Forest Products Research Institute, 1 Matsunosato, Tsukuba 305-8687, Japan

Sang-Chul Shin (Chapter 5)
Department of Forest Diseases and Insect Pests, Korea Forest Research Institute, 207 Cheongnyangni 2-dong, Dongdaemun-gu, Seoul 130-172, Republic of Korea

Susumu Shiraishi (Chapter 34)
Laboratory of Silviculture, Graduate School of Bioresource and Bioenvironmental Science, Kyushu University, 6-10-1 Hakozaki, Higashi-ku, Fukuoka 812-8581, Japan

Etsuko Shoda-Kagaya (Chapter 18)
Insect Ecology Laboratory, Forestry and Forest Products Research Institute, 1 Matsunosato, Tsukuba 305-8687, Japan

Rina Sriwati (Chapter 28)
Laboratory of Environmental Mycoscience, Graduate School of Agriculture, Kyoto University, Kitashirakawa Oiwake-cho, Sakyo-ku, Kyoto 606-8502, Japan
Nematology Laboratory, Plant Pests and Diseases Department, Agricultural Faculty, Syiah Kuala University, Banda Aceh 23111, Indonesia

Jack R. Sutherland (Chapter 3)
1963 St. Ann Street, Victoria B.C. V8R 5V9, Canada

Shuhei Takemoto (Chapter 12)
Laboratory of Environmental Mycoscience, Graduate School of Agriculture, Kyoto University, Kitashirakawa Oiwake-cho, Sakyo-ku, Kyoto 606-8502, Japan
Environmental Biofunction Division, National Institute for Agro-Environmental Science, 3-1-3 Kannondai, Tsukuba 305-8604, Japan

Yuko Takeuchi (Chapter 23)
Laboratory of Environmental Mycoscience, Graduate School of Agriculture, Kyoto University, Kitashirakawa Oiwake-cho, Sakyo-ku, Kyoto 606-8502, Japan

Katsumi Togashi (Chapters 15, 17, 19)
Laboratory of Forest Zoology, Graduate School of Agricultural and Life Science, The University of Tokyo, Yayoi, Bunkyo-ku, Tokyo 113-8657, Japan

Paulo C. Vieira (Chapter 6)
NemaLab-ICAM, Department of Biology, University of Évora, 7002-554 Évora, Portugal

Fuyuan Xu (Chapter 33)
Forestry Academy of Jiangsu Province, Nanjing 211153, People's Republic of China

Toshihiro Yamada (Chapters 22, 25)
University Forest in Chiba, The University Forests, The University of Tokyo, 770 Amatsu, Kamogawa 299-5503, Japan

Bo Guang Zhao (Chapters 4, 24, 27)
Department of Forest Protection, Nanjing Forestry University, Nanjing 210037, People's Republic of China

Color Plates

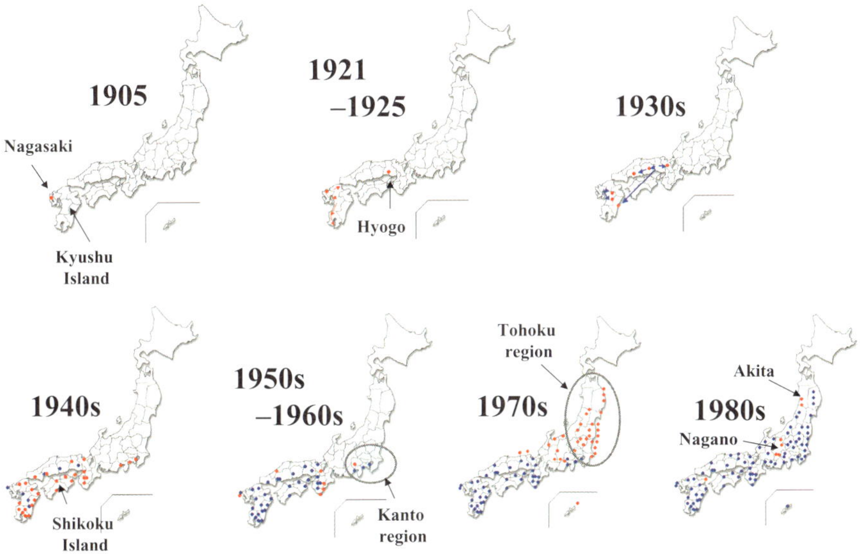

Fig. I.1 (p. 6) History of pine wilt spreading over Japan

Fig. I.3 (p. 9) Limited numbers of Japanese red pine trees are surviving on the ridge of a mountain after serious devastating by pine wilt disease

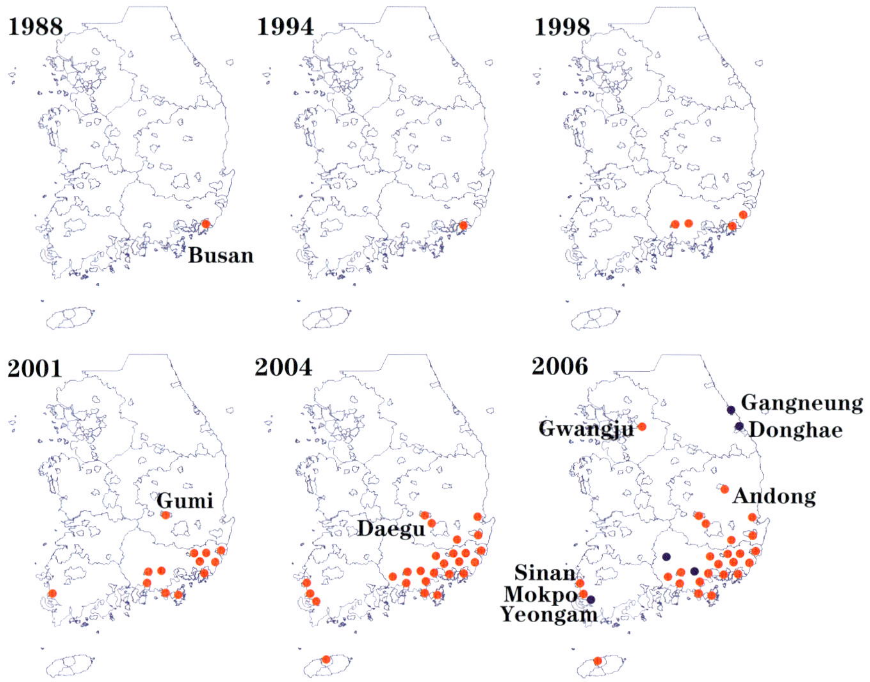

Fig. I.10 (p. 27) Spatial dispersal of pine wilt disease in Korea from 1988 to 2006. *Red dots* indicate the areas in which trees affected by pine wilt disease were observed and *blue dots* indicated the areas in which affected trees by pine wilt disease were found until 2005, but where no pine wilt disease has been observed since 2006

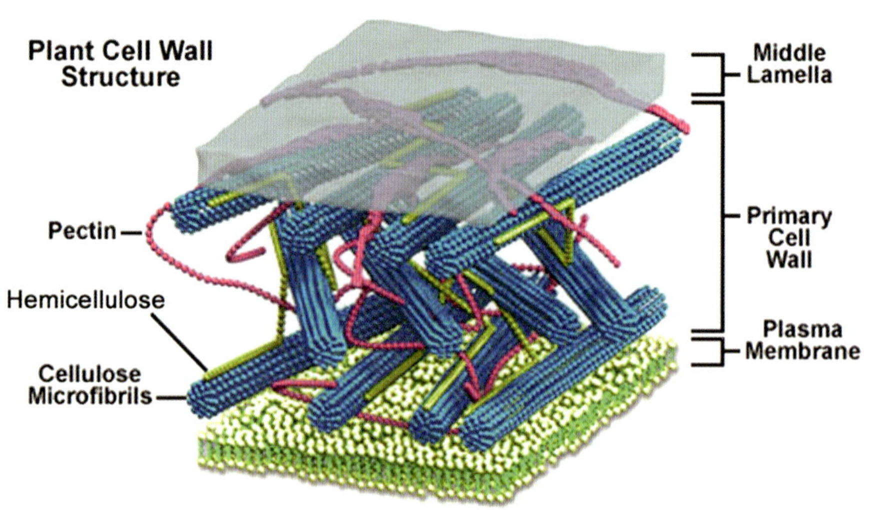

Fig. II.11 (p. 72) Structure of primary plant cell wall, from http://micro.magnet.fsu.edu/

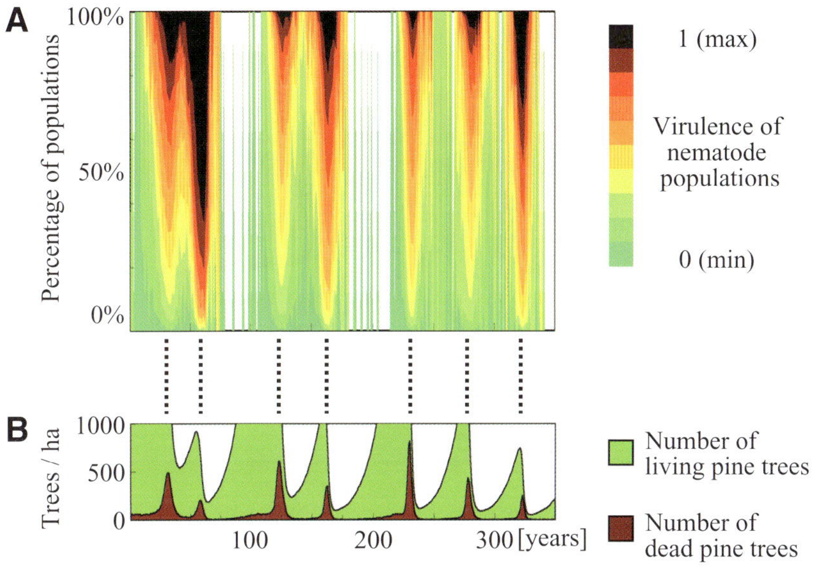

Fig. II.36 (p. 120) Fluctuation in the virulence of *Bursaphelenchus xylophilus* populations synchronized to the dynamics of the host population. **A** Change in the virulence of *B. xylophilus* populations. **B** Changes of the number of living and dead trees

Fig. II.38 (p. 125) *Monochamus alternatus* larva in the pupal chamber. A pine log infested with beetle larvae was chopped in two with a hatchet to observe the pupal chamber. Larva bores a tunnel into the xylem while plugging it with frass to make a pupal chamber. The *arrow* shows the entrance of the tunnel (*scale bar* = 2 cm)

Fig. II.41 (p. 130) Abdominal spiracles of *Monochamus alternatus* adult. The left elytron was removed to observe the spiracles. Seven pairs of spiracles are arranged in pairs on seven abdominal segments. The first abdominal spiracle (*circle*) is the largest among the spiracles and a great number of fourth-stage dispersal juveniles enter the trachea through here

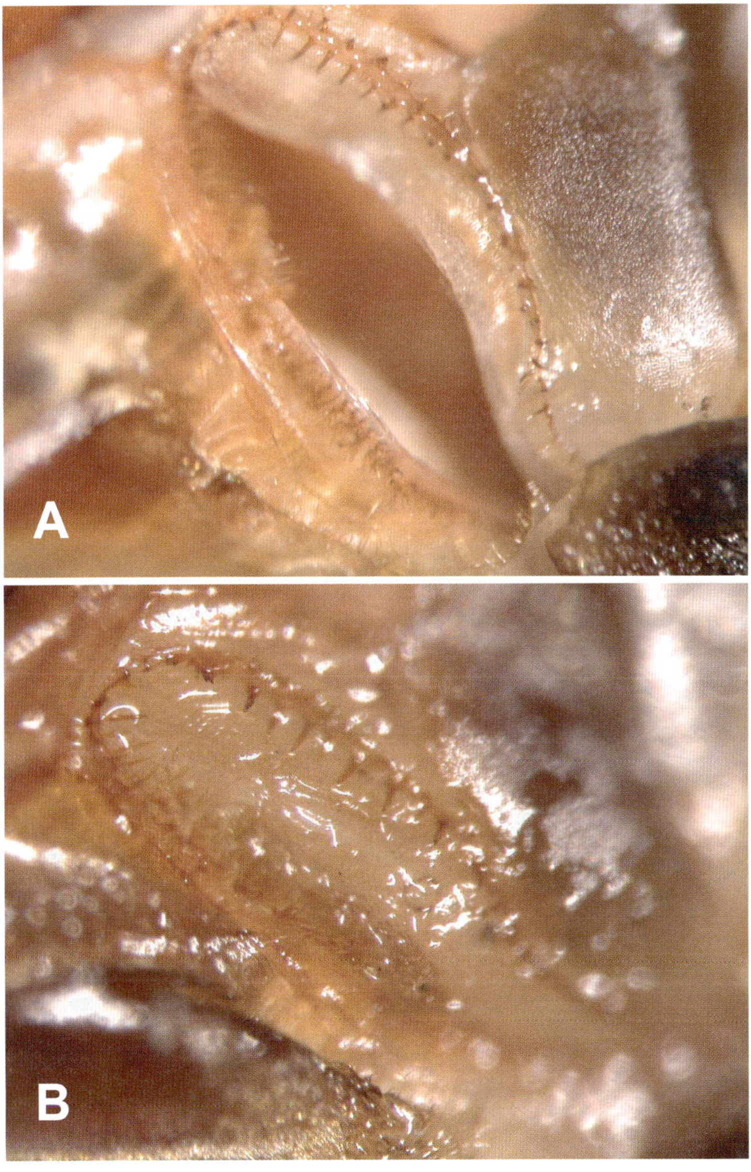

Fig. II.42 (p. 131) Atrium of the first abdominal spiracle of *Monochamus alternatus* adult. **A** The atrium of a beetle infested with no fourth-stage dispersal juvenile (J_{IV}). The cavity of the atrium can be clearly observed. **B** The atrium of a beetle infested with numerous J_{IV}s. The cavity of the atrium cannot be observed as the result of being packed with many J_{IV}s. Modified after Aikawa (2006), with permission

Fig. II.43 (p. 132) Fourth-stage dispersal juveniles (J_{IV}s) in the trachea of the hind leg of *Monochamus alternatus* adult. **A** The ventral area of a beetle whose hind legs were removed. A trachea connected to the hind leg is exposed (*arrow*). **B** The trachea contains many J_{IV}s. Modified after Aikawa (2006), with permission

Fig. II.44 (p. 133) Fourth-stage dispersal juveniles (J_{IV}s) crawling on the body surface of *Monochamus alternatus* adult. After J_{IV}s exited from the spiracles, they move to the abdominal terminal to depart from the beetle's body

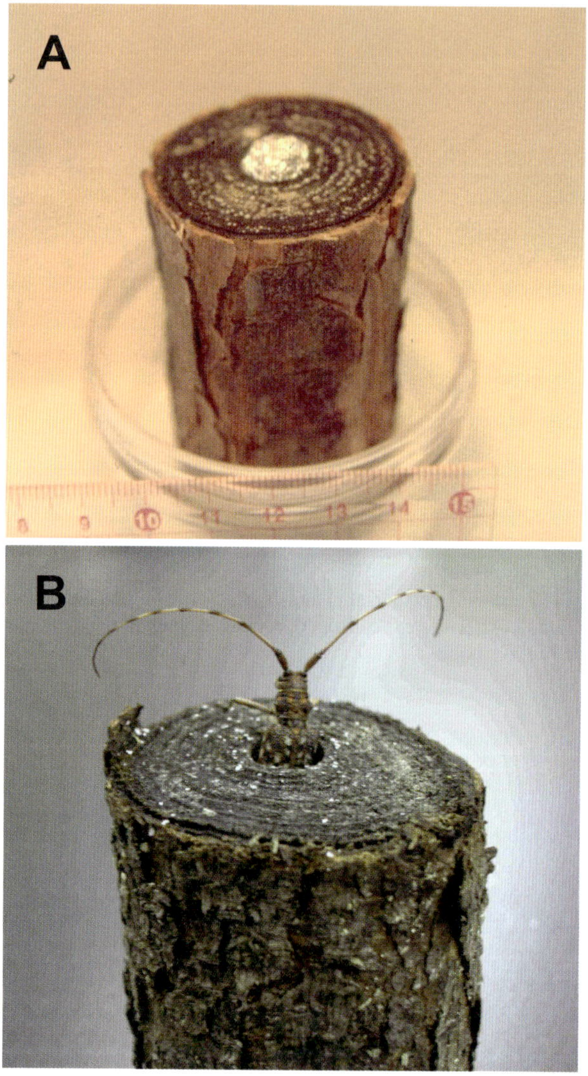

Fig. II.45 (p. 137) Pine bolt for loading *Monochamus alternatus* adult with fourth-stage dispersal juveniles (J_{IV}s) of *Bursaphelenchus xylophilus*. A hole 1.05 cm across and 5 cm deep was drilled in the center of the xylem at the cut end of the pine bolt (7.5-cm long and about 4-cm diameter) as an artificial pupal chamber for *M. alternatus* larva. The bolt was set upright in a layer of quartz sand in a polycarbonate container. The container with the pine bolt was autoclaved and then the bolt was inoculated with *Ophiostoma minus* under sterilized conditions. Two weeks later, 5,000 propagative nematodes suspended in 0.5 ml water and a fourth-instar *M. alternatus* larva were put into the hole at the same time and the hole was plugged with aluminum foil (**A**). During 30–40 days after inoculation of the nematodes and larva, adult beetle infested with a great number of J_{IV}s gnawed through the aluminum plug and emerged from the bolt (**B**). A beetle infested with more than 150,000 J_{IV}s has so far been obtained by this method (T. Aikawa, unpublished data)

Fig. III.2 (p. 147) Life cycle of *Monochamus alternatus*. **A** female adult coming out of an emergence hole (*arrow*), **B** feeding wounds by adult beetles on a *Pinus thunbergii* twig, **C** oviposition scars (indicated by the *arrows*), **D** an egg laid in the inner bark, **E** a feeding larva and frass under the bark, **F** a mature larva in a pupal chamber (*arrows* show the entrance hole)

Fig. IV.1 (p. 205) Symptom development of *Pinus thunbergii* sapling inoculated with pine wood nematode, *Bursaphelenchus xylophilus*: **A** healthy 5-year-old sapling; **B** start of old-needle discoloration and drooping of apical needles 3 weeks after inoculation. Trees indicated by *arrows* are the same tree of different dates

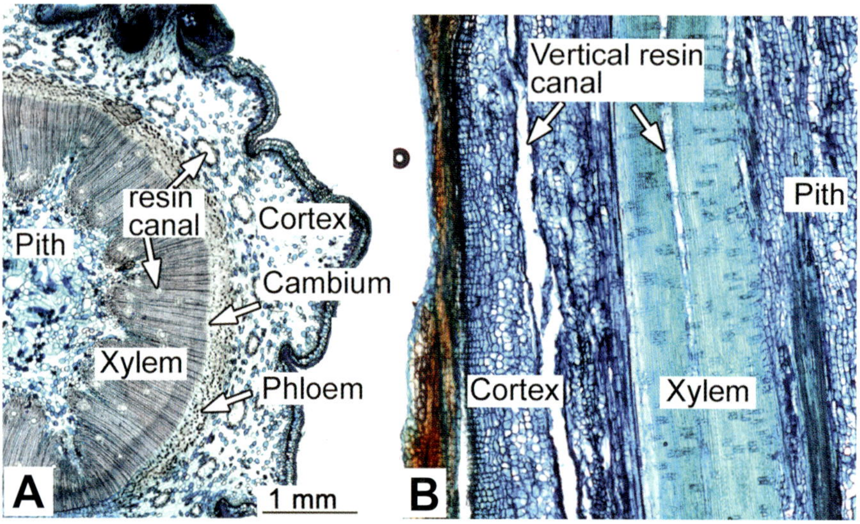

Fig. IV.2 (p. 206) Structure of a current year *Pinus densiflora* shoot. **A** Cross section, **B** radial section. Stained with nile blue

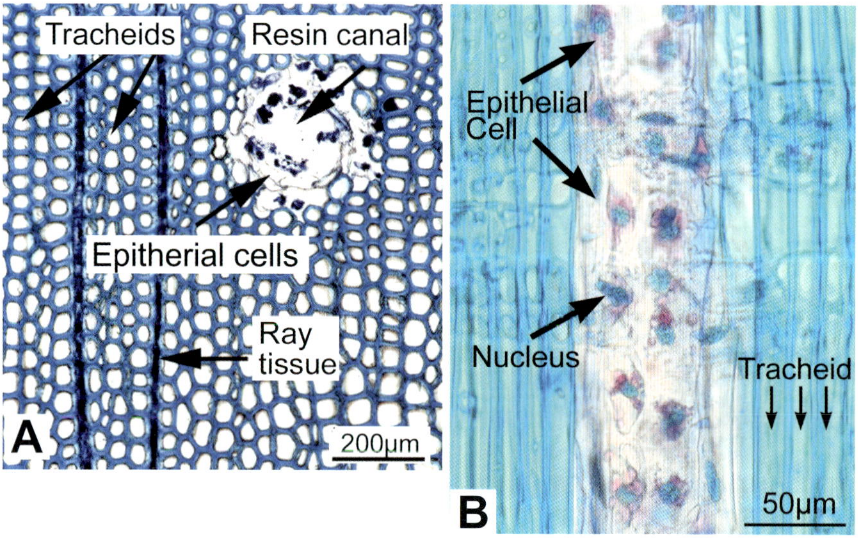

Fig. IV.3 (p. 206) Xylem tissue of *Pinus* species. **A** Cross section of *Pinus thunbergii*. **B** Radial section of a vertical resin canal in the xylem of *P. densiflora*. The diameter of the resin canal in *P. densiflora* is thinner than that in *P. thunbergii*. Stained with nile blue

Fig. IV.5 (p. 209) Ray parenchyma cells in the xylem of *Pinus thunbergii* stained with nile blue. **A** Contents of healthy cells (*radial section*). **B** Exudates (*arrow*) and changes in the cell contents in shape and stainability are observed at the developing phase of the disease (*tangential section*)

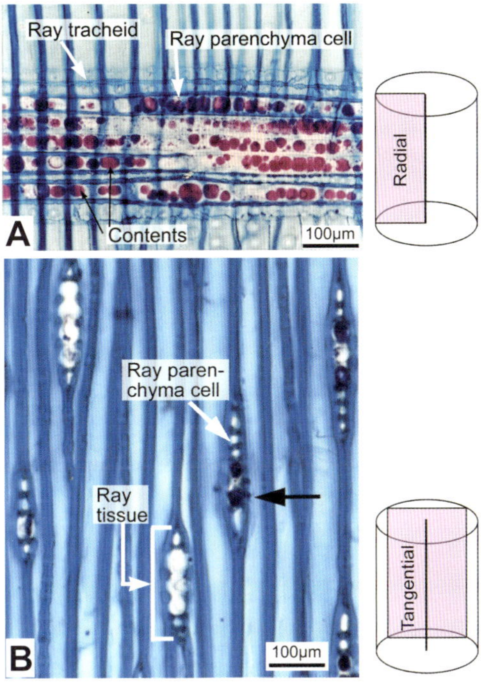

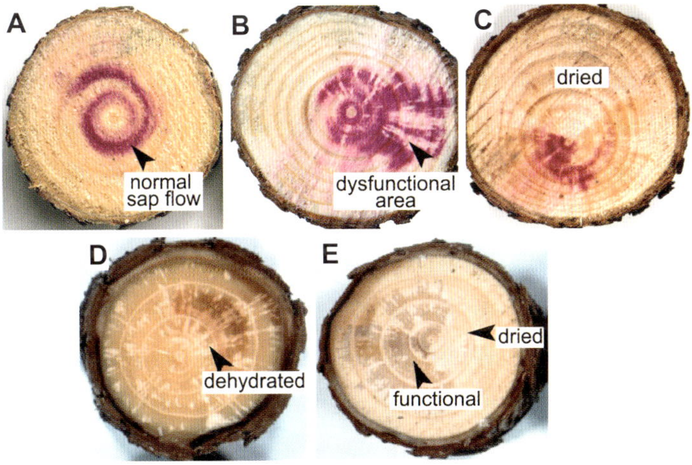

Fig. IV.7 (p. 212) Development of the disturbance and decrease of sap flow after inoculation of *Pinus thunbergii* with the pine wood nematode. An acid fuchsine solution was injected in the base of trunks **A** to **C** before the tree was cut down. **A** Normal spiral sap flow in healthy condition before inoculation of the pine wood nematode. **B** Three weeks after inoculation. Sap flow is disturbed by the partial blockage of sap ascent. **C** Four weeks after inoculation. Sap flow decreases extensively, and necrosis is occurring in the cambium. **D** *White patches* of dehydrated areas are observed in the early phase (2 weeks after inoculation). **E** Desiccation of the xylem is progressing at the start of old-needle discoloration in the developing phase (3 weeks after inoculation)

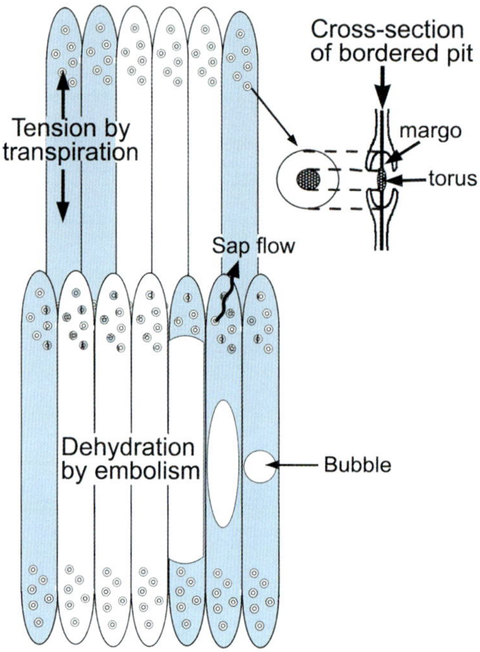

Fig. IV.8 (p. 213) Embolism occurring in the tracheids of pine sapwood infected with the pine wood nematode and structure of a bordered pit. Bubble formed in the tracheids immediately enlarges and dehydrates the tracheid by strong tension induced by transpiration. The dehydrated area spreads to adjacent tracheids in a chain reaction

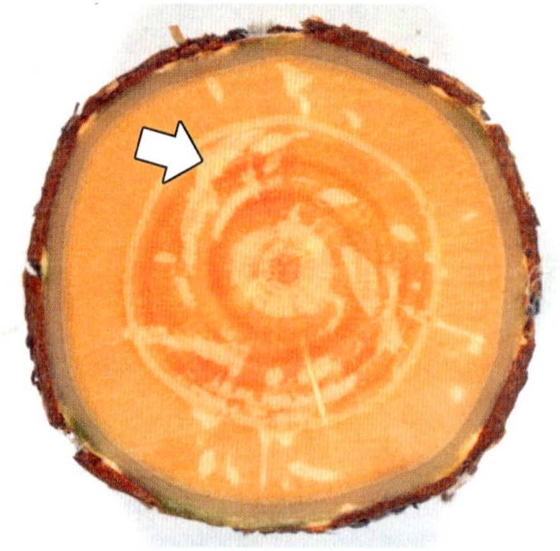

Fig. IV.10 (p. 218) Cross section of a *Pinus taeda* stem 5 weeks after inoculation with pine wood nematode. White patches of the dehydrated area (*arrow*) are narrower than those in *P. thunbergii* (Fig. IV.7)

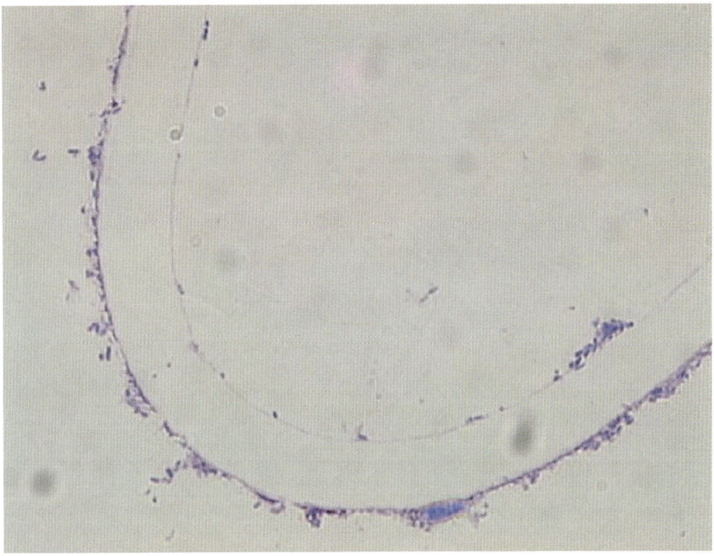

Fig. IV.22 (p. 256) Bacteria carried by pine wood nematode under a light microscope. The nematode was isolated from a naturally diseased Japanese black pine (stained with crystal violet)

Fig. V.6 (p. 287) Blue-stained *Pinus densiflora* logs (triangular areas)

Fig. V.7 (p. 287) *Cryptoporus volvatus* mushrooms on *Pinus densiflora* logs

Fig. V.9 (p. 291) Blue-stain fungus on the elytra of the Japanese pine sawyer, *Monochamus alternatus*

Fig. VI.8 (p. 337) Inoculation of Japanese black pine to determine resistance to the pine wood nematode

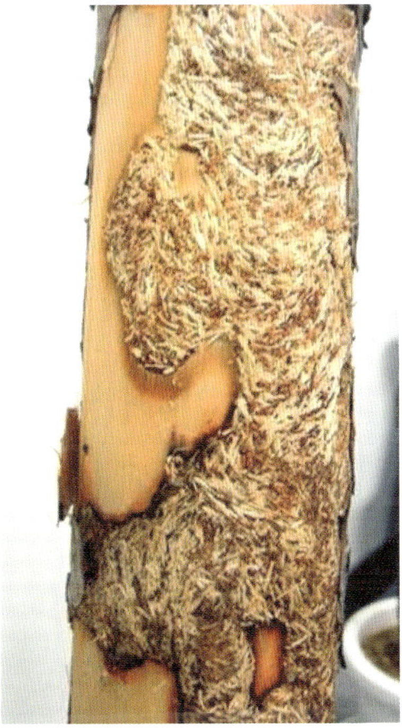

Fig. VI.23 (p. 378) Sawdust produced by larvae of the Japanese pine sawyer, *Monochamus alternatus*, feeding on sapwood where entomopathogenic nematodes can migrate

Fig. VI.24 (p. 379) Second to fifth stage (*left* to *right*) larvae of the Japanese pine sawyer beetle, *Monochamus alternatus*, killed by the entomopathogenic nematode *Steinernema carpocapsae*

Part I
Historical Overview

1 Introduction

1.1 Discovery of the Pine Wilt Disease Pathogen

Prior to the pine wood nematode (PWN) being shown to be the causal agent of pine wilt disease, the widespread death of pines in Japan had been attributed to bark and wood boring insects, especially coleopteran insects in the families Cerambycidae, Curculionidae, and Scolytidae. However, during a national project (1968–1971) to find a control for the problem it was shown that such insects, which supposedly had been responsible for killing the trees, could not lay their eggs in healthy pine trees, and also that the trees were wilting prior to insect attack. Consequently, the goals of the project had to be changed from focusing just on insects as causing the mortality to looking at other possible causes such as microorganisms, edaphic, tree physiological, and climatic factors.

In early autumn, 1968, Dr. Y. Tokushige, a tree pathologist member of the National Project at the Kyushu Branch of the Forestry and Forest Products Research Institute (FFPRI), while looking at possible causes of the problem happened to notice something in a Petri dish that caught his eye. The pile of Petri dishes on his table contained microorganisms that had originated from the wood of dead pines. The Petri dish that caught his eye contained not only immobile microorganisms but also squirming nematodes. Consequently, Dr. Tokushige asked his colleague, Dr. Kiyohara, a nematologist working in Tokushige's laboratory, to identify the nematode. After careful examination, Dr. Kiyohara identified the nematode as a species of *Bursaphelenchus* (Tokushige and Kiyohara 1969). In subsequent studies the two scientists collected many wood samples from dead pine trees in various districts on Kyushu Island, and confirmed the ubiquitous presence of the nematode in the dead trees, that is, they fulfilled Koch's first postulate by demonstrating that the organism was consistently associated with diseased plants. Most species of the genus *Bursaphelenchus* are mycophagous and can be propagated on fungi. Next, to fulfill Koch's second postulate they reared the nematodes on some fungi isolated from dead pine trees. They then identified the potential pathogens and inoculated them into pine seedlings to determine their pathogenicity. Although they were skeptical about the ability of the nematode to kill trees, they also inoc ulated *Bursaphelenchus* nematodes into eight, pine trees growing outdoors at the Institute. Surprisingly, most of the trees receiving the nematodes showed acute and well-defined wilting symptoms, that is, the researchers had fulfilled Koch's postulate No. 3. Thus, they had fulfilled Koch's postulates for the newly found *Bursaphelenchus* nematode. After the unexpected results of their initial work, they made a well-planned, series of inoculation trials consisting of eight experiments. In 1970 they made many inoculation tests and clearly confirmed the pathogenicity of the *Bursaphelenchus* nematode (Kiyohara and Tokushige 1971).

1.2 New Control Tactics and Discouraging Results

About 40 species of *Bursaphelenchus* nematodes had been described before PWN was found with of them having a phoretic relationship with insects, that is, they are vectored by insects. Using this knowledge an intensive search for vector insects of the PWN was conducted with emphasis on the insects related to wilting and dead pine trees. Ultimately, a species of sawyer beetle, *Monochamus alternatus*, was found to be the sole vector of the PWN (Mamiya and Enda 1972; Morimoto and Iwasaki 1972), and after exhaustive research the disease cycle was determined (Mamiya 1975). Based on this new information, spraying pine trees with an insecticide such as fenitrothion was implemented to prevent maturation feeding of *Monochamus* beetles. This new control tactic received government support in 1978. Such spraying seemed to be effective and reasonable, but its use over vast areas of pine forests raised environmental concerns which have always restricted its use. Such restrictions have led to poor overall results. Thus, as the result of pine wilt disease Japanese pine forests have suffered losses of over 46 million cubic meters of trees in the last 50 years.

1.3 Spread of Pine Wilt Disease to Other Countries

Pine wilt disease spread from Japan to neighboring East Asian countries such as China and Korea in 1982 and 1988, respectively. Both countries have made enormous efforts to eradicate the disease by employing similar control methods as used in Japan, but there too the results have never been successful (see Chaps. 4 and 5). After the discovery of PWN in wood chips imported from Canada and the USA, Nordic governments placed a strict ban on the importation of wood products from these and other countries where PWN occurs. The European and Mediterranean Plant Protection Organization (EPPO) listed both the PWN and the vector *Monochamus* beetles as quarantine pest A1, and implemented strict inspection of imported wood products. In 1999 pine wilt disease was found in Portugal (see Chap. 6), thus this epidemic disease has become a widespread threat to forests.

1.4 The Purpose of This Textbook

Because of the serious damage that pine wilts causes many scientists have tried numerous techniques for its control. So far, more than 3,000 papers on pine wilt disease have been published in Japan, and as well many papers have been published in China and Korea. Since most of these papers have been written in the local language it is difficult for foreigners to access them. To solve this problem, Dr. Y. Kishi made an extensive review of the Japanese literature, which he published in English (Kishi 1995). Since then, however, research on pine wilt disease has progressed at a rapid rate and innumerable papers have been published. Consequently, in 2006 Dr. B.G. Zhao and Dr. K. Futai agreed to prepare this new textbook, which includes both basic information about pine wilt and updates recent findings on the disease.

2
Pine Wilt in Japan: From First Incidence to the Present

Kazuyoshi Futai

2.1 Historical Overview

Before pine wilt disease brought devastation to the majority of pine forests in Japan, most countryside forests, especially those in southwestern Japan, were dominated by Japanese red pine, *Pinus densiflora*. Actually, vegetation ecologists classified the flora of the southwestern region of Japan as a *P. densiflora–Quercus serrata* zone. Recently, however, it has become very difficult to find healthy pine trees in our surrounding mountains. The tiny, 1-mm-long, pine wood nematode (PWN) has dramatically changed our familiar flora.

As shown in Fig. I.1, the first incidence of pine wilt disease in Japan was reported in Nagasaki City on Kyushu Island in 1905 (Yano 1913). Since then, intensive efforts have been made to control this epidemic disease, and the first outbreak of pine wilt disease was stamped out by 1915. Pine wilt disease, however, recurred in a harbor town 50 km from Nagasaki City in 1925, and then gradually spread into the surrounding areas.

Pine wilt disease spread to the mainland in 1921, and old pines planted in a shrine in a harbor town in Hyogo Prefecture began to wilt, and the number of diseased trees increased year after year. In the 1930s, pine wilt disease spread gradually into neighboring prefectures both on Kyushu Island and the mainland. In the 1940s, life in Japan was harsh because of World War II and the forests were largely left unattended. As such, pine wilt disease devastated many pine forests. The disease spread rapidly, not only to surrounding regions but also to remote regions such as Shikoku Island and the Kanto districts.

After World War II, the General Headquarters (GHQ) of the Allied occupation military became seriously concerned about the spread of the devastation that was

Laboratory of Environmental Mycoscience, Graduate School of Agriculture, Kyoto University, Kitashirakawa Oiwake-cho, Sakyo-ku, Kyoto 606-8502, Japan

Tel.: +81-75-753-2266, Fax: +81-75-753-2266, e-mail: futai@kais.kyoto-u.ac.jp

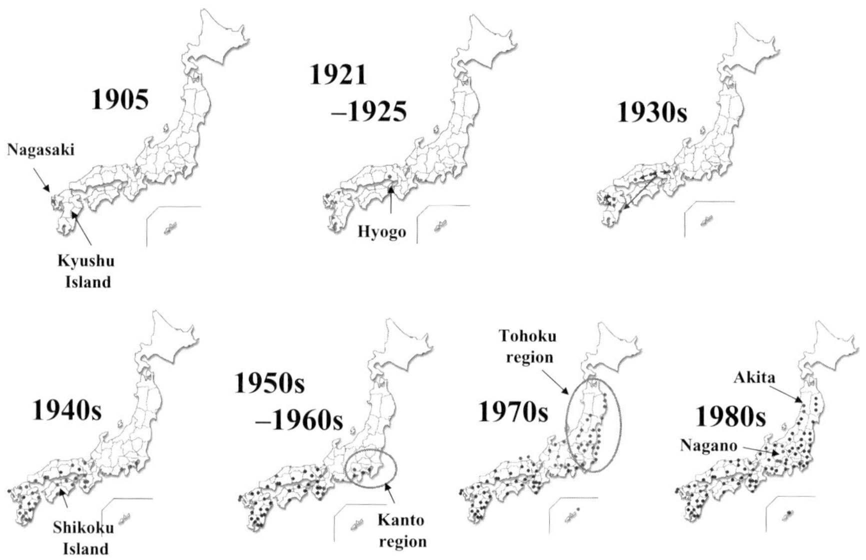

Fig. I.1 History of pine wilt spreading over Japan (see Color Plates)

occurring in pine forests throughout Japan. So much so that they asked Dr. R. L. Furniss, a forest entomologist, to inspect the many pine forests being affected by pine wilt disease. After intensive inspection he submitted two reports, in which he recommended very simple control methods "felling and burning" of dead pine trees. The GHQ implemented this recommendation and urged the Japanese government to start control methods. The extensive control efforts following Furniss' recommendations, together with the plentiful labor available then, succeeded in reducing the damage from pine wilt disease. Thus, the spread of pine wilt disease slowed down in the 1950s to 1960s.

At the beginning of the 1970s, a governmental project found the PWN as the causal agent of pine wilt disease and *Monochamus* beetles as its vector, and established new control tactics based on the newly discovered pine wilt disease cycle. The Japanese government pushed ahead with these policies by enacting a new law, "the Special Law in Force for PWN control", which recommended aerial spraying of insecticides to prevent mature feeding of *M. alternatus* and thereby suppress nematode infection. The annual loss of pine trees resulting from pine wilt disease, however, could not be reduced, but instead increased quickly, and the dry and hot summer of 1978 resulted in a marked expansion of pine wilt disease toward northern regions such as Kanto and Tohoku, and increased the annual loss up to two million cubic meters in 1978 until 1981.

In the 1980s, Nagano Prefecture and Akita Prefecture, both of which had remained free from pine wilt disease, were newly invaded. Thus, pine wilt disease prevailed throughout Japan except for the northernmost two prefectures, Aomori and Hokkaido.

2.2 Possible Factors Influencing Pine Wilt Disease Spread

The spread of pine wilt disease is influenced not only by climatic conditions such as temperature and precipitation, but also by edaphic, topographic, biological, and human factors.

2.2.1 Meteorological Conditions and Flight Ability of Vector Beetles

The flight ability of *Monochamus* beetles is an important factor affecting pine wilt disease spread, and is reported to be 50–260 m during a beetle's life span (Togashi 1990), but the combination of flight and wind can often carry beetles several kilometers.

Temperature determines both the rate of development and the mobility of PWNs and those of vector beetles, all of which influence pine wilt spread, while precipitation determines the host tree's water status. When infected by PWNs, pine trees show wilting symptoms, but the velocity of symptom development varies depending upon the host tree's water status. Thus, precipitation influences the spread rate of pine wilt disease.

2.2.2 Soil Eutrophication Adversely Affects Pine Trees

Based on the conservative and empirical idea that fully grown trees must be more resistant to disease, some plant ecologists supposed that pine trees would acquire resistance to pine wilt if they were fertilized. From the viewpoint of the mycorrhizal relationship, however, pine trees are healthier when growing in nutrient poor soils, even though their growth may be less. To examine which idea is most possible, my colleagues and I carried out a field experiment in a pine stand on a coastal sand dune in Tottori Prefecture, placing two 20 × 20 m plots side by side in the stand. One plot was left as the control, and received no treatment. The other plot was fertilized at a rate of 0.5 kg m^{-2} with a slow-release nitrogen fertilizer. The rate of pine wilt disease spread in the control plot was slower than in the fertilized plot. At the end of the experiment, many healthy pine trees remained in the control plot, while there were fewer surviving trees in the fertilized plot (Fig. I.2). Thus, the application of fertilizer, and therefore soil eutrophication, seemed to speed up pine wilt disease spread over the 3 years from 1997 to 2000.

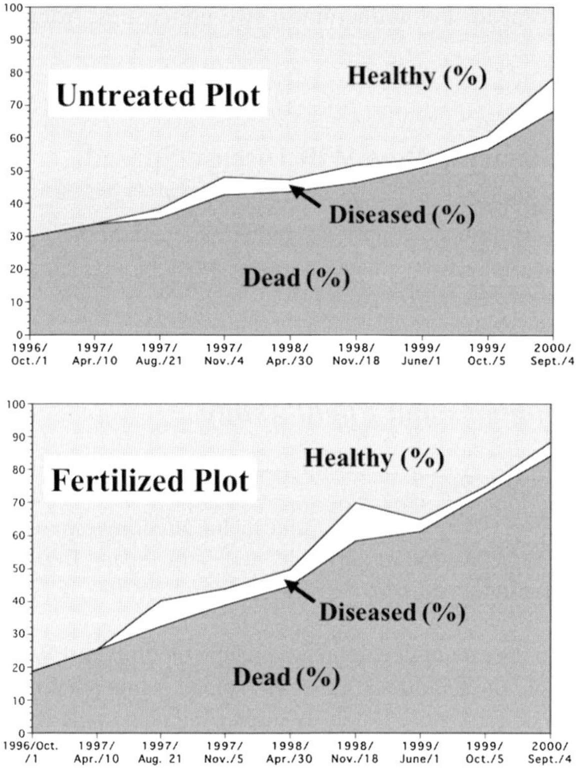

Fig. I.2 The effect of soil eutrophication on the spread of pine wilt disease in two small plots of Japanese black pine, *Pinus thunbergii*; one is fertilized and the other is non-fertilized plot

2.2.3 Topographic Conditions and the Role of the Mycorrhizal Relationship

When a pine stand on a slope is devastated by pine wilt, some healthy pine trees tend to survive at the mountain ridge top (Fig. I.3). The situation prevails to the present. This suggests that pine trees survive better on the upper part of a slope than on the lower part. To test this hypothesis my colleagues and I carried out a field survey from 1993 to 1998 at the experimental station of Kyoto University Forests in Yamaguchi Prefecture, located in the westernmost part of the mainland Japan. In a pine stand at the station, about 4,000 pine seedlings of 23 families of Japanese black or red pines were planted in 1973. The stand is ca. 1.4 ha, and is located on a 25° inclined slope with a height of about 50 m. All the pine families were planted in rows along the slope.

Since 1979, pine wilt has spread into this stand. By the end of 1993, the pine wilt damage had become severe, and more than 70% of the pine trees had been killed (Fig. I.4). However, by 1993 some of the pine families were surviving in quite high numbers. Using two families of such Japanese black pine, shown as the

Fig. I.3 Limited numbers of Japanese red pine trees are surviving on the ridge of a mountain after serious devastating by pine wilt disease (see Color Plates)

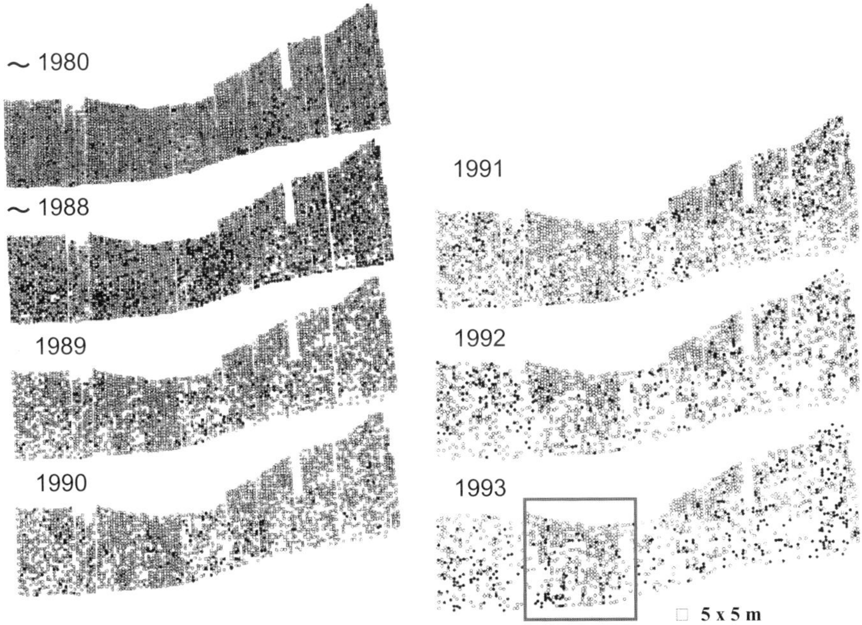

Fig. I.4 Decrease in surviving pine trees since 1980–1993 at a low hill in Tokuyama experimental forest station, Kyoto University Forest (after Nakai et al. 1995)

framed area in the figure, we compared the survival along different heights of the slope. Table I.1 shows the survival of Japanese black pine along different heights of the slope. In both pine families the survival was higher at the upper part of the slope. To test a possible reason as to why this occurred we examined mycorrhizal development on trees at different heights along the slope. The results showed that, the number of fine tap roots and the mycorrhizal ratio (=the number of mycorrhized root chips / the total number of root chips examined) were higher on trees at the upper parts of the slope than lower on the slope (Fig. I.5).

In the summer in Japan, most plants are exposed to a dry and hot climate; however, pine trees planted at the higher part of a slope might well suffer less water stress as the result of water being supplied by the mycorrhizal symbiosis. Supposedly, the pines planted on the lower part of a slope could not obtain enough water to avoid such stress. We concluded that PWN affected pine trees growing on the lower part of such slopes must be far more vulnerable to the disease.

Table I.1 The survival ratio of Japanese black pine at different heights of a slope

	No. 241	No. 236
Upper part	74	65
Middle part	54	39
Lower part	35	17

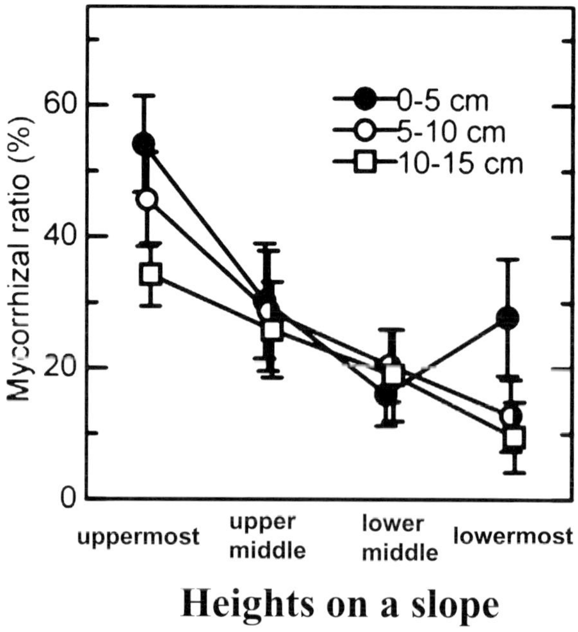

Fig. I.5 The proportion of mycorrhizae to the total of fine tap roots and mycorrhizae by fresh weight in each depth and at different heights on a slope. Samples were obtained from a small area of the slope shown in Fig. I.4. The area is enclosed in red line in the figure (after Akema and Futai 2005)

2.2.4 Asymptomatic Carrier Trees Ensure Continuity of Pine Wilt Disease—A Chain Infection Model

It is well known among forest workers that pine wilt disease often recurs in the vicinity of the stump of a pine tree that was killed the previous year, even after thorough eradication of the dead tree. Togashi (1980) called this phenomenon "hysteresis". To determine the mechanism of "hysteresis", the mode of pine wilt disease spread across a pine stand must be followed from the beginning of the incidence. A small stand consisting of 72 trees of 45-year-old Korean pine, *P. koraiensis*, provided an ideal opportunity to study this problem as inoculation tests have shown *P. koraiens* is to be one of the most susceptible species to pine wilt disease (Futai and Furuno 1979). Interestingly, this stand had been free from pine wilt disease until late autumn 1990 when two pine trees in this stand were killed by pine wilt. This situation provided an ideal opportunity to study the spread of the disease in this stand. The two dead trees (including their trunks, branches, and twigs) were removed from the stand in early spring, 1991. By early autumn, however, four more trees located near the stumps of the trees killed in the preceding year were killed by pine wilt disease. The next year, 1992, all the dead trees were again removed from the stand before spring. In early summer, however, *Monochamus* beetles visited some trees surrounding the stumps of dead trees and laid their eggs. Thus, pine wilt disease spread from tree to tree, and killed more than 40 trees until the end of the study in 1994. To determine the health status of all the pine trees in the stand, the resin exudation ability of the trees was examined once or twice a month, 58 times in total for 4 years, that is, from December 1990 to August 1994. The resin exudation data showed that some of the newly killed pine trees had already been infected in the previous year or earlier and then had survived without any visible symptoms. Such asymptomatic trees may play an important role in pine wilt epidemics.

Based on data from the above study of *P. koraiensis*, a chain infection model was proposed (Fig. I.6) to explain the mode of pine wilt spread, that is, where an asymptomatic carrier tree releases volatiles and attracts vector beetles. The conclusion is that, these asymptomatic trees play an important role in ensuring the continuity of pine wilt disease.

Recently, my colleagues and I established a new molecular method to detect PWNs in pine wood. This new method revealed that the number of asymptomatic trees in the field was far greater than expected (see Part IV). Thus, the role of asymptomatic carrier trees in spreading pine wilt disease may be more crucial than previously thought.

2.2.5 World-Wide Trade Could Facilitate the Invasion of Other Continents by Pine Wilt Disease

Since the beginning of the twentieth century, various nursery seedlings and natural products have been transported among continents. Many pests accompany such

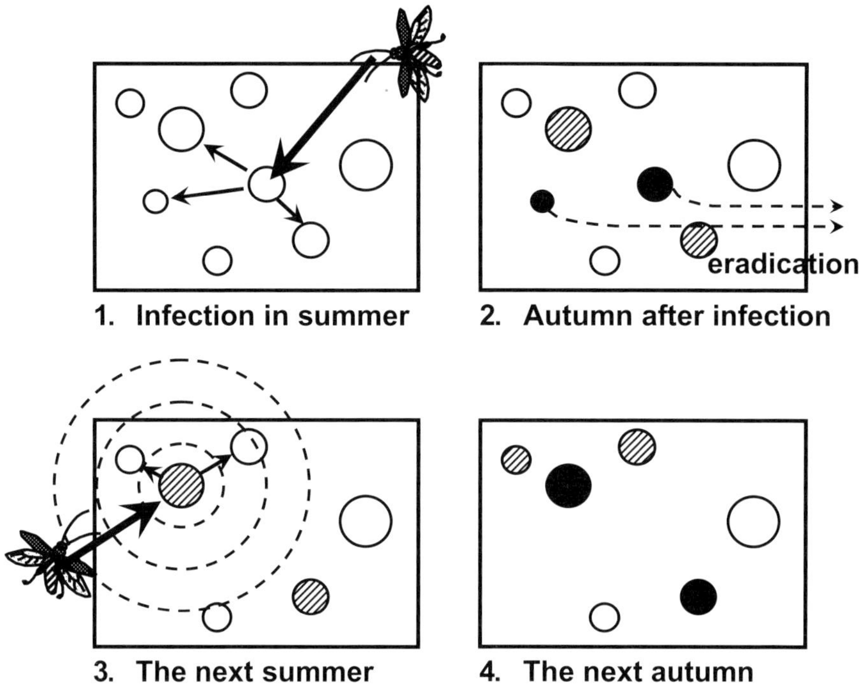

Fig. I.6 A chain infection model for the spread of pine wilt mediated by an asymptomatic carrier tree, proposed by Futai (2003a). This model was based on a study in a stand of *Pinus koraiensis* in Kamigamo Experimental Station, Kyoto University Forests, where dead pine trees were thoroughly eradicated before the next season of the pine wilt began. *Solid* and *shaded circles* represent dead and asymptomatic carrier trees, respectively. Immigration from outside the stand and translocation within the stand of vector beetles are represented as large and *small arrows*, respectively. *Long thin arrows* in the *top right* (2) mean the removal of dead trees from the stand. *Dotted circles* represent the diffusion of volatiles from diseased trees

products and plants are thus spread to other continents. At the beginning of the twentieth century pine wilt disease was also spread in this manner. Pine wood and wood products often carry PWNs or vector beetles, or both. For example, pine wood infested with vector beetles and PWNs was transported to the Okinawa Islands for use as construction material in 1973, and thereafter many Ryukyu pines (*P. luchuensis*) were killed by this disease. The PWN and its vector beetle could be concealed in these wood packing products, and thereby spread to other countries. Thus, global trade and human transportation enable pine wilt to move to previously uninfested regions.

3
A Brief Overview of the Pine Wood Nematode and Pine Wilt Disease in Canada and the United States

Jack R. Sutherland

3.1 Introduction

The purpose here is not to review all the relevant literature on pine wilt disease, the pathogen, that is, the pine wood nematode (PWN), *Bursaphelenchus xylophilus*, host pines, or the nematode's insect vector, species of *Monochamus*, in North America. Instead, it is to briefly summarize for North America what has already been cited in several review articles that have been published on these topics, especially on the pathogen, and disease, and where relevant, the possible and known vectors in the USA (Dwinell and Nickle 1989) and Canada (Bowers et al. 1992). Then follows a narrative on the conditions that limit the importance of pine wilt in Canada and the USA. Additional information on pine wilt, the pathogen, vectors, effects of climatic conditions on the two, hosts and so on can be found in the review articles by Mamiya (1983) and Dwinell (1997) and the book entitled "Pathogenicity of the Pine Wood Nematode" (Wingfield 1987a). Although the PWN has been reported in Mexico (Dwinell 1985a), essentially nothing else is known about it or the disease there; consequently, the discussion here is confined to Canada and the USA.

3.2 Historical Aspects

The history of pine wilt disease follows a long series of research findings beginning with the publication of Steiner and Buhrer (1934) who described *Aphelenchoides xylophilus*, n. sp., which they had isolated 5 years earlier from blue-stained, longleaf pine, *Pinus palustris*, logs at a sawmill in Louisiana, USA. Dwinell and Nickle

1963 St. Ann Street, Victoria, BC V8R 5V9, Canada

Tel.: +1-250-5984033, e-mail: jacksutherland@shaw.ca

(1989) give point by point landmarks in the history of PWN research which started some 35 years after Steiner and Buhrer's work when Tokushige and Kiyohara (1969) reported the occurrence of a *Bursaphelenchus* sp. in the wood of dead pine trees in Japan. The next year, Nickle (1970) transferred *A. xylophilus* to the genus *Bursaphelenchus*. Kiyohara and Tokushige (1971) in Japan demonstrated the pathogenicity of a *Bursaphelenchus* sp. by inoculating 25-year-old Japanese red pine, *P. densiflora*. Later in Japan, Mamiya and Kiyohara (1972) described the PWN as *B. lignicolus* while that same year, Mamiya and Enda (1972) reported transmission of this nematode during maturation feeding by *Monochamus alternatus*. Seven years later, pine wilt disease was reported in Missouri, USA (Dropkin and Foudin 1979), and Nickle et al. (1981) placed *B. lignicolus* as a synonym of *B. xylophilus*. Wingfield (1983) showed transmission of *B. xylophilus* during oviposition of *Monochamus* vectors. Transmission during maturation feeding is commonly referred to as primary transmission while transmission during oviposition is secondary transmission. In 1983, *B. xylophilus* was found associated with dwarf mistletoe-infested Jack pine, *P. banksiana*, trees in Manitoba, Canada (Knowles et al. 1983). In 1984, Rautapaa (1986) reported that pine, *Pinus* spp., wood chips imported into Finland from the USA and Canada contained the nematode. Subsequently, in 1986, as outlined by Bowers et al. (1992) and references therein, the European Plant Protection Organization (EPPO) concluded that the PWN met all the criteria of a class "A-1" pest, that is, potentially important (Anonymous 1986). In 1985 Finland restricted the importation of conifer chips and timber from countries where the PWN is present because wood chips shipped from the USA and Canada were infested by the nematode (Rautapaa 1986). Other Nordic countries acted similarly and the EPPO recommended that Europe ban softwood products, except kiln-dried lumber, from countries with the PWN. Up until then, the nematode and the disease, which had only affected a few, mostly exotic, trees in the southern USA (Dwinell and Nickle 1989) had not caused serious economic consequences. Prior to that, the serious losses that had occurred in Japan (Mamiya 1983), while of interest to North Americans, were viewed mainly as a serious problem that occurred in Japan and nearby countries; however, these EPPO recommendations which also affected North American products resulted in an accelerated interest in the pathogen and the disease in both Canada and the USA. Since then, both the PWN and pine wilt disease have been found in Portugal (Mota et al. 1999).

3.3 Overview of the PWN in Canada

In Canada, work on the PWN began shortly after the EPPO restrictions were imposed and included surveys to determine the geographic and host distribution of the nematode in Canada and possible insect vectors. As well, research was done in British Columbia at the Pacific Forestry Centre, Forestry Canada, in Victoria, and at Simon Fraser University in Burnaby.

3.4 Nationwide Survey of the Nematode, Hosts and Potential Insect Vectors

Initially, in the early 1980s, several province- or area-specific surveys were made to determine the presence and status of the nematode; however, in 1985, the Forest Insect and Disease Survey of Forestry Canada initiated a nationwide survey to determine the distribution of the nematode and its potential insect vectors. The detailed protocols used in this survey and the survey results are given by Bowers et al. (1992). Briefly, the survey methods prioritized and emphasized the sampling of both host tree and vector species most likely to harbor the PWN. Consequently, the results represented the worst-case scenarios. Some 3,706 trees were sampled during the 5-year-long survey, of which 2,773 were dead or dying, and 1,294 *Monochamus* beetles, which were considered as having the greatest potential of containing the PWN, were assayed to establish the frequency of the nematode within potential host trees and insect vectors. Another 4,325, forest-inhabiting insects were assayed for the nematode.

The results showed that the PWN was present in: (1) all provinces except Prince Edward Island, (2) isolated groups of one or two trees, (3) both the mucronated (m), the most common and widespread, and round-tailed (r) forms were present, (4) low, for example Ontario, to rare, for example British Columbia, frequencies, and (5) in low numbers in six species of pines, that is, Scots, *P. sylvestris*, jack, *P. banksiana*, lodgepole, *P. contorta*, red, *P. resinosa*, Ponderosa, *P. ponderosa* and eastern white, *P. strobus*. As well, (6) the "r" form was only found in pines while the "m" form was more prevalent in *Abies* and *Picea* spp. (7) the "r" form occurs in New Brunswick, Nova Scotia, Quebec, Ontario, Manitoba and Alberta, plus (8) the PWN was present in low numbers in balsam fir, *Abies balsamea*, white spruce, *Picea glauca*, black spruce, *P. mariana*, red spruce, *P. rubens*, tamarack, *Larix laricina* and Douglas fir, *Pseudotsuga menziesii*. Almost all trees in which the nematode was found were weakened by other factors. Regarding the insect assays, of the 1,519 insects assayed for the PWN, including 1,294 *Monochamus* spp., only 1 specimen of *Monochamus clamator* yielded the nematode, while no other vectors contained PWNs, including possible vectors from logs which had been inoculated with massive numbers of the nematode.

Based on the survey results, it was concluded that the risk of importing the PWN with Canadian lumber and logs, or indirectly through vectors, is considered to be very low, especially with continued programs to eliminate bark and grub holes.

3.5 Research on the PWN in Canada

Canadian research on the nematode and the disease was done at the Pacific Forestry Centre, Canadian Forest Service, in Victoria, and at Simon Fraser University in Burnaby, both in British Columbia. The Victoria work was headed up by

J. Sutherland and J. Webster was the lead researcher at Simon Fraser. In Victoria, several researchers such as T. Forge, T.S. Panesar, K. Futai, B. Yang, F.G. Peet, and T.S. Sahota carried out studies to determine PWN host range, mainly by inoculating conifer seedlings with Canadian PWN isolates, the pathogenicity of "m" and "r" form nematodes, host attractiveness, movement of the nematode through wood bark and PWN survival in wood chips. At Simon Fraser, research was done on topics such as on PWN pathogenicity by E. Riga and J.M. Webster, and T.A. Rutherford and J.M. Webster on the effect of temperature on pine wilt disease occurrence.

3.6 Overview of the PWN in the United States

Although numerous papers have been written about the PWN and pine wilt disease in the USA, the most recent overview is that of Dwinell and Nickle (1989). Much of the historical information given in the introduction of this chapter comes from that publication. Not only have they reviewed their own work, but they have also summarized the work of others in the USA and elsewhere, for example, related to pathogenicity, hosts, and insect vector relationships of the pathogen.

Three prerequisites are needed for disease development, that is, presence of (1) a pathogen (and for pine wilt disease also a vector), (2) a susceptible host, and (3) of environmental conditions favorable for disease development. Dwinell and Nickle (1989) reviewed the PWN and pine wilt disease situation in the USA, especially the first two points, and made the best case as to why pine wilt disease is only of minor importance in the USA and Canada.

3.7 Relationship of *Bursaphelenchus* with *Monochamus*, that is, Presence of the Pathogen and Vectors

In the USA, PWN dauerlarvae have been found associated with five *Monochamus* species (pine sawyers), that is, *M. carolinensis*, *M. scutellatus oregonensis*, *M. titillator*, *M. mutator* and *M. notatus*. In Canada, 14 species of potential PWN vectors are known. The biologies of Asian, North American, and Euro-Siberian *Monochamus* are similar; however, species differ in their geographic distributions, hosts, oviposition site preference, and so forth, including the length of the life cycle. Adult pine sawyers are attracted to newly dead or dying trees and freshly killed timber (including logs) for breeding. The cause of tree death is not crucial, for example, it can vary from inclement weather, death or injury from other insects or diseases or fire damage. In more northern parts, 2 years are required to complete the life cycle. In the presence of the callow adult of the vector, the third larval dispersal stage of PWN moults to the fourth larval stage, that is, the transmission stage or dauerlarva. These are transmitted to the host tree during maturation feeding by the beetles (primary transmission) or during oviposition in weakened trees (secondary transmission). Thus, both in Canada (see above) and the USA, the pathogen and one or more suitable vectors are present; however, even under these conditions pine wilt disease is rare.

3.8 Pathogenicity of the PWN to North American Pines: Presence of a Susceptible Host

Dwinell and Nickle (1989) state that "Much of the information about pathogenicity of the PWN in North America is based on seedling pathogenicity tests". They then point out that this information may have contributed to the impression that pine wilt is epidemic in North America. In fact, field pathogenicity tests in which the inoculation protocols were similar to the natural infection process, and fulfilled Koch's postulates, have only been done with Scots, *P. sylvestris*, slash, *P. taeda* and Japanese red, *P. densiflora* and Japanese black, *P. thunbergii*, pines (see references cited by Dwinell and Nickle 1989). Of these, only slash pine is native to North America and in the field, pine wilt occurrence on this species is extremely rare. Based on these scientific reports, Dwinell and Nickle (1989) conclude that native, North American conifers are either immune or highly resistant to pine wilt disease. In North America, isolation of the PWN most likely results from the trees being weakened by other factors followed by secondary transmission of the nematode during oviposition.

Rutherford and Webster (1987) reviewed the relationship of temperature and pine wilt disease. They concluded that in North America and Japan where the PWN and its vectors occur, pine wilt in susceptible pines only occurs where mean air temperatures are above 20°C for long periods. In these warm areas, susceptible pines grow disease free only at high elevations while pines resistant to pine wilt transcend the 20°C temperature threshold without becoming diseased. Finally, even in the presence of PWN and its vectors there are no reports of susceptible pines dying from the disease in those regions where mean summer temperatures are less than 20°C. Such high temperatures seldom occur in either Canada or the USA.

3.9 Conclusions

1. Surveys show that PWNs, and often-suitable vectors, occur throughout much of Canada and the USA; however, pine wilt disease is of only minor importance in these countries.
2. The reasons for this are that most indigenous North American conifers are resistant to the disease and high summer temperatures are of too short a duration to favor the pathogen and pine wilt development.

4
Pine Wilt Disease in China

Bo Guang Zhao

4.1 Introduction and Spread of the Disease

Pine wilt disease was first discovered in People's Republic of China in 1982 in Nanjing City, Jiangsu Province (Cheng et al. 1986). That year only 256 dead trees were found in the city. Subsequently, the disease has spread to 10 provinces and a city: Jiangsu, Zhejiang, Anhui, Guangdong, Shandong, Jiangxi, Hubei, Hunan, Yunnan, Guizhou Provinces and Chongqing City (Fig. I.7). The affected areas have reached nearly 80,000 ha and 50,000,000 trees have been killed by the disease. In recent years, the disease has spread quickly. The infected forest areas and the number of dead trees per year since 1982 are shown in Figs. I.8 and I.9. The disease threatens Huangshang and many other famous scenic spots and places of World Natural or Cultural Heritages; or both, and so, the disease has become a serious forest problem affecting local economies.

Since 1982, when the disease was identified in Jiangsu Province, it has spread in the next 10 years to several locations in Nanjing and Zhenjiang City and Wuxi City. In 1996, the disease was found in Suzhou City and, in 1998, in Changzhou City and Yangzhou City. Up to now, the disease has infested forests in 23 prefectures or cities in Jiangsu Province.

In Anhui Province, the disease was first found in 1988 in Maanshan City, He Prefecture and Mingguang City. Now occurs in the forests of 22 prefectures or cities, including Dangto, Chaqozhou, Xuancheng, Ningguo, Guangde, Chuzhou, Mingguang, Dingyuan, Laian, Quanjiao and Huainan Prefectures or cities in Anhui Province. At present damage occurs on 5,000 ha.

In Zhejiang Province the disease first appeared in a forest in Xiangshan Prefecture. Since then it has spread to 21 prefectures or cities, such as Dinghai, Diemu,

Department of Forest Protection, Nanjing Forestry University, Nanjing 210037, People's Republic of China

Tel.: +86-25-85427302, Fax: +86-25-85423922, e-mail: zhbg596@126.com, boguangzhao@yahoo.com

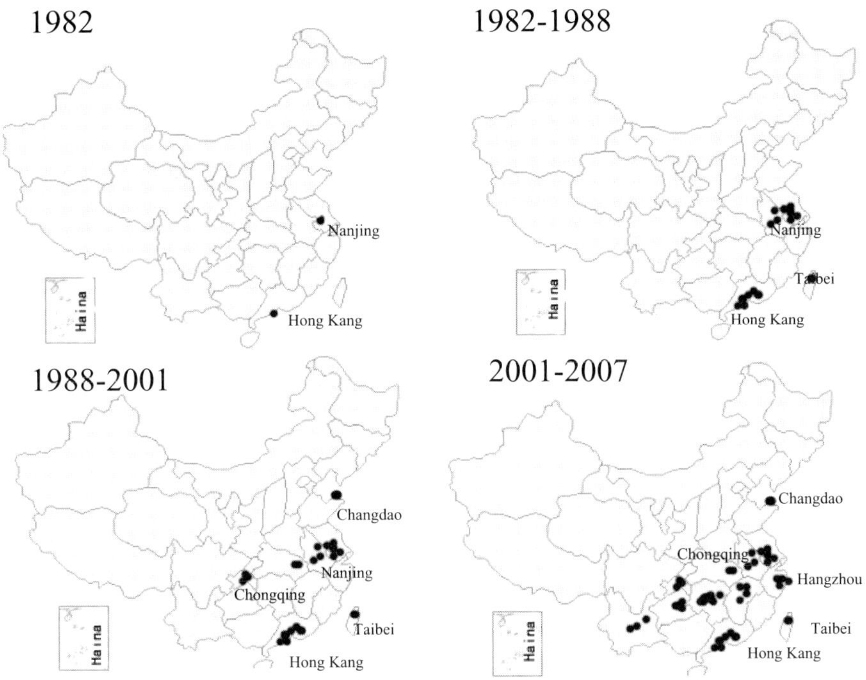

Fig. I.7 Distribution changes of pine wilt disease in China

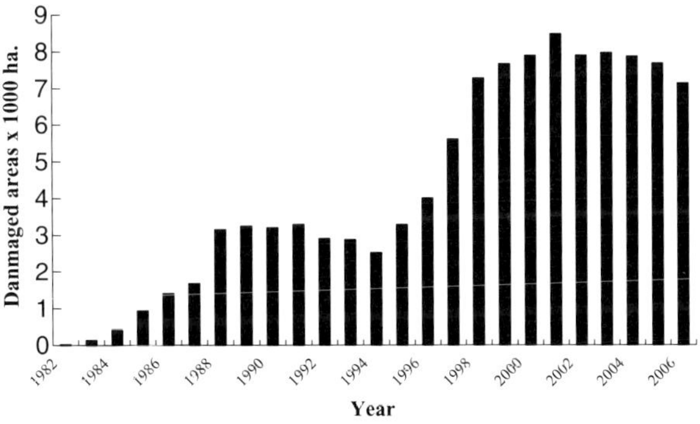

Fig. I.8 The annual areas damaged by pine wilt disease in mainland China (based on information from the Chinese Ministry of Forestry)

Ningbo, Ninghai and Hangzhou. In 2006, the total damaged areas rcached 23,000 ha of forest and 53,540,000 trees have died of the disease in Zhejiang Province.

Pine wilt was first discovered in Guangdong Province in 1988. Up to now, damaged forests have been found in 15 prefectures, including Huizhou City, Dongguan City, Shenzhen City and its prefectures. The total affected area has now increased to 1,600,000 ha.

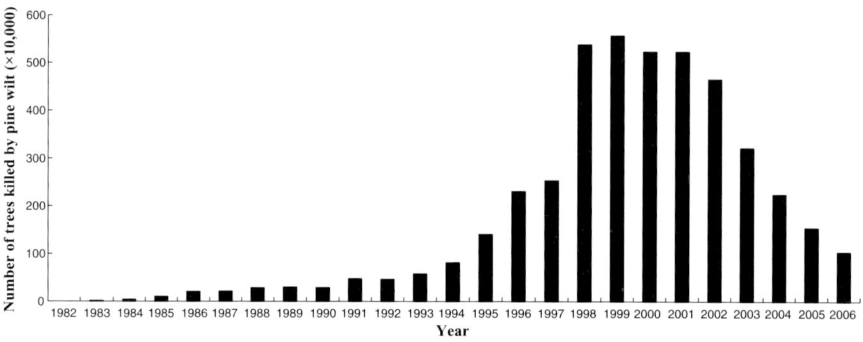

Fig. I.9 The annual number of trees killed by pine wilt disease in mainland China (based on information from the Chinese Ministry of Forestry)

In recent years, the disease has been found in several other provinces, Hubei Province in 2000, Chungqing in 2001, Jiangxi Province, Hunan Province, Guizhou Province in 2003, and Yunnan Province in 2004. The total damaged areas amount to 11,800 ha (all of these data have been provided by the Chinese Ministry of Forestry).

From the 1970s in Taiwan province, China, in the Northern Forest of the island, pines began to show wilt symptoms. In 1983, pine wilt disease was finally identified in Shimen Village of Taibei City. The damaged area has now reached 3,263 ha.

Five key factors determine occurrence of pine wilt disease: (1) yearly mean temperature, (2) the mean temperature in July, June and August, (3) latitude, (4) days above 25°C and (5) precipitation. Based on these criteria the southeast part of mainland China, is a potentially suitable area for occurrence of the pine wood nematodes (PWN), *Bursaphelenchus xylophilus* (Lu et al. 2005; Zhang et al. 2006). If a line were drawn from Beijing to Yunnan Province the southeast part of China below that line is potentially suitable for pine wilt occurrence.

4.2 Strategies to Control Pine Wilt Spread

The main control measures taken after the finding of pine wilt in China in 1982 are described below.

4.2.1 Quarantine Measures

After the disease was found in Nanjing in 1982, the Chinese government strengthened quarantine measures at Chinese ports of entry and established quarantine stations within China to prevent the movement of infected logs, lumber and wood products from infected to uninfested areas. These measures played an important role in stopping and delaying the spread of the disease (Li and Zhao 2006).

4.2.2 Monitoring the Disease

Traps with artificial attractant lures for monitoring the pine sawyer were developed and used in the disease management program. Several Synthetic formulations of lures for *Monochamus alternatus* have been developed and are used for controlling the disease (Liu et al. 2003; Wang et al. 2006).

4.2.3 Chemicals Used to Control the Disease

In some provinces, insecticide sprays or trunk injection for trees in affected forests has resulted in some success; however, such insecticides have resulted in harmful side effects to the environment. For example, the populations of certain birds and some beneficial insect parasites and predators have been reduced by the sprays. These side effects led to the government decision in the 1980s to stop spraying chemical insecticides for controlling pine wilt in Nanjing. Several new formulations of insecticides were developed, such as the spray formulation, Baosongling, and a microcapsule formula of insecticides, Green Mines, and the injection formulation Chongxianqing (Huang 2001). Because they are more environment friendly and high efficiency, they are used operationally in some severely affected forests to control pine sawyers.

4.2.4 Removal of Dead and Weakened Trees in the Affected Forests and Establishment of Isolation Belts Around Such Forests

Upon entry of the disease into China the main control measure was to remove dead trees from infested forests. In 1999, the Chinese government launched a project to manage pine wilt disease. The measures taken were mainly to treat and remove the dead trees in the infested forests and to monitor the pine sawyer populations. In Guandong Province 449,286 trees were killed in 1997 before these measures were implemented, however use of these practices resulted in fewer trees being killed by 35.2% in 2003 (Ye et al. 2005). Removing dead trees from affected forests and fumigation or burning the dead lumber to kill the pine sawyer larvae have been also very effective. In some severely damaged forests all the pine trees were clear cut. Those measures resulted in the reduction of both dead trees and infested forest areas in the two provinces (Pan 2000). By the turn of the century, the Chinese

Ministry of Forestry had taken strong measures to remove and treat dead trees in wilt-affected forests. These measures have led to a substantial reduction in numbers of diseased trees (Fig. I.9).

To protect parks, areas around historic buildings and so forth the government decided that cultural sites, such as Huangshan, a World Natural and Cultural Heritage site, to establish a 4-km wide, 100-km long pine tree-free strip around the mountain. Within this belt, all the pine trees were felled (Zhang 2005). The belt has had a major effect in protecting the forests inside the area from the disease; however, it could not stop PWN resulting from human activity.

There are 33,330,000 ha of forests in South-east of China of which 80,000 has been damaged, that is, 0.24% of the total forest area, the task of protecting the remaining healthy forest is still tough.

4.3 Research on PWN and its Vector Insect, *Monochamus alternatus*

After pine wilt was discovered in China, considerable time and research has been devoted to the pathogen and its insect vector. Over 400 papers on pine wilt disease have been published in Chinese (Ning et al. 2005). That information is summarized below.

There are 24 species or subspecies of the genus *Monochamus* in China, the most common being *M. alternatus*, *M. saltuarius* and *M. sutor* (Ning et al. 2004).

Research on the control *M. alternatus* by using biological agents (Table I.2) focused on, the parasite wasp, *Scleroderma guani*, and this parasite has been used operationally (Song et al. 1998; He et al. 2006; see Part VI). Techniques such as artificial rearing in forests have been studied and protocols established (Yang et al. 2005; Zhou et al. 2005). In the subtropical forests, the wasp can be established locally and it has long-lasting benefits (Song et al. 1998; Zhou et al. 2005).

Other biological control agents, which have been studied in detail include *Beauveria bassiana*, and *Metarrhizium anisopliae*, *Syncephalastrum racemosum* and *Hypomyces* sp., which produces metabolites that inhibit the development of plant nematodes including PWN (Mo et al. 2002; Ning et al. 2005). Strains of *B. bassiana* have been selected for their pathogenicity to *M. alternatus* (Zhang et al. 2000). The fungus in mixture with other agents has been used for a long time to control forest insect pests in China (Zhang et al. 2000; Lai et al. 2003).

Plant secondary metabolites were also examined for nematicidal chemicals for use against the PWN. *Sophora alopecuroides*, a desert shrub that grows in north-west China, contains high amounts of quinolizidine alkaloids. Zhao (1996) reported that aloperine, one of the alkaloids from the plant, was very toxicity to the PWN at LC50, 2.63×10^5 g. The molecular mechanism of the high toxicity of aloperine was also hypothesized (Zhao 1999) and later demonstrated experimentally (Li et al. 2000). Tests of the chemical showed that treated pines could be effectively protected (Zhao et al. 1998).

Table I.2 New control measures against *Monochamus alternatus* in China

	Species	Effects	Application	References
Parasite insects	*Scleroderma guani*	Wasps parasitize larvae of beetles	In the first year, 15,000–18,000 wasps hm^{-2} were released in Guangdong, and mortality was reduced upto 85.16–92.24% the next year.	Song (1998) and Zhang and Yang (2006)
	Dastarcus longulus	Pupae and larvae of *M. alternatus* can parasitize beetle	Large-scale field tests have not been done yet.	Yang (2004) and Zhang and Yang (2006)
Fungi	*Beauveria bassiana* *B. brongniatii* *B. tenella* *Metarhizium* sp. *Isaria farinose* *Aspergillus flavus* *Verticillium* spp. *Acremonium* sp.	Both adults and larvae can be infected with them, causing death	Spray the spores in forests	Zhang et al. (2000)
Insecticide	Green mine	Contact-breaking release microcapsules of pesticides	Spray	Tang et al. (1999) and Yan et al. (1999)

Other extracts from certain plants were also studied (Table I.3); however, no other active anti-PWN compounds have not been isolated or identified before 2003 (Liu et al. 2003).

The mechanism of pine wilt disease has been one of the main research projects supported by the Natural Science Foundation of China. Although some scientists still think PWN is the only pathogen causing wilting and tree death, this one-pathogen theory has been challenged by a new hypothesis that the disease is caused by PWN and the symbiotic pathogenic bacteria that it carries (Zhao et al. 2003). The new theory was proposed followings a series of inoculation experiments (Zhao et al. 2003). According to the new hypothesis bacteria carried by the nematode play an important role in the disease; therefore, control of those bacteria should result in reducing PWN pathogenicity. Experiments in which Japanese black pine or Masson pine seedlings were treated with antibiotics showed that such treatments delayed wilting (Zhao et al. 2000a; Chen et al. 2003).

Table I.3 Natural compounds in scientific research to control pine wood nematode (PWN) in China

	Species	Compounds	Toxicity	References
Fungi	*Syncephalastrum racemosum*	Water soluble	All PWNs died when cultured with the fungus	Sun (1997), and Zhou et al. (2005)
	Lampteromyces japonicu *Pleurotus spodoleucus* *P. corticatus* *P. ferulae* *P. memberancens*	Unknown toxins	>90% PWNs died when cultured with these fungi	Dong et al. 2000
	Streptomyces	Ramification of peptide acylcytosine nucleotides	PWN mortality 100% within 4 h at 144 μg ml^{-1}	Song et al. (2000)
	Basidiomycetes YL14	Unknown toxins	>90% PWNs died when cultured with the fungus	Li and Zhang (2001)
Antibiotics		Mixture of antibiotics	Inhibitory effects on symptoms of the disease	Zhao et al. 2000 Chen et al. 2003
Plants	*Sophora alopecuroide*	Aloperine	LC50 = 2.63×10^{-5} g ml^{-1}	Zhao (1996, 1999), and Zhao et al. (1998)
	Derris elliptic *Cephalotaxus fortune* *C. sinensis* *Paeonia suffruticosa* *Sophora viciifolia* *Dendranthema indicum*	Unknown toxins	Inhibitory effects on reproduction of PWN	Wen et al. (2001)

To identify dead trees or lumber from pine wilt killed trees a method has been devised for checking the pH of wood. The technique based on the observation that the pH of tissue from trees that are killed by the disease is significantly lower than that of healthy trees (You et al. 1994; Wang et al. 2001). Also several methods have been studied for identifying PWN. These include use of DNA probes, however, other techniques including PCR identification methods, including PCR–RAPD, PCR–RFLP and PCR–SSCP techniques and satellite DNA techniques have been developed (Wang 2004), but some problems still exist in using these techniques on a large scale, for example, their cost and accuracy are among the main obstacles to use in disease management.

Many fungi have been tested for PWN control in China, including some that produce nematicidal chemicals (Liu et al. 2003); however, most of their active components have not been isolated and identified. More research is needed in this area.

4.4 Research on Resistance of *Pinus massoniana* to Pine Wilt Disease

Because Masson pine, *Pinus massoniana*, grows well in the arid, sandy soil and the dry climate of certain areas of southeast China it is a major forest species in those areas. Before 1982, when pine wilt was first found in China some scientists thought that Masson pine was resistant to the disease; however, subsequently it has been shown to be only moderately resistant, and many Masson pines have been killed by the disease. Research showed that when Masson pines reached age older than 16 years that they begin to gradually lose their resistance to the disease (Wang et al. 2006). Recently, trees younger than 10 years old were also found to be killed by pine wilt. This indicates that the resistance of Masson pine or virulence of the pathogens are changing. *P. massoniana* resistance to the disease has been related to the tree growth rate, some metabolites, like phenolic compounds and less content of free amino acids in the resistant tree species (Xu et al. 2000).

Plant hormone, calcium, salicylic acid and ammonium were tested to increase *P. massoniana* wilt resistance. Applying calcium alone or combined with ammonium were shown to be the best treatments for increasing resistance to the disease (Ge et al. 1999).

Inoculation of trees form 40 Masson pine strains collected from South-east China showed that all of three resistant strains originated from Guangdong and Guangxi Provinces along China's-southern coast (Xu et al. 2000). Interestingly, the most virulent PWN strains were also reported to come from Guangdong Province (Gao et al. 2005).

5
Pine Wilt Disease in Korea

Sang-Chul Shin

5.1 Introduction

In Korea, pine trees are both culturally and spiritually important. According to the fourth forest resource survey from 1996 to 2005, pines occur widely on some 1,507,118 ha of land representing 23.5% of Korea's forest area and 15.1% of the country's land mass (Kwon 2006). Pines have been the dominant tree species in Korean forests even after the attack by the pine caterpillar in 1970s, the outbreak of pine needle gall midge in 1980s, and the occurrence of black pine blast scale in 1980s and 1990s. Because of the serious losses from pine wilt disease, which was first reported in Busan in 1988, this disease is a serious threat to Korea's pine forests (Yi et al. 1989). Despite the strenuous efforts to manage the disease over the last 19 years, pine wilt has spread to Mokpo, Sinan, and Yeongam in Jeonnam Province (west), Daegu, Gumi, and Andong in Gyeongbuk Province (inland), and Gangneung and Donghae in Gangwon (northeast) (Fig. I.10). Since 2006, at least in the southern Korea, spread of the disease has slowed as the result of intensive management (Fig. I.11). Until 2005 Japanese red pine (*Pinus densiflora*) and Japanese black pine (*P. thunbergii*) were reported as natural hosts of pine wilt disease in Korea; however, in 2006 Korean white pine (*P. koraiensis*) was found as being affected by pine wilt disease. This section gives a brief history and management strategy for pine wilt disease in Korea.

5.2 Pine Wilt in Korea

Like pine wilt disease in other countries, the pine wood nematode (PWN) *Bursaphelenchus xylophilus* causes pine wilt disease in Korea. This ca. 1-mm long nematode cannot be seen with the naked eye. A similar nematode, *B. mucronatus*,

Department of Forest Diseases and Insect Pests, Korea Forest Research Institute, 207 Cheongnyangni 2-dong, Dongdaemun-gu, Seoul 130-172, Republic of Korea

Tel.: +82-2-9612651, Fax: +82-2-9612679, e-mail: shinsc99@foa.go.kr

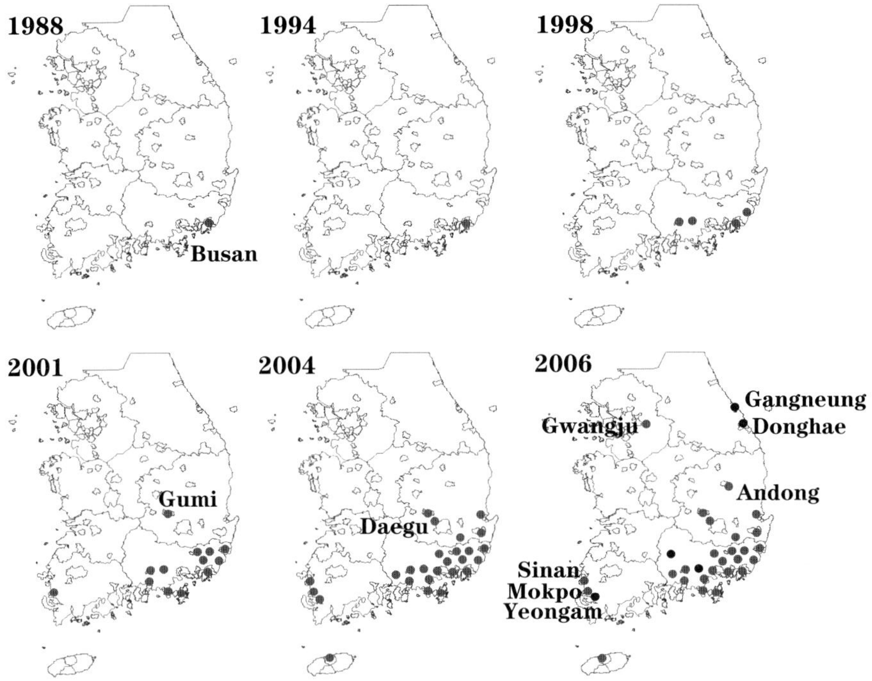

Fig. I.10 Spatial dispersal of pine wilt disease in Korea from 1988 to 2006. *Red dots* indicate the areas in which trees affected by pine wilt disease were observed and *blue dots* indicated the areas in which affected trees by pine wilt disease were found until 2005, but where no pine wilt disease has been observed since 2006 (see Color Plates)

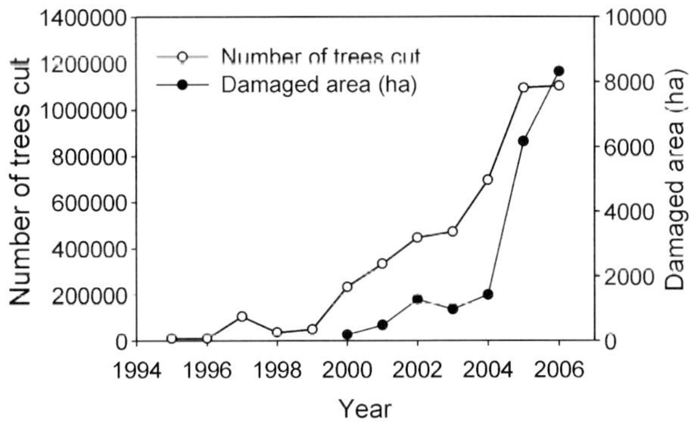

Fig. I.11 Sequential changes in the number of pine trees felled and areas damaged by pine wilt disease from 1994 to 2006

also occurs in pine trees. Recently, a European strain of *B. mucronatus* was identified by ribosomal ITS DNA sequence analysis (Han et al. 2008). As the nematode cannot move among trees by itself, it needs an insect vector. The Japanese pine sawyer, *Monochamus alternatus* and another pine sawyer, *M. saltuarius* are the vectors for red pine and probably for white pine, too. These two beetles have different distributions, probably resulting from differences in heat adaptation. *Monochamus alternatus* mainly occurs in southern Korea whereas *M. saltuarius* is found in the north of the country (Kwon et al. 2006). As well, the emergence periods for the two beetles differ with *M. alternatus* emerging from mid-May to early August while *M. saltuarius* emerges from early May to early June (Han et al. 2007). Pine wilt symptoms differ in the two host species with wilt symptoms being systematic in red pine while white pine wilts partially or systemically and has green needles even after infection. In infested forests, red pine mortality is 80% in the first year and the remaining 20% the next year; that is, red pine mortality reaches 100%, while in white pine mortality is lower even 2 years after wilt occurrence. In addition to these *Pinus* species, pine wilt disease is also naturally found in Douglas fir (*Pseudotsuga menziesii*), Eastern and European larch (*Larix* spp.) and blue spruce and white spruce (*Picea* spp.) in North America (Suzuki 2002).

5.3 Outlook and Status of Damage

Pine wilt disease was first found in Korea in Busan in 1988, and the affected area has increased continuously. Since 2000, newly infested areas have been found around the nation, especially in 15 cities and counties of 5 provinces such as Gangneung City in Gangwon Province, and Buk-gu in Daegu City. In 2005, another 15 cities and counties were reported as infested areas and the pine wilt disease was assumed to have spread naturally, in other words without any human activity, to 6 counties, districts, and cities (Jung-gu and Dong-gu in Ulsan City, Cheongdo County in Gyeongbuk, Uiryeong and Namhae Counties in Gyeongnam), and by human activity such as movement of logs infested with PWN to other nine districts, counties, and cities. From 1988 to 2005, the disease was reported from 53 districts, counties, and cities in 8 cities and provinces. In 2005 damaged occurred on 7,811 ha and the number of diseased trees removed were 138,441, 200,637, and 862,542 from 2003 to 2005 with yearly disease management costs of 6.9, 7.8 and 9.5 million US dollars per year, respectively. The average yearly cost was 8 million US dollars and this has been increasing annually. Consequently, a more effective and economic management strategy is required.

The spread of pine wilt disease by its insect vector, *M. alternatus*, is about 1.4 km per year (maximum 5 km). So, it is suspected that 38% of the increase in the size of the affected area was the results from spread attributable to human activity such as the movement of infested logs. However, control methods such as fumigation, aerial application of pesticides, injection of nematicides into tree trunks and use of the vector's natural enemies could reduce disease spread.

Since the disease has received much attention by the media, most Koreans know about pine wilt. As well, the government gives a reward to people who report the occurrence of pine wilt disease, and the transfer of pine trees from infected areas is restricted by a special law established in 2005. This law should restrict spread of the disease to new areas.

5.4 Control Methods and Problems

Control methods such as fumigation and burning of infested trees have been employed against pine wilt in Korea. Injection of nematicides into tree trunks and aerial application of pesticides helped in the battle against the disease. These control methods have changed according to research findings and social demand. Until 2001, burning of diseased trees was the main control method, but this changed to fumigation in 2002. In 2005, other control methods were developed such as year-round cutting of both wilt-affected and healthy pines in diseased forest.

5.4.1 Felling and Fumigation

Until now felling and fumigation of affected trees has been the most effective control method. Infected trees are felled and cut into 1–2 m^3 sections. Metham-sodium at 1 l m^{-3} is sprayed onto the tree sections, and then the logs are covered with a 0.1-mm thick vinyl sheet. The mortality for the beetle and the PWN is 100% after 7 days. Although the most effective method is to collect and fumigate whole trees and twigs more than 2 cm in diameter, it is, because of logistic problems, impossible to do this in the mountains.

5.4.2 Trunk Injection of Nematicides

This method has been used for prevention. A nematicide such as morantel tartrate is applied to the pine tree trunk; however, the cost is so high that application is restricted to nursery or ornamental trees. In 2005, abamectin and emamectin benzoate were registered for trunk injection use. Two nematicides (abamectin and emamectin benzoate) are being used in 2006; however, they are difficult to apply in large areas, and nematicides should be injected every 2 years because they are known to be effective at least within 2 years.

5.4.3 Felling and Crushing

Special machines are used to grind pine wilt diseased trees into sawdust or chips smaller than 1.5 cm in size, preventing survival of the pine sawyer beetle larvae.

An advantage of this procedure is that the sawdust and chips from the diseased trees can be used, but the end product is expensive since it is a labor-intensive procedure and the machines must be moved in the forests.

5.4.4 Felling and Burning PWN-Infested Pine Trees

Felling and burning infected trees is an easy, low cost and highly efficient method to control PWN, but its use is restricted to periods when forest fire risk is low. Also, heat from the fires can damage other pine trees.

5.4.5 Aerial Spraying of Insecticides

To prevent dispersal of the pine sawyer beetle, aerial spraying of insecticides is a highly efficient method that can be applied over a wide area. When adult beetles emerge, from May to July, beetle dispersal can be reduced by aerial spraying of fenitrothion or thiacloprid before the beetle is able to do maturation feeding on pine twigs. Concern about such insecticides causing environmental contamination is a major complaint of NGOs.

5.4.6 Environmentally Friendly Control Methods

Since some of the above control methods may lead to environmental problems, environmentally friendly control methods such as use of parasitoids, predators, entomopathogenic micro-organisms and non-pathogenic nematodes have replaced the use of pesticides. In 2005, parasitoids were imported from The People's Republic of China and are now being tested for field application. However, environmentally friendly control methods require a long time to kill pine sawyer beetles and the PWN. This is a major problem because Korean governmental policy calls for quick suppression of pine wilt disease.

5.4.7 Legal Methods

Most likely pine wilt disease was brought to Korea from Japan and long-distance dispersal of pine wilt disease within the country is attributable to human-related activities. This means that the introduction of pine wilt disease from overseas went undetected during quarantine inspections and the dispersal of pine wilt disease in Korea is attributed to commercial movement of infested pine wood. As such, government restrictions on the movement of such wood are needed to prevent the spread of pine wilt disease. In May 2005, the Korean government established a "special law to control pine wilt disease" and it came into effect in September 2005.

5.5 Inter-Disciplinary Research

To develop new pine wilt controls the Korean Forest Research Institute (KFRI) reorganized its research organization and projects, founding on 9 September 2005, "the Pine Wilt Disease Research Center". The Forest Disease and Insect Pests division of the KFRI focuses on basic research and the Southern Forest Research Center deals with control operations in the field. Other KFRI divisions concentrate on activities such as genetic studies, ecological analyses, the forest environment, the burning of infested materials, and aerial surveys. As well, another group consisting of universities, NGOs, research institutes and pesticide companies has been established.

5.6 Aims of the Research

Until now, Korean researcher dealt with *Bursaphelenchus* taxonomy (Lee et al. 1990), mortality rate of affected pine trees per month and timing of pine sawyer beetle emergence (Kim et al. 2003). Too, studies have been made on the pine sawyer beetle biology (Chung et al. 2003; Lee et al. 2004), fumigation methods using metham-sodium (Lee et al. 2003a). Other studies have centred on selecting appropriate pesticides for aerial spraying (Lee et al. 2003b) and tree injection, discovery of potential nematicides from plant essential oils (Park et al. 2005), and PWN diagnosis methods using RT-PCR (real-time polymerase chain reaction). In spite of the progress it is still necessary to develop new, more efficient and more environmentally friendly controls. From 2004 to 2006 work started on integrated research on pine wilt disease, development of control technologies and a pine wilt disease research center was established.

Over 6 million US dollars during 3 years is being spent on pine wilt-disease research and researchers in many universities and institutes have joined the projects. Cooperative research with China and Japan are in progress or planned; all of which should result in the development of practices for pine wilt disease control.

5.7 Conclusions

In the 19 years since pine wilt disease was first found in Korea the disease has spread, probably because initial eradication of the disease failed. However, the Korean government has established a "special law to control pine wilt disease" and put the law into operation, and researchers have concentrated their efforts on developing new control technologies. Hopefully the results will reduce the spread and damage resulting from the disease. We now know that pine wilt disease is a lethal disease in pine forests and can induce catastrophic damage. As such interest in and research on pine wilt disease is necessary if we want to maintain healthy pine forests in Korea.

6 Pine Wilt Disease in Portugal

Manuel M. Mota* and Paulo C. Vieira

6.1 Introduction

In Europe, species of the nematode genus *Bursaphelenchus* have been known and studied for a long time (Fuchs 1937; Rühm 1956). Earlier, except from a purely biological or ecological point of view, no particular interest was paid to this group of mycophagous nematodes. In 1979, however, a study conducted in southwestern France showed that the nematode *Bursaphelenchus lignicolous* was associated with declining pines (Baujard et al. 1979). This report caused alarm in Europe, since *B. lignicolous* is a synonym *of B. xylophilus*; the nematode in question was later identified as *B. mucronatus* (de Guiran and Boulbria 1986), which had been described as a new species that year. In 1984, a shipment of wood from North America to Finland was found to carry the pine wood nematode (PWN), *B. xylophilus* (Rautapaa 1986). This important interception prompted European authorities to develop more rigorous inspections at sea ports, and in particular of wood products coming from North America. However, no equivalent emphasis was placed on such products coming from East Asia. Between 1996 and 1999, an EU-funded project (RISKBURS) resulted in an updated survey of the *Bursaphelenchus* species in Europe. For an updated situation on the species distribution in the EU, see Braasch (2001). In 1999, the PWN, the causal agent of pine wilt disease, was first detected in the European Union (EU), in Portugal (Mota et al. 1999; Mota 2004), and this immediately prompted several national and EU governments to assess the extent of the nematode's distribution, and to restrict *B. xylophilus* and its insect vector (*Monochamus galloprovincialis*) to an area with a 30-km radius in the Setúbal Peninsula, 20 km south of Lisbon (Rodrigues 2008). The origin of the population

NemaLab-ICAM, Department of Biology, University of Évora, 7002-554 Évora, Portugal

*Tel.: +351-266-760800, Fax: +351-266-760914, e-mail: mmota@uevora.pt

of PWN found in Portugal remains unknown, although recent research indicates that it originated from Eastern Asia (Vieira et al. 2007). Several hypotheses have been suggested on how it entered the country, namely from North America or from Japan or China. World trade of wood products such as timber, wooden crates, and palettes play an important role in the potential dissemination of the PWN (Evans et al. 1996). In fact, human activities involving the movement of wood products may be the single most important factor in PWN spread. Despite the dedicated and concerted actions of government agencies, both the PWN and pine wilt disease continue to spread. In 2006 in Portugal, forestry and plant quarantine authorities (DGRF and DGPC) announced a new strategy for managing the problem. The plan is to establish a phytosanitary strip, 3-km wide, devoid of *Pinus pinaster*, surrounding the affected area, for the control and ultimately the eradication of the nematode, under the coordination of the national program for the control of the PWN (DGRF 2006). Research on the bioecology of the nematode and its insect vector, new detection methods, for example, involving real-time PCR, tree ecology and pathology, and control methods, has been underway since 1999. As well there are two major ongoing projects for the EU: PHRAME (http://www.forestresearch.gov.uk/website/forestresearch.nsf/ByUnique/INFD-63KGEF) and PortCheck (http://www.portcheck.eu.com/index.cfm). This research has been instrumental in helping to understand the scientific aspects of pine wilt disease. The objective of the present paper is to highlight the progress made in Portugal and the EU. International agreements (GATT, WTO) and sharing of scientific information is of paramount importance for achieving effective control of the nematode and its vector, and in turn protection of our forest ecosystems and forest economies.

6.2 General Surveys

With the discovery of PWN in Portugal in 1999, all EU governments were prompted to conduct general surveys for the nematode in their pine forests. Although some surveys had been conducted prior to 1999 (Braasch et al. 2000), it was not until then that new records of *Bursaphelenchus* species plus descriptions of new species began to appear (Braasch 2001; Escuer et al. 2004; Kulinich 2004; Magnusson et al. 2004; Michalopolous-Skarmoutsos et al. 2004; Penas et al. 2004, 2006a, b). These activities, funded mainly by national governments and the European Community, have been extremely useful in monitoring for the presence of PWN and also in gaining a better understanding of the diversity of the genus *Bursaphelenchus*. The results of these surveys are assessable on the EU web page on the PWN. In Portugal, such surveys have been undertaken in the affected area on the Setúbal Peninsula, near Lisbon and outside the affected area (Mota and Vieira 2004; Rodrigues 2008). These results are available on the PROLUNP web site (DGRF 2006). To date the PWN has only been found within a well-defined area in Portugal, but not elsewhere in the EU.

6.3 PWN Taxonomy and Biology

The genus *Bursaphelenchus* was established in 1937 by Fuchs, with *B. piniperdae* as the type species. To date, more then 80 valid *Bursaphelenchus* species have been described worldwide (Ryss et al. 2005; Vieira et al. 2006; see Chap. 9), mainly in the northern hemisphere. However, after 1999 26 new species have been described, frequently from Asia. Although 28 *Bursaphelenchus* species are known from Europe (Braasch 2001), no research had been done on this genus in Portugal before detection of the PWN in 1999. Since then, several studies have been made to determine the species' diversity and distribution within Portugal. Although the survey has focused mainly on maritime pine (*P. pinaster*), nine other *Bursaphelenchus* species, besides *B. xylophilus*, have been identified and characterized based on morphological characters and by using molecular biology tools (ITS-RFLP, 18S and 28S rDNA D2/D3 domain sequences, RAPD-PCR) (Mota et al. 1999; Penas et al. 2004, 2006a; Vieira et al. 2007). The *Bursaphelenchus* species, namely *B. hellenicus*, *B. leoni*, *B. mucronatus*, *B. pinasteri*, *B. sexdentati*, *B. terastospicularis*, *B. tusciae* (Penas et al. 2004) mainly occur in north and central Portugal, coinciding with the occurrence of the major pine forests. Recently, a new *Bursaphelenchus* species, *B. antoniae*, has been described (Penas et al. 2006a). To determine the insect vectors of *Bursaphelenchus* spp. associated with *P. pinaster*, bark- and wood-boring insects belonging mainly to the families Cerambycidae, Scolytidae, Buprestidae and Curculionidae (Coleoptera) were captured from specific locations in Portugal, and their associated nematodes identified using ITS-RFLP analysis of dauer juveniles and morphological identification of the adults (Penas et al. 2006b). Several insect-nematode associations have beed identified such as *B. teratospicularis* and *B. sexdentati* associated with *Orthotomicus erosus*; *B. tusciae*, *B. sexdentati* and/or *B. pinophilus* with *Hylurgus ligniperda* and *B. hellenicus* with *Tomicus piniperda*, *Ips sexdentatus* and *H. ligniperda*. Other nematode genera besides *Bursaphelenchus*, which were found associated with the insects included two species of *Ektaphelenchus*, and one species each of the genera *Parasitorhabditis*, *Parasitaphelenchus*, *Contortylenchus*, and other unidentified nematodes. Of major interest is the molecular characterization of the PWN isolates within the affected area, using rDNA genes, and in particular the ITS regions (ITS-1 and ITS-2) of rDNA. Intra-specific variability using RAPD–PCR techniques has proven very useful for evaluating genetic distances and for helping develop phylogenies and pathway analysis of world populations of the PWN (Burgermeister et al. 2005b; Ye et al. 2007; Metge and Burgermeister 2008). Recent studies (Vieira et al. 2007) have focused on the genetic diversity of *B. xylophilus* in Portugal using RAPD–PCR techniques for evaluating intra-specific variation. The results show that there is no significant genetic variation among PWN isolates from the pine wilt disease affected area. Also, the data seems to suggest a recent and single PWN introduction into Portugal; however, since no correlation could be made between the genetic and geographic matrices, it seems that this technique is not useful in such studies. Nonetheless, the studies have shown that the Portuguese isolates display a much lower genetic distance to Asian isolates, especially from China as compared to North American isolates (Vieira et al. 2007). Recently the use of satDNA as a marker for *B. xylophilus* has been developed in the EU (François et al. 2007; Castagnone-Sereno et al. 2008); and it may be a useful tool for intraspecific analysis of the PWN and, together with mtDNA, provide additional information on the geographic spread of he nematode. Such information would supply important information for pathway analysis (P. Vieira, personal communication). Also, PCR using a "real-time" technique (Leal 2007) has been developed and may be very useful when field equipment becomes available for use by quarantine authorities. This is a main goal of the EU "PortCheck" project mentioned above; initial ring-testing, utilizing a species-specific probe based on the *Msp*I satellite DNA family, has already been done in Portugal. The results have demonstrated great precision in discriminating *B. xylophilus*, however the issue of a reliable sampling method from the tree remains to be solved. Regarding PWN biology, and specifically cytogenetics, progress has been made in clarifying the basic chromosome number of the species ($n = 6$) and also the possible sex determination mechanism (XX–XY) (Hasegawa et al. 2006). Ongoing research should hopefully provide more detailed information on PWN cell division and developmental biology processes, and also the possibility of triploid populations (K. Hasegawa, personal communication).

6.4 Insect Vector, *Monochamus* spp.

Initial surveys done in Portugal (Sousa et al. 2001, 2002) demonstrated that the cerambycid beetle *Monochamus galloprovincialis* was the only insect vector for the PWN. Elsewhere, studies have focused on surveying insects for the presence of nematodes, tree-insect acoustics, and molecular taxonomy (Garcia-Alvarez et al. 2008; Vincent et al. 2008; C. Tomiczek, personal communication). Research on the bioecology of the insect vector in Portugal has provided much data such as how the nematode enters the vector, PWN population dynamics inside the vector (Naves et al. 2006a), feeding and oviposition (Naves et al. 2006b, 2007), flight patterns, effectiveness of lures and traps, reproduction (Naves et al. 2006c), and also on natural enemies such as parasitoids (Naves et al. 2005). The latter paper includes a review of parasitoids from East Asia and North America. The reproductive potential of *M. galloprovincialis* in Portugal seems to be less than elsewhere (Naves et al. 2006c). Studies on feeding and oviposition demonstrate that in Portugal *Pinus sylvestris* and *P. halepensis* are also good hosts for the PWN (Naves et al. 2006b). Laboratory studies on PWN transmission to *P. pinaster* by *M. galloprovincialis*, have provided valuable information including the fact that the critical period for nematode infection to the tree is within the first few weeks of beetle emergence. This type of information is particularly useful for control measures (Naves et al. 2007).

6.5 Disease Modeling and Climate

Earlier, a pest risk analysis (PRA) had been done regarding EU concerns about wood shipments arriving in Europe (Evans 1996). However, this PRA was done without PWN actually being present in Europe. After 1999 the need arose for a new PRA with the inclusion of new perspectives and parameters. The EU funded an eight partner (six nation) consortium, in 2002, designated "PHRAME" (Development of improved pest risk analysis techniques for quarantine pests, using the PWN in Portugal as a model system) which was supposed to, among other things, establish a predictive model and sub-models, based on numerous parameters related to climate, the nematode, the soil, the insect vector, for risk evaluation in the EU. One goal within this project was "process-based modeling" (PBM) on the plant system and which considered a mechanistic approach to analysing the disease progress, based on water potential and carbon availability (Evans et al. 2008). Another approach has been to use spatial modeling, based on a set number of eco-climatic variables (Pereira and Roque 2008). The results from these models are still being tested, both in the laboratory and the field, and will hopefully continue during the coming years. The recent concerns over global warming and climatic changes may constitute an increased threat to pine forests in Portugal (J. Corte-Real, personal communication).

6.6 Pathogenicity

In Portugal, *P. pinaster* is most likely the only known host as it is the preferred pine upon which the insect vector does maturation feeding. However, *P. sylvestris* and *P. halepensis*, are also fed upon by *M. galloprovincialis* (Naves et al. 2006b), but their abundance and distribution is relatively limited in Portugal. In nature, *P. pinea* is not affected by PWN since the vector, *M. galloprovincialis*, does not feed upon or colonize the tree. Recently, preliminary results (M.M. Mota et al., unpublished data) have demonstrated that PWN can invade, multiply in, infect and kill *P. pinea*, but less quickly than for *P. pinaster*. Inoculation studies using Portuguese and Japanese PWN isolates on *P. thunbergii* (Mota et al. 2007) show more virulence (i.e., greater mortality) for one Portuguese isolate (T) to Japanese black pine. Other recent studies have dealt with PWN distribution and migration behavior when inoculated into a susceptible host (Daub et al. 2008). Until now, and although studies have been made on the development of disease and of resistance factors, no single report has been made on the plant expression (mRNA, proteins) when in the presence of the PWN. This information is essential for understanding the pathogenicity of PWN on pine species and also for making decisions on reforestion practices. In 2007–2008, under a COST program (COST 872: http://www.cost.esf.org/index.php?id=181&action_number=872), it is expected that some progress may be made in this area, in Portugal and elsewhere in the EU.

6.7 Other Pine Wilt Disease Research

A deluge of information is a common feature of research in any scientific area. Today it is extremely difficult to keep abreast of all the relevant literature in a particular field. Also, it is important to establish new taxonomic keys, based mainly on computer technology and using specific algorithms. Regarding taxonomical databases, much effort has been put forward during the past few years to access all published descriptions of all *Bursaphelenchus* species, in pdf format, making it less time consuming to access such documents (Vieira et al. 2004, 2006; Eisenback et al. 2008). Also, programs such as EndNote or RefManager have become increasingly helpful in organizing bibliography information. Regarding taxonomical keys, a new system, based on the simultaneous use of several morphological and morphometric characters (polytomous key) has been developed for the identification of *Bursaphelenchus* species (Ryss et al. 2004).

Another important issue concerns quarantine and international borders (Braasch and Enzian 2004). Joint collaborations between the EU and neighboring countries such as Turkey and Russia have been instrumental in helping prevent the spread of the PWN and its insect vectors (Kulinich 2004; Akbulut et al. 2006).

6.8 Future Outlook and Conclusions

1. The PWN continues to pose a threat to forests worldwide (Mota and Vieira 2008). International trade and globalization, despite its great economic benefits, carries a certain amount of risk in introducing new pathogens and pests to countries and continents.
2. In recent years, the EU has particularly felt this risk with the introduction of the PWN into Portugal. We still do not know exactly where the nematode originated, although preliminary scientific results point to an East Asian source.
3. It is only through the mutual cooperation and exchange of scientific and technical information that these issues can be minimized or resolved. Also, the basis for decision-making must always be based on scientific information.

7
Concluding Remarks

Part I of this book gives an overview of the history of pine wilt disease in Japan, China, Korea, and North American and Europe. The first author, K. Futai, gives the history of pine wilt disease in Japan starting early in the twentieth century. As well he describes the spread of the disease in Japan and some factors that may have influenced the spread. Various factors such as the role of asymptomatic carriers, which to date have been largely unstudied, may be very important in the persistence and spread of pine wilt disease.

In Chap. 3, J.R. Sutherland briefly summarizes the historical events in North America, and reviews both the PWN situation in Canada and the United States, and some research programs carried out in these countries. He lists three prerequisites needed for disease development, and explains why pine wilt disease is only of minor importance in the USA and Canada. Based on the report of Dwinell and Nickle (1989), he concludes that North American conifers are either immune to or highly resistant to pine wilt disease, and high summer temperatures are of too short a duration to favor the pathogen and pine wilt development.

In Chap. 4, B.G. Zhao describes the invasion and spread of pine wilt disease throughout China from two points of view, that is, the size of the area damaged and the number of trees killed by the disease. Also, he gives the control strategies adopted by the Chinese government including using within China quarantine measures to stop and delay pine wilt spread. As well he recommends newly developed insecticides, sanitation in diseased forests and the use of cutting all trees in belts around affected areas to protect surrounding, unaffected forests. He then reviews the progress of Chinese research on PWN and its vector, and the resistance of native Chinese pines.

At the beginning of Chap. 5, S.-C. Shin points out both the spiritual and cultural importance of pines in Korea. As is the case in China and Japan, pine trees are also an important forest resource in Korea. Shin then reviews the historical development of pine wilt disease in Korea and the control procedures that the Koreans have adopted after evaluating the results and the problems associated with the procedures. Next, the author covers the development of pine wilt research in Korea and stresses the importancc of cooperative work between countries, governments institutes and universities.

In Chap. 6, M.M. Mota and P.C. Vieira report historical pine wilt disease events in Europe, and introduce how EU authorities have developed quarantine measures to keep the nematode and vector away from the EU. Regardless of such efforts, however, the PWN was detected in Portugal in 1999. Since then, two major projects, PHRAME and PortCheck were established to understand the scientific basis of pine wilt disease and provide effective control of the disease. After highlighting the progress achieved in Portugal and in the EU by these projects, authors emphasize the importance of international agreements and sharing of scientific information to effectively control the nematode and its vector.

Part II
Pine Wood Nematode

8
Introduction

Pine wilt disease results from a multitude of complicated, biological organisms, that is, the pathogen, a host and an insect vector and climatic conditions. The general disease cycle for pine wilt disease in the Asian temperate zone is as follows. In early summer, the vector of the disease, the Japanese pine sawyer, *Monochamus alternatus*, which harbors the dispersal fourth-stage (dauer) juveniles of the pathogenic pine wood nematode, *Bursaphelenchus xylophilus* (PWN) in its tracheal system, emerge from dead pine trees and feed (maturation feeding) on the twigs of healthy pine trees. PWNs invade the healthy host tree through the feeding wounds made at this time. They then feed on the host's cells and multiply and eventually kill the host tree. After the death of the tree, the PWNs feed on fungi growing in the dead tree and maintain their population until the next year, while mature vectors lay eggs in recently killed trees, and the larvae which hatch from the eggs grow and become vectors the next year.

The above disease cycle seems uncomplicated; however, it includes many complicated factors. For example, PWNs kill their host trees, but their pathogenicity, (virulence), varies drastically among PWN isolates, and the resistance of host trees also varies among individuals and among species. As for dispersal of pine wilt disease, the vector preference of PWN is very narrow, that is, the beetles *Monochamus alternatus* as the primary vector and sometimes also *M. saltualius*, and dauer juveniles are formed in response to signals produced by the appropriate vectors. Certain issues such as how the PWN kills trees, the origin of their pathogenicity, how they choose vectors and what is the cue for dauer induction, remain unclear.

In this part, the disease cycle is explained, focusing on the "nematode". Basic information about the PWN, its taxonomic and systematic details and related nematode species and genera are described in Chap. 9. The genus *Bursaphelenchus*, to which the PWN belongs, varies morphologically and ecologically, and its generic definition is very wide. The evolutional process of the life history and morphology of the PWN and its relatives is discussed based on a comparison of the morphology, biology and molecular phylogeny of *Bursaphelenchus* spp. The difficulties and problems of morphological observation and morphological taxonomy of this genus are also described.

In Chap. 10, studies on the "genes" of PWN are introduced. Currently, an expressed sequence (EST tag) project is ongoing in Japan, and other EST project groups are working in other countries. To date, 13,000 EST of the PWN and 4,000 EST of *B. mucronatus*, the sister species of the PWN, are available on line. From these EST analyses, some interesting results have been found, for example, horizontal cellulase gene transmission from fungi to nematode. In this section, recent genetic information about the PWN and the future use of this information is discussed.

In Chap. 11, the life cycle of the PWN is described at the embryogenetic level. The cascade of PWN development, that is, developmental processes from sperm and egg cells to adults, is basically the same as that of *Caenorhabditis elegans*, a biological model organism; however, several characteristic patterns have been observed for the PWN. Some of the characteristic developmental features of PWN found during fertilization to early embryonic development are explained in detail.

The population ecology of the PWN within the host tree is described in Chap. 12. The PWN reproduces by feeding on the tissue of living trees and fungi growing in the dead wood. When the PWN invades healthy pine trees, whether or not they can obtain their first feeding resource, plant tissue, is determined by the balance between PWN pathogenicity and host resistance, while if the nematode enters a tree which is already dead, pathogenicity may not affect their fate. Also, there may be competition, for example, competition for food, among different populations of the PWN and among different nematode species. Regardless, pathogenicity can be a powerful tool to obtain the best feeding resources and habitat. Here, transition of the genetic structure of the PWN population during propagation is explained, focusing on the evolution of pathogenicity at the population level.

In Chap. 13, the biological (ecological) traits of the PWN are described, focusing on the vector of the PWN. There are two routes by which pine wilt disease spreads, that is, long-distance spread resulting from human transportation of PWN-infested logs and short-distance spread by vector beetles. In this section, the behavior of PWN when entering vector beetles, departure from such beetles and the invasion of host trees are noted as key factors in both kinds of disease spread. Several hypotheses about the controlling factors of nematode behavior, which are scientifically interesting, are also explained.

9 Taxonomy and Systematics of the Nematode Genus *Bursaphelenchus* (Nematoda: Parasitaphelenchidae)

Natsumi Kanzaki

9.1 Introduction

Nematodes in the genus *Bursaphelenchus*, which are mycophagus or plant parasitic, or both, have been considered a potential risk to cultivated plants, especially conifers, since the end of the 1970s. The reason for this is that the genus contains two virulent plant pathogens, the pine wood nematode (PWN), *B. xylophilus*, and the red ring nematode, *B. cocophilus*. To date almost 90 *Bursaphelenchus* species have been described (Hunt 1993; Ryss et al. 2005; Kanzaki 2006; see Table II.1), especially from Europe (Rühm 1956; Braasch 2001) and the USA (Massey 1974), as associates of coleopteran beetles; however, because of the finding of the PWN in Portugal (Mota et al. 1999), the practical importance of the taxonomy of this genus has been re-evaluated worldwide. Recently, probably because of the global interest in this nematode genus, the number of newly described species from Asian countries such as China, Thailand and Japan, where in the past only a few *Bursaphelenchus* nematodes have been reported, has increased (Braasch and Braasch-Bidasak 2002; Braasch et al. 2005; Gu et al. 2005, 2006a, b; Kanzaki 2006; Kanzaki and Futai 2007).

The main purposes of this chapter is: (1) to bring together information on the taxonomy and systematics of the genus *Bursaphelenchus*, that is, the taxonomic status of the genus within the family Parasitaphelenchidae and the superfamily Aphelenchoidoidea, (2) review the morphological characteristics for identification and molecular systematics, and (3) discuss the morphological and ecological evolution of the genus.

Forest Pathology Laboratory, Forestry and Forest Products Research Institute, 1 Matsunosato, Tsukuba 305-8687, Japan

Tel.: +81-29-829-8246, Fax: +81-29-874-3720, e-mail: nkanzaki@affrc.go.jp

The information in this chapter is an updated version of a review article by N. Kanzaki (2006) Taxonomy and systematics of *Bursaphelenchus* nematodes. J Jpn For Soc 88:392–406, originally published in Japanese.

Table II.1 Species list of the genus *Bursaphelenchus*

Species name	Original description	Ryss et al. (2006) grouping	Molecular profiles[h]
B. aberrans	Fang et al. (2002a)	aberrans	—[i]
B. abietinus	Braasch and Schmutzenhofer (2000)	piniperdae	RS
B. abruptus	Giblin-Davis et al. (1993)	xylophilus	RS
B. africanus	Braasch et al. (2006d)	africanus[e,f]	RS
B. anamurius	Akbulut et al. (2007)	piniperdae[f]	R
B. anatolius	Giblin-Davis et al. (2005)	hunti[f]	S
B. antoniae	Penas et al. (2006a)	piniperdae	RS
B. arthuri	Burgermeister et al. (2005a)	hunti[f]	RS
B. baujardi	Walia et al. (2003)	xylophilus	—
B. bestiolus	Massey (1974)	piniperdae	—
B. burgermeisteri	Braasch et al. (2007)	africanus[f]	RS
B. borealis	Korenchenko (1980)	borealis	RS
B. chitwoodi	Rühm (1956)	piniperdae	—
B. cocophilus	Cobb (1919)	hunti	S
B. clavicauda	Kanzaki et al. (2007)	piniperdae[f]	RS
B. conicaudatus	Kanzaki et al. (2000)	xylophilus	RS
B. conjunctus[a]	Fuchs (1930)	—[g]	—
B. conurus[a]	Steiner (1932)	—[g]	—
B. corneolus	Massey (1966)	piniperdae	S
B. crenati	Rühm (1956)	xylophilus	—
B. cryphali[a]	Fuchs (1930)	borealis	—
B. curvicaudatus	Wang J et al. (2005)	piniperdae[f]	R
B. debrae	Hazir et al. (2007)	hunti[f]	S
B. digitulus	Loof (1964)	eidmanni	—
B. dongguanensis	Fang et al. (2002b)	hunti	—
B. doui	Braasch et al. (2005)	xylophilus[f]	RS
B. eggersi	Rühm (1956)	piniperdae	RS
B. eidmanni	Rühm (1956)	eidmanni	—
B. elytrus	Massey (1971a)	aberrans	—
B. eproctatus	Sriwati et al. (2008)	piniperdae	—
B. eremus	Rühm (1956)	piniperdae	RS
B. eroshenkii	Kolossova (1997)	xylophilus	—
B. erosus	Kurashvili et al. (1980)	eidmanni	—
B. eucarpus	Rühm (1956)	piniperdae	—
B. fraudulentus	Rühm (1956)	xylophilus	RS
B. fuchsi	Kruglik and Eroshenko (2004)	piniperdae	—
B. fungivorus	Franklin and Hooper (1962)	hunti	RS
B. georgicus	Devdariani et al. (1980)	piniperdae	
B. gerberae	Giblin-Davis et al. (2006a)	piniperdae[f]	S
B. glochis	Brzeski and Baujard (1997)	piniperdae	—
B. gonzalezi	Loof (1964)	hunti	—
B. hellenicus	Skarmoutsos et al. (1998)	piniperdae	RS
B. hildegardae	Braasch et al. (2006b)	piniperdae[f]	RS
B. hofmanni	Braasch (1998)	piniperdae	RS
B. hunanensis	Yin et al. (1988)	piniperdae	—
B. hunti	Steiner (1935)	hunti	—
B. hylobianum	Korenchenko (1980)	piniperdae	RS
B. idius	Rühm (1956)	aberrans	—
B. incurvus	Rühm (1956)	piniperdae	—

(continued)

Table II.1 (Continued)

Species name	Original description	Ryss et al. (2006) grouping	Molecular profiles[h]
B. kevini	Giblin et al. (1984)	hunti	S
B. kolymensis	Korentchenko (1980)	xylophilus	—[j]
B. leoni	Baujard (1980)	borealis	—
B. lignophilus[b]	Körner (1954)	—[g]	—
B. lini	Braasch (2004)	piniperdae	RS
B. luxuriosae	Kanzaki and Futai (2003a)	xylophilus	RS
B. maxbassiensis	Massey (1971b)	piniperdae	—
B. minutes	Walia et al. (2003)	piniperdae	—
B. mucronatus	Mamiya and Enda (1979)	xylophilus	RS[j]
B. naujaci	Baujard (1980)	piniperdae	—
B. newmexicanus	Massey (1974)	piniperdae	—
B. nuesslini	Rühm (1956)	piniperdae	—
B. paracorneolus	Braasch (2000)	piniperdae	RS
B. parvispicularis	Kanzaki and Futai (2005)	piniperdae	R[k]S
B. pinasteri	Baujard (1980)	piniperdae	RS
B. piniperdae	Fuchs (1937)	piniperdae	—
B. pinophilus	Brzeski and Baujard (1997)	piniperdae	R
B. pityogeni	Massey (1974)	piniperdae	—
B. platzeri	Giblin-Davis et al. (2006b)	hunti[f]	S
B. poligraphi	Fuchs (1937)	piniperdae	RS
B. rainulfi	Braasch and Burgermeister (2002)	piniperdae	RS
B. ratzeburgii	Rühm (1956)	piniperdae	—
B. ruehmi[c]	Baker (1962)	—[g]	—
B. sachsi	Rühm (1956)	piniperdae	—
B. scolyti	Massey (1974)	piniperdae	—
B. seani	Giblin and Kaya (1983)	hunti	RS
B. sexdentati	Rühm (1960)	piniperdae	RS
syn. *B. bakeri*	Rühm (1964)		
B. silvestris	Lieutier and Lamond (1978)	borealis	—
B. sinensis	Palmisano et al. (2004)	aberrans	RS[i]
B. singaporensis	Gu et al. (2005)	xylophilus[f]	R[l]S
B. steineri	Rühm (1956)	eidmanni	—
B. sutoricus	Devdariani (1974)	piniperdae	—
B. sychnus	Rühm (1956)	piniperdae	—
B. talonus	Thorne (1935) and Kaisa (2003)[d]	piniperdae	—
B. teratospicularis	Kakuliya and Devdariani (1965)	eidmanni	—
B. thailandae	Braasch and Braasch-Bidasak (2002)	piniperdae	RS
B. tritrunculus	Massey (1974)	piniperdae	—
B. tusciae	Ambrogioni and Palmisano (1998)	borealis	RS
B. typographi	Kakuliya (1967)	piniperdae	—
B. uncispicularis	Zhou et al. (2007)	borealis	—
B. vallesianus	Braasch et al. (2004)	piniperdae	RS
B. varicauda	Thong and Webster (1983)	piniperdae	—
B. wakuae	Kurashvili et al. (1980)	piniperdae	—
B. wilfordi	Massey (1964)	piniperdae	—

(continued)

Table II.1 (Continued)

Species name	Original description	Ryss et al. (2006) grouping	Molecular profiles[h]
B. willi	Massey (1974)	piniperdae	—
B. willibaldi	Schönfeld et al. (2006)	hunti[f]	RS
B. xerokarterus	Rühm (1956)	piniperdae	—
B. xylophilus	Steiner and Buhrer (1934)	xylophilus	RS
syn. *B. lignicolus*	Mamiya and Kiyohara (1972)		
B. yongensis	Gu et al. (2006a)	piniperdae[f]	RS

[a]"Species inquirendae" in Hunt (1993)
[b]"Species incertae sedis" in Hunt (1993)
[c]"Species indeterminatae" in Hunt (1993)
[d]Redescription
[e]New group proposed by Braasch et al. (2006d)
[f]Described after Ryss et al. (2005)
[g]Excruded from the genus by Ryss et al. (2005)
[h]R: ITS-RFLP profiles are reported in Burgermeister et al. (2005); S: DNA sequence(s) is (are) available from GenBank Database
[i]"*B. aberrans*" in Burgermeister et al. (2005) was corrected to *B. sinensis* by Kanzaki and Futai (2007)
[j]*B. kolymensis* might be a junior synonym of *B. mucronatus* European type
[k]Shown in Kanzaki and Futai (2005)
[l]Shown in Gu et al. (2005)

9.2 Taxonomic Status of the Genus *Bursaphelenchus*

The genus *Bursaphelenchus* was established by Fuchs (1937) with the type species, *B. piniperdae*. According to Hunt (1993), the genus is a member of the Family Parasitaphelenchidae, Superfamily Aphelenchoidoidea, Suborder Aphelenchina, Order Aphelenchida. Although the Superfamily Aphelenchoidoidea contains some predatory nematodes, obligate plant parasites and insect parasites, most *Bursaphelenchus* species are free-living mycophagus species inhabiting soil or dead plant material, including dead wood. Many species are also known as entomophilic (phoretic) nematodes. The families and genera belonging to the superfamily are listed below.

Currently, the taxonomy of the phylum Nematoda is changing drastically, and the order Aphelenchida, to which the genus *Bursaphelenchus* belongs, is now considered as belonging to the Superfamily Aphelenchoidea, Infraorder Tylenchomorpha, Order Rhabditida based on molecular phylogenetic analyses conducted by Blaxter et al. (1998) and De Ley and Blaxter (2002). However, the construction of the new taxonomic system has not been completed; consequently, the widely accepted system proposed by Hunt (1993) is used here. The main feeding habits are indicated after each family name.

Order Aphelenchida
Superfamily Aphelenchoidoidea

Family Aphelenchoididae: mycophagus, entomoparasitic, plant parasitic
  Subfamily Aphelenchoidinae
    Genera: *Aphelenchoides*, *Laimaphelenchus*, *Megadorus*, *Ruehmaphelenchus*, *Schistonchus*, *Sheraphelenchus*, *Tylaphelenchus*
  Subfamily Anomyctinae
    Genus *Anomyctus*
Family Seinuridae: predator
  Subfamily Seinurinae
    Genera *Seinura*, *Aprutides*, *Papuaphelenchus*, *Paraseinura*
Family Ektaphelenchidae: entomoparasitic, mycophagus (?)
  Subfamily Ektaphelenchinae
    Genera *Ektaphelenchus*, *Cryptaphelenchus*, *Cryptaphelenchoides**, *Ektaphelenchoides*
Family Acugutturidae: entomoparasitic
  Subfamily Acugutturinae
    Genus *Acugutturus*
  Subfamily Noctidonematodae
    Genera *Noctidonema*, *Vampyronema*
Family Parasitaphelenchidae: entomoparasitic, mycophagus, plant parasitic
  Subfamily Parasitaphelenchinae
    Genus *Parasitaphelenchus*
  Subfamily Bursaphelenchinae
    Genera *Bursaphelenchus*, *Rhadinaphelenchus***
Family Entaphelenchidae: entomoparasitic, mycophagus (?)
  Subfamily Entaphelenchidae
    Genera *Entaphelenchus*, *Peraphelenchus*, *Praecocilenchus*, *Roveaphelenchus*

Note: *Cryptaphelenchoides** and *Rhadinaphelenchus*** are now considered synonyms of *Ektaphelenchus* and *Bursaphelenchus*, respectively (Baujard 1989; Giblin-Davis et al. 1989; Ryss et al. 2005; Ye et al. 2007; Hunt 2008), and the possibility of mycophagy of the families Ektaphelenchidae and Entaphelenchidae, followed by "?", has been suggested, but has not been confirmed.

Hunt (1993) noted the genus *Bursaphelenchus* as a "home to a considerable assemblage of species, some of which have been placed in separate genera", and stated the necessity of further taxonomic work. Actually, the morphological definition of this genus has been extended many times by several authors (e.g., Braasch 2004; Kaisa 2005; Ryss et al. 2005), and overlaps with the other genera belonging to the same superfamily. Several important features are excerpted from the current generic morphological definition (Hunt 1993), and are listed below. The features are also illustrated in Fig. II.1. Some exceptions are shown in parentheses.

Genus *Bursaphelenchus*
Body: slender, ventrally arcuate when killed with heat, 0.3–1.4 mm in length, lateral field with 2–6 incisures present but not described in several species.
Lip: well-developed, separated in six equal-sized lips (exceptions: *B. lini* and *B. eproctatus*, two lateral lips are narrower than the other four), constricted clearly at posterior end (several species, e.g., *B. platzeri*, have weak constriction).

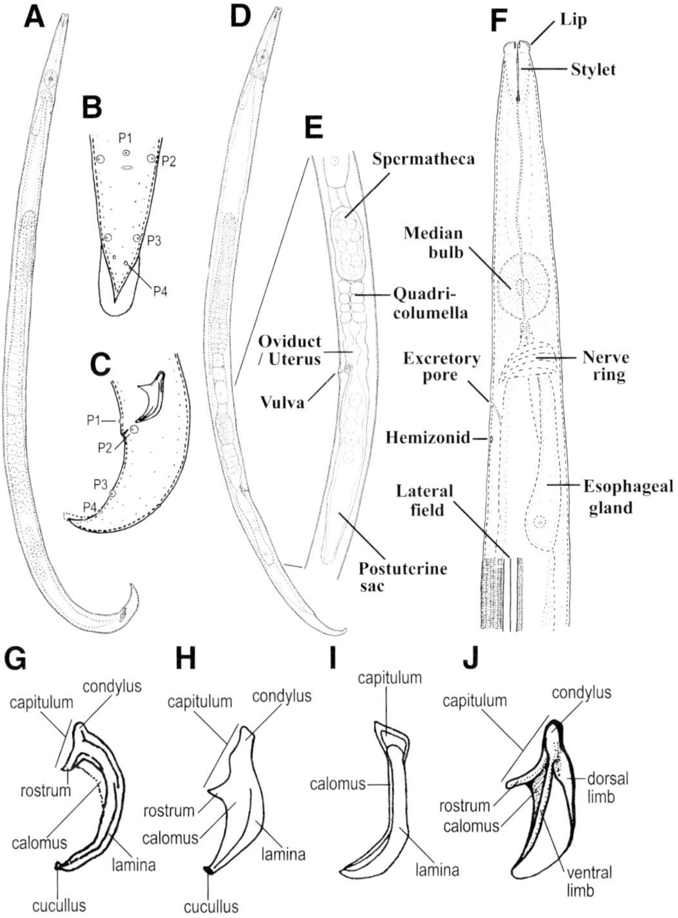

Fig. II.1 General morphology of *Bursaphelenchus* nematodes. Adult male (**A**), ventral view of male tail (**B**), lateral view of male tail (**C**), adult female (**D**), female reproductive organ (**E**), anterior part of female (**F**) and spicules of *B. conicaudatus* (**G**), *B. corneolus* (**H**), *B. aberrans* (**I**) and *B. seani* (**J**), modified after Kanzaki (2006), with permission

Stylet: well-developed, 12–20 µm in length but sometimes reaches 26 µm, basal swelling called "basal knob" usually present (several species, e.g., *B. lini*, lacking the knob).

Metacorpus: well-developed, muscular, spherical- or oval-shaped.

Excretory pore: usually conspicuous, located at various positions, that is, anterior to metacorpus to posterior to nerve ring, but usually located at the same level as or just posterior to the metacorpus.

Female reproductive organ: monodelphic, *V* value usually 70–80%, vulva various, that is, without any flap, with short (=side) flap or long (=real) vulval flap, most species probably possess a pair of three-celled structures at the junction of the uterus and postuterine sac; postuterine sac is usually well-developed, 3–6 body diameter. Length, occupying more than 50% of the vulva–anus distance, but very short in some species.

Female tail: varies from short to elongated, rounded to pointed tip, mucronated in some species.

Male spicule: usually separated (but fused in some species), variable in size and shape, usually arcuate.

Male bursa: present, variable in shape.

Male caudal papillae: 4 (2 pairs) to 11 (single ventral one +5 pairs) have been reported.

As suggested in the above parentheses, there are many exceptions in generic definitive characteristics. In particular, the distinction between *Bursaphelenchus* and its sister taxa, *Parasitaphelenchus*, is very vague. These two genera are separated mainly by the presence or absence of the parasitic juvenile stage, body length, *V* value and fused or separated male spicule; however, some *Bursaphelenchus* nematodes have a *Parasitaphelenchus*-like parasitic juvenile stage, for example, *B. hylobianum* (Korentchenko 1980), a large *V* value, for example, *B. dongguanensis* (*V* value = 86–92; Kaisa 2005), and fused spicule, for example, *B. platzeri* (Giblin-Davis et al. 2006b), and some *Parasitaphelenchus* species have a small body, for example, *P. oldhami* has a body length of ca. 1 mm (Hunt and Hague 1974). Thus, an integrated generic revision of these two taxa is needed in the future.

9.3 Methods of Taxonomy Identification and Systematics of the Genus *Bursaphelenchus*

9.3.1 Morphological Taxonomy and Identification

Many authors have attempted to systematize the genus *Bursaphelenchus*, which contains many species. Tarjan and Baeze-Aragon (1982) and Yin et al. (1988) proposed a pictorial key for species based on morphological characteristics; however, their main purpose was construction of the pictorial key, and they did not propose any taxonomic system to divide the genus into subsets. After their pictorial keys, as the results of efforts to organize the taxonomic system by several authors, currently, the genus is divided into subsets called "groups", which is not a formally accepted taxonomic unit, but is roughly equivalent to "subgenus". The "group" is defined by spicule morphology, and species belonging to the same "group" are distinguished by the other morphological traits, for example, arrangement of caudal papillae, vulval structure, female tail shape and morphometrics.

The original concept of the "group", the assemblage of species, which share characteristic spicule morphology, was proposed and tested by Giblin and Kaya (1983), but they did not apply this concept to all species. Braasch (2000) expanded this concept, taking the number of lateral incisures, the number and arrangement of caudal papillae, morphology of male bursa and female vulval structure into consideration, and divided 28 European conifer-inhabiting species into 10 groups, while Ryss et al. (2005) used spicule morphology as the most important feature, because only spicules are described in all species evenly. They also proposed detailed morphometrics about spicule morphology, for example, ratio of spicule length and maximum width, capitulum length and ratio of capitulum and spicule length, and divided the genus into six groups according to systematic analysis based on their morphometrics and other morphological traits. This system proposed by Ryss et al. (2005) could be considered as a well-constructed integrated system based on the "group" concept (Giblin and Kaya 1983) and systematic analysis (Tarjan and Baeze-Aragon 1982); however, their major purpose was the construction of an "identification system", thus the system is still too typological to systematize the genus. Hence, as the authors remarked, their system still contains some arbitral clades, because they constructed the system based on original descriptions, which contain many misinterpretations in morphological observation, and did not evaluate the weight (= importance in evolutionally systematics) of each characteristic. Nevertheless, their efforts could serve as a starting point for integrated discussion, because they listed all nominal species at that point, and proposed the digitalization and generalization of morphological traits, especially spicule morphology.

9.3.2 Molecular Systematics and Identification

Molecular techniques have many advantages in identification compared to morphological methods, for example, they do not require special training in morphological observation and the methods are applicable to juveniles, which do not have specific diagnostic morphological characteristics. Also the cost and labor involved in molecular techniques is now becoming reasonable.

Many of the molecular techniques that have been developed and used with other *Bursaphelenchus* nematodes have also been used to identify the PWN, for example, RFLP (Webster et al. 1990; Abad et al. 1991; Beckenbach et al. 1992; Tarés et al. 1992), satellite DNA probe (Tarés et al. 1993, 1994; Harmey and Harmey 1994), species specific PCR (Matsunaga and Togashi 2004; Takeuchi et al. 2005), RAPD–PCR (Braasch et al. 1995), PCR–RFLP of ITS rDNA (Hoyar et al. 1998; Iwahori et al. 1998, 2000; Burgermeister et al. 2005b) and DNA sequencing (Iwahori et al. 1998, 2000; Kanzaki and Futai 2002a; Ye et al. 2007). Among these molecular techniques, PCR–RFLP profiles of ITS rDNA (=ITS–RFLP) and DNA base sequencing of several genetic loci have recently been widely employed.

The PCR-RFLP technique has been applied to many nematode species to identify them at species or strain level (Harris et al. 1990; Ferris et al. 1993; Ibrahim et al. 1994; Orui 1996). This technique was introduced for *Bursaphelenchus* nematodes by Hoyar et al. (1998) and Iwahori et al. (1998) to identify the isolates of *B. xylophilus* and *B. mucronatus* using ITS rDNA. Reference profiles of more than 30 *Bursaphelenchus* species have been provided by several authors (e.g.,

Burgermeister et al. 2005; Kanzaki and Futai 2005; details given in Table II.1). The advantages of ITS–RFLP is its simplicity and speed and relatively low cost; however, it is difficult to apply the technique to systematic analyses, and reference patterns to compare with the query profile are required every time. Thus, ITS–RFLP is less general than DNA sequencing and its utilization may be limited to identify certain species or strains (isolates).

DNA sequencing may become the standard tool for molecular taxonomy and identification, as suggested by Ye et al. (2007), because of the expansion of DNA sequence databases, for example, GenBank. Databases also enable us to compare query sequences with all other sequences stored in the database using a computer system, and to estimate the phylogenetic relationships if proper genetic loci and analytical algorithms are chosen. The cost of sequencing, the largest disadvantage of this technique, is now becoming lower, similar to that for ITS–RFLP.

Several ribosomal DNAs, that is, 18S rDNA (SSU) (Kanzaki and Futai 2002a; Ye et al. 2007), ITS region (Iwahori et al. 1998, 2000; Kanzaki and Futai 2002a) and 28S rDNA (D2/D3 LSU) (Kanzaki and Futai 2007; Ye et al. 2007), and mitochondrial COI (Kanzaki and Futai 2002a, 2002b; Ye et al. 2007) have been applied to molecular systematic analyses of *Bursaphelenchus* nematodes at various levels. Each molecular region has its own substitution rate and inherent characteristics, for example, mitochondrial DNA is inherited maternally and has a relatively high substitution ratio, thus DNA sequences are applicable to various levels of comparison if a proper genetic loci, for example, stable loci for higher taxa and variable loci for lower taxa, are chosen for analysis. Kanzaki and Futai (2002a) and Ye et al. (2007) compared the features of those genetic loci, and defined the applicable range of each locus. The characteristics of each locus are summarized as follows. The sequences of universal primers are summarized in Fig. II.2.

ITS Region

Internal transcribed spacer (ITS) region consists of ITS 1, 5.8S ribosomal DNA and ITS 2. ITS 1 and 2 are kinds of "intron" sequences located between small and large subunits of ribosomal RNA coding regions. Sequence mutations accumulate easily in this region, therefore, the ITS region is suitable for analyses of intraspecific, that is, isolate group phylogeny (Iwahori et al. 1998, 2000; Kanzaki and Futai 2002a, b); however, it is not applicable for interspecific phylogeny because sequence divergence is too high in this locus (Kanzaki and Futai 2002a). The sequence length, that is, length of PCR products amplified with universal primer sets, of this region varies so much among species, ranging from 0.7 to 1.2 kbps.

D2D3 LSU

D2D3 LSU, which consists of highly variable D2 and relatively stable D3, is a part of 28S ribosomal DNA (large subunit ribosomal RNA). This locus is applicable to

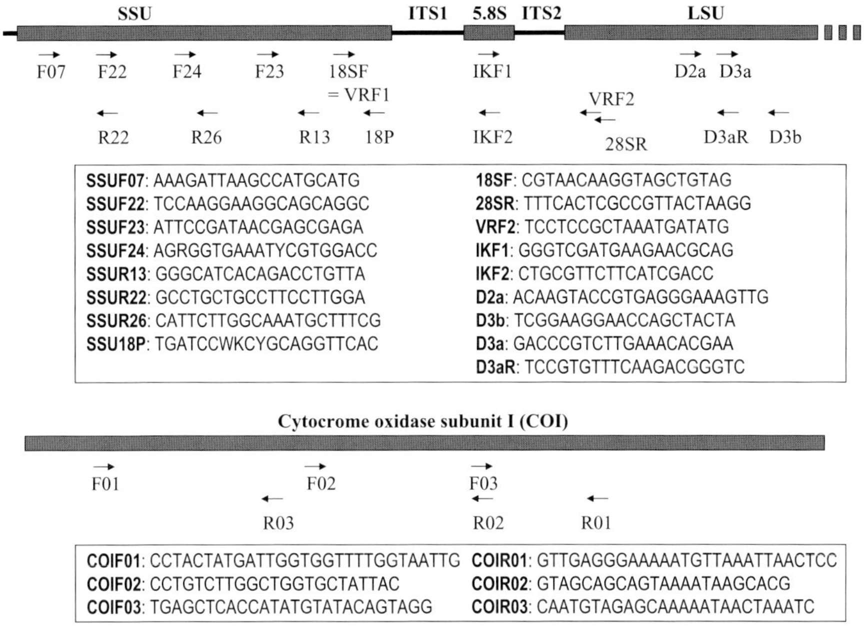

Fig. II.2 Sequences and locations of universal PCR primers applicable to *Bursaphelenchus* nematodes. *Arrows* indicate the direction and position of primers

relatively wide-ranging phylogenetic analyses, that is, among closely related species to among genera (Ye et al. 2007). To date only Ye et al. (2007) have applied this region to the genus *Bursaphelenchus*, however, this region seems very effective to analyze interspecific variation within the genus, and is expected to be a standard region for molecular profiles of *Bursaphelenchus* spp. The sequence length of this region is about 0.75 kbps.

18S rDNA

The 18S ribosomal DNA is a coding region of 18S ribosomal RNA (SSU: small subunit). This locus is highly stable, and is used for molecular systematics of higher taxa, for example, class and order level (Blaxter et al. 1998; Floyd et al. 2003). In the genus *Bursaphelenchus* and related nematodes, this region is very useful for molecular systematics among groups, genera and families, because of its stability. Many molecular sequences of this region are also available in databases (Blaxter et al. 1998; Giblin-Davis et al. 2005; Kanzaki and Futai 2005; Ye et al. 2007); however, the SSU is not applicable in the analyses of intraspecific variations because this region contains few intraspecific variations (N. Kanzaki, unpublished data). Many authors (e.g., De Ley et al. 2002) have developed universal primers for this region, and many of those primer sequences are available at Prof. Dr. Blaxter's website (http://www.nematodes.org/barcoding/sourhope/nemoprimers.html). The sequence length of this region is relatively long, ca. 1.7 kbps.

Mitochondrial COI Gene

Cytochrome oxidase subunit I (mtCOI) gene, a part of the mitochondrial genome, is suitable to analyze intraspecific variations and variations among closely related species, because of its high sequence diversity (Kanzaki and Futai 2002a, b). This gene is transcribed to protein, thus, the amino acid sequences of this gene could be available for higher level analysis, for example, among groups (Kanzaki and Futai 2002a, 2003a). About 1.0 kbps of DNA sequence is available with the primer set developed by Kanzaki and Futai (2002a).

9.4 Comparison of Molecular Phylogeny, Morphology and Life History of *Bursaphelenchus* Nematodes

DNA sequences of SSU and D2D3 LSU of several *Bursaphelenchus* species downloaded from the GenBank database were compared using maximum likelihood analysis. The morphological and life history traits of the nematodes were then plotted on phylograms (Figs. II.3, II.4, II.5).

About 30 species available for this comparison fell into three large clades (Figs. II.3, II.4):

Clade I contains only one species, *B. abruptus*, which is outside of the phylogram, and is distinct from the other clades. Morphologically, *B. abruptus* is similar to "xylophilus" group species, belonging to clade III (subclade III-d), because of its very characteristic spicule shape, relatively large spicule possessing narrow lamina and calomus, and well-developed condylus and rostrum, and a long female vulval flap (Giblin-Davis et al. 1993). Also, the biological traits of *B. abruptus*, associated with a soil-dwelling bee, *Anthophora abrupta*, are similar to those of *B. anatolius*, *B. kevini* and *B. seani*, which belong to clade III (subclades III-b and III-c) (Giblin and Kaya 1983; Giblin et al. 1984; Giblin-Davis et al. 1990, 1993, 2005). However, Giblin-Davis et al. (1993), who identified this species, pointed out the possibility of morphological and biological convergence, because *B. abruptus* has a unique lip structure, lacking head annulations and possessing a circular oral depression, which is clearly different from other *Bursaphelenchus* nematodes. Molecular analysis based on SSU and D2D3 LSU suggests that these morphological and biological traits of clade I (=*B. abruptus*) are convergent characteristics, which occurred independently from those of clade III. *B. abruptus* may be separated from the other *Bursaphelenchus* nematodes. Generic or subgeneric reconstruction might be considered following detailed morphological observations.

Fig. II.3 Phylogenetic relationship among 29 *Bursaphelenchus* nematodes based on small subunit. *Aphelenchus avenae* and *Aphelenchoides fragariae* served as outgroup species. *Bursaphelenchus* spp. #209 assumed to be similar to *B. eremus* and *B. yongensis* is now being identified by the author

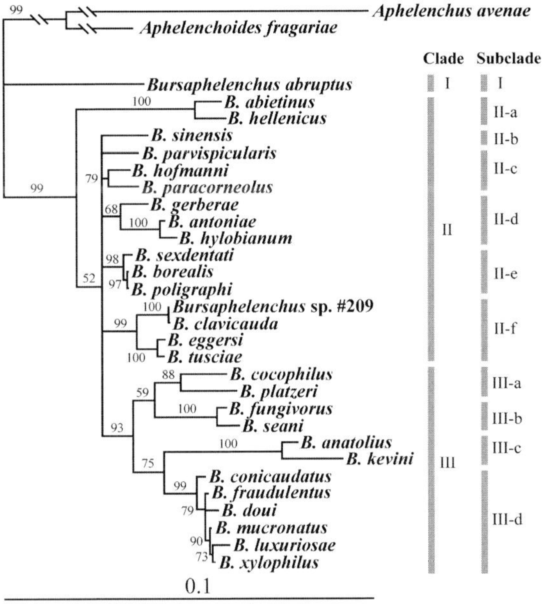

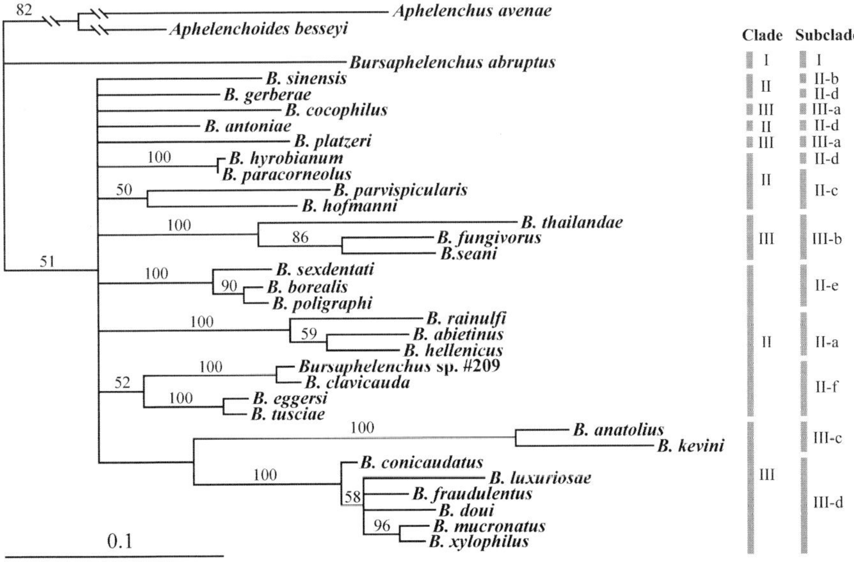

Fig. II.4 Phylogenetic relationship among 31 *Bursaphelenchus* nematodes based on D2/D3 expansion segment of large subunit. *Aphelenchus avenae* and *Aphelenchoides fragariae* served as outgroup species. *Bursaphelenchus* spp. #209 assumed to be similar to *B. eremus* and *B. yongensis* is now being identified by the author

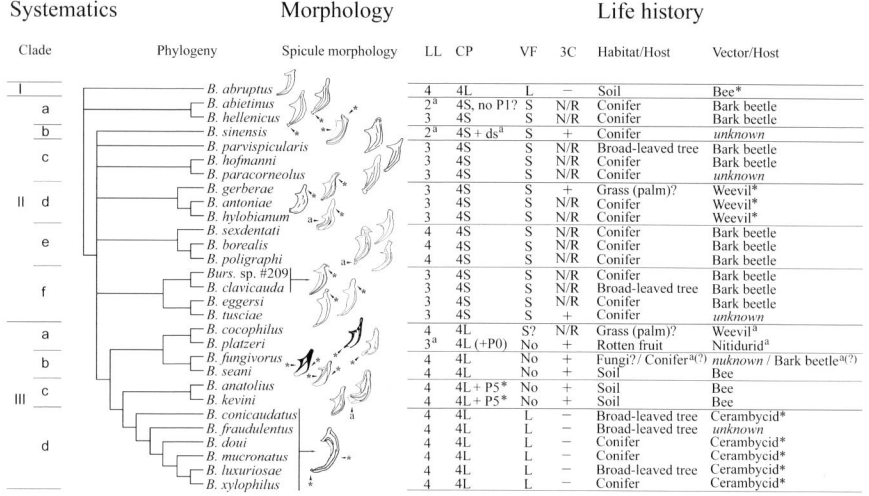

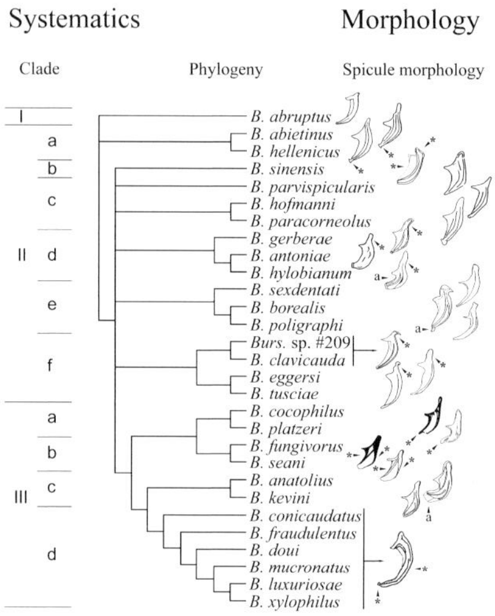

Life history

Clade	Species	LL	CP	VF	3C	Habitat/Host	Vector/Host
I	*B. abruptus*	4	4L	L	–	Soil	Bee*
II a	*B. abietinus*	2[a]	4S, no P1?	S	N/R	Conifer	Bark beetle
II a	*B. hellenicus*	3	4S	S	N/R	Conifer	Bark beetle
II b	*B. sinensis*	2[a]	4S + ds[a]	S	+	Conifer	*unknown*
II c	*B. parvispicularis*	3	4S	S	N/R	Broad-leaved tree	Bark beetle
II c	*B. hofmanni*	3	4S	S	N/R	Conifer	Bark beetle
II c	*B. paracorneolus*	3	4S	S	N/R	Conifer	*unknown*
II d	*B. gerberae*	3	4S	S	+	Grass (palm)?	Weevil*
II d	*B. antoniae*	3	4S	S	N/R	Conifer	Weevil*
II d	*B. hylobianum*	3	4S	S	N/R	Conifer	Weevil*
II e	*B. sexdentati*	4	4S	S	N/R	Conifer	Bark beetle
II e	*B. borealis*	4	4S	S	N/R	Conifer	Bark beetle
II e	*B. poligraphi*	4	4S	S	N/R	Conifer	Bark beetle
II f	*Burs.* sp. #209	3	4S	S	N/R	Conifer	Bark beetle
II f	*B. clavicauda*	3	4S	S	N/R	Broad-leaved tree	Bark beetle
II f	*B. eggersi*	3	4S	S	N/R	Conifer	Bark beetle
II f	*B. tusciae*	3	4S	S	+	Conifer	*unknown*
III a	*B. cocophilus*	4	4L	S?	N/R	Grass (palm)?	Weevil[a]
III a	*B. platzeri*	3[a]	4L (+P0)	No	+	Rotten fruit	Nitidurid[a]
III b	*B. fungivorus*	4	4L	No	+	Fungi?/ Conifer[a(?)]	*nuknown* / Bark beetle[a(?)]
III b	*B. seani*	4	4L	No	+	Soil	Bee
III c	*B. anatolius*	4	4L + P5*	No	+	Soil	Bee
III c	*B. kevini*	4	4L + P5*	No	+	Soil	Bee
III d	*B. conicaudatus*	4	4L	L	–	Broad-leaved tree	Cerambycid*
III d	*B. fraudulentus*	4	4L	L	–	Broad-leaved tree	*unknown*
III d	*B. doui*	4	4L	L	–	Conifer	Cerambycid*
III d	*B. mucronatus*	4	4L	L	–	Conifer	Cerambycid*
III d	*B. luxuriosae*	4	4L	L	–	Broad-leaved tree	Cerambycid*
III d	*B. xylophilus*	4	4L	L	–	Conifer	Cerambycid*

Fig. II.5 Comparison among phylogeny based on small subunit, morphology and biological traits of 29 *Bursaphelenchus* nematodes. *LL* number of lateral lines, *CP* arrangement of caudal papillae, *4L* large (conspicuous) P4 papillae, *4S* small P4 papillae, *ds* dot-like sensilla, *P0* an extra pair at pre-anal, *P5* an extra pair around tail tip, *VF* vulval flap structure, *3C* a three-celled structure at uterus/post-uterine sac junction; *N/R* not reported, + present, – absent. Common traits shared within each subclade are indicated by an *asterisk*. Derived character is indicated as *a* (= alternation)

The species belonging to Clade II have relatively small and stout spicules, tiny P4 papillae and a short vulval flap (side flap: see Giblin-Davis et al. 2006a). This clade is divided into six subclades, II-a to II-f (Fig. II.5). Morphologically, clade II consists of the "aberrans" group, "borealis" group and "piniperdae" group sensu Ryss et al. (2005) or "abietinus" group, "eggersi" group, "hofmanni" group and "sexdentati" group sensu Braasch (2001); however, neither of these classifications corresponds to the phylogenetic clades clearly, mainly resulting from the polyphyogeny of "piniperdae" group sensu Ryss et al. (2005) and "hofmanni" group sensu Braasch (2001). As Braasch (2001) pointed out, the species belonging to clade II are difficult to classify clearly, because they have relatively small spicules similar to each other and few conspicuous morphological characteristics. The following morphological characteristics of each clade can be mentioned: lack of condylus, rostrum and cucullus in spicule (II-b: *B. sinensis*); relatively stout spicule and three lateral lines (II-c, II-d); relatively stout spicule with more or less recurved condylus and four lateral lines (II-e); relatively stout spicule with recurved and more or less pointed condylus (II-f); however, clear diagnostic morphological characteristics to distinguish these molecular phylogenetic clades have not been identified so far.

Interestingly, three alternations on the number of lateral lines and a convergence of spicule morphology are found in clade II. The numbers of lateral lines are two in *B. abietinus* and three in *B. hellenicus* (subclade II-a), two in *B. sinensis*, four in *B. aberrans* (subclade II-b: *B. aberrans* was not available for this molecular comparison, but based on their very unique spicule shape, there is no doubt that *B. aberrans* and *B. sinensis* are sister species.), three in *B. platzeri* and four in *B. cocophilus* (Fig. II.5). These alternations occurred within each phylogenetic clade, while with the spicule shape, a large and dorsally recurved condylus with a pointed end, which was considered as a key characteristic of the "borealis" group sensu Ryss et al. (2005), was shared with two phylogenetic subclades, that is, *B. borealis* (II-e) and *Bursaphelenchus* sp. #209 which is close to *B. eremus* and *B. yongensis* and is now being identified (N. Kanzaki et al., unpublished data). This characteristic morphology may be a convergent characteristic or an ancestoral characteristic of a common ancestor of subclades II-e and II-f. Detailed re-observation to figure out clade-specific characteristics is necessary in the future.

The biological traits of clade II species are similar to each other. All were isolated from various kinds of dead wood, that is, conifers or broad-leaved trees, or coleopteran insects, or both, inhabiting shallow (=beneath the bark) dead wood. There was no clear preference for tree species. Vector preferences seems to correspond to phylogenetic clades, that, tree species may be explained by host preferences of the vector beetles. Species in subclade II-d, *B. gerberae*, *B. antoniae* and *B. hylobianum*, have been isolated from weevils, and the others arre associated with bark beetles (family Scolytidae), although vector insects have not been identified for *B. sinensis*, *B. paracorneolus*, *B. tusciae* and *B. fraudulentus*. The weevil associate is probably derived from a vector-switching event, which occurred in an ancestor of subclade II-b. Further, *B. hylobianum*, which was originally described as a member of the genus *Parasitaphelenchus*, has a parasitic juvenile stage, which is very similar to that of *Parasitaphelenchus* spp. Parasitism of vector weevil may have evolved independently in this species or as the re-emergence of an ancestoral characteristic.

Clade III is very variable in morphology and life history, thus, it is difficult to identify the common traits of this clade morphologically and/or biologically. The only common morphological trait is conspicuous P4 caudal papillae. This clade, containing "hunti" group sensu Ryss et al. (2005) and "xylophilus" group sensu Braasch (2001), is divided into four subclades. Subclades III-a (*B. cocophilus* and *B. platzeri*), III-b (*B. fungivorus* and *B. seani*), III-c (*B. anatolius* and *B. kevini*) and III-d ("xylophilus" group sensu Braasch 2001). The morphological characteristics of each subclade are as follows: III-a has a fused and semi-circle-shaped male spicule and lacks a real female vulval flap, III-b has a broad spicule with conspicuous ventral and dorsal limbs, and is totally lacking a female vulval flap; III-c has a relatively broad spicule and extra (small P5 pair) caudal papillae in males and very short post uterine sac in females, and totally lacks a female vulval flap; III-d has a long, slender and strongly arcuate spicule with well-developed condyles and rostrum in males and long vulval flap in females. Morphologically, subclades III-a, III-b and III-c are similar in their spicule shape, thus, spicule morphology of the "xylophilus" group is assumed to be a derived characteristic occurring in the ancestor of this subclade, and the ancestoral spicule morphology of clade III may be semi-circular.

The biological traits of clade III are also very variable. Common biological traits within subclade III-a and III-b are not identified clearly. In subclade III-a, *B. cocophilus*, the red ring nematode, is vectored by a species of palm weevil, *Rhynchophorus palmarum*, and inhabits and feeds on palm tissue. This species has a unique feeding habitat, obligate plant parasite, and entomoparasitism is also suspected (Griffith 1987; Gerber and Giblin-Davis 1990), while *B. platzeri*, another member of III-a, is vectored by a nitidulid beetle, *Carpophilus humeralis*, and inhabits rotten fruit, feeding on many species of fungi (Giblin 1985; Giblin-Davis et al. 2006b). Both of these two species have different hosts, vectors and feeding habitat preferences from each other and from other *Bursaphelenchus* species. Thus, the ancestoral characteristics and origins of these unique biological traits are still unknown. *B. fungivorus* and *B. seani*, members of III-b, are morphologically similar, but *B. seani* is associated with a soil-dwelling bee, *Anthophora bomboides*, and inhabits the vector's nest, feeding on fungi (Giblin and Kaya 1983), while *B. fungivorus* was described from a species of broad-leaved tree, *Gardenia* sp. affected by *Botrytis cinerea* in a greenhouse (Franklin and Hooper 1962), and was recently isolated from a species of bark beetle, *Orthotomicus erosus* emerging from a dead pine tree (Arias et al. 2005). Although *B. fungivorus* may consist of several cryptic species, which are morphologically identical but genetically different, if *B. fungivorus* is one species, it may have flexible habitat and vector preferences. The biological traits of *B. anatolius* and *B. kevini* (III-c) are also similar to each other: both are associates of soil-dwelling bees, *Halictus* spp., and inhabit their vector's nest, feeding on various fungi (Giblin et al. 1984; Giblin-Davis et al. 2005). The life histories of "xylophilus" group species (III-d) are similar, and are unique to the genus, thus, the characteristics are assumed to have occurred in the common ancestor of this subclade. They inhabit relatively deep wood of dead or dying trees, feeding on various fungi. The species in the "xylophilus" group have an unique dauer stage, fourth-stage dispersal juvenile, that is, most of *Bursaphelenchus* nematodes have third-stage dauers, and dauer juveniles are vectored by longicorn beetles of the tribe Lamiini, entering the vector's tracheal system (Kanzaki and Futai 2003b).

The ancestors of the genus *Bursaphelenchus* may be soil-inhabiting mycophagus nematodes such as species of *Aphelenchus* and *Aphelenchoides*, the outgroup species of the phylogram. A comparison of the morphological and biological traits and the phylogenetic relationship suggest many radiations and convergences within the genus (Fig. II.5).

Regarding vector preference, bee associations occurs at least twice, that is, in clade I and III (the origins of the associations in subclade III-b and III-c are not specified as the same or different). Weevil associations also occur at least twice, that is, II-d and III-a. Bark beetle association, which is widely distributed through clade II, longicorn beetle association ("xylophilus" group = subclade III-d) and nitidulid association (*B. platzeri*: subclade III-a) may have occurred at least once. Regarding morphology, there are several species-specific alternations on spicule morphology, which has been the primary taxonomic characteristic of the genus *Bursaphlenchus* (e.g., Ryss et al. 2005). A strongly bent lamina/calomus complex of *B. hylobianum*, a distal projection (cucullus?) of *B. borealis* and a sac-like structure of *B. anatolius* may be species-specific morphologies. In the present analysis, the other morphological and biological characteristics, that is, structure of caudal papillae and vulval flap and associated vectors, also seem to correspond to phylogenetic clades.

Currently, about a third of nominal *Bursaphelenchus* nematodes are available for molecular analysis. To construct an integrated taxonomic system, many more species should be added to the molecular analysis, and the morphological and biological characteristics must be re-evaluated.

9.5 Future Taxonomic Issues

9.5.1 Old Descriptions without Type Material

Type material is very important for taxonomic studies; however, the type material for many *Bursaphelenchus* species are not available, because in the original descriptions type specimens were not designated for some old species and the type material of other species have been lost or have limited availability due to problems at collection institutions.

Species lacking type materials should be re-isolated and re-described, and neotypes should be designated. Re-isolation, identification and new type designation may be possible for clearly described species; however, it may be almost impossible to identify old and poorly described species. For example, Hunt (1993) considered *B. pinasteri* (Baujard 1980; with type materials) as a junior synonym for *B. sachsi* (Rühm 1956; without type materials = without type designation), while Ryss et al. (2005) treated both of them as valid species. In this case, the description of *B. sachsi* is not sufficient, although it was sound enough at the time of description; therefore, a conclusion, the same or different, cannot be reached. In a similar case, *B. kolymensis* is suspected to be a junior synonym for *B. mucronatus* (Magnusson and Kulinich 1996; Braasch et al. 2005). In this case, type materials of both species are available. Further, many isolates of *B. mucronatus* are available as cultures, and biological information about both species is described very well (Mamiya and Enda 1979; Korentchenko 1980). Although the morphological re-observation by Magnusson and Kulinich (1996) did not provide a taxonomic conclusion, re-isolation of *B. kolymensis* followed by molecular analysis and hybridization tests may be possible, and the relationship between these two species could be clarified in the future.

Species descriptions often contain misinterpretations. If a description without type designation contains misinterpretations, the situation becomes complicated. These misinterpretations allow fictional species to remain in the species list and pictorial and text keys, and may cause misidentification of synonyms. Continuous efforts to re-isolate and re-describe old species, culture preservations at a reliable institute, and rules for specimen voucher are needed. It is also necessary to obtain and accumulate molecular profiles.

The morphological traits easily misinterpreted are as follows:

Spicule morphology. Spicule morphology is one of the most important taxonomic characteristics, because the spicule has been described for all nominal species within the genus; however, spicule morphology is three-dimensional, and its shape seems to differ depending on the microscopic focal plane and direction (angle) of the spicule. Furthermore, there are some, usually slight, morphological variations among individuals; so, it is becoming difficult to understand spicule shape based on just one drawing or photograph. Figure II.6 shows morphological variations of spicules within *B. parvispicularis*, *B. gerberae* and *B. clavicauda*. Also, almost identical spicule shapes are sometimes interpreted as different. Figure II.7 shows the

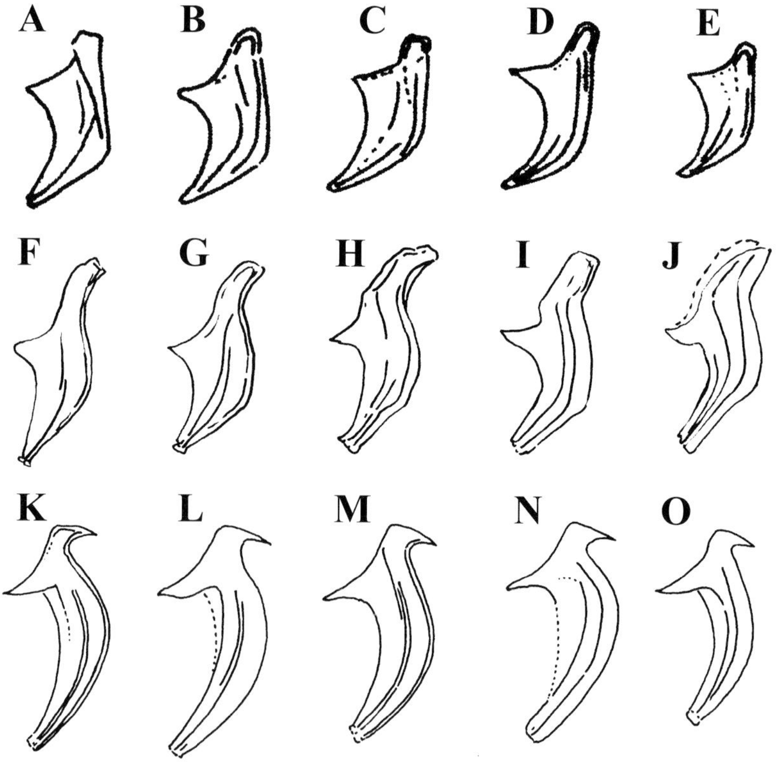

Fig. II.6 Intraspecific variation of spicule morphologies. **A–E** *Bursaphelenchus parvispicularis*; **F–J** *B. gerberae*; **K–O** *B. clavicauda*. Modified after Kanzaki and Futai (2005) (**A–E**), Giblin-Davis et al. (2006b) (**F–J**); and **K–O** Kanzaki et al. (2007)

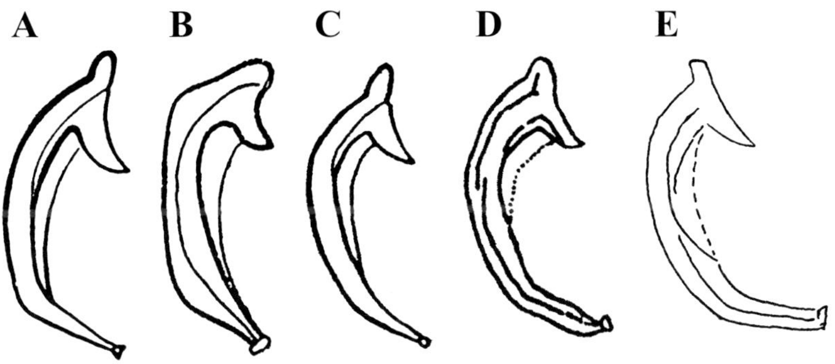

Fig. II.7 Spicules of five (four?) "xylophilus" group species. **A** *B. xylophilus*; **B** *B. kolymensis* (=*B. mucronatus*?); **C** *B. mucronatus*; **D** *B. conicaudatus*; **E** *B. luxuriosae*. The spicule morphologies of these five species are basically the same; however, they are somewhat different from each other in rostrum and condylus morphologies. Modified after Mamiya and Kiyohara (1972) (**A**), Korentchenko (1980) (**B**), Mamiya and Enda (1979) (**C**), Kanzaki et al. (2000) (**D**) and Kanzaki and Futai (2003a) (**E**)

descriptions of *B. xylophilus*, *B. kolymensis* (synonym for *B. mucronatus*?), *B mucronatus*, *B. conicaudatus* and *B. luxuriosae*. These spicule shapes are almost identical; however, the drawings appear to be different. Similarly, the spicule of *B. eremus*, drawn by Rühm (1956) in the original species description, seems totally different from that of the re-isolated culture reported by Braasch et al. (2006c).

Female vulval flap. Generally, the vulval flap is described as "present" or "absent", and when present, "long" or "short"; however, the structure of the vulval flap (or vulva) is roughly classified into three types (Fig. II.8). The species belonging to subclades III-a, III-b and III-c lack a vulval flap, and the anterior and posterior vulval lips seem a little protuberant (Fig. II.8). The ventral view of the vulva on SEM micrograph looks like a simple horizontal slit (Fig. II.8), while the vulval flap of the "xylophilus" group and *B. abruptus* is obviously long and conspicuous; however, other species have a short vulval flap, which is referred to as a "side flap" (see Giblin-Davis et al. 2006a). These species have a dome-shaped expansion just posterior to the vulva, and the flap covers both sides of the vulva, but not the central part (Fig. II.8). This flap sometimes seems like a short flap in the lateral view (Fig. II.8). Re-observation and confirmation are necessary for the species where a short flap was described. For example, the short vulval flaps of *B. hylobianum*, *B. paracorneolus*, and *B. hofmanni* were confirmed as side flaps by Giblin-Davis et al. (2006a).

The number and arrangement of male caudal papillae. Generally, males of *Bursaphelenchus* nematodes have seven caudal papillae, P1 to P4 (Fig. II.1). Within these papillae, P2 and P3 are observed easily with a light microscope; however, P1 is sometimes located at the same level as P2, thus it is often masked by the P2 pair when observed in lateral view, or confused with the cloacal structure or spicule, or

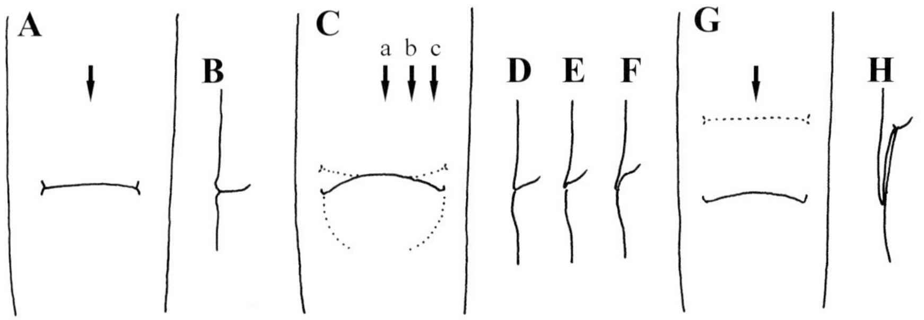

Fig. II.8 Variation of female vulval flap structure. Without the vulval flap, ventral view (**A**) and lateral view (**B**). Short vulval flap (=side flap) in ventral view (**C**) and lateral views (**D**–**F**). Long vulval flap in ventral view (**G**) and lateral view (**H**). *Arrows* in ventral view indicate the focal plane of corresponding lateral views. The *a*, *b* and *c* in **C** correspond to **D**, **E** and **F**, respectively

both, when observed in the ventral view. Therefore, P1 is sometimes missed in the species description, and is confirmed by re-observation with a high-resolution microscope (e.g., Kanzaki and Futai 2002c; Giblin-Davis et al. 2006a). P4 is also missed or easily confused. The structure of the P4 pair is different among phylogenetic clades, that is, the species in clades I and III have relatively large and distinctive P4 located at almost the same level as P3 (e.g., Kanzaki and Futai 2002c), and those in clade II have a small P4 pair, referred to as glandpapilla in Ryss et al. (2005), located on a rounded square cuticle raise plate (see Giblin-Davis et al. 2006a, b). Thus, the P4 pair of clades I and III is sometimes confused with P3 and missed in the lateral view in the original description (e.g., Kanzaki et al. 2000, *B. conicaudatus*; Mamiya and Kiyohara 1972, *B. lignicolus* = *B. xylophilus*; Rühm 1956, *B. fraudulentus*). In clade II, P4 paired papillae are located close to each other and asymmetrically on the right and left. Thus, a P4 pair is sometimes confused as two pairs in the lateral view. Also, the edges of the cuticle plate are sometimes misinterpreted as papillae (Fig. II.9). Actually, this characteristic was corrected in several species by re-observation of type materials and/or living cultures using SEM or a high-resolution light microscope. The number, arrangement and structure of caudal papillae are very important taxonomic characteristics and should be observed carefully, and the possibility of misinterpreting old descriptions should be taken into account.

Besides the basic pattern (seven: P1–P4), various numbers and arrangements of caudal papillae have been reported: four (P2 and P3: many species); six (P2–P4: many species); eight [P2–P5: *B. georgicus* (Devdariani et al. 1980); *B. gonzalezi* (Loof 1964)]; nine [P1–P5: *B. hylobianum* (Korentchenko 1980); *B. kevini* (Giblin et al. 1984); *B. anatolius* (Giblin-Davis et al. 2005)] and 11 [P1–P6: *B. piniperdae* (Fuchs 1937) and *B. poligraphi* (Rühm 1956)]. Within this variation, most of "four" and "six" are assumed to have missed P1 and P4. Only *B. lini* and *B. eproctatus* were confirmed to have just four papillae using SEM (Braasch et al. 2006a; Sriwati et al. 2008). While most of "8" to "11" may have been confused with a cuticle plate

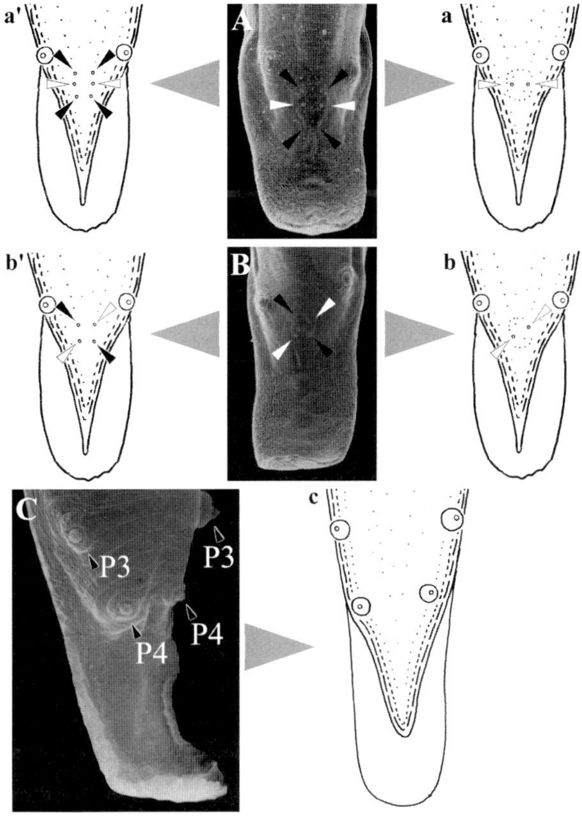

Fig. II.9 Caudal papillae structure. A pair of small P4 papillae (A, B, a, b) easily misinterpreted as three pairs (a′) or two pairs (b′) of papillae with misinterpretation of the edge of cuticular raised plates (A, B) and conspicuous P4 paired papillae. *Solid arrow* edge of cuticular plate, *white arrow*: P4 papillae. Modified after Giblin-Davis et al. (2006a) (A, B, a, b, a′, b′) and Giblin-Davis et al. (2006b) (C and c′)

as mentioned above. Only two species, *B. kevini* and *B. anatolius* have been shown to have an extra pair on photographs (Giblin et al. 1984; Giblin-Davis et al. 2005). Also, some individuals of *B. platzeri* have an extra pair of caudal papillae just anterior to P1 (Giblin-Davis et al. 2006b).

Number of lateral lines. Generally, *Bursaphelenchus* nematodes have a longitudinal tape-like structure on both sides of the lateral body, which is called the "lateral field". The lateral field usually has several incisures called "lateral lines". The number of lateral lines are usually two to four, and a few species have more than five (Rühm 1956; Braasch 2001). The number is mostly difficult to count with a light microscope, and is sometimes missing from the species description. In some old descriptions, only the presence of the lateral field is mentioned without giving the number of lines, or the lateral field is totally missing from the description. The

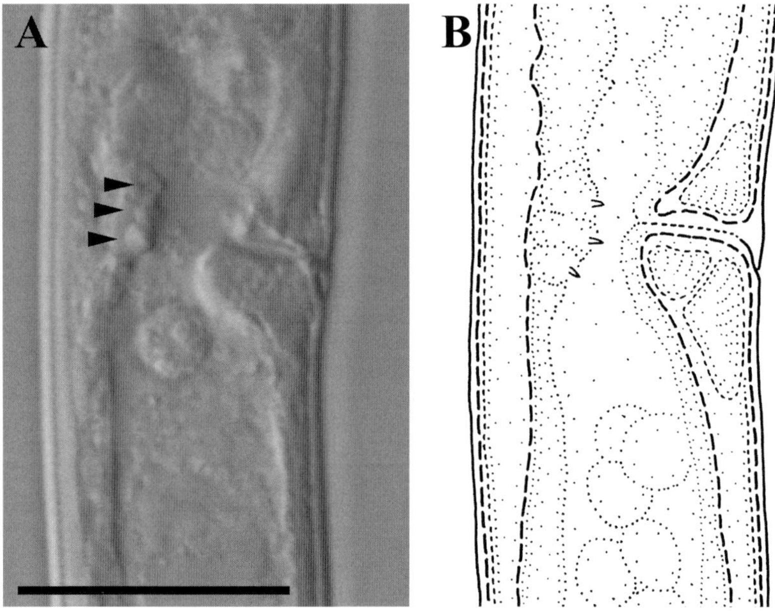

Fig. II.10 A pair of three-celled pronged structures of *Bursaphelenchus sinensis*. **A** light microscopy; **B** drawing. The structure is indicated by *arrows*. Modified after Kanzaki and Futai (2007)

number may correspond relatively well to the phylogenetic relationship, although several alternations are suspected (Figs. II.3, II.4, II.5), but at present is difficult to use as a taxonomic characteristic.

A three-celled structure at uterus-postuterine sac junction. In several species, a pair of three-celled structures are reported at the uterus-post uterine sac junction (Fig. II.10). The presence of the structure is known in just a few species; however, this structure may be present in most *Bursaphelenchus* nematodes, because it has been found in many phylogenetic groups (Kanzaki and Futai 2007). The function of the structure has not been clarified, but it probably has some role concerning oviposition. Kanzaki and Futai (2007) reported that this structure was absent in subclade III-d. Probably, the structure was lost, or altered morphologically in a common ancestor of the "xylophilus" group.

9.5.2 Species Morphologically Overlapping with Other Genera and Families

Several members of the genus *Bursaphelenchus* diverge from general generic definitions. Some have very unique characteristics and others overlap with other genera or families in their essential morphological features. The systematic positions of these species within the genus, or a suitable home of these species has not determined so far. For example, Baujard (1989) and Giblin-Davis et al. (1989) considered *B. cocophilus* as a member of the genus *Bursaphelenchus*, based on its seven caudal papillae and rounded bursa, while Hunt (1993) placed this species in another genus, *Rhadinaphelenchus*, based on its obligate plant parasitism, extraordinarily slender body and fused male spicule. Similarly, *B. hylobianum* was originally described as a member of the genus *Parasitaphelenchus*, although the adult morphological traits fit the generic definition of *Bursaphelenchus*, because this species has a special parasitic stage, which has a head hook and a tail hook, characteristic of the parasitic third juvenile of *Parasitaphelenchus* species (Hunt 1993; Korentchenko 1980).

In the case of *B. aberrans*, Fang et al. (2002a) reported an elongated median bulb, similar to those of *Ektaphelenchus* spp., and Braasch and Braasch-Bidasak (2002) reported a "parasitic adult" of *B. aberrans*, which is very similar to *Ektaphelenchus* species, that is, the "parasitic adult" of *B. aberrans* has a flattened lip and stylet with a wide lumen. As *B. sinensis*, a sister species of *B. aberrans*, clearly belongs to the genus *Bursaphelenchus* (Figs. II.3, II.4, II.5; Kanzaki and Futai 2007), *B. aberrans* is assumed to belong to the genus; however, the origin of this *Ektaphelenchus*-like parasitic stage is still unknown, that is, as to wheather it developed independently from *Ektaphelenchus* or is derived from a common ancestor.

Fortunately, living material is available for the above three species, and they were confirmed to be included clearly in the inner clades of the genus, based on their molecular phylogenetic positions; however, no living material is available for the other species with unique characteristics. *B. dongguanensis* and *B. digitulus* share several adult morphological traits with the genus *Parasitaphelenchus* although a parasitic juvenile has not been reported for these species, for example, large (more than 85%) *V* value (Loof 1964; Fang et al. 2002b; Kaisa 2005) and weak curvature of the male tail when killed by heat (Loof 1964; Kaisa 2005). Besides *B. aberrans*, *B. lini* and *B. eproctatus* also share several essential characteristics with the family Ektaphelenchidae, that is, elongated (long-oval) median bulb (common in *Ektaphelenchus*), lacking anus and rectum (common in Ektaphelenchidae) and four (two pairs) caudal papillae in males (common in *Cryptaphelenchus*) (Braasch 2004; Braasch et al. 2006a; Sriwati et al. 2008). Braasch (2004), who described *B. lini*, mentioned these features, unique in *Bursaphelenchus* and common in Ektaphelenchidae, and placed this species in the genus *Bursaphelenchus* because this species has a clear male bursa. In the generic description of the *Cryptaphelenchus* (Ekitaphelechidae), Rühm (1956) stated that the bursa is absent in this genus; however, he drew a bursa-like flap at the male tail tip in his illustration. Up to the present, 19 species have been described as members of the genus *Cryptaphelenchus*, but unfortunately, there is no type material available for them; therefore, the presence or absence and structure of the bursa-like flap drawn by Rühm (1956) is still unknown.

The phylogenetic affiliation of those species, as well as the integrated definition, diagnoses and phylogenetic relationship among *Bursaphelenchus*, *Parasitaphelenchus*, *Ektaphelenchus* and *Cryptaphelenchus* remain important taxonomical subjects. Re-isolation, followed by molecular analyses, are needed for these confusing species without type materials.

9.6 Concluding Remarks on the Taxonomy and Systematics

The present situation of the taxonomy, systematics and evolutional hypotheses of the genus *Bursaphelenchus* are summarized in this chapter. Similar to other nematode groups, the taxonomic system of the genus *Bursaphelenchus* is still incomplete, and the border of the genera is still unclear, because of the many problems remaining in morphological and molecular taxonomic systems. These problems are mainly the result of old and unclear descriptions, and the lack of type materials; however, in the future these issues will be addressed individually.

Since the finding of the PWN in Portugal (Mota et al. 1999), which warned the world about the PWN threat, the importance of the taxonomy and identification of *Bursaphelenchus* nematodes has increased rapidly. Further, the pathogenicity of several *Bursaphelenchus* nematodes on *Pinus* spp. has been reported recently (e.g. Skarmoutsos and Michalopoulos-Skarmoutsos 2000), and that has increased the importance of the genus for plant quarantine. This trend may help to solve the taxonomical problems of the genus.

There are two important issues to reconstruct the generic taxonomic system. Primary, old and unclear species are to be organized and verified again, that is, correct the mistakes occurring in the original descriptions and designations of lectotype or neotypes, and secondly, a standard method of description and deposition is necessary.

To reconstruct the taxonomic system, the following procedure is proposed, as described in the last half of this chapter. First, construction of a temporal taxonomic system based on detailed observation of type materials or living specimens, or both, and molecular analysis. Here, morphological and partial biological information may be plotted for phylogenetic groups. Then, re-isolation of unclearly described species based on their biological information, for example, host, habitat or locality, ascription or correction of their morphological traits and plotting on the phylogenetic tree. At present, a temporal system seems to be under construction by several research groups (Braasch 2001; Ryss et al. 2005; Kanzaki 2006; Ye et al. 2007), and several corrections of old species have been proposed (Braasch et al. 2006c; Giblin-Davis et al. 2006a, b; Kanzaki and Futai 2007). The reconstructed generic taxonomy is expected to be proposed in the near future.

Conversely, to standardize the description, although there are general rules suggested in the International Code of Zoological Nomenclatures, besides a proper deposition system for morphological specimens, a system for molecular specimens (=DNA vouchers) is also essential. At present, neither a molecular voucher system nor molecular barcode region has been standardized. The establishment of molecular or culture vouchers may be an important area in the future.

The manuscript of this chapter was written in August 2007, and updated in December 2007. After the last update, taxonomic framework of the superfamily Aphelenchoidea, formerly called "order Aphelenchida", was updated by Hunt (2008). The latest taxonomic system and species list are provided in the paper; Hunt DJ (2008) A check list of the Aphelenchoidea (Nematoda: Tylenchina). J Nematode Morph Syst 10:99–135

10
Parasitism Genes of the Pine Wood Nematode

Taisei Kikuchi

10.1 Introduction

Although the life history and some behavioral ecological aspects of *Bursaphelenchus xylophilus*, the pine wood nematode (PWN) are now well documented, little is known about the molecular basis of the nematodes' biology and host-parasite interaction.

Molecular biology has allowed for the details of many aspects of nematode biology and host-parasite interactions to be worked out at a far deeper level than would once have been thought possible. One of the most useful applications of these techniques has been the development of a detailed and accurate phylogeny for the Phylum Nematoda. A lack of clearly homologous characteristics and the absence of a fossil record that would allow the evolution of nematodes to be studied prevent interpretation of the deeper phylogenetic relationships within the Phylum. On the basis of comparisons of the small subunit ribosomal DNA sequences of nematodes, a detailed phylogeny of the Phylum has been drawn up and the interrelationships of the major nematode groups have been established (Blaxter et al. 1998; Dorris et al. 1999). Nematodes can be divided into five major clades, all of which include parasitic species, indicating that parasitism of both animals and plants has arisen multiple times during evolution (Blaxter et al. 1998; Dorris et al. 1999).

In contrast, another technique—analysis of expressed sequence tags (ESTs)—has been of paramount importance in developing an understanding of the proteins produced by nematodes that allow them to parasitize plants. To date, more than 700,000 EST sequences from nematodes, including free-living species as well as animal and plant parasites, are available in databases (Parkinson et al. 2003, 2004). Many parasitism genes have been identified and detailed analysis of such genes present in the EST dataset has been undertaken. One of the most remarkable

Forest Pathology Laboratory Forestry and Forest Products Research Institute, 1 Matsunosato, Tsukuba 305-8687, Japan

Tel.: +81-29-873-3211, Fax: +81-29-873-1543, e-mail: kikuchit@affrc.go.jp

findings of the EST analysis was the identification of genes, which encode proteins that are likely to be important for plant parasitism (parasitism genes).

Previous studies on the parasitism genes have been largely restricted to economically important Tylenchid, cyst and root-knot nematodes. There have been few studies on plant parasitism genes in other nematodes that are phylogenetically or ecologically distinct from cyst and root-knot nematodes. *Bursaphelenchus xylophilus* is part of the same clade (IVb) as cyst/root-knot nematodes (Blaxter et al. 1998; Meldal et al. 2007); however, *Bursaphelenchus* spp. are not directly related to these nematodes but form a distinct grouping with other mycophagus nematodes including *Aphelenchoides* spp. (Meldal et al. 2007). Studies on the parasitism genes of *B. xylophilus* will undoubtedly provide clues to understanding the mechanisms underlying parasitism in *B. xylophilus*, and will also further our understanding of the evolution of plant parasitism in nematodes.

In this chapter, I present an overview of EST analysis of *B. xylophilus* and detailed characteristics of some parasitism genes in *B. xylophilus* identified during this EST project.

10.2 EST Analysis

The *Caenorhabditis elegans* genome sequence was completed several years ago (The-Caenorhabditis-elegans-Sequencing-Consortium 1998) and since then, substantial annotation of the sequence has taken place. In addition, whole-genome scale gene expression and RNA interference (RNAi) studies have allowed detailed functional analysis of the biological role of many genes (Maeda et al. 2001). Although genome sequencing projects are well underway for several other nematode species, including *C. briggsae*, *Haemonchus contortus* and *Brugia malayi*, such resources are not available for many other nematodes. Consequently, many parasitic nematode genomes are being explored using ESTs. Analysis of ESTs by single-pass random sequencing of cDNA libraries is a powerful tool for rapid and cost-effective gene discovery. High-throughput projects involving more than 30 nematode species have generated nearly 500,000 ESTs from parasitic nematodes, including datasets from animal parasites and plant parasites (Table II.2) (Parkinson et al. 2003, 2004). Including the sequences from *C. elegans* and *C. briggsae*, there are currently over 700,000 nematode ESTs in the publicly accessible dbEST database. EST analysis has been a powerful tool for the identification of plant parasitic nematode genes, which have a possible role in parasitism.

Most of this analysis has been performed on the economically important Tylenchid nematodes. Almost all plant parasitic nematodes in Table II.2 are Tylenchid, with the exception of several groups of ectoparasitic nematodes such as species of *Xiphinema*, *Trichodorus* and *Longidorus*, which are found in clades I and II of the Phylum. By contrast, previous characterization of the *Bursaphelenchus* species genome has been limited to a very few sequences, which were used only for phylogenetic analysis and diagnostic purposes.

Table II.2 Nematode expressed sequence tag (EST) projects

Nematode species	Description	Submitted ESTs[a]	Clades[b]	Major EST sources[c]
Bursaphelenchus xylophilus	Pine wood nematode	13,340	IVb	1
Bursaphelenchus mucronatus	Pine wood nematode	3,193	IVb	1
Globodera pallida	Potato cyst nematode	4,378	IVb	2, 3, 4
Globodera rostochiensis	Potato cyst nematode	11,851	IVb	2, 3, 4
Heterodera glycines	Soy bean cyst	24,444	IVb	2
Heterodera schachtii	Sugar beat cyst nematode	2,818	IVb	2
Meloidogyne arenaria	Root-knot nematode	5,018	IVb	2
Meloidogyne chitwoodi	Root-knot nematode	12,218	IVb	2
Meloidogyne hapla	Root-knot nematode	24,452	IVb	2
Maloidogyne incognita	Root-knot nematode	20,334	IVb	2, 4
Meloidogyne javanica	Root-knot nematode	7,587	IVb	2
Meloidogyne paranaensis	Root-knot nematode	3,710	IVb	2
Pratylenchus penetrans	Cobb's lesion nematode	1,928	IVb	2
Pratylenchus vulnus	Plant lesion nematode	5,812	IVb	2
Radopholus similis	Plant migratory endoparasite	1,154	IVb	2
Xiphinema index	California dagger nematode	9,351	I	2, 3
Ancylostoma caninum	Dog hookworm	9,618	V	2
Ancylostoma ceylanicum	Human hookworm	10,651	V	2
Angiostrongylus cantonensis	Rat lung worm	1,279	V	10
Ascaris suum	Swine gut parasite	40,771	III	2, 5
Ascaris lumbricoides	Human gut parasite	1,822	III	2
Anisakis simplex	Marine mammal parasite	475	III	11
Brugia malayi	Human lymphatic parasite	26,215	III	2, 9
Dirofilaria immitis	Canine heart worm	4,005	III	2
Haemonchus contortus	Sheep gut parasite	21,967	V	2, 5
Litomosoides sigmodontis	Mouse filarial worm	2,699	III	5
Necator americanus	Human hookworm	5,032	V	5, 7
Nippostrongylus brasiliensis	Rat gastrointestinal parasite	8,238	V	2, 5
Onchocerca volvulus	Human filarial parasite	14,974	III	9
Ostertagia ostertagi	Cattle gut parasite	7,006	V	2
Parastrongyloides trichosuri	Possum gut parasite	7,963	IVa	2
Strongyloides ratti	Rodent gut parasite	14,761	IVa	2
Strongyloides stercoralis	Human gut parasite	11,392	IVa	2

(continued)

Table II.2 Continued

Nematode species	Description	Submitted ESTs[a]	Clades[b]	Major EST sources[c]
Teladorsagia circumcincta	Sheep gut parasite	4,313	V	5, 7
Toxocara canis	Canine gut parasite	4,889	III	2, 5
Trichinella spiralis	Human muscle parasite	11,568	I	2
Trichuris muris	Mouse threadworm	2,714	I	5, 7
Trichuris vulpis	Dog whipworm	3,063	I	2
Wuchereria bancrofti	Human bancroftian filariasis	4,847	III	2
Caenorhabditis briggsae	Free living	2,424	V	2
Caenorhabditis elegans	Free living	346,046	V	2, 6, 7, 8
Caenorhabditis remanei	Free living	20,292	V	2
Heterorhabditis bacteriophora	Insect associated bacteriovore	7,514	V	2
Pristionchus pacificus	Free living	14,663	V	2
Zeldia punctata	Free living	391	IVb	2
Total		746,647		

[a]ESTs deposited as at 10 October 2004
[b]The phylum Nematoda has previously been defined into five clades (Dorris et al. 1999)
[c]1, Forestry and Forest Products Research Institute, Japan; 2, Genome Sequencing Center, Washington University, USA; 3, Scottish Crop Research Institute, UK; 4, Wageningen University, the Netherlands; 5, University of Edinburgh, UK; 6, National Institute of Genetics, Japan; 7, The Wellcome Trust Sanger Institute, UK; 8, The Institute for Genome Research, USA; 9, The Filarial Genome Network, UK; 10, Chang Gung University, Taiwan; 11, Pusan University, Korea

The EST project for the PWN was done at the Forestry and Forest Products Research Institute, Japan from 2004 to 2006. In this project, over 13,000 ESTs from *B. xylophilus* and, by way of contrast, over 3,000 ESTs from a closely related species that does not as readily parasitize plants; *B. mucronatus*, were produced (Kikuchi et al. 2007). Four libraries from *B. xylophilus* and one library from *B. mucronatus* were constructed and used to generate ESTs. Sixty-nine percent of the total *B. xylophilus* ESTs were from a mixed-stage library derived from nematodes feeding on fungi, 11% were from a library made from nematodes feeding on plant material and 20% were from two dauer-like larvae libraries (Table II.3). A variety of proteins potentially important in the parasitic process of *B. xylophilus* and *B. mucronatus*, including proteins important in fungal feeding as well as proteins that break down various components of the plant cell wall, were identified in the libraries. As well, several gene candidates potentially involved in dauer entry or maintenance were also identified in the EST dataset.

The 13,327 *B. xylophilus* ESTs grouped into 6,487 clusters and the 3,193 *B. mucronatus* ESTs formed 2,219 clusters. Assuming 19–20,000 total genes as in *C. elegans*, these clusters are likely to represent 30% of all *B. xylophilus* genes and 11% of *B. mucronatus* genes (Kikuchi et al. 2007).

Table II.3 ESTs generated from *Bursaphelenchus* cDNA libraries

Species	Library name	Strain	Stage	Description	ESTs	Average length (bp)
B. xylophilus	K1	Ka-4	Mixed stage	Vigorously growing on fungi	9,194	578
	KP	Ka-4	Mixed stage	Growing on plants	1,476	550
	KDw	Ka-4	Dauer larvae (JIV)[a]	Separated from wood	658	442
	KDi	Ka-4	Dauer larvae (JIV)[a]	Separated from insect	1,999	455
					13,327	550
B. mucronatus	U1	Un1	Mixed stage	Vigorously growing on fungi	3,193	564

[a] JIV: dispersal fourth stage juvenile

The availability of these sequences and further bioinformatic analysis, including functional categorization and detailed comparative analysis of the ESTs, will provide useful information to investigate the biology, pathogenicity and evolutionary history of *B. xylophilus*.

10.3 Parasitism Genes

Plant parasitic nematodes are mainly biotrophic root parasites, and can be sedentary or migratory, and ectoparasites or endoparasites (Gheysen and Jones 2006). Migratory nematodes feed on plant cells, frequently causing cell death, and then move to another cell to repeat the feeding. Sedentary endoparasites, including cyst and root-knot nematodes, feed from a single cell or a group of cells for a prolonged period of time. For this sustaining feeding, sedentary parasites have the ability to dramatically modify root cells into elaborate feeding cells.

All plant parasitic nematodes have evolved a hollow, called a stylet, and use it to penetrate the wall of a plant cell, to remove plant cell contents during feeding and to introduce nematode secretion into plant tissue.

Secretions from the stylet, which are produced in esophageal gland cells, play important roles in plant parasitism of nematodes. Plant parasitic nematodes have two sets of these gland cells, dorsal and sub-ventral, and these gland cells enlarged considerably as nematodes evolved from bacterial-feeding nematodes to parasites of higher plants. Tylenchid nematodes have two sub-ventral gland cells and one dorsal gland cell, and the products of the gland cells are developmentally regulated (Gheysen and Jones 2006).

Bursaphelenchus xylophilus have a stylet and the same number of esophageal gland cells as other plant parasitic nematodes although it is difficult to distinguish each cell because the three-esophageal gland cells of *Bursaphelenchus* dorsally overlap and all connect to similar positions in the large median esophageal bulb.

However, *B. xylophilus* is unique compared to the major plant parasitic nematodes as it is a parasite of aboveground parts of trees, and does not enter the soil but migrates through plant tissues. Furthermore, *B. xylophilus* is basically a fungal feeder and uses an insect as a transmitting vector. This implies that *B. xylophilus* should have a set of parasitism genes distinct from the major plant parasitic nematodes.

10.3.1 Cell Wall-Degrading Enzymes

The plant cell wall is the primary barrier faced by most plant pathogens and the production of enzymes able to degrade this cell wall is of critical importance for plant pathogens including plant parasitic nematodes. The plant cell wall is a complex but highly organized composite of polysaccharides and protein (Fig. II.11). Plant cell wall-degrading enzymes had previously only been shown to be produced by plants, bacteria and fungi, with no clear reports of their production by animals. Although the ability of animal to hydrolyze cellulose had been the subject of various studies, it is difficult to establish whether the enzyme is synthesized by an animal or by associated microorganisms without isolating the corresponding gene. The first animal cellulase genes were described in plant parasitic cyst nematodes in 1998. Since then many plant cell wall-degrading enzymes have been identified

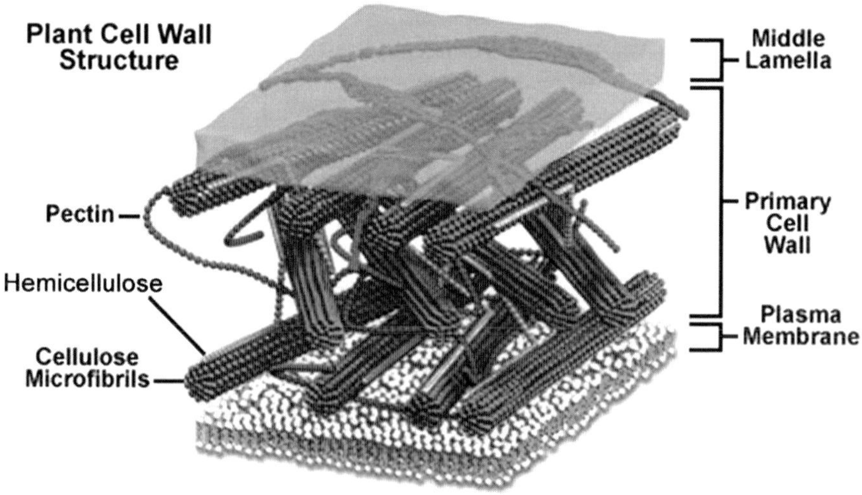

Fig. II.11 Structure of primary plant cell wall, from http://micro.magnet.fsu.edu/ (see Color Plates)

in Tylenchid nematodes, including endo-β-1,4-glucanases (cellulases) (Rosso et al. 1999; Smant et al. 1998), pectate lyase (Doyle and Lambert 2002; Popeijus et al. 2000), polygalacturonase (Jaubert et al. 2002) and xylanase (Dautova et al. 2001), as well as proteins that disrupt non-covalent bonds in plant cell walls (Qin et al. 2004).

1 Cellulase

Cellulose is a major component of plant cell walls (Fig. II.11) and consequently, cellulases (endo-β-1,4-glucanases) are produced by many plant pathogens including bacteria and fungi (Barras et al. 1994; Walton 1994). Endogenous cellulase genes have been identified from plant parasitic nematodes, including *Heterodera*, *Globodera* (cyst nematode) and *Meloidogyne* (root-knot nematode) species (Smant et al. 1998; Rosso et al. 1999; Goellner et al. 2000). These cellulases are produced within the esophageal gland cells of these nematodes and secreted through the nematode stylet into plant tissues (de Boer et al. 1999). They are therefore likely to facilitate the penetration and migration of nematodes into root tissues during parasitism. In addition, their removal using RNAi prevents successful invasion of plant roots (Chen et al. 2005). The proteins encoded by these genes belong to the glycosyl hydrolase family (GHF) 5 and are far more similar to bacterial than to eukaryotic cellulases. It has therefore been suggested that these genes have been acquired via horizontal gene transfer from bacteria (Yan et al. 1998). GHF5 cellulase genes have also been found in one migratory endoparasitic nematode *Pratylenchus penetrans* (Uehara et al. 2001) that is related to cyst and root-knot nematodes.

Cellulase genes of the PWN were identified by screening the EST dataset. Surprisingly, the cellulases showed high similarity with GHF45 cellulases from fungi (Fig. II.12). The catalytic domain of *B. xylophilus* cellulases shows 62–66% overall amino acid identity with cellulases from two fungi, *Scopulariopsis brevicaulis* and *Rhizopus oryzae*. This extremely high similarity between *B. xylophilus* cellulases and fungal cellulases, together with the absence of sequences resembling GHF45 cellulases from other nematodes, including *C. elegans* and *C. briggsae* for which full genome sequences are available, suggests that *B. xylophilus* cellulases might have been acquired via horizontal gene transfer from fungi.

GHF45 cellulase genes were shown to exist as a multiple gene family in the *B. xylophilus* genome. The enzymatic activity of the protein was confirmed by heterologous expression in *Escherichia coli* and the endogenous nature of the genes was confirmed by Southern blotting. The presence of predicted signal peptide sequences at the N-termini of the proteins encoded by these genes coupled with the specific localization of the transcripts imply that these cellulases could be secreted into plant tissues to help nematodes feed and migrate into plants (Fig. II.13). Localization of the protein by immunofluorescence confirmed this (Fig. II.13).

The biochemical properties were examined using purified recombinant proteins expressed in *Pichia*. These analyses suggested that *B. xylophilus* cellulases act not only on cellulose but also on hemicellulose. Among celluloses, these enzymes act

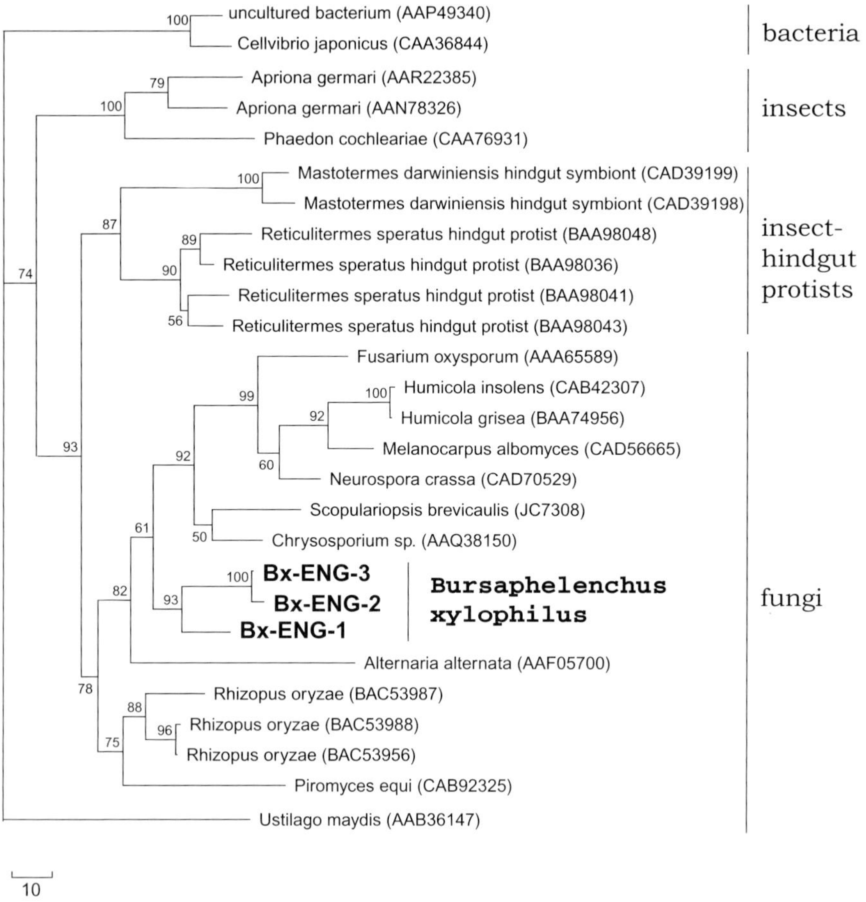

Fig. II.12 Unrooted, phylogenetic tree of GHF45 cellulases. *Scale bar* represents 10 substitutions per 100 amino acid positions

largely on amorphous cellulose but hardly hydrolyze crystalline cellulose, indicating that the nematode does not use cellulases for complete digestion of the cell walls. Once a nematode invades a pine tree, it migrates primarily through the resin canals in the tree and feeds on parenchyma cells surrounding the canals. The cell wall of the parenchyma cell consists of a primary wall, without a thick and rigid secondary wall. In the primary walls, cellulose exists as elementary fibrils that form a complex with hemicellulose.

Before the identification of the cellulase genes, cellulase enzymatic activity was reported in homogenates and secretions of *B. xylophilus* (Odani et al. 1985; Yamamoto et al. 1986). Moreover, close observations of pine tissues infected with *B. xylophilus* suggested that the destruction of pine cells might be a result of cell

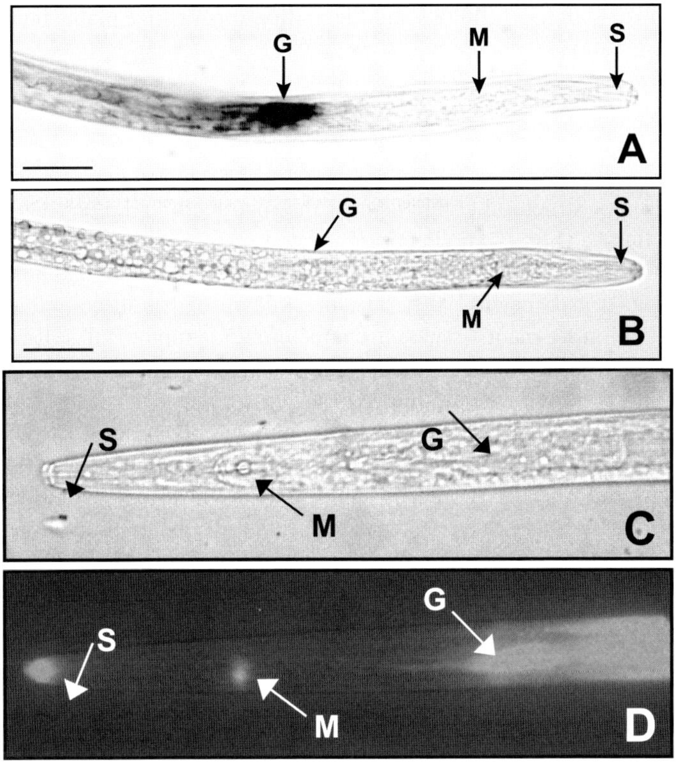

Fig. II.13 Localization of the *Bursaphelenchus xylophilus* cellulase 1 (Bx-ENG-1) transcript and protein. **A**, **B** localization by in situ hybridization of Bx-eng-1 transcripts in oesophageal gland cells of *B. xylophilus* adult female with antisense (**A**) and sense (**B**) Bx-eng-1 digoxigenin-labelled cDNA probes. Expression is restricted to oesophageal gland cells (*bar*, 20 μm); **C**, **D** immunofluorescence localization with antiserum against recombinant Bx-ENG-1, showing that protein is present in the oesophageal gland cells of the nematode and on the exterior of the nematode's head; **C** illustrates the bright-field image, whereas **D** illustrates the same specimen viewed under fluorescence optics; *G* oesophageal glands; *S* stylet, *M* metacarpus

wall-degrading enzymes such as cellulase (Ishida ct al. 1997; Ichihara et al. 2000a). Thus, these results taken together suggest that *B. xylophilus* secretes cellulases to act on the cellulose–hemicellulose complex, resulting in weakening of the mechanical strength of the cell wall.

It was shown that *B. xylophilus* cellulases are secreted through the nematode's stylet and, like the cellulases of other plant parasitic nematodes, may soften the plant cell wall to facilitate their feeding and migration; however, the *B. xylophilus* cellulases showed most similarity to fungal cellulases and were classified into the glycosyl hydrolase family (GHF) 45, while cellulases of cyst/root-knot nematodes belong to GHF5 and are most similar to bacterial cellulases. It was therefore proposed that cyst/root-knot nematodes and *Bursaphelenchus* spp. might have evolved both the ability to digest cellulose and the ability to parasitize plants independently (Kikuchi et al. 2004).

2 Pectate Lyases

Pectin is a major structural component of the plant cell wall along with cellulose and hemicellulose (Fig. II.11). Pectin is located mainly in the middle lamella and primary cell wall, and functions as a matrix anchoring cellulose and hemicellulose fibers (Carpita and Gibeaut 1993). The breakdown of pectin consequently leads to the maceration of plant tissues, a characteristic symptom of soft-rot diseases (Lietzke et al. 1994). Pectin degradation requires the combined action of several enzymes. These can be divided into two groups, namely pectin esterases, which remove the methoxyl groups from pectin and depolymerases (hydrolases and lyases) that cleave the backbone chain (Tamaru and Doi 2001). Pectate lyase (pectate transeliminase, EC 4.2.2.2), which catalyzes the cleavage of internal α-1,4-linkages of unesterified polygalacturonate (pectate) by beta-elimination, is known to play a critical role in pectin degradation (Barras et al. 1994).

Pectate lyases are widely distributed among bacterial and fungal plant pathogens and have been the focus of several studies that have aimed to ascertain their function as virulence factors (Barras et al. 1994). They are used by plant pathogens to degrade host cell walls in order to allow penetration and colonization. Plant parasitic cyst nematodes and root-knot nematodes are known to secrete pectate lyases. Genes encoding pectate lyases have been cloned from several species of plant parasitic nematodes including *Heterodera*, *Globodera* (cyst nematode) and *Meloidogyne* (root-knot nematode) species (Popeijus et al. 2000; de Boer et al. 2002; Doyle and Lambert 2002; Huang G et al. 2005). These pectate lyases are produced in esophageal gland cells and are secreted from the stylet of the nematode. They are thought to play an important role in the infection and parasitism of plants.

Two pectate lyase genes have been cloned from *B. xylophilus*. Like cellulases, the pectate lyases were shown to be secreted from the stylet of the nematode (Kikuchi et al. 2006). *Bursaphelenchus xylophilus* is required to migrate within plant tissue during its lifecycle. Once the nematode invades the pine tree, it migrates primarily through resin canals of the tree and feeds from parenchyma cells surrounding the canals (Mamiya 1983). Although pectin substrates are scarce in the xylem of woody plants, they are present in the primary cell wall of the cambium and parenchyma cells of woody plants, including pine trees (Hafren et al. 2000; Westermark et al. 1986). As such it is likely that pectate lyases are secreted from the nematode stylet and help the nematode to migrate and feed within the tree.

Phylogenetic analysis of pectate lyases, including those from bacteria, fungi and nematodes, resulted in a tree in which the nematode sequences were not monophyletic (Fig. II.14). Although it is difficult to determine conclusively from this analysis whether nematode pectate lyase genes have an ancient, common origin, it seems

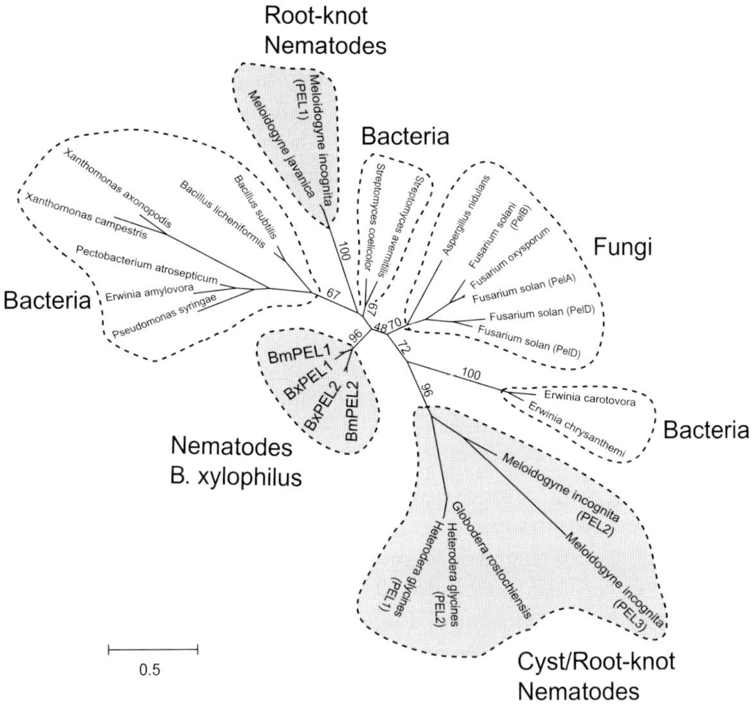

Fig. II.14 Unrooted phylogenetic tree of pectate lyases belonging to polysaccharide lyase family 3. Numbers on the node represent bootstrap support percentages. The *scale bar* represents 50 substitutions per 100 amino acid positions

likely that this is the case. Further support for this idea comes from observation of the conserved position of the intron. Pectate lyases are likely to be widely distributed in plant parasitic nematodes.

3 1,3-Glucanases

β-1,3-Glucanases are widely distributed among bacteria, fungi and higher plants. β-1,3-Glucanases catalyze the hydrolysis of β-1,3-D-glucosidic linkages in β-1,3-D-glucan. This polymer is a major component of fungal cell walls and a major structural and storage polysaccharide (in the form of laminarin) in marine macroalgae (Hong et al. 2002). The physiological functions of β-1,3-glucanases are distinct and depend on their source. Although they have the same hydrolytic activity, bacterial enzymes are classified into glycosyl hydrolase family 16 (GHF16), whereas most plant and fungal enzymes are grouped in GHF17, on the basis of differences in their amino acid sequences (Henrissat and Bairoch 1993).

In the animal kingdom, functionally characterized β-1,3-glucanases are restricted to *B. xylophilus* and marine invertebrates. Genes encoding β-1,3-glucanases have been cloned from the sea urchin *Strongylocentrotus purpuratus* (Bachman and McClay 1996) and the bivalve mollusk *Spisula sachalinensis* (Kozhemyako et al. 2004). They are classified into GHF16 and are thought to be involved in the digestion of algal food. In addition, many β-1,3-glucanase-like proteins have been isolated, and the encoding genes cloned, from insects (Dimopoulos et al. 1997; Kim et al. 2000; Ma and Kanost 2000; Ochiai and Ashida 2000; Zhang et al. 2003) and other invertebrates (Seki et al. 1994; Beschin et al. 1998; Lee et al. 2000; Sritunyalucksana et al. 2002). Although these sequences contain regions that are very similar to the activation region of GHF16 β-1,3-glucanases, they have not been shown to exhibit glucanase activity. These proteins bind specifically to β-1,3-glucan, which is found on the cell surface of microbes but is absent in the host, and they have been shown to play a role in the innate immune system by recognizing foreign material.

Most *Bursaphelenchus* species are solely fungal feeders and all species rely on fungi as a food source at some stage in their life cycle. Many *Bursaphelenchus* species feed on fungi colonizing dead trees. Many *Bursaphelenchus* species, including *B. xylophilus*, therefore have a close association with fungi. Since β-1,3-glucan is the main structural component of fungal cell walls, it seems likely that β-1,3-glucanases play an important role in the life cycle of these nematodes.

The first nematode β-1,3-glucanase genes were identified from *B. xylophilus* during an EST project (Kikuchi et al. 2005). The *B. xylophilus* gene is expressed in oesophageal gland cells and, like the previously characterised cellulases, the enzyme is secreted from the nematode stylet. However, in contrast to cellulases, glucanase is most similar to bacterial enzymes. It is therefore suggested that the *B. xylophilus* β-1,3-glucanase was acquired by horizontal gene transfer from bacteria.

10.3.2 Other Parasitism Genes

Analysis of the EST datasets revealed that other proteins potentially important for parasitism were present, including chitinase and expansin (Kikuchi et al. 2007). Chitinase was described from the cyst nematode *Heterodera glycines*, where it was found to be expressed in subventral oesophageal gland cells, suggesting a role in parasitism, although the precise nature of its potential role remains uncertain (Gao et al. 2002). While a similar function for chitinase may be possible for *B. xylophilus*, this nematode uses fungi as a food source and a beetle as a transmission vector, whose structures contain chitin, and it is possible that chitinase plays a role in these processes. Expansins have been described from cyst nematodes (Qin et al. 2004) and it is thought that these proteins disrupt non-covalent bonds in the plant cell wall, enhancing the activity of other enzymes such as cellulases.

10.4 Conclusions

In the EST project of the PWN, over 13,000 ESTs from *B. xylophilus* and over 3,000 ESTs from a closely related species, *B. mucronatus*, were generated. This project demonstrated that EST generation is an effective method for the discovery of new genes in plant parasitic nematodes. Previous characterization of the *Bursaphelenchus* species genome has been limited to a very few sequences which were used only for phylogenetic analysis and diagnostic purposes. The EST sequences will provide a solid base for future research to investigate the biology, pathogenicity and evolutionary history of this nematode.

One of the most remarkable findings of the EST analysis was the identification of genes, which encode cell wall-degrading enzymes that are likely to be important for plant parasitism. Distinct types of cell wall-degrading enzyme genes were identified from the EST dataset: cellulase, β-1,3-glucanase, pectate lyase. Molecular characterization of these genes showed that they are endogenous nematode genes and are each present as gene families. In situ hybridization showed that all these enzymes are produced in esophageal glands and are therefore extremely likely to be secreted from the nematode through the stylet to the external environment. Further evidence for this was obtained from experiments in which enzyme activity was detected in stylet secretions collected from nematode samples. It has therefore been suggested that these nematodes use a mixture of enzymes to attack the plant or fungal cell wall. These secreted enzymes have been suggested to help in feeding and migration of the nematode.

Cell wall-degrading enzymes of cyst/root-knot nematode are suggested that they were acquired by horizontal gene transfer (HGT) from bacteria (Jones et al. 2005; Scholl et al. 2003) because they are not found in other nematodes or almost any other animals, and are most similar to bacterial genes. *Bursaphelenchus xylophilus* also showed that at least two independent HGT events, one each from bacteria (Kikuchi et al. 2005) and fungi (Kikuchi et al. 2004), have occurred during the evolution of the *Bursaphelenchus* group and that these events have helped shape the evolution of two feeding strategies (fungal feeding and plant parasitism) within this group.

It seems clear therefore that horizontal gene transfer has played an important role in the evolution of plant parasitism in at least two major groups of plant-parasitic nematodes. ESTs have also been obtained from a representative of a third group, *Xiphinema index* (Table 11.2). Analysis of these ESTs suggests that cellulase-like genes may also be present in this nematode and that these are most similar to GHF12 cellulases. Further studies on these genes are required and are currently in progress (J. Jones, personal communication) but if these genes are confirmed as genuine *X. index* genes, this will provide another example of the role of genes acquired by HGT in plant-nematode interactions. Studies on other nematodes, particularly the less intensively studied ectoparasites such as *Trichodorus* and fungal-feeding nematodes belonging to groups different from *Bursaphelenchus* such as *Aphelenchus* and *Tylencholaimus*, would be useful to determine whether the presence of cellulases or other cell wall-degrading enzymes is a requirement for nematode parasitism of plants and whether HGT has driven the evolution of plant parasitism in other nematode groups.

As more genome sequences are obtained from a wider range of nematodes and as EST datasets are compared and analyzed in more detail, it is possible that other horizontally acquired genes may be identified and the role that this process plays in the evolution of nematodes will be fully appreciated.

RNAi (RNA interference) is a powerful tool for the analysis of gene function that has been used extensively for model organisms such as *C. elegans*. This technique exploits the fact that exposure of an organism to double-stranded RNA (dsRNA) from a gene of interest causes silencing of the endogenous gene and allows the null phenotype to be mimicked (Fire et al. 1998). RNAi has been used for genomic scale studies in *C. elegans* (Maeda et al. 2001). In recent years, many groups have been working to transfer this technology to plant parasitic nematodes. To date, successful application of RNAi has been published for plant parasitic cyst and root-knot nematodes (Urwin et al. 2002; Bakhetia et al. 2005; Chen et al. 2005; Fanelli et al. 2005; Rosso et al. 2005). This technique will surely be helpful in determining the detailed functions not only of the genes encoding cell wall-degrading enzymes but also other interesting genes in *B. xylophilus*.

11 Embryology and Cytology of *Bursaphelenchus xylophilus*

Koichi Hasegawa[1,2] and Johji Miwa[1,3*]

11.1 Introduction

Because of their small size, availability, simple and transparent body, and reproducible development nematodes have been popular organisms for studying animal development since the nineteenth century. The many important events in embryogenesis, such as fertilization, meiosis, pronuclear meeting, and chromosome-based sex-determination were discovered as the result of observations of the horse roundworm *Parascaris equorum* (*Ascaris megalocephala*) embryo, and the concepts of individuality and physical continuity of the chromosomes were established and generalized as biological phenomena by nineteenth century nematologists (Triantaphyllou 1971). A variety of chromosome numbers, reproduction modes, and early developmental patterns have been discovered and studied in both parasitic and free-living nematodes.

Since the initial use of the free-living soil nematode *Caenorhabditis elegans* as a model organism to study genetic animal development and behavior (Brenner 1974), various novel biological phenomena have been found and an enormous amount of sophisticated knowledge has been accumulated through research on this nematode. Completion of the embryonic and postembryonic cell lineages of *C. elegans* has established a strong foundation for the molecular understanding of animal development (Sulston and Horvitz 1977; Sulston et al. 1983). The

[1]Institute for Biological Function, Chubu University, 1200 Matsumoto, Kasugai 487-8501, Japan

[2]Laboratory of Environmental Mycoscience, Graduate School of Agriculture, Kyoto University, Kitashirakawa Oiwake-cho, Sakyo-ku, Kyoto 606-8502, Japan

[3]Graduate School of Bioscicncc and Biotechnology, Chubu University, 1200 Matsumoto, Kasugai 487-8501, Japan

[*]Tel.: +81-568-51-6218, Fax: +81-568-51-6218, e-mail: miwa@isc.chubu.ac.jp

connection and wiring diagram of all neuronal cells (White et al. 1986) has made it possible to analyze animal responses to environmental stimuli on cellular and molecular terms. The genome of *C. elegans* has been sequenced with about 19,000 predicted gene-coding regions in 100 million pairs of nucleotides (The *C. elegans* Sequence Consortium 1998). From the WormBase homepage, all published *C. elegans* data are retrievable by personal computer (WormBase, http://elegans.swmed.edu/).

In this chapter, taking advantage of the knowledge accumulated for *C. elegans*, we introduce the embryology and cytology of the pine wood nematode (PWN) *Bursaphelenchus xylophilus* (Nematoda: Parasitaphelenchidae), the causal agent of pine wilt disease in East Asia (Mamiya 1983), and Europe (Mota et al. 1999). Although parasitic in nature, *B. xylophilus* is easily maintained in the laboratory where usually it is cultured on the fungus *Botrytis cinerea*, or sometimes on callus cultures of several plants (Iwahori and Futai 1990). Under laboratory conditions, the entire life cycle from fertilization to mature adult takes 5 days at 25°C (Fig. II.15). *Bursaphelenchus xylophilus* reproduces gonochoristically (male and female sexes), with a large brood of about 100–800 offspring (Bolla and Boschert 1993). Embryonic development takes about 25 h at 25°C from fertilization through such events as cell division, differentiation, and one molt in the eggshell to L2 hatchee (Hasegawa et al. 2004). If enough food is available, postembryonic development takes about 4 days from L2-hatched larva through 3 molts to adult (Mamiya 1975a). In nature, *B. xylophilus* invades a living pine tree via feeding wounds made by the

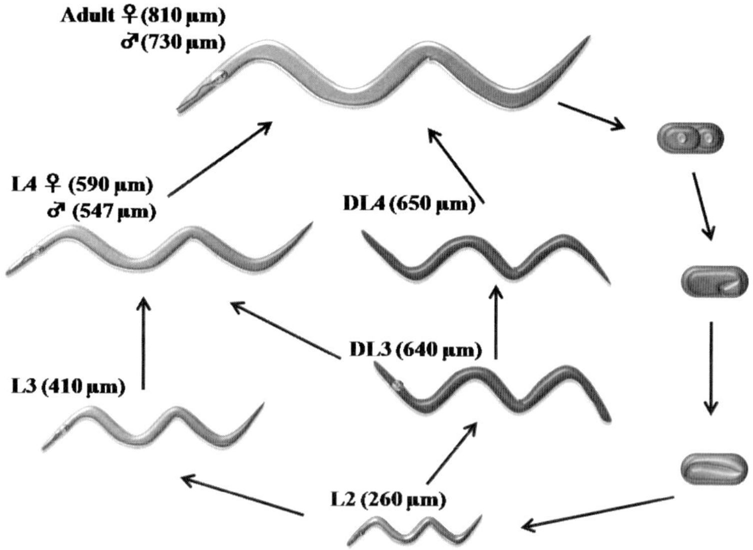

Fig. II.15 Schematic images of the *Bursaphelenchus xylophilus* lifecycle. Body sizes are adapted from Mamiya (1975). *DL3* third-stage dispersal larva, *DL4* fourth-stage dispersal larva

vector beetle *Monochamus alternatus*, feeds on fresh live pine tissues at an earlier stage of invasion, and then feeds on fungi that grow in dead pine trees. In the absence of food or under unfavorable conditions, the nematode larva enters the dispersal third stage (L_{III}, DL3), which is specialized for surviving adverse conditions and is biologically significant since this stage precedes the dauer larval stage (Ishibashi and Kondo 1977; Kondo and Ishibashi 1978). Moreover, the dispersal third-stage larva (L_{III}, DL3), stimulated by the *M. alternatus* pupa, molts to become the dispersal fourth-stage larva (L_{IV}, DL4) which is adapted for being carried by the vector beetle (Maehara and Futai 1996). After the dispersal fourth-stage larva (L_{IV}, DL4) transfers from the vector into the pine tree, it develops into the adult stage and starts reproducing (Mamiya 1975a).

11.2 Reproduction

Nematodes employ a variety of reproductive strategies. (1) Asexual (apomictic) reproduction by which an organism creates a genetically identical copy of itself without a contribution of genetic materials from another individual. This type of reproduction includes (a) meiotic parthenogenesis, in which the haploid chromosomes in a polar body fuse with the haploid chromosomes of an oocyte to maintain diploidy, (b) mitotic parthenogenesis, in which a diploid egg produced without meiosis and fertilization divides and develops normally to a full-fledged diploid nematode. (2) The second is sexual amphimictic reproduction by which organisms create descendants that have a combination of genetic materials from male and female members of the species. Sexual amphimictic reproduction includes meiosis and fertilization whereby two successive cell divisions occur to form ovum or sperm of haploid chromosomes, which fuse to become a zygote. (3) Some other nematode species, however, employ another sexual reproduction called hermaphroditism, in which both sperm and ovum are produced in one individual, that is, a hermaphrodite. (4) Yet another curious reproductive strategy exists and is called pseudogamy, in which eggs develop normally by mere sperm stimulation with no introduction of genetic materials (Goldstein 1981). The PWN, *B. xylophilus*, employs amphimictic sexual reproduction (Aoyagi and Ishibashi 1983; Hasegawa et al. 2004, 2006).

The *B. xylophilus* reproductive system consists of somatic gonads that provide the structure and environment for fertilization and egg laying and the germline gonad, which eventually produces gametes. Both male and female *B. xylophilus* have one tubular gonad, which extends straight forward from the cloaca (in males) or vulva (in females) (proximal region) toward the anterior (distal region). Figure II.16 shows both an adult *B. xylophilus* male and female. The nematode shown in Fig. II.17A and B has a monodelphic gonad; one tubular gonad extends from the vulva toward the head. An adult germline gonad has distal–proximal polarity, and the distal region forms the syncytium with nuclei dividing and proliferating continuously by mitosis before the nuclei enter meiosis. The spermatheca, where sperm

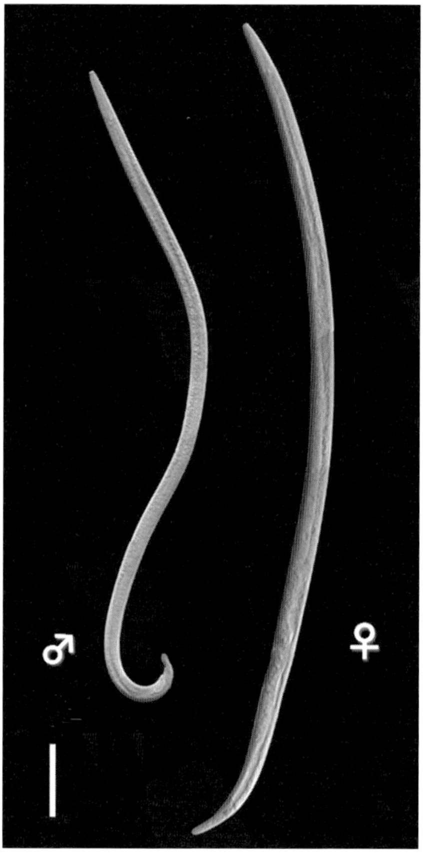

Fig. II.16 Nomarski differential interference (DIC) images of *Bursaphelenchus xylophilus* adult male and female. Top is head. *Scale bar* 100 μm

from the male are stored, is positioned between the oviduct and uterus. A healthy young adult female in the reproductive stage usually has one fertilized embryo in the uterus. A post-vulval sac located posterior to the vulva contains no germ cells.

Although almost all cell nuclei in the distal region marked by the distal tip cell (DTC) contain condensed chromosomes visible typically as dots with DAPI staining, a few undergo mitotic division (Fig. II.17C). Moving from the distal region toward the proximal region, cell nuclei undergo meiosis, and chromosomes in the pachytene stage look tangled up like "Japanese ramen noodles before cooking" (Fig. II.17D). In spermatogenesis, germ cells complete the first and second meiotic divisions to form haploid sperm. During this spermatogenesis, bivalent chromosomes are clearly visible as six dots (Fig. II.17E). In oogenesis, germ cell division is arrested in the diakinetic stage until fertilization. Female germ cell nuclei are enclosed by a membrane and enlarge as they move toward the proximal region of the gonad. Six bivalent chromosomes are seen in the mature oocyte (Fig. II.17F).

The *C. elegans* DTC functions to lead gonad elongation during development and inhibits meiotic division, while it promotes the mitotic division of germ cells near

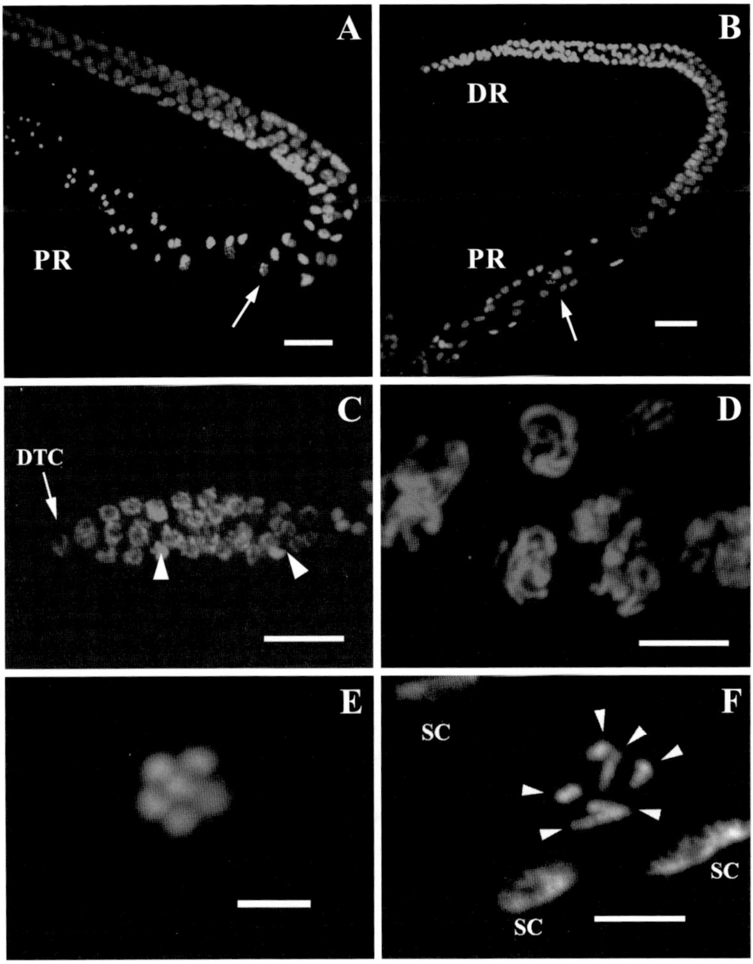

Fig. II.17 Chromosomal structures of *Bursaphelenchus xylophilus* adult male and female germ cells, stained with DAPI and viewed by confocal laser scanning microscopy (LSM). **A** Male germ line, *arrow* indicates six bivalent chromosomes; **B** female germ lines, *arrow* indicates bivalent chromosomes in diakinesis; **C** germ nuclei at the distal region of the male gonad arm, arrowheads indicate mitotically dividing cell nuclei; **D** male germ cells in the meiotic pachytene stage. *DR* distal region. **E** High magnification of bivalent chromosomes in the male germ cell shown in **A**; **F** high magnification of the six bivalent chromosomes in the female germ cell shown in **B**. *PR* proximal region, *DTC* distal tip cell, *SC* somatic seam cell. *Scale bar* **A**, **B** 20 μm; **C** 10 μm; **D**, **F** 5 μm; **E** 2 μm (2006, Nematology, Brill, with permission)

DTC (Hedgecock et al. 1987; Kimble et al. 1998). Although a *C. elegans* male has a single-armed gonad (monodelphic gonad), a hermaphrodite has a didelphic U-shaped gonad, which is rotationally symmetrical (Kimble and Hirsh 1979). *Caenorhabditis elegans* hatches (L1 animals) possess the two primordial germ cells Z2 and Z3 flanked by the two somatic gonad precursors Z1 and Z4. In the hermaphrodite, each of the Z1 and Z4 cells gives rise to a DTC, which promotes the growth and migration anteriorly right or posteriorly left to form the two gonad arms (didelphic gonad) (Kimble and Hirsh 1979). In the *C. elegans* male, the Z1 and Z4 cells generate two DTCs in one end of the somatic gonad, which regulate germline development and one linker cell (LC) in the other end, which regulates asymmetric gonad elongation and formation of the J-shaped gonad (Kimble and Hirsh 1979). The *B. xylophilus* L2 hatchee (*B. xylophilus* molts once in the eggshell and hatches as a L2 larva) has also the same configuration of somatic and germ precursor cells. Following the *C. elegans* cell nomenclature, the two larger primordial germ cells Z2 and Z3 are flanked by the two smaller somatic gonad precursors Z1 and Z4. Z1 and Z4 develop asymmetrically, as the posterior gonad precursor cell Z4 divides and elongates posteriorly to form the spermatheca, uterus, and post-vulval sac (or post vulvar uterine branch) in females, and seminal vesicle and vas deferens in males (Ishibashi et al. 1978).

11.3 Fertilization Process

Bursaphelenchus xylophilus adult females cannot lay either fertilized or unfertilized eggs unless they have mated. Before mating, there are no sperm in the spermatheca, uterus, or post-uterus branch, and mature oocytes in the most proximal region of the gonad remain there until sperm are supplied into the spermatheca (Fig. II.18A, B). After mating, many sperm are supplied into not only the spermatheca, but also the uterus as well as the post-vulval sac (Fig. II.18C, D). As described above, the post-vulval sac, positioned posterior to the vulva, has no ability to make germ cells, and might function only to store sperm (Fig. II.18C, D). Sperm supply triggers the oocyte to move from the oviduct to the uterus. In *B. xylophilus*, the spermatheca is located between the oviduct and the uterus and has only one opening toward the uterus, and so sperm supplied from the male should enter the spermatheca from this opening. As the oocyte moves from the oviduct toward the uterus through a very narrow path between them, it squeezes out some sperm from the spermatheca, although only one sperm fertilizes the oocyte (Fig. II.19A). Some unfertilized sperm seem to have a chance to go back to the spermatheca again for the next fertilization, but others may be pushed out from the vulva or into the post-vulval sac by the moving fertilized egg (zygote). The zygote stays in the uterus undergoing successive events such as resumption of meiosis, breakdown of the germinal vesicle, and eggshell formation (Fig. II.19B). The process from fertilization to egg laying takes about 20 min at 25°C.

11.4 Pronuclear Meeting and Axis Polarity

The *B. xylophilus* embryo is long and slender, about 25-μm wide and 60-μm long. As with *C. elegans*, the first cell division is unequal and produces the larger AB cell and the smaller P_1 cell, which are destined to a certain fate, although they are

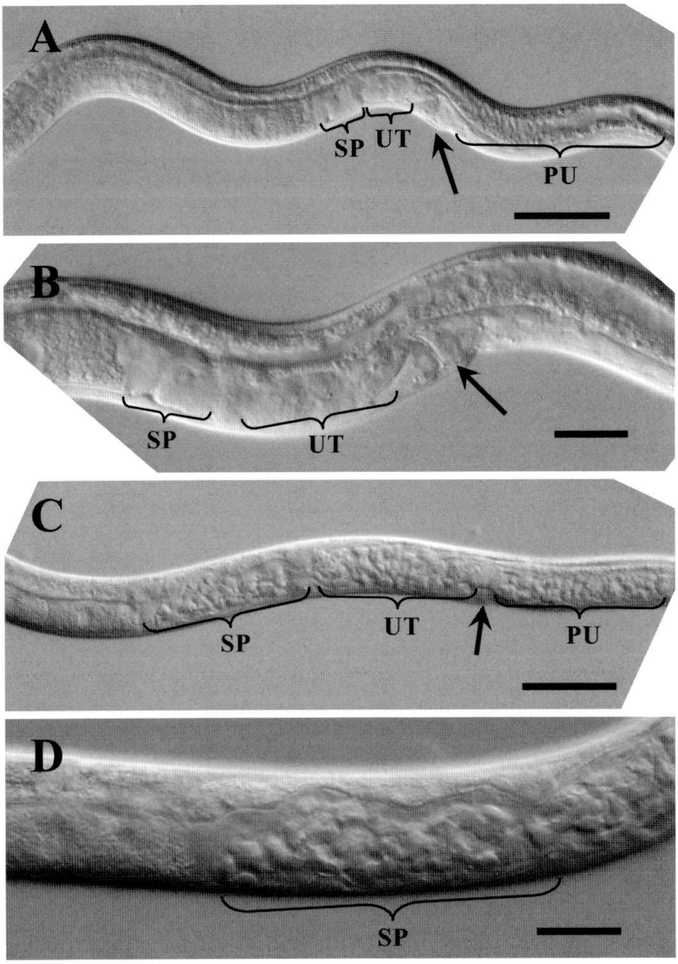

Fig. II.18 Adult female, before or after mating: **A**, **B** before mating, there are no sperm in the spermatheca, uterus, or post-uterus branch. A mature oocyte in the proximal-most oviduct remains there until sperm is supplied into the spermatheca. **C**, **D** After mating, spermatheca and uterus are expanded and filled with sperm. Sperm is also supplied into the post-vulval sac. *SP* spermatheca, *UT* uterus, *PU* post-vulval sac. *Arrow* indicates vulva. Anterior is left, dorsal is top. *Scale bar* **A**, **C** 200 μm; **B**, **D** 20 μm

still reprogrammable at this stage (Gönczy and Rose 2005). This division reveals the anterior–posterior (A–P) axis of a nematode in that the AB and P_1 cells indicate, respectively, the future anterior and the posterior sides (Hasegawa et al. 2004). In this section we describe the *B. xylophilus* embryonic cell division pattern from pronuclear meeting to the two-cell embryo, whereby the cell cytoskeleton forms and the anterior–posterior axis of the embryo is determined, following certain predictable behavior of the male and female pronuclei.

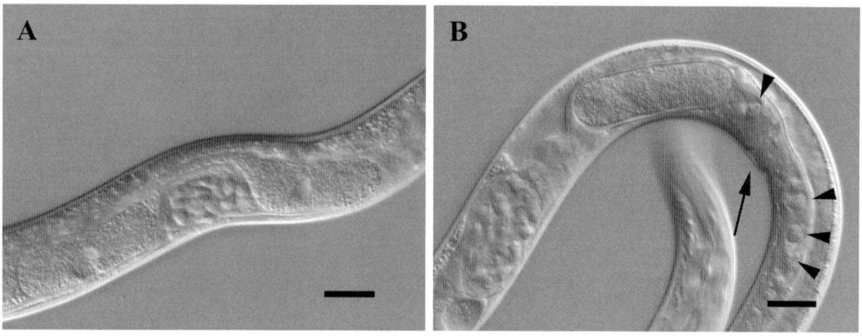

Fig. II.19 Fertilization processes, viewed by DIC imaging: **A** mature oocyte moves from oviduct to the uterus, squeezing out some sperms from the spermatheca; **B** fertilized zygote stays in the uterus for a while. *Arrow* indicates vulva, arrowheads indicate sperm in the post-vulval sac. Left is anterior, top is dorsal. *Scale bar* 10 μm

The cortical membrane of a newly laid egg is rumpled and active, and seemingly random cytoplasmic streaming is visible (Fig. II.20A). At this time, a single DAPI-stained sperm is seen as a dot positioned at the future anterior side of the embryo (Figs. II.20A, II.21A), and six chromosomes are clearly visible inside the first polar body (Fig. II.21B, C). Meiosis is propelled by the small seemingly acentriolar meiotic spindle positioned at the lateral mid-point position of the embryo (Fig. II.20A). The first and second polar bodies are extruded out as the vesicular male and female pronuclei are reconstituted (Figs. II.20B, II.21D–F). The male pronucleus is slightly bigger than that for the female and appears at the future anterior pole. The female pronucleus emerges by the second polar body. Duplicated centrosomes are on the surface of the male pronucleus, interacting with the cortical actin (Fig. II.20B). In *C. elegans*, the fertilizing sperm is known to bring not only the haploid genome set but also the centrosome into the oocyte. And it is also thought to determine the anterior–posterior axis of the embryo such that the sperm-entry point on the embryo marks the future posterior end (Goldstein and Hird 1996). In contrast, the sperm entry point appears to become the future anterior end of the *B. xylophilus* embryo, although the *B. xylophilus* sperm also brings a centrosome into the oocyte. Chromosomes in each of male and female pronuclei are dispersed throughout the pronucleus (Fig. II.21D–F). The rumpled cortical membrane becomes smooth, and the male and female pronuclei move toward each other and eventually meet (Fig. II.20C). Chromosomes become condensed in the prophase stage, and six chromosomes are visible in each pronucleus (Fig. II.21G, H). They move to the center where they rotate 90° (Figs. II.20D, II.21I, J), and fuse to become one (Fig. II.20E). The pair of male and female pronuclear-derived chromatids are fused and aligned along the metaphase plate (Figs. II.20E, II.21K). The pair of chromatids is arranged parallel to each other, and the longitudinal axes of the chromatids are perpendicular to the anterior–posterior axis of the embryo (Fig. II.21L, M). Two centrosomes are located at the center of the embryo along its longitudinal axis (Fig. II.20E). In the

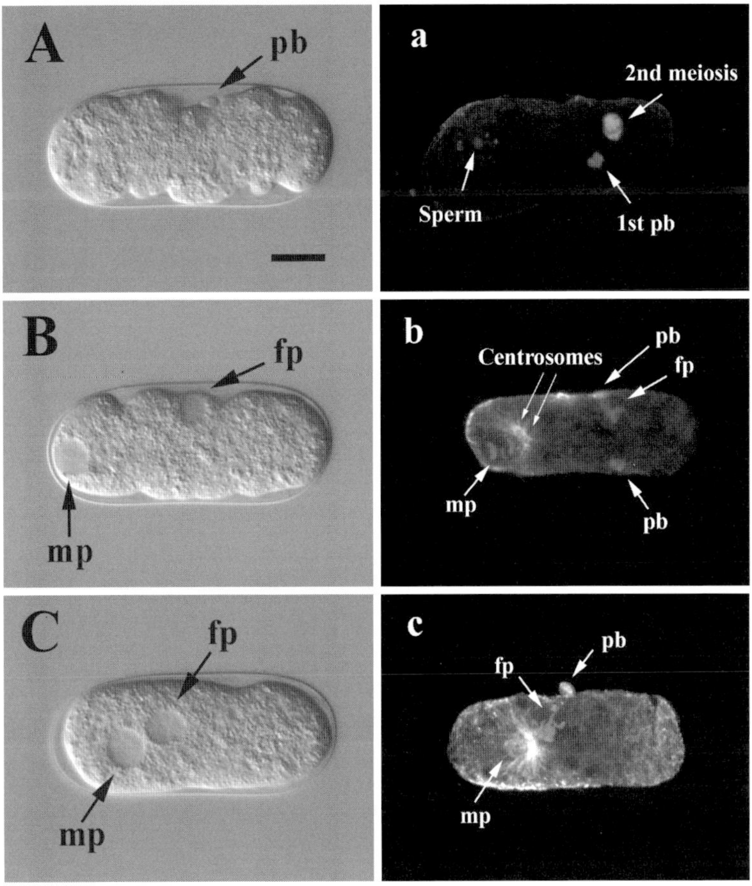

Fig. II.20 Embryonic cell divisions of *Bursaphelenchus xylophilus* from fertilization to the two-cell stage. **A–G** Nomarski differential interference contrast images, **a–g** confocal laser-scanning microscope images, microtubule and actin cytoskeleton are visualized with antibody, and DNA with DAPI. **A** Just after oviposition, the entire cortical region is ruffling and one of the polar bodies is visible. **a** At this stage, the sperm is visible as a faint dot, and second meiosis starts with microtubules elongated. **B** Soon after the completion of meiosis, the male pronucleus appears at the future anterior pole of the embryo, and the female pronucleus emerges at a lateral mid-point position. **b** Duplicate centrosomes are visible at the surface of the male pronucleus, interacting with cortical actin. **C** Pronuclear meeting. **c** Two centrosomes are nucleating. **D**, **d** Juxtaposed pronuclei move to the center and rotate 90°. **E** Two pronuclei are fusing. **e** Metaphase-stage embryo, two centrosomes are located in the center of the embryo along the longitudinal axis. **F** Anaphase-stage embryo. **f** Posterior centrosome moves posteriorly to pull the chromosomes posteriorly. **G** Two-cell-stage embryo. **g** Two-cell-stage embryo. *pb* polar body, *mp* male pronucleus, *fp* female pronucleus. Anterior is left. *Scale bar* 10 μm

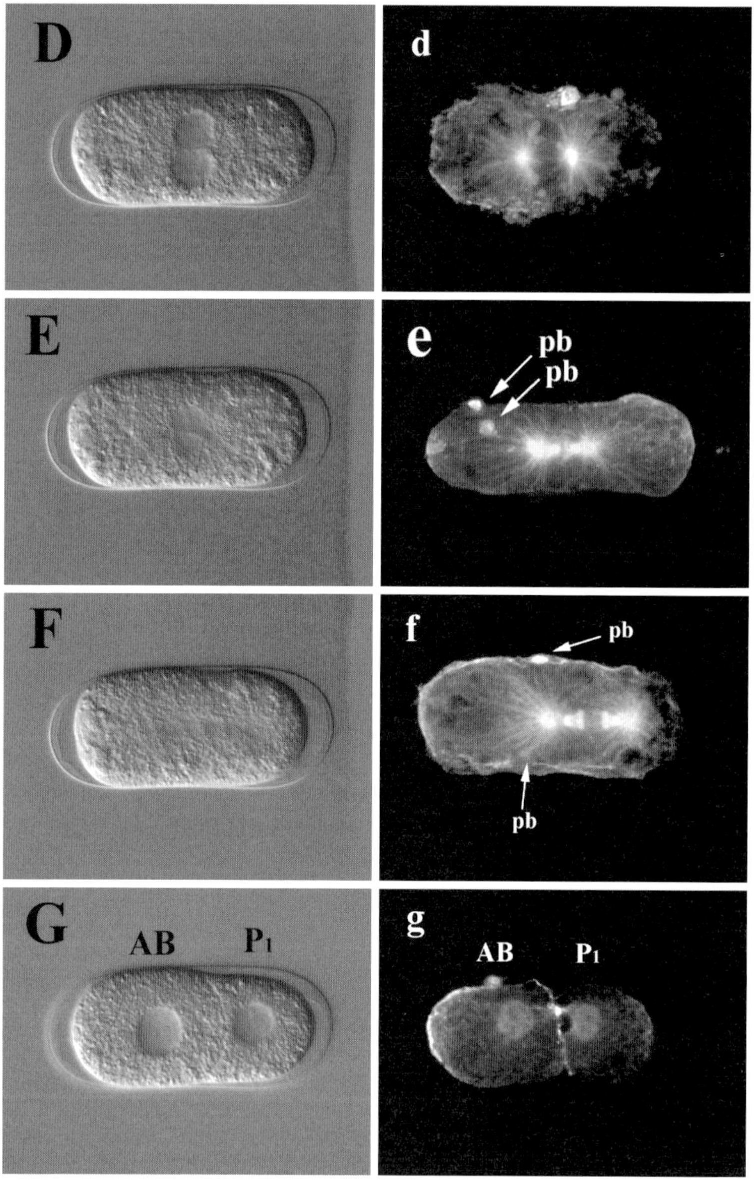

Fig. II.20 (Continued)

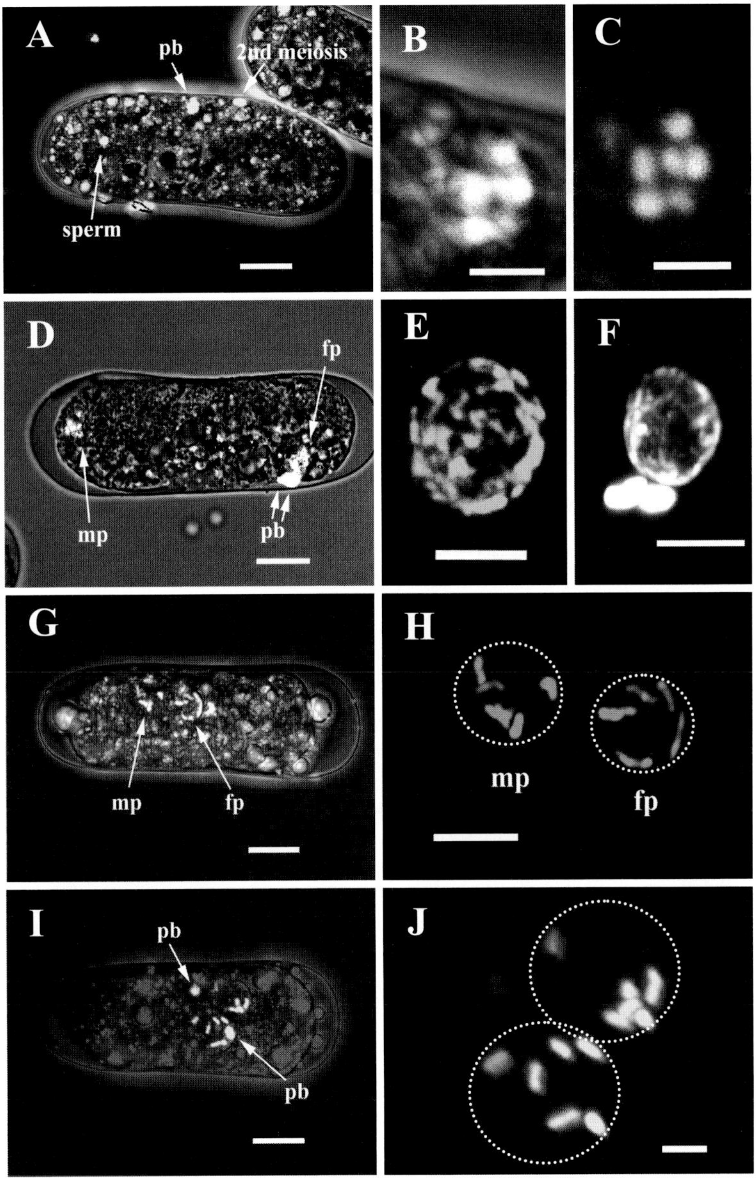

Fig. II.21 Chromosome behavior from fertilization to the two-cell stage. **A** Fertilized egg, just after extrusion from the vulva (oviposition). The sperm is seen as a dot at the future anterior end of the embryo (*left* side in this photograph). **B** High magnification of the first polar body shown in **A**. Merged view of the bright field and fluorescence images. **C** Fluorescence-only image of the 1st polar body, six chromosomes are seen. **D** Rearrangement of male (*mp*) and female (*fp*) pronuclei. **E** High magnification of male pronucleus and **F** female pronucleus. The first and second polar bodies are seen next to the female pronucleus. **G** Pronuclear meeting. **H** High magnification of two pronuclei. **I** Rotation of two juxtaposed nuclei. **J** High magnification of juxtaposed nuclei shown in **I**. Each pronucleus contains six haploid chromosomes.

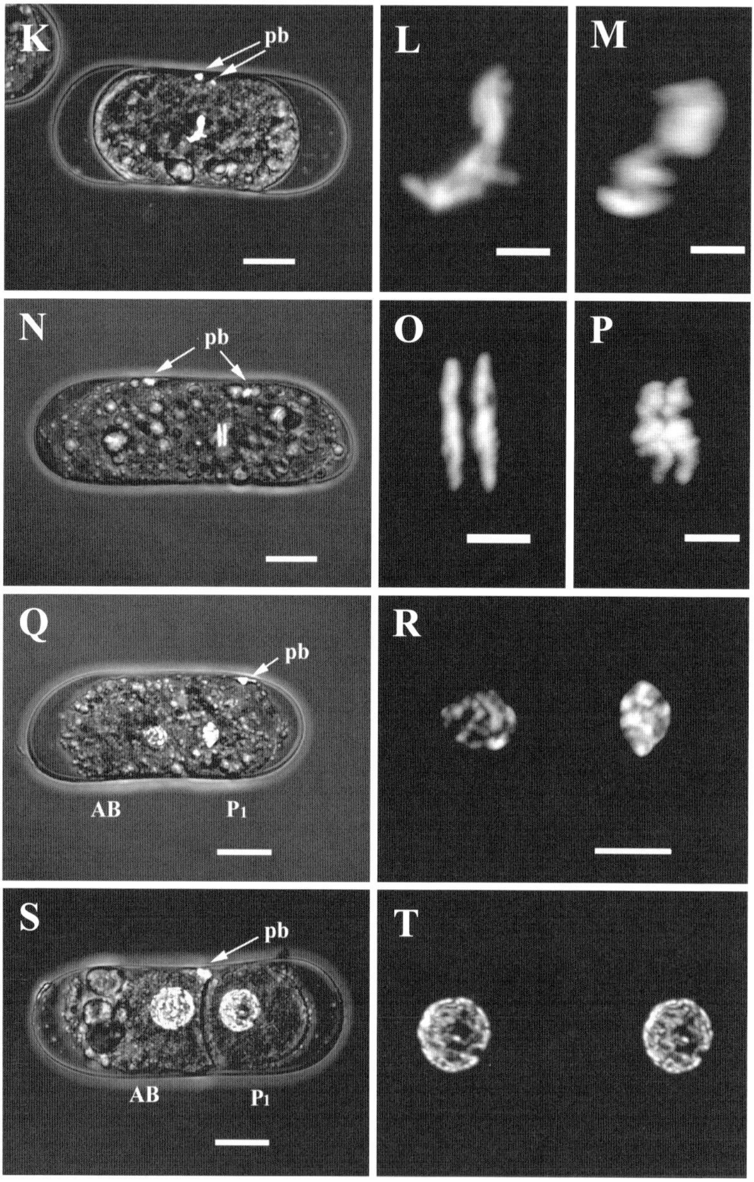

Fig. II.21 (Continued) **K** Mitotic metaphase after fusion of the two pronuclei. **L** High magnification of chromosomes arranged along the metaphase plate. **M** The same image as in **L** seen from 30°-forward diagonal. **N** Mitotic anaphase, segregating chromosomes. **O** High magnification of segregating chromosomes in **N**. **P** The same image as in **O** from 30°-forward diagonal. **Q** End of mitotic telophase, with cytokinesis completed. **R** High magnification of **Q**. AB cell nucleus is on the left and that for P_1 cells is on the right. Nuclear membranes are reconstructed in sister blastomeres. **S** 2-cell-stage embryo. **T** Stereo image of P_1 nucleus: *pb* polar body, *mp* male pronucleus, *fp* female pronucleus. Anterior is left. *Scale bar* **A**, **D**, **G**, **I**, **K**, **N**, **Q**, **S** 10 µm; **E**, **F**, **H**, **R** 5 µm; **B**, **C**, **J**, **L**, **M**, **O**, **P** 2 µm. Adapted from Hasegawa et al. (2006, Nematology, Brill, with permission)

anaphase stage, the posteriorly positioned centrosome moves posteriorly to separates the chromatids antero-posteriorly while the anteriorly positioned centrosome remains near the center of the embryo (Figs. II.20F, II.21N–P). Subsequently, the embryo divides unequally (Fig. II.21Q, R) to form the larger anterior AB cell and the smaller posterior P_1 cell (Figs. II.20G, II.21S, T), thereby determining the anterior–posterior axis. It takes 52.6 ± 7.7 min ($N = 7$, at 25°C) from oviposition to the two-cell stage.

From the shape of the anaphase chromosomes, *B. xylophilus* mitotic chromosomes are judged to be holocentric or polycentric. This means that kinetochores are formed along the entire length of the chromosomes (Fig. II.21O, P).

Polarity formation is an important feature for establishing many different cell types through several rounds of asymmetric cell divisions. The initial asymmetric cues such as sperm entry (Goldstein and Hird 1996) establish the cell polarity followed by cytoskeleton reorganization and polarized localization of several cortical proteins in the *C. elegans* oocyte. A mature oocyte passing through the spermatheca is fertilized by sperm there (Ward and Carrel 1979). Compared with sperm, the oocyte is much larger and contains more cellular factors necessary for cell division, cell-fate determination, and morphogenesis. Although almost all factors required for early embryogenesis are supplied maternally (Miwa et al. 1980), paternal factors such as chromosomes, centrosome, and cytoplasm also constitute essential and important contributions (Singson 2001). Fertilization triggers completion of the final stage of oocyte meiosis and promotes rearrangement of the cortical actin cytoskeleton that generates cytoplasmic flows anteriorly in the cortical region and posteriorly in the interior of the zygote to distribute cell-fate determinants (Munro et al. 2004). Asymmetric distribution of cell-fate determinants should eventually lead the first cell division to produce the larger anterior AB cell and the smaller posterior P_1 cell that are programmed to have different fates (Sulston et al. 1983; Miwa 1986).

The relation of sperm entry to body axis determination is reported in some other nematode species, which belong to the phylogenetic clades III, IV and V (Blaxter et al. 1998; Goldstein et al. 1998). *Acrobeloides* sp. (PS1146, PS156), *Meloidogyne incognita*, *Panagrellus redivivus* (PS1163), *Panagrolaimus* sp. (PS443), and *Tubatrix aceti* all employ male/female (amphimictic) sexual reproduction. In these species, however, sperm entry seems to have no apparent influence in establishing the embryonic polarity (Goldstein et al. 1998). Some parthenogenetic species, *Acrobeloides nanus*, *Acrobeloides* sp. (ES501), *Cephalobus oryzae* (PS1165), *Cephalobus* sp. (PS1215), *Chiloplacus minimus*, *Panagrolaimus* sp. (PS1159), and *Zeldia punctata* (PS1153), were examined for polarity determination. Quite obviously they can not use sperm entry as a cue to determine the antero–posterior (AP) body axis, but seem to utilize some other cues like maternally supplied materials, environmental conditions, or/and random decision (Goldstein et al. 1998). By anaesthetizing adult nematodes with 0.03–0.05% NaN_3 or 1% 1-phenoxy 2-propanol, fertilized eggs develop in the uterus without being laid, and so we can observe the relation of the embryonic axis formation in the uterus to the orientation of the maturing germ cells or of their mother. In the parthenogenetic *A. nanus*, 98% of embryos express a defined AP polarity, whereas in another parthenogenetic species, *Diploscapter coronatus*, AP polarity is expressed at random (Lahl et al. 2006). The AP-axis of a *B. xylophilus* embryo when laid is usually oriented opposite to that of its mother (Hasegawa et al. 2004). That is, the future anterior end of the embryo is ejected first from the vulva facing toward the posterior end of the mother (Hasegawa et al. 2004). In *Acrobeloides* sp. (PS1146), the sperm does not seem to be involved in determining the AP-axis, as the centrosomes are not associated with the sperm pronucleus (Goldstein et al. 1998).

Actin and several other cortical molecules such as PAR-1 to -6 and PKC-3 play important roles in the AP determination in *C. elegans* (Kemphues and Strome 1997; Tabuse et al. 1998). After fertilization, the distribution of these proteins is first established and then maintained; the PAR-3/PAR-6/PKC-3 complex is located at the anterior cortex, PAR-4 and PAR-5 throughout the cytoplasm and the cortex, and both PAR-1 and PAR-2 at the posterior cortex (Fig. II.22) (Schneider and Bowerman 2003). The accurate distribution of these pilot proteins is necessary for correct displacement of the posterior centrosome (Fig. II.22B) and the polarized distribution of cell-fate determinants along the AP-axis. Homologues of the *C. elegans par* genes are reported to also be essential for establishing the cell polarity in different cell types like neuroblast cells and sensory organ precursor cells, and vertebrate epithelium cells and in the *Drosophila* embryo (Suzuki and Ohno 2006). Although the AP-axis determination of the *B. xylophilus* embryo is entirely opposite to that of the *C. elegans* embryo, many questions remain. Does the *B. xylophilus* embryo also use the same types of molecules to generate AP polarity as in the *C. elegans* embryo? And, furthermore, what happens during the first cell division of parthenogenetic nematodes? A homologue of the serine/threonine kinase gene *par-1* has been isolated from the animal parasitic nematode *Haemonchus contortus*, and its structure and expression patterns were analyzed (Nikolaou et al. 2002, 2004). *Hc-par-1* expresses at least four alternative splicing forms, and their expression is higher in the first- and fourth-stage larvae than in the larvae of the other developmental stages (Nikolaou et al. 2002, 2004). The cloned *Bx-par-1* encodes a deduced protein of 967 amino acids with a conserved serine/threonine kinase domain in the N-terminus and a kinase associated domain at the C-terminus. Its amino acid sequence shares 40% homology with that of *C. elegans* (K. Hasegawa and J. Miwa, personal communication).

11.5 Divisions from Two Cells to Four Cells

The AB cell divides equally and antero-posteriorly, followed by the antero-posterior and unequal P_1 cell division to enter the four-cell stage (Fig. II.23A–C). First, the two centrosomes are positioned dorso-ventrally in the AB cell (Fig. II.23A). Because the space in the eggshell is too small, the dividing axis is gradually inclined to become parallel to the AP-axis (Fig. II.23b). The two centrosomes in the P_1 cell are positioned anterior-posteriorly, and the P_1 cell divides as it forms the EMS and P_2 cells (Fig. II.23c). The ABp cell is positioned at the dorsal side of the future

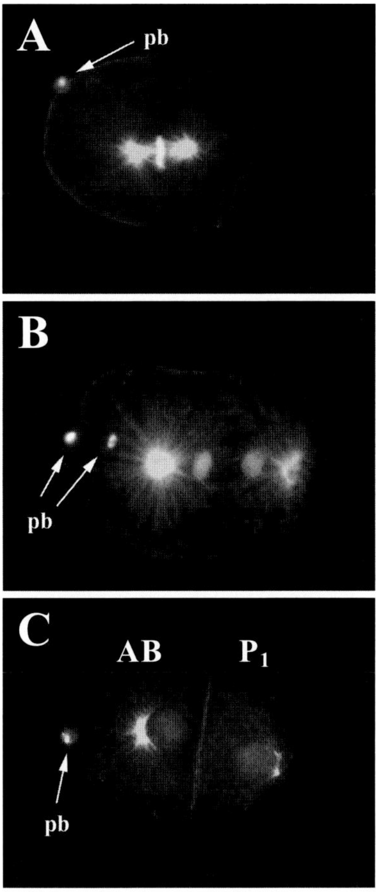

Fig. II.22 LSM image of *Caenorhabditis elegans* embryo from one-cell metaphase to two-cell stage, microtubules and PKC-3 are visualized with antibodies, and DNA with DAPI. **A** Metaphase stage, bivalent chromosomes are arranged at the metaphase plate, and the two centrosomes are positioned symmetrically along the anterior–posterior axis. PKC-3 proteins are restricted to the anterior cortex. **B** Anaphase stage, the posterior centrosome moves posteriorly. **C** Two-cell stage. PKC-3 is uniformly present at the cortex of the AB cell and the boundary of AB and P_1 cells

dorso-ventral (DV) axis, and the consequent right and left divisions of both ABa and ABp should determine the right–left (RL) axis. It takes 78 ± 7 min ($N = 8$, at 25°C) from oviposition to the 4-cell stage. These events are similar to those of *C. elegans*, in which the PAR proteins control the centrosome orientation and the P_1 asymmetric division in two-cell to four-cell embryos (Boyd et al. 1996).

Based on the DNA sequence analysis of a small nuclear subunit (18S) RNA gene, the phylum Nematoda is grouped into five clades (I–V) and four classes, Rhabditia (Clades III, IV, V), Chromadoria (the others), Enoplia (Clade II), and Dorylaimia (Clade I) (Blaxter et al. 1998; De Ley and Blaxter 2002). *Parascaris equorum* (Clade III), which displays a round-shaped embryo, initially forms a T-shaped four-cell stage and then the cells rearrange to assume a rhomboid shape (Malakhov 1994; Hope 2002). Like embryos of Tylenchida and Aphelenchida, the *B. xylophilus* embryo is long and slender. The AB cell divides anterior-posteriorly,

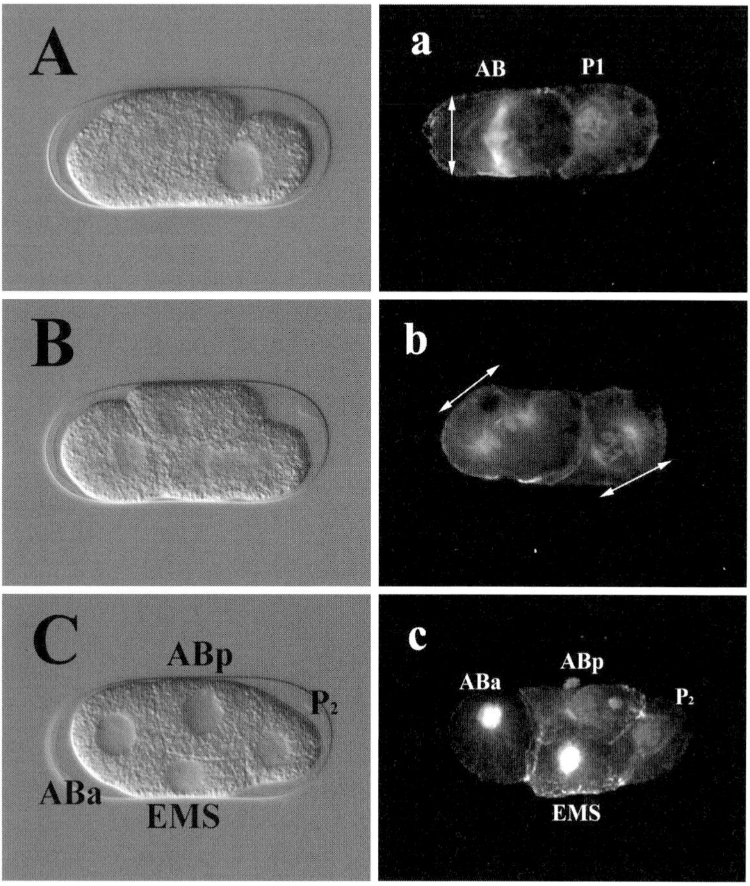

Fig. II.23 Cell divisions from two-cell to four-cell stage. **A**, **a** Two centrosomes in the AB cell dividing dorso-ventrally (*arrow*). **B**, **b** Anterior–posterior division of the AB cell followed by anterior-posterior division of the P1 cell (*white arrow*). **C**, **c** Four-cell stage

and the embryo eventually arranges itself to reach the rhomboidal four-cell stage (Fig. II.23). There are some additional arrangements of four-cell embryos, such as linear and zigzag arrangements (Dolinski et al. 2001; Hope 2002).

Cell positions are very important in the cell–cell interactions and communication, which play major roles in specifying cell fates. The P_2 cell in the *C. elegans* four-cell embryo has direct contact with ABp and EMS cells, but not with ABa. The ABp cell is destined to assume the dorsal side of the worm via APX-1 and GLP-1/Notch signals, and EMS is polarized via a Wnt signal and SRC-1/MES-1 signals. All of these signals are derived from the P_2 cell (Gönczy and Rose 2005).

11.6 Cell Lineage

The first cell division of the *B. xylophilus* embryo is asymmetric to produce the larger AB cell and the smaller P_1 cell. The AB cell line divides equally and synchronously for several cycles of cell division, whereas the P_1 cell line continues a few cycles of unequal and asymmetric cell division. In this process, the six founder cells of AB, MS, E, C, D, and P_4 are generated from the unequal P-cell divisions, and their respective daughters divide synchronously (Hasegawa et al. 2004). ABa and ABp cells in the 4-cell embryo divide along the (R–L) axis, producing ABar and ABal and ABpr and ABpl, respectively; these two are the only R–L divisions observed before the 46-cell stage. Both of the left-side daughters (ABal and ABpl) are slightly shifted anteriorly to the right-side daughters (ABar and ABpr). By the time of the 16-cell embryo, the C cell has just divided and the E cell has divided to produce the Ea and Ep cells, which guide the embryo into the first step of gastrulation. During the 5th cycle of the AB-cell division, 2C cells start to divide. By the 46-cell embryo, the 4C cells are positioned asymmetrically in the posterior dorsal-right region, and the Ea and Ep cells have just divided to produce their respective daughters after journeying into their present interior position from the posterior-ventral inner-surface of the embryo (Fig. II.24). The P_4 cell always stays in close proximity to the Ep cell, and they move inward side-by-side during gastrulation. At 30°C, it takes approximately 250 min for the *B. xylophilus* embryo to reach the 46-

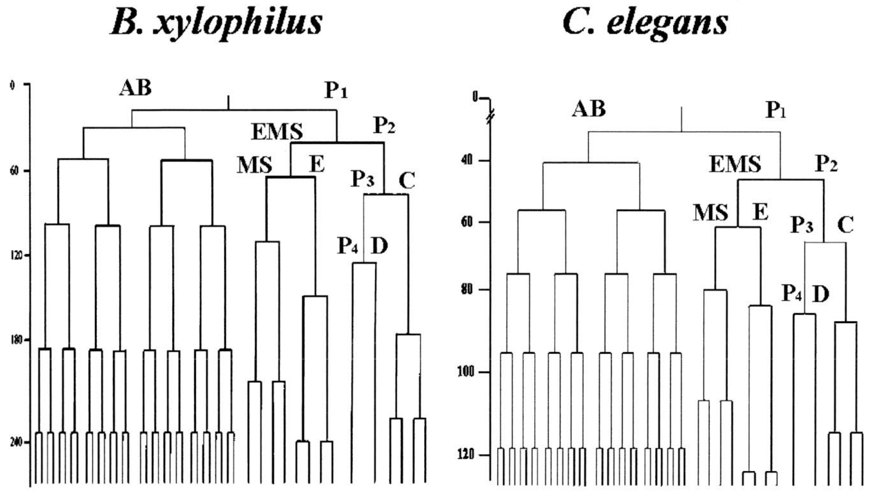

Fig. II.24 Cell lineage trees of *Bursaphelenchus xylophilus* (*left*) and *Caenorhabditis elegans* (*right*). Vertical axes indicate time (minutes). *B. xylophilus* cell division followed at 30°C (Hasegawa et al. 2004). *C. elegans* cell lineage tree is adapted and modified from Schierenberg et al. (1980)

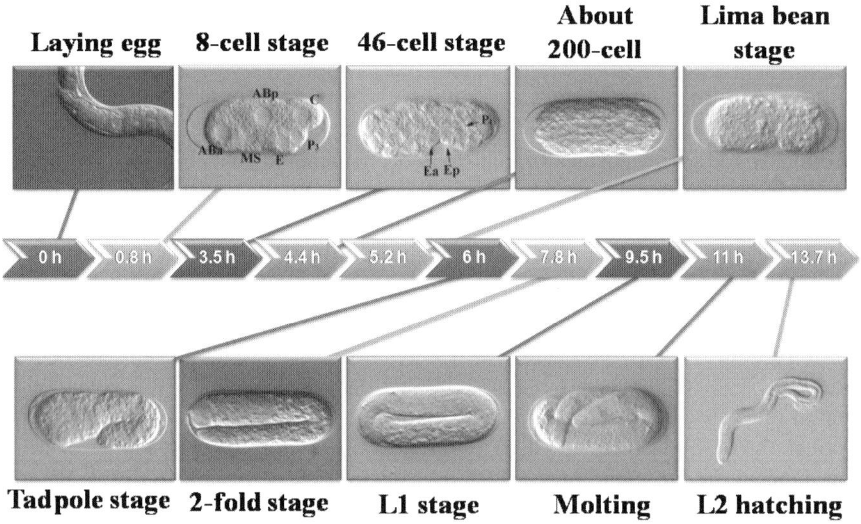

Fig. II.25 Key stages of the *B. xylophilus* embryo. Adapted from Hasegawa et al. (2004)

cell stage after fertilization (Figs. II.24, II.25), almost twice as long as the 130 min required for the *C. elegans* embryo at 25°C (Schierenberg et al. 1980).

Patterns of early embryogenesis are similar in Rhabditia (the phylogenetic clades III, IV, V) (Sulston et al. 1983; Malakhov 1994; Borgonie et al. 2000; Hope 2002; Hasegawa et al. 2004). The complete embryonic cell lineage is finished in *Pellioditis marina* (Clade V), and is compared with *C. elegans* (Clade V) for the timing of cell division and terminal differentiations (Houthoofd et al. 2003). Although the lineage homology between these two nematodes is 95.5%, the cell fate homology is only 76.4% (Houthoofd et al. 2003). Embryogenesis of the nematodes that belong to the phylogenetic clades I and II is prominently different from that for the other nematodes; because the early embryonic cell divisions of all cells in *Enoplus brevis* (Clade II) and *Prionchulus punctatus* (Clade I) are synchronous and because blastomeres are identical to each other by size and position (Malakhov 1994; Voronov and Panchin 1998; Borgonie et al. 2000; Hope 2002). In these nematodes, the endoderm cell (E cell) is a daughter of the anterior blastomere in the two-cell embryo, although the E cell in Rhabditia is derived from EMS next to the posterior P_2 cell (Hope 2002). In another free living nematode, *A. nanus* (Clade IV), neighboring cells can adopt the developmental program of eliminated blastomeres (Wiegner and Schierenberg 1999), and in *E. brevis*, the future fates of early blastomeres are not predetermined (Voronov and Panchin 1998). These features are quite different from the mosaic nature of *C. elegans* embryogenesis (Sulston et al. 1983). Different cell-fate determinants, different cell positions, and different cell communications orchestrate diverse patterns of embryonic cell divisions in the Phylum Nematoda.

11.7 Developmental Stages

Figures II.25 and II.26 show the developmental rates of the *B. xylophilus* embryo from egg laying to hatching, which are times needed to reach different key stages at 20, 25, 30, and 35°C. At 20°C, all of the embryos took more than 30 h before hatching, and their developmental rates varied very little. Also, all of the embryos developed essentially at the same rate at 25 or 30°C. The frequency of successful hatching at 25 and 30°C was high at 95.7 and 100%, respectively. At 35°C, however, the hatching frequency fell to 61.5%, and both the developmental rate and the hatching frequency varied greatly. In addition, at 35°C, of the five dead embryos, two died before the first division, one was arrested between the 200-cell and tadpole stages, and the remaining two arrested at the L1 stage. Nevertheless, the average developmental rate from fertilization to hatching was nearly the same at both 30 and 35°C (Fig. II.26).

Early developmental events, such as oocyte maturation, meiosis, pronuclear rearrangement, and cytoplasmic streaming necessary to allocate and distribute cytoplasmic determinants correctly, may be sensitive to temperatures higher than 30°C. It is also possible that maternal products necessary for early embryogenesis are not produced properly at the higher temperatures.

The *B. xylophilus* embryogenesis from fertilization to hatching is conveniently divided into three distinct phases. The first is the cell proliferation phase from one cell to the lima bean stage, which includes pronuclear meeting, cell cleavages, gastrulation, and some degree of cell differentiation when blue autofluorescence becomes visible in gut cells (Fig. II.27). The fluorescing particles in the gut cells

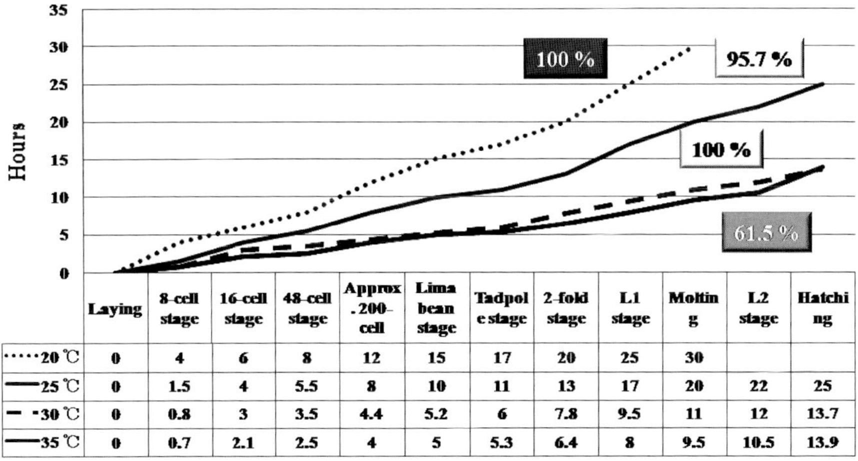

	Laying	8-cell stage	16-cell stage	48-cell stage	Approx. 200-cell	Lima bean stage	Tadpole stage	2-fold stage	L1 stage	Molting	L2 stage	Hatching
·····20 °C	0	4	6	8	12	15	17	20	25	30		
——25 °C	0	1.5	4	5.5	8	10	11	13	17	20	22	25
– -30 °C	0	0.8	3	3.5	4.4	5.2	6	7.8	9.5	11	12	13.7
——35 °C	0	0.7	2.1	2.5	4	5	5.3	6.4	8	9.5	10.5	13.9

Fig. II.26 Developmental rate from egg laying to hatching at 20, 25, 30, and 35°C. *X*-axis shows key embryonic stages, and *Y*-axis indicates time in hr. Adapted and modified from Hasegawa et al. (2004)

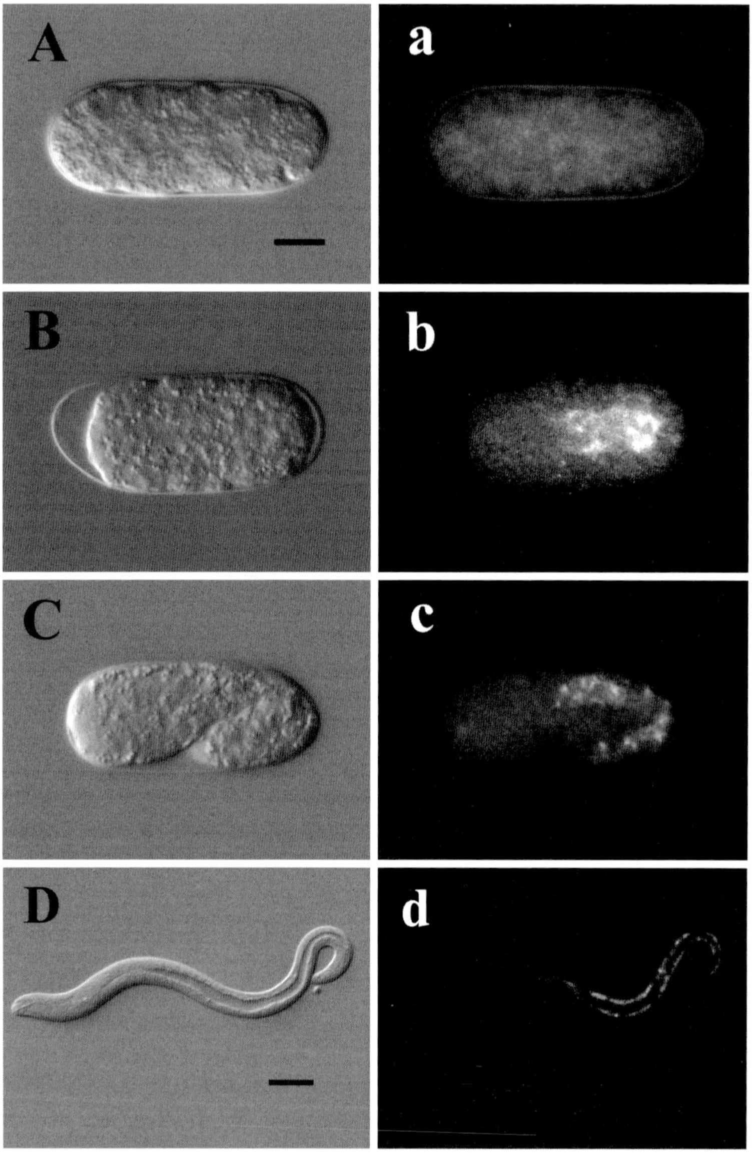

Fig. II.27 Gut differentiation during development. **A–D** DIC images and **a–d** fluorescence images. **A**, **a** 46-cell stages gut cell differentiation is not visible. **B**, **b** At about the 100-cell stage, a fluorescence signal is detected from the posterior-center of the embryo. **C**, **c** Lima bean stage, gut cells are arranged along the center of the tail. **D**, **d** L1 hatchee

are products from tryptophan catabolism (Babu 1974). The second is the morphogenesis phase from the lima bean to L1 stage, which covers morphogenesis and organogenesis that execute most of the body plan, although the stylet and central valve remain to be completed. The third is the L1 phase from the L1 to L 2 hatchee stage, including the first molting and the final stage of organogenesis wherein the stylet and the central valve apparatus are completed and the worm starts to move vigorously. At 25°C the cell proliferation phase lasted for about 10 h, the morphogenesis phase for about 7 h, and the L1 phase for about 4.2 h. At 30°C, these three phases lasted for approximately 5.2, 4.3, and 4.2 h, respectively.

11.8 Cytology and Sex Determination

The haploid chromosome number detected in all observed *B. xylophilus* male and female germ cells is six, and all of the six chromosomes appear identical to one another in all the observed male and female germ cells (Fig. II.17), pronuclei of the one-cell stage embryo, and nucleus of the early embryo (Fig. II.21). In the absence of any reports that the male to female ratio in the *B. xylophilus* population is distorted in the populations extracted from either pine trees or the vector beetle *M. alternatus* as well as in the population treated with high-temperature, *B. xylophilus* exhibits an essentially one to one ratio of males to females.

The sex determination system in *B. xylophilus* does not appear to be of the XO type as in *C. elegans*. It might be the XX/XY system as in all plant parasitic nematodes or other multiple sex chromosome systems examined, such as that for *Ascaris suum* (Goldstein 1981). Although *C. elegans* adopts the XX/XO system for sex determination, sex regulating genes are scattered over various chromosomes, and any one of the autosomes except for chromosome I (LGI) has the potential to become a sex chromosome by mutation (Hodgkin 2002). Although, in general, the male or female sex determination depends on a chromosome-based mechanism in almost all animal species, environmental conditions play a role in a few cases represented by the arthropod parasite Mermithidae (Haag 2005) and the vertebrate parasite *Strongyloides ratti* (Haevey and Viney 2001).

Some *B. xylophilus* isolates are reported to have different chromosome numbers; their haploid number varies within $N = 3, 5, 6$ (Bolla and Boschert 1993), and some mating among these aneuploid isolates could occur (Bolla and Boschert 1993). Some of these mating combinations have a curious preference for sexuality, and produce viable or inviable progeny depending on which isolate takes a female or male sex; for example, the combinations US12 ($N = 6$) male with US1 ($N = 3$) female, and US10 ($N = 3$) male with US12 ($N = 6$) female can both produce viable progeny, but reciprocal male–female combinations cannot (Bolla and Boschert 1993). We found a few cases of triploid embryos ($3N = 18$) in a *B. xylophilus* Portugal isolate. Their chromosome number in somatic blastomeres was 18, and 3 pairs of ribosomal clusters were recognized (Fig. II.28). This might have resulted from a spontaneous fusion of two oocytes before fertilization, and the two female

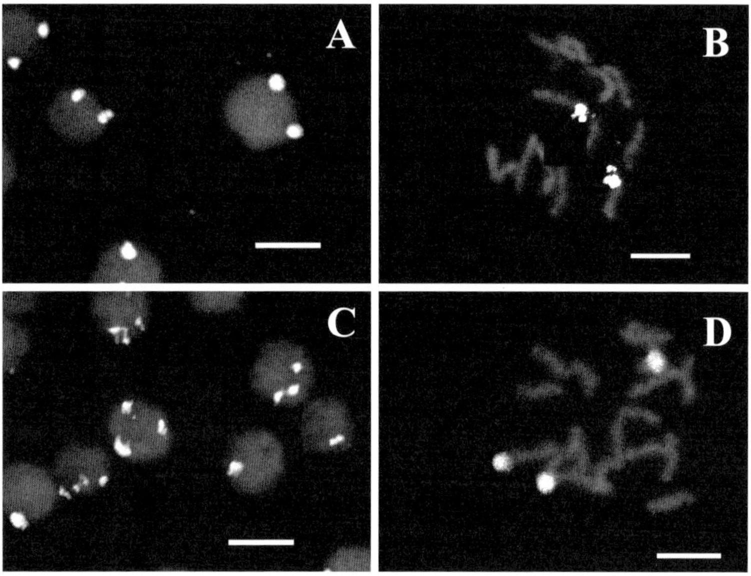

Fig. II.28 Chromosomes hybridized *in situ* with the ribosomal ITS probe and visualized with FITC. Signals are detected in ribosomal RNA cluster regions. **A** Interphase nuclei in early embryo of the S-10 isolate. Two fluorescence signals are detected in one nucleus. **B** Metaphase chromosomes in early embryo of the S-10 isolate. $2N = 12$. **C** Interphase nuclei in early embryo of the Portugal isolate T1. Three fluorescence signals are detected in one nucleus. **D** Metaphase chromosomes in early embryo of the Portugal isolate T1. $3N = 18$. *Scale bar* 5 μm (2006, Nematology, Brill, with permission)

pronuclei met one male pronucleus to form one embryo (Fig. II.29). Although this triploid embryo divided and hatched normally to become a giant L2 larva (not shown), we could not follow the postembryonic growth and fecundity of this *B. xylophilus*. Irle and Schierenberg (2002) reported that artificially constructed *C. elegans* triploid embryos (from two female pronuclei fertilized with one male pronucleus) showed several phenotypes; some embryos divided abnormally or developed and hatched normally as L1 but were sterile, and others produced fertile individuals. The standard number of the haploid chromosomes in *B xylophilus* seems to be 6, and a triploid population may be produced by chance such as oocyte fusion. And such a population might be maintained in some percentage in several *B. xylophilus* isolates. We have just begun to seek answers to the questions about sex determination mechanisms and cytology of this nematode. We have no knowledge of either $N = 3$ or $N = 5$ *B. xylophilus* isolates. Are they really *B. xylophilus*? We may have to start out with determining the chromosome numbers of these isolates and carefully observe the chromosomal behaviors of eggs from prefertilization to at least several embryonic divisions in parental as well as in F1 hybrids.

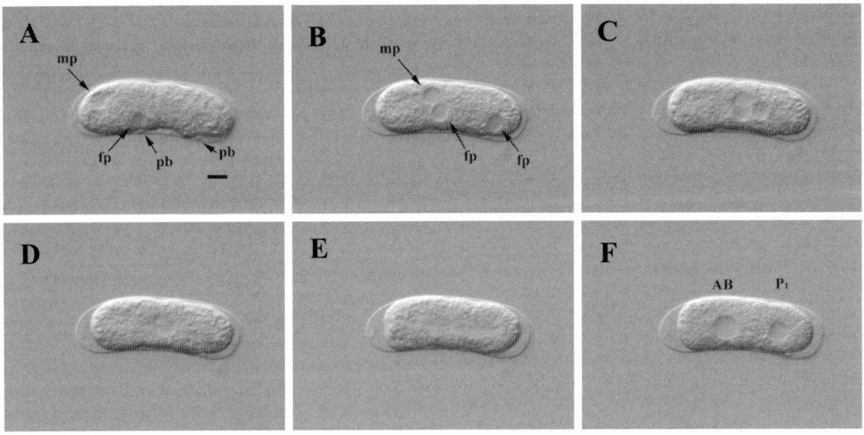

Fig. II.29 Meeting of two female and one male pronuclei. DIC images. **A** The male pronucleus appears at the future anterior pole of the embryo, and the female pronucleus emerges at a lateral position. **B** The male and one female pronuclei meet and move to the center, and the other female pronucleus appears. **C** Three pronuclei positioned at the center of the embryo. **D** Three pronuclei are fusing. **E** Anaphase. **F** Two-cell stage embryo. *pb* polar body, *mp* male pronucleus, *fp* female pronucleus. Anterior is left. *Scale bar* 10 μm

11.9 Concluding Remarks

The sophisticated methodology and vast amount of accumulated knowledge available about *C. elegans*, together with comparative research on developmental patterns among divergent species in the Phylum Nematoda, should shed light on the evolutionary history and changes in the ways of "making animals." Although the culturing and handling of *B. xylophilus* is far easier than it is for the other plant parasitic nematodes, we have yet to establish useful and effective molecular methods to analyze genetic schemes programmed and deployed by *B. xylophilus*. And so, our understanding of this nematode is still at its infancy and only exists at the descriptive level. Although classical genetics (forward genetics) provides the most powerful tool to understand most biological phenomena, very few organisms offer this wonderful tool. Recently discovered RNAi (RNA mediated interference) is now considered by many to be a "wonderful" substitutive of genetics in non-model organisms in which classical genetics is not amenable, and is sometimes referred to as "reverse genetics," since we first start with materials like RNA and then a phenotype, but not vice versa (Grishok 2005). The phenomenon of RNAi was discovered first in *C. elegans* (Fire et al. 1998). The RNAi-based "reverse genetics" in *C. elegans* offers three different ways for obtaining the interfered phenotype induced by tailor-made double stranded (ds) RNA material: by (1) microinjection of dsRNA into the nematode body (Fire et al. 1998), (2) feeding *E. coli* that expresses dsRNA (Timmons and Fire 1998), and (3) soaking nematodes in dsRNA

(Tabara et al. 1998; Maeda et al. 2001). For feeding and soaking RNAi, dsRNA is absorbed from the gut and transported to other parts of the body. Such systemic effects of RNAi do not appear to occur in other animals (Tijsterman et al. 2004). Successful results of RNAi in parasitic nematodes have been reported; dsRNA was delivered by soaking (e.g., Urwin et al. 2002; Fanelli et al. 2005), or by electroporation (Issa et al. 2005). Because *B. xylophilus* is very slender, it is quite difficult to perform microinjection. And systemic RNAi might be suppressed in *B. xylophilus*, and thus soaking RNAi is not effective (K. Hasegawa and J. Miwa, personal communication).

We need breakthrough technology to analyze functional aspects of *B. xylophilus* genes and to understand the evolution of developmental systems adopted by *B. xylophilus* and its kin.

12
Population Ecology of *Bursaphelenchus xylophilus*

Shuhei Takemoto[1,2]

12.1 Introduction

Population ecology is the study of biological populations, and it includes diverse subfields: population dynamics concerns the size and structure of a population and examines the effects of environmental factors on it. In the study of population genetics, the changes of allele frequencies within populations and their distribution are especially interesting. The evolutionary change of populations may be studied using some biological characteristics.

This chapter focuses on the population biology of the pine wood nematode (*Bursaphelenchus xylophilus*). The chapter begins by reviewing the intraspecific phylogeny of *B. xylophilus*, and continues by clarifying the population dynamics within a single pine tree, then gene flow among populations and their spread pattern. Finally, the process of virulence evolution over a time-scale longer than the lifetime of host trees is discussed.

Knowledge on the population biology of *B. xylophilus* should prove useful for planning or improving the strategies to manage this destructive forest tree disease. As well, it is quite an exciting challenge to explain how populations of *B. xylophilus* behave in the midst of the interrelations among the populations and in relation to other organisms.

[1]Laboratory of Environmental Mycoscience, Graduate School of Agriculture, Kyoto University. Kitashirakawa Oiwake-cho, Sakyo-ku, Kyoto 606-8502, Japan

[2]Current affiliation: Environmental Biofunction Division, National Institute for Agro-Environmental Sciences, 3-1-3 Kannondai, Tsukuba 305-8604, Japan

Tel.: +81-29-838-8269, Fax: +81-29-838-8269, e-mail: ts1@affrc.go.jp

12.2 Intraspecific Phylogeny

As mentioned in Chap. 9, *B. xylophilus* is closely related to *B. mucronatus*. Since the two species usually colonize dead conifers such as pine trees where their host ranges overlap, there is the possibility that they hybridize in nature. In fact, Matsunaga et al. (2004) gave molecular evidence for introgressive hybridization in a natural pine stand. However, many other studies have indicated that these two nematodes are genetically differentiated from each other (Webster et al. 1990; Beckenbach et al. 1992, 1999; Riga et al. 1992; Bolla and Boschert 1993; Iwahori et al. 1998). Consequently, in the following discussions xylophilus × mucronatus hybrids are not covered, that is, the discussion is limited to *B. xylophilus*.

Bursaphelenchus xylophilus is distributed widely throughout the northern hemisphere. When we consider the worldwide distribution, it is natural for us to expect the existence of "geographic strains", which are differ from each other. According to the work to date, Asian and Portuguese populations are considered to be small samples from the North American, their native origin. Iwahori et al. (1998) showed that three Japanese virulent isolates (populations in culture) and one Chinese isolate examined were identical to an American isolate in their PCR–RFLP (polymerase chain reaction–restriction fragment length polymorphism) banding patterns of the ITS region (internal transcribed spacer region) of ribosomal DNA. Beckenbach et al. (1999) indicated that one Japanese avirulent isolate should be assigned to an intermediate position between the Japanese virulent isolate and Canadian isolates in their phylogenetic tree based on ITS sequences. Mota et al. (2006) showed that two Portuguese isolates were identical to some Japanese virulent isolates in their ITS sequences. Recently, Metge and Bugermeister (2006) characterized 30 *B. xylophilus* isolates with 530 ISSR (inter simple sequence repeat) and 611 RAPD (random amplification of polymorphic DNA) markers and showed that the Asian and Portuguese isolates, except for one Japanese avirulent isolate, formed a monophyletic group as a subclade of the North American clade in their phylogenetic study. They further mentioned that the Portuguese populations were most likely introduced from East Asia, not directly from North America.

The *B. xylophilus* isolates contain somewhat diverse phenotypic traits other than the molecular traits mentioned above. One of the most outstanding examples is the dimorphism of female tail tip, that is, the "R (round)" and "M (mucronate)" forms. Interestingly, some M-form isolates from *Abies balsamea* in Canada specifically showed pathogenicity against *A. balsamea* (Wingfield et al. 1983). Bolla and Boschert (1993) found that interbreeding occurred in the laboratory between some M and R forms, and that the chromosome number was inconsistent among isolates within each of the forms. It appears that, the morphology of the tail tips seems not to reflect phylogeny. Bolla et al. (1986) found two pathotypes, one of which was specifically pathogenic to *Pinus sylvestris* and the other to *P. strobus*. Using RFLP analysis Bolla et al. (1988) demonstrated that the two pathotypes were genetically different, but there was no reproductive isolation between them (Bolla and Boschert 1993). Phylogenetic data on these R and M forms are still incomplete because few isolates are available.

Sexual incompatibility may occasionally enhance such phenotypic and molecular differentiation among isolates. Bolla and Boschert (1993) showed that pair mating among *B. xylophilus* isolates did not produce a fertile F_1 generation in some cases, even among isolates that had the same origin, the USA. On the other hand, Riga et al. (1992) found that all combinations among three Canadian, one USA and one Japanese isolates of *B. xylophilus* produced fertile F_1. Kiyohara et al. (1998) showed that all combinations among six isolates collected from different countries generated fertile progeny. Mota et al. (2006) also reported that two Portuguese isolates freely hybridized with two Japanese isolates. Until now, sexual incompatibility among *B. xylophilus* isolates has been found only in the case of North American isolates (Bolla and Boschert 1993). This suggests that the speciation of sibling species is about to start in the native, North American origin of *B. xylophilus,* owing to the long history of divergence there. Detailed studies are needed on the sexual incompatibility among North American populations in relation to their geographic origin and phylogeny.

Among many *B. xylophilus* populations collected in Japan, several from different geographical origins have only slight or no virulence to host pine trees (e.g., Kiyohara and Bolla 1990), and are generally called "avirulent" isolates. Takemoto and Futai (2007), examined the PCR–RFLP patterns of heat shock protein 70A gene (*hsp70A*) of 29 *B. xylophilus* isolates, including some isolates well known for their virulence: C14-5 and OKD1 have considerably low virulence while Ka4, NS [a subculture of the isolate "Shimabara" in Ikeda et al. (1994)], S10, S6–1 and T4 have rather high virulence against host pine species (e.g., Kiyohara and Bolla 1990; Ikeda et al. 1994; Aikawa et al. 2003a). According to the PCR–RFLP patterns of *hsp70A*, two isolates, C14-5 and OKD1, were classified into one group, and five isolates, Ka4, NS, S10, S6-1, and T4 into another (Takemoto and Futai 2007). Aikawa et al. (2003a) also reported that a PCR–RFLP of the ITS region enabled the discrimination of five virulent isolates (Ka4, S10, T4, S6-1 and No. 375) from two avirulent isolates (C14-5 and OKD1). Based on these results, we should reject the possibility that there is no relationship between their (an isolate's) virulence and PCR–RFLP patterns with a significance level of $P = 0.0476$ (Fischer's exact probability test) for both of the two loci, *hsp70A* and ITS, since those isolates are random samples of Japanese *B. xylophilus* populations. This strongly infers that avirulent and virulent isolates were derived from different ancestral populations and are characterized with alleles unique to each of them. These data support the hypothesis that the virulent and avirulent isolate groups should be genetically differentiated from one another (Iwahori et al. 1998), though no reproductive isolation seems to have been developed between the two groups (e.g., Kiyohara and Bolla 1990; Aikawa et al. 2003a; Mota et al. 2006).

There has been no report on the collection of avirulent *B. xylophilus* isolates since 1984, when OKD1 was isolated (Kawazu et al. 1996a); however, populations that had both alleles specific to the virulent and avirulent isolate groups were found in the two areas around Kasama City and the southern part of the Kujukurihama coast. The persistence of the avirulent-type allele in these areas suggests that avirulent ancestral populations may have made a certain contribution to the gene pool of Japanese *B. xylophilus* by introgressive hybridization with virulent ancestral populations.

Avirulent isolates seldom kill pine trees. Therefore, vector beetles rarely transmit them since vector beetles lay their eggs exclusively on dying or recently killed pine trees (e.g., Kojima 1960; Kobayashi et al. 1971). This inferior transmissibility of avirulent isolates is apparently disadvantageous if they compete with virulent isolates. Recent studies on the interaction between avirulent and virulent populations are mentioned later in this section.

12.3 Population Dynamics within a Tree

The expanding process of pine wilt disease includes two incomparably different dimensions. One is the propagation process within a single tree and the other is the transmission process among trees. For convenience, we should consider them separately. The discussion here initially focuses on the process within a tree.

12.3.1 Population in Culture

It might be useful to know the population dynamics in vitro before considering the population dynamics within a tree growing in a natural stand. Unlike most other plant parasitic nematodes, *B. xylophilus* can be cultured on fungus mycelium growing on culture media such as PDA (potato dextrose agar) or MEA (malt extract agar), that is, *B. xylophilus* is mycophagous. *Bursaphelenchus xylophilus* feeds on various fungi such as blue stain fungi, which colonize wilt-killed pine trees (e.g., Kobayashi et al. 1974; see also Chaps. 28, 29). Using a modified Baermann funnel method the nematodes can be easily extracted from wood chips collected from wilt-killed trees. Nematodes are effectively (70–80%) extracted within 2 days at 10°C or higher (Mamiya 1975b). It is very important to remove microbial contamination from the surface of nematodes before starting the culture. The following is a practical method for surface sterilization routinely used in the Forest and Forest Products Research Institute, Japan (Mamiya et al. 2004): after rinsing the nematodes three times with sterilized water in a 10-ml centrifuge tube, remove the supernatant and transfer the nematodes into another sterilized tube. Add 6–8 ml of sterilized 3% lactic acid aq. and centrifuge at 1,500 rpm for 30 s. The nematodes must not be left in the lactic acid aq. for longer than 3 min. Rinse the nematodes once again with sterilized water. Remove the supernatant and surface-sterilized nematodes are suspended in a droplet at the bottom of the tube. *Bursaphelenchus xylophilus* are usually conveniently reared on a *Botrytis cinerea* culture that has lost the ability to produce spores. Such populations in culture are commonly denoted as "isolates". Note that an "isolate" of *B. xylophilus* is possibly composed of genetically diverse individuals and is not necessarily homogenous in contrast to an "isolate" in the research field of mycology.

The population of *B. xylophilus* increases rapidly in culture. Logistic curves were well fitted to the population growth (Dozono and Yoshida 1974). Mamiya (1975a) reported that it took only 3 days for *B. xylophilus* to complete its lifecycle at 30°C, and 4–5 days at 25°C, respectively, though developmental disorder was observed at 28°C or higher. Futai (1980a) reported in his comparative study between *B. xylophilus* and *B. mucronatus* that the developmental zero point and heat units required for *B. xylophilus* to complete one generation were 10.0 and 63.7 ± 3.4, respectively, which resulted in a higher developmental rate of *B. xylophilus* than that of *B. mucronatus*. He also mentioned that the sex ratio of adults started at a low value with faster maturation of females. As well as on culture medium, the population of *B. xylophilus* rapidly grows on a pine segment inoculated with a *Ceratocystis* sp. fungus as its food source, reaching its maximum density of ca. 25,000 g^{-1} dry wood 4 weeks after nematode inoculation (Fukushige 1991).

12.3.2 Population Growth and Migration within a Single Tree

The process within a single tree begins with the invasion of dispersal fourth-stage juveniles (dauer larvae, J_{IV}s) via feeding wounds made by vector beetles on young branches of the tree (see Chap. 13 for further information on the relationship between *B. xylophilus* and vector beetles). It has been found that *B. xylophilus* J_{IV}s leave intermittently from the body of a vector beetle over its life span, generally, with a peak around 2–3 weeks after the beetle has emerged from a dead tree (e.g., Morimoto and Iwasaki 1972; Nakane 1976; Mineo 1983; Togashi 1985; Shibata 1985); therefore, the timing and pathway of transmission seems to be quite limited. Furthermore, a considerable portion of the nematodes that have left the beetle body fail to invade the tissue of healthy pine trees. For example, Kishi (1995) examined the number of nematodes transmitted from vector beetles onto pine twigs for over a 7-year period, and reported that the invasion rate of the nematode was 12.1–35.0% (23.2% on average). Asai and Futai (2006) reported that the invasion rate of virulent PWN to branch segments of *P. thunbergii* placed on agar plates was 7.2–40.0% (25.6% on average).

As mentioned above, the population growth of *B. xylophilus* within a tree should begin with few founders. Although most of the nematodes remain at the initial invasion site, others rapidly disperse throughout the tree. Togashi et al. (2003) inoculated twigs of *P. densiflora* with a virulent PWN isolate, T4, and determined the migration pattern of the nematodes. They suggested that nematodes that have entered a pine branch from a wound can be divided into two groups: one which stays near the inoculum court (beetle feeding wound), and the other which migrates at random throughout the branch away from the wound. The proportion of non-migratory nematodes and the migration distance of migratory nematodes were estimated as 45.7% and 7.2 cm day^{-1} on average, respectively, when 1,000 nematodes were introduced onto the upper end of 25-cm-long twigs. The inhibitory effect of pine branches to nematode migration varied greatly among individual trees, the positions of branches on a tree and the timing of sampling in a day (Matsunaga and Togashi 2004). In a review on the population dynamics of *B. xylophilus* in relation to its pathogenesis, Hashimoto (1975) mentioned that nematodes could spread throughout an 8-m-tall tree in 10 or 14 days after invasion and their maximum migration distance was estimated at 20–80 cm day^{-1}. Kuroda and Ito (1992) found that nematodes could migrate 150 cm day^{-1} through the main stem of a 7-year-old *P. thunbergii* tree. Since the population density of nematodes over the stem usually remains quite low until 1 week after inoculation (Hashimoto 1975; Mamiya 1990a), both of these studies (Hashimoto 1975; Kuroda and Ito 1992) incubated wood samples for 1 month before detecting *B. xylophilus*.

In spite of the random migration of nematodes over a single branch (Togashi et al. 2003), their distribution in the bole of a tree is uneven during early infection (e.g., Kuroda and Ito 1992). This phenomenon might result from some inhibitory effect against nematode migration or the uneven distribution of a preferable micro-environment. Futai et al. (1986) analyzed the change in the vertical distribution of nematodes within pine logs, and inferred that the population dynamics of *B. xylophilus* follows this pattern: high density and uniform distribution → as the population gradually shrinks, the distribution becomes clumped → the population continues to diminish and the distribution becomes random → the distribution becomes clumped again possibly due to the aggregation of nematodes around *M. alternatus* pupal chambers as Mamiya (1972) pointed out. Sriwati et al. (2006) also showed that *B. xylophilus* was more abundant around pupal chambers and tunnels of *M. alternatus* than elsewhere in the wood (see Chap. 28).

Monochamus alternatus lays its eggs preferably on dying or recently killed trees, and does not utilize rotten pine trees as a food and habitat resource (e.g., Kojima 1960; Kobayashi et al. 1971); therefore, there is only one chance for nematodes in a dead pine tree to be transmitted to another tree. Sometimes one beetle carries up to 250,000 PWNs (Mamiya 1975c); however, the nematode load varies greatly. Only 20% of 958 vector beetles carried 1,000 nematodes or more in their body, which was as much as 95% of the total number of nematodes carried by the entire beetles captured (Kobayashi 1975). Based on these studies, it is thought that only a limited number of all nematodes that aggregate around a pupal chamber will be successfully transmitted to another host tree, and that this process likely has a significant bottlenecking effect on nematode populations.

12.3.3 Factors Affecting Population Growth

Cohabiting fungi significantly affect the growth of *B. xylophilus* populations on culture media (e.g., Kobayashi et al. 1974) and on pinewood (e.g., Fukushige 1991). As well as fungi, a recent study indicated that bacteria are also an important factor (Zhao et al. 2007). Since the details of the relationship between microorganisms and *B. xylophilus* are discussed in Part V, in this section our recent unpublished study about the influence of high temperature on *B. xylophilus* is introduced.

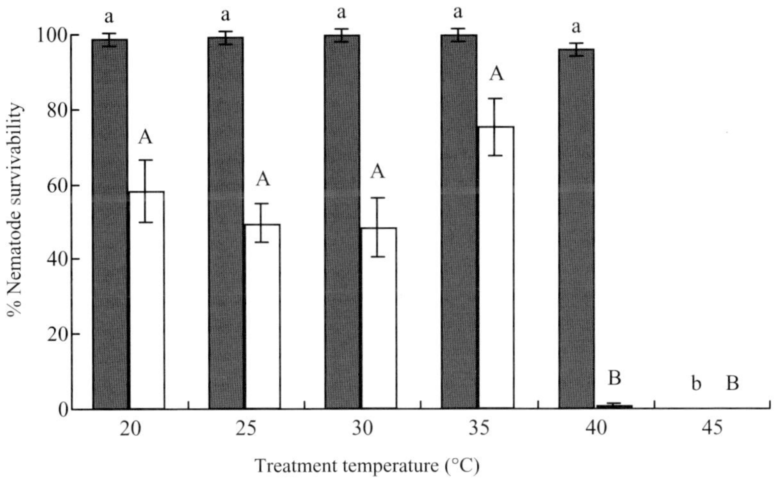

Fig. II.30 The influence of thermal treatments on the survivability of *Bursaphelenchus xylophilus*. *Bars* indicate standard errors. *Shaded* and *open columns* are 1- and 7-day treatments, respectively. There is no significant difference between treatments with *identical letters* (Tukey–Kramer multiple comparison test, $P > 0.05$)

Previous researchers reported that *B. xylophilus* cannot reproduce at 40°C and thus die out within a few months (Tomminen 1991; Panesar 1994), and that even a few minutes exposure to temperatures above 50°C is lethal to *B. xylophilus* (Kinn 1986; Nickle and Coburn 1988; Dwinell et al. 1994). In our recent study, *B. xylophilus* exposed to 40°C for 7 days were almost destroyed (Fig. II.30), and those exposed to 45°C for 8 h were completely killed, which agrees with previous reports. However, it has also been shown that exposure to 40°C for 1 day does not kill *B. xylophilus* (Fig. II.30).

Nickle and Coburn (1988) and Dwinell et al. (1994) studied the reproductivity of heat-treated *B. xylophilus* to evaluate the sterilizing effect of heat treatments on infested wood chips. Giblin-Davis and Verkade (1988) also evaluated the effect of soil solarization using the reproductivity of *B. seani* as an indicator. Although these researchers examined the effect of heat treatments on the reproductivity of these nematodes, they mentioned only about nematodes' reproductuvity, and did not show data on the rate of reproduction. Kanbe et al. (2000) immersed *B. xylophilus* in warm water and showed that thermal treatments reduce the reproductive rates. Also, in our experiment, *B. xylophilus*, on culture medium plates, was exposed to several temperatures prior to measuring its reproductive rates. The results showed that higher temperatures and longer treatments resulted in a lower rate of reproduction.

A temperature higher than 28°C affects PWN development (Mamiya 1975a), and the embryonic development of *B. xylophilus* is disturbed at 35°C (Hasegawa et al. 2004). My colleagues and I also confirmed that PWN reproduction was suppressed

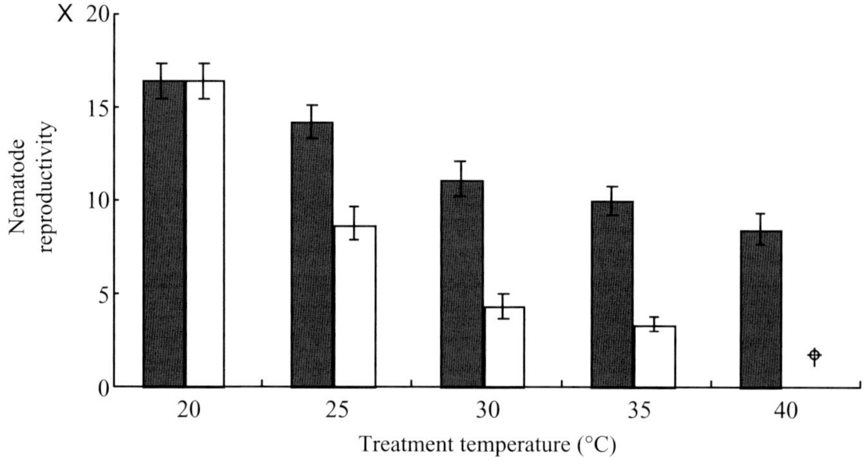

Fig. II.31 The influence of thermal treatments on the reproductivity of *Bursaphelenchus xylophilus*. *Bars* stand for standard errors. *Shaded* and *open columns* are 1- and 7-day treatments, respectively. *Dagger* datum is not shown because of insufficient replication of measurement

after treated at temperatures above 35°C. Conversely, it is known that *B. xylophilus* reproduces well at 20–30°C (Dozono and Kiyohara 1971; Futai 1980a); in our study, however, exposure of *B. xylophilus* to even such moderate temperatures without a food fungus resulted in low reproductivity, and the suppressive effect was positively correlated with temperature (Fig. II.31). Heat stress affects the sex ratio of some animals, for example, Bearded dragon *Pogona vitticeps* (Quinn et al. 2007). In our case, though, the sex ratio of *B. xylophilus* adults was not affected by treatment temperature. Therefore, their low reproductivity could not be attributed to the alteration of the sex ratio that would reduce the chance of mating. Possibly, some damage to their sexual system or exhaustion caused the low reproductive rate.

Air temperature in central and eastern Japan ordinarily exceeds 30°C in the summer, when infected pine trees are dying; therefore, it was thought that the boles (trunks) of dead trees heat up to rather high temperatures. In fact, the bole temperature was usually over 25°C, even at the pith, and frequently exceeded 35°C (Fig. II.32). Based on these results it was suggested that high temperatures within the boles of dead trees possibly acts as a suppressive factor on the reproduction of *B. xylophilus* therein.

12.3.4 Intraspecific Competition

The above-mentioned avirulent isolates seem to be at an apparent disadvantage if they compete with virulent isolates. Avirulent isolates have less potential to invade the bark tissue of pine shoots (Asai and Futai 2006), and less able to disperse

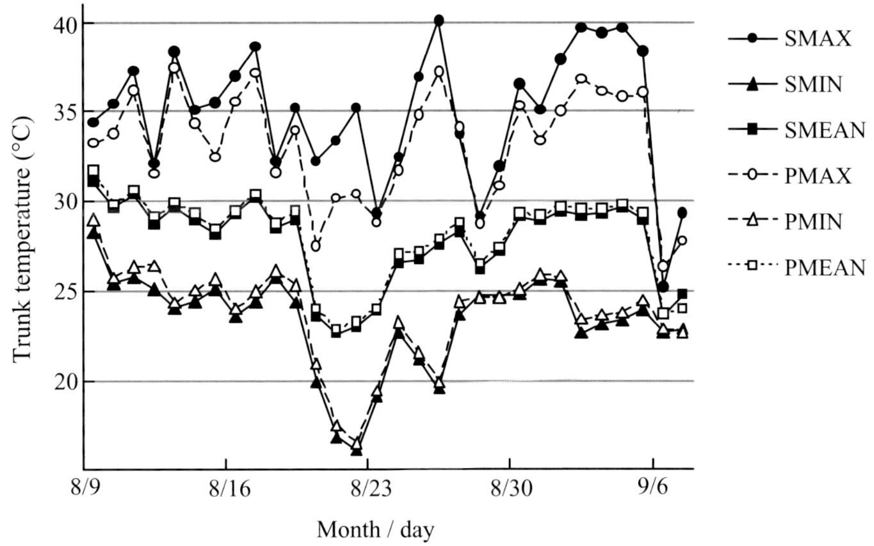

Fig. II.32 Trunk temperature of a dead tree. SMAX, SMEAN and SMIN are the daily maximum, daily mean and daily minimum temperatures of the surface of the trunk, respectively. PMAX, PMEAN and PMIN are the daily maximum, daily mean and daily minimum temperatures at the pith of the trunk, respectively

(Ichihara et al. 2000a) and propagate (Kiyohara and Bolla 1990) within healthy pine trees. Avirulent isolates rarely kill pine trees, and thus, vector beetles seldom transmit them. Aikawa et al. (2003b) reported that the number of avirulent nematodes carried by a vector beetle was far fewer than that of virulent isolates.

Since the virulence of a *B. xylophilus* population varies among killed pine trees and also among vector beetles even within a single pine stand (Kiyohara and Bolla 1990), a pine tree should be infected with nematode populations having a diverse level of virulence; however, Kiyohara and Bolla (1990) cultured five isolates from two wilt-killed *P. thunbergii* trees and showed that virulence did not differ among the isolates that originated from a single tree regardless of the height at which the wood samples were taken. This implies that a population within a single pine tree should be composed of genetically homogeneous individuals resulting from a strong bottle-necking or selection force, or that a population should interbred freely throughout a single tree, though it might still be heterogeneous.

To explain the mechanism by which such uniformity among isolates from a single tree develops, Aikawa et al. (2006) inoculated single *P. thunbergii* trees with two *B. xylophilus* isolates that could be distinguished from each other with specific PCR–RFLP markers, and analyzed the competition between the two isolates based on the population structure of the nematodes. When a virulent and an avirulent isolate were introduced into a single pine tree, the virulent isolate always became dominant irrespective of the order of introduction and the avirulent one was nearly out-competed. When two virulent isolates were introduced into a single pine tree, the PCR–RFLP pattern of the isolate that had been first introduced into the tree became dominant, though the first colonizer did not out-compete the second. Nematodes with a hybrid PCR–RFLP pattern between the two virulent isolates were frequently observed, suggesting that the two interbred with each other.

These results (Aikawa et al. 2006) clearly indicate that the avirulent isolate makes only a small contribution to the population in a single pine tree, due to its inferior competitiveness, that is, inferior ability to enter and propagate within the living tree before hybridizing with the virulent population. However, these results also show that the avirulent population can interbreed with the virulent population, resulting in a hybridized population within a single tree, though its genetic contribution is quite small. Virulence evolution within and among such hybridized populations is discussed later.

Normally, different genetic loci are under the influence of different selection pressures depending on their own function or that of neighboring loci; therefore, we should examine the various traits individually. Here, a study that clarified the behavior of the alleles of *hsp70A* in a *B. xylophilus* population in vitro is presented. Takemoto et al. (2005) prepared mixed populations of *B. xylophilus* from two parental populations, S10 (virulent isolate) and C14-5 (avirulent isolate), as follows: A 500-μl suspension containing ca. 10,000 second-stage juveniles of S10 or C14-5, or both, were pipetted onto a culture of *B. cinerea* growing in a 50-ml conical flask. Each cultural line started at various proportions of S10 (i.e., 100, 99, 90, 70, 50, 30, 10, 1 or 0% of the initial population). These lines were maintained at 20°C with subculturing once a month. In these experiments, 1-month-old mycelium of *B. cinerea* grown on barley grain medium (unthreshed barley grain 10 ml; tap water 10 ml, autoclaved at 121°C for 20 min) in 50-ml conical flasks served as the food source for the nematodes. When the initial S10/C14-5 ratio was adjusted to 1 : 1, the frequency of the virulent-type allele of the *hsp70A* gene was 90–96% from 20 to 30 days after inoculation, which differed from the results of Aikawa et al. (2003a). Aikawa et al. (2003a) inoculated 100 individuals of each of two *B. xylophilus* isolates, OKD1 (avirulent isolate) and Ka4 (virulent isolate), onto the same Petri dish, and found that three types of PCR–RFLP patterns (i.e., virulent–virulent homo-, virulent–avirulent hetero- and avirulent–avirulent homozygote type) were present for the ITS region in the ratio of 1 : 2 : 1 on day 25 after inoculation. These differences in the results may result from the isolates and the loci being different in the two studies.

The *hsp70A* alleles were maintained polymorphic in the S10/C14–5 mixtures. One of the best examples of such an allelic polymorphism is human sickle-cell anemia studied by Allison (1954). In this case, the anemia allele causes a lethal effect on a homozygote, but provides resistance against malaria to a heterozygote. Due to malaria resistance, the relative fitness of a heterozygote is even higher than that of a non-anemia homozygote. This over dominance explained why the anemia locus was maintained in a polymorphic state. Takemoto et al. (2005) applied this overdominance model to analyze the convergence of the allele frequency observed in this study and estimated a selection coefficient for the homozygote whose *hsp70A* alleles were the same as those of C14-5. Except for a few cases, the predictions from an overdominance model with a constant selection coefficient were well fitted to the observations, irrespective of the initial proportion of S10 in the mixed cultures. As a result, the frequency of the virulent-type allele converged at ca. 0.7.

12.4 Gene Flow and Population Structure

Earlier in this chapter, I discussed the population growth of *B. xylophilus* within a single tree. Now, we will consider the process of transmission and the resulting population structure of *B. xylophilus*. For more details about the relationship between *B. xylophilus* and the vector, *M. alternatus*, see Chaps. 13 and 17.

Kiyohara and Bolla (1990) showed that the virulence of *B. xylophilus* isolates varied significantly among pine trees within a stand, implying that there should be genetic variation among them. Takemoto and Futai (2007) directly indicated the variation of the allele frequency of *hsp70A* among *B. xylophilus* subpopulations taken from different individual pine trees in a local area. An in vitro experiment revealed that the frequency of the virulent-type allele of *hsp70A* converged at ca. 0.7 within 200–300 days after starting the culture of a virulent isolate mixed with an avirulent one (Takemoto et al. 2005); in the field, however, the frequency of the virulent-type allele deviated from the proportion of 0.7 and varied among populations (Takemoto and Futai 2007). Since virulent- and avirulent-type alleles are identical at the amino acid level (Takemoto et al. 2005), perhaps the allele frequency of *hsp70A* in the field populations was determined not mainly by natural selection forces, but was significantly influenced by random factors such as genetic drift.

Zhou et al. (2007), regarding PWNs extracted from a single pine tree as a subpopulation, analyzed the genetic structure of *B. xylophilus* populations inhabiting three distant pine forests (Tanashi forest, Tokyo Metropolis; Tsukuba forest, Ibaraki Prefecture; Chiba forest; Chiba Prefecture) with four microsatellite (SSR) markers, and also suggested that there should be a strong bottlenecking event based on the fact that heterozygocity observed was significantly less than expected. The populations in the Tanashi forest were quite homogeneous, sharing almost the same monomorphic genotype. The populations in the Chiba and Tsukuba forests were less homogeneous than those in Tanashi, but their degree of polymorphism was still low compared to other organisms such as *Cenococcum geophilum* (Wu et al. 2005), *Pinus densiflora* (Watanabe et al. 2006) and *Parus major minor* (Kawano 2003). In their phylogenetic analyses, using a model-based clustering method and a neighbor-joining method, Zhou et al. (2007) indicated that the Chiba and Tsukuba populations were intermingled, and that the Tanashi populations were genetically isolated from either of them. They inferred that the genetic isolation of Tanashi populations had been maintained, probably as the result of the geographic isolation of the Tanashi forest by the large residential area surrounding it, and that their monomorphism suggests that they began with quite a few founders.

12.5 Virulence Evolution

Takemoto and Futai (2007) showed that the frequency of *hsp70A* alleles specific to virulent and avirulent isolates varied among field populations within a local area. Though this *hsp70A* locus is not considered to have a linkage with virulence gene, the genetic variation observed suggests that the populations in the area might contain alleles derived both from virulent and avirulent ancestral populations in various proportions. Therefore, it is thought that Japanese populations of *B. xylophilus* potentially have a genetic background that is heterogeneous in terms of virulence.

12.5.1 Performances of Heterogeneous Populations of **Bursaphelenchus xylophilus**

In the case of pine wilt disease, symptoms, or damage to a host plant are caused by a population that is more or less heterogeneous in the natural condition. This is also often the case with experimental studies in the nursery or laboratory, unlike many of fungus pathogens. Taking this into account, unfortunately, analyses of the virulence of *B. xylophilus* sometimes seem to be difficult and imprecise. This is why we should recognize the heterogeneity and observe the performances of such a heterogeneous population.

Takemoto and Futai (2006) used a virulent isolate (S10), which was hybridized with an avirulent isolate (C14-5), and examined its virulence on 1- and 2-year-old *P. thunbergii* seedlings. They found that introgressive hybridization between a virulent and avirulent population resulted in various levels of virulence according to the genetic contribution of S10 to the inoculum populations (Fig. II.33; Takemoto et al. 2006).

Takemoto et al. (2006) also analyzed $\boldsymbol{T_0}$ (tolerance limit) and $\boldsymbol{R_{mi}}$ (rate of mortality increase) of seedlings of different ages inoculated with different inocula using the method of Asai and Futai (2005). Asai and Futai (2005) proposed the rate of mortality increase (originally called mortality velocity, $\boldsymbol{S}$, by Asai and Futai 2005), $\boldsymbol{R_{mi}}$, as an index that reflects disease progress in a seedling population and the tolerance limit, $\boldsymbol{T_0}$, as the critical value of load (cumulative damage) necessary to kill the first seedling. Takemoto et al. (2006) suggested that there was no relationship between $\boldsymbol{T_0}$ of *P. thunbergii* seedlings and the virulence of nematode populations; while there was a significant difference in the average $\boldsymbol{T_0}$ among different aged seedlings: 2-year-old seedlings 71.39 ± 1.56 (mean ± S.E.) and 1-year-old seedlings 49.70 ± 1.14 (Takemoto and Futai 2006). Asai and Futai (2005) found a lower $\boldsymbol{T_0}$ (26.47) for 4-month-old *P. thunbergii* seedlings. They also found that $\boldsymbol{T_0}$ varied among seedling groups that had a different pretreatment with simulated acid rain. These results imply that $\boldsymbol{T_0}$ depends on the physiological status of seedlings rather than the characteristics of the inoculum. Adding to the support of this idea is the

Fig. II.33 Relationship between the genetic contribution of S10 (a virulent isolate) to a hybridized population and the rate of mortality increase ($\boldsymbol{R}_{mi}$) of the *Pinus thunbergii* seedlings inoculated. Horizontal axis is the frequency of the allele of *hsp70A* specific to the parental virulent population (*S*-allele) in inoculum. *Closed* and *open circles* are 2- and 1-year-old seedlings, respectively

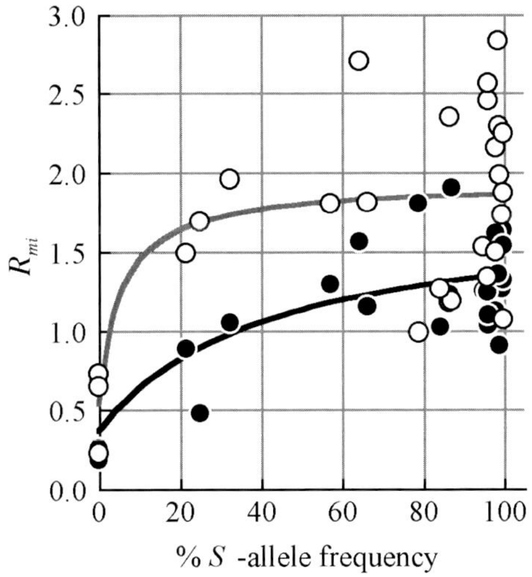

fact that, older seedlings show a higher tolerance limit (Takemoto and Futai 2006). Spearman's rank correlation coefficients between the mortality of the seedlings (2 months after inoculation) and the rate of mortality increase ($\boldsymbol{R}_{mi}$) were 0.659 for the 2-year-old seedlings and 0.755 for the 1-year-old seedlings, respectively. This infers a close relationship between seedling mortality and $\boldsymbol{R}_{mi}$.

This analytical model of the mortality process with $\boldsymbol{T}_0$ and $\boldsymbol{R}_{mi}$ (Asai and Futai 2005) is an empirical one, which makes it difficult to draw physiological meaning from the analysis. Therefore, I emphasize that the model still needs to be rationalized with experiments: we need more data on how the physiological interaction between the nematode and the host plant influences the mortality process.

12.5.2 Virulence Evolution

It is important to determine how the virulence of *B. xylophilus* populations evolves and to predict the evolutionary process. Kiyohara (1976) isolated a *B. xylophilus* population from a dead *P. thunbergii* sapling previously inoculated with a *B. xylophilus* population that had partly lost its virulence during long subculture. Then he inoculated 5-year-old saplings with the re-isolated and the original population, and found no clear difference in the mortality rate between them, 0/10 and 1/10 (dead/inoculated). This means that the virulence of the attenuated population was not regained during the infection and disease development on the host plants. Possibly, this is because the population was too homogeneous to put selection pressure on the virulence.

In a recent study (Takemoto and Futai 2006), inoculated hybridized PWN populations (between a virulent and an avirulent isolate) into healthy pine seedlings and

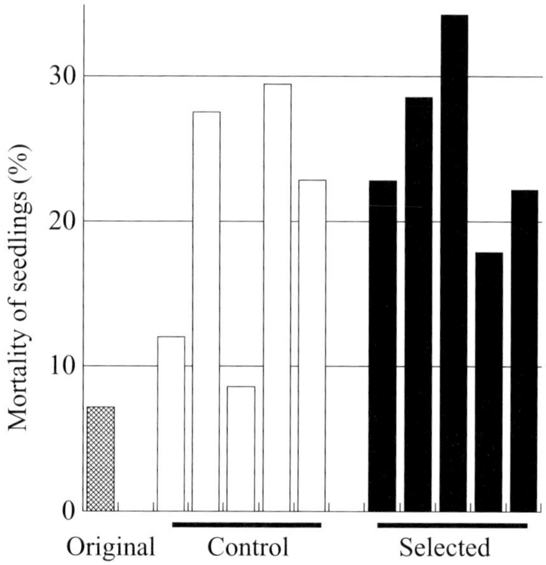

Fig. II.34 Comparison of the virulence of *Bursaphelenchus xylophilus* populations that experienced different substrates. Vertical axis is the mortality of 3-month-old *Pinus thunbergii* seedlings on the day 45 after inoculation. "Original" is a population that had just subcultured on fungal mycelia without experiencing a host plant. "Selected" is a group of populations that had once inoculated into healthy pine seedlings, and then, re-isolated from those seedlings. "Control" is a group of populations that had once inoculated into pine seedlings previously killed with boiling water, and then, re-isolated from those seedlings. All of them were originated from one hybridized population derived from virulent and avirulent parental populations

into the pine seedlings previously killed with hot water, and then *B. xylophilus* was re-isolated from these seedlings a month after inoculation. The populations re-isolated from the former and the latter were designated as the "selected group" and the "control group". Both the selected and control populations killed more seedlings than the original population that had been maintained in culture on a fungus (Fig. II.34). This shows that a population may increase its virulence after inoculation into a host plant, regardless of host condition; healthy or dead, though this tendency was not statistically significant. In contrast, the selected group showed a higher rate of invasion into a healthy pine seedling than the control group, although the difference was insignificant ($P = 0.147$, two-level nested ANOVA for arcsine-transformed values) between the selected and the control group. This suggests that *B. xylophilus* individuals that had high invasive ability were possibly selected during the invasion into a healthy pine seedling. Although the difference was not significant, the mortality of the seedlings inoculated with the selected group was slightly higher than that of the seedlings inoculated with the control group on average ($P = 0.584$, Mann–Whitney U test), and the control and selected groups were more virulent than the original population that had been subcultured on a

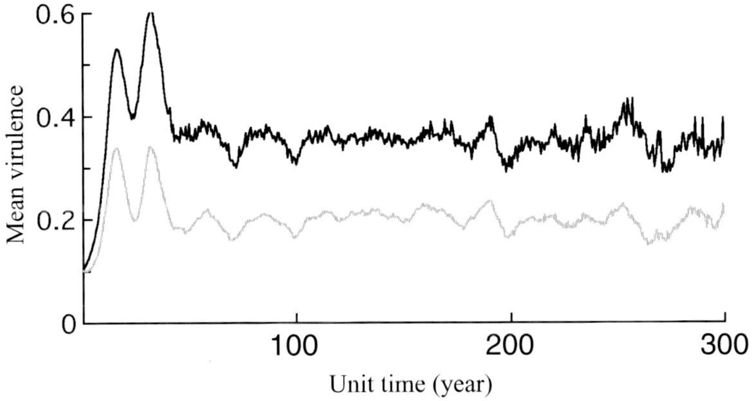

Fig. II.35 Evolution of virulence among *Bursaphelenchus xylophilus* in a numerical simulation with a lattice model. This shows a summary of 250 simulations starting with a few pioneer *B. xylophilus* populations with virulence $\alpha = 0.1$, transmission rate $\beta = 0.001$. Solid and dotted lines indicate *B. xylophilus* populations on wilt-killed trees and infected but still living trees, respectively

fungus ($P = 0.182$, Mann–Whitney *U* test), suggesting that increased virulence may help *B. xylophilus* individuals survive and multiply within a host tree, and that the beneficial effect is, possibly, slightly greater for nematodes inhabiting a living tree than for those inhabiting a dead tree.

So, how does virulence evolve among populations? In a recent study, Takemoto (2007) attempted to mathematically analyze the virulence evolution of *B. xylophilus*. The algebraic model showed that higher virulence always evolves during the progress of pine wilt disease in a pine stand; however, if the stochastic process is considered in the lattice model, moderate mean virulence was maintained (Fig. II.35). Boots and Sasaki (1999) considered a lattice-structured population of the host and also predicted that moderate virulence and transmission rates evolve when infection occurs locally, and that extremely high virulence and transmission rates may evolve when the infection occurs globally. They argued that if infection occurs locally, the low transmission rate and low virulence of the pathogen allowed (1) infected host individuals to reproduce, and thus, tended to increase the local density of susceptible host individuals around the infected hosts, and (2) the pathogen to persist on a cluster of host individuals. In the lattice model of the study by Takemoto (2007), if one simulation run is regarded as one local cluster of pine trees, moderate mean virulence generated in the summary of the 250 results of simulation (Fig. II.35), which may be explained by the balance between the evolution of higher virulence and the local extinction of highly virulent populations, as Boots and Sasaki (1999) mentioned.

If we focused on one result of the simulations (Takemoto 2007), it was shown that mean virulence declines when the population of host trees become sparse due to the disease (Fig. II.36). This may be because highly virulent *B. xylophilus*

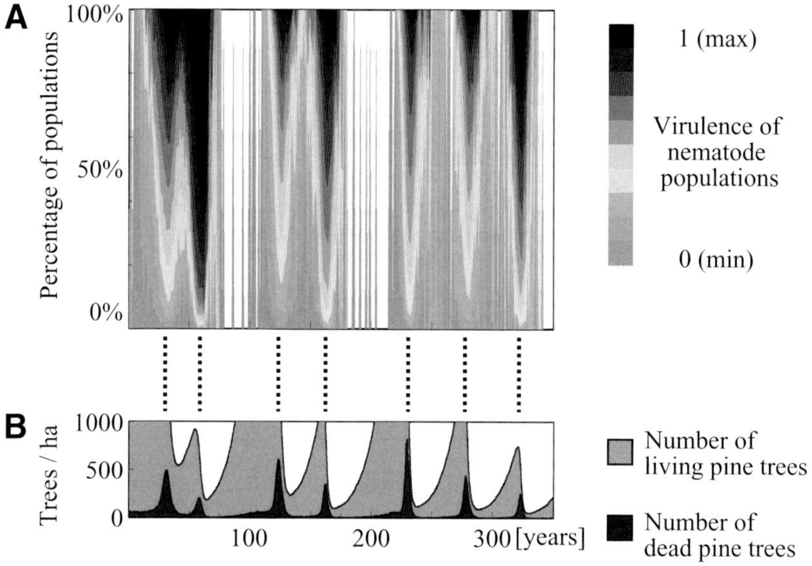

Fig. II.36 Fluctuation in the virulence of *Bursaphelenchus xylophilus* populations synchronized to the dynamics of the host population. **A** Change in the virulence of *B. xylophilus* populations. **B** Changes of the number of living and dead trees (see Color Plates)

populations are rapidly removed from such a sparse population of host trees, and only slightly virulent *B. xylophilus* populations in latent carriers survived by chance until the host population flourished again. The latency is biologically realistic because some researchers have reported that *B. xylophilus* can persist in living *P. sylvestris* (Halik and Bergdahl 1994) and *P. koraiensis* trees (Futai 2003a) without inciting pine wilt disease (see also Chap. 23). The present results well explain the high virulence of *B. xylophilus* populations isolated from a natural pine stand under expanding injury (Akiba et al. 2003) and the lower virulence of *B. xylophilus* populations isolated at the end of the epidemic (Togashi et al. 2002).

Bursaphelenchus xylophilus is transmitted to healthy and dying trees via feeding and oviposition wounds by the vector beetle, respectively. Togashi (2001) pointed out that the two pathways should be separately considered, and described the infection cycle of *B. xylophilus* as a recurrence formula including these two pathways. He supposed the existence of a positive relationship between the transmission rate via feeding wounds and the virulence, and a negative relationship between the transmission rate via oviposition wounds and virulence, referring to the three transmission systems of *Bursaphelenchus* nematodes vectored by sawyer beetles. He concluded that if dying trees are frequently supplied, the lowest virulence possibly maximizes the basic reproductive rate of the nematode populations depending on the form of the trade-offs between the virulence and the transmission rates. Although the analyses by Takemoto (2007) explained the fluctuation of virulence without assuming these explicit trade-offs, further studies are needed to establish whether these trade-offs could stabilize the low virulence when the host population becomes sparse. It is also important to examine the form and intensity of the supposed trade-offs.

Supposedly high virulence would always evolve among *B. xylophilus* populations during the spread of pine wilt disease in a pine stand, because the process of virulence evolution in a single population of *B. xylophilus* is unlikely to cancel the evolution of higher virulence among populations (Takemoto and Futai 2006). If we allow populations of *B. xylophilus* to be transmitted from dead to living trees, they will increase their virulence, and the destruction of the pine stand will accelerate. So, I emphasize the importance of the removal of dead trees to block the infection cycle to slow down the virulence evolution of *B. xylophilus* populations. To examine the effect of the removal of dead trees on the evolutionary process, both field survey and mathematical studies are needed.

12.6 Conclusion and Perspectives

Firstly, we saw that *B. xylophilus* is widely distributed across the Northern Hemisphere, in which the Asian and Portuguese populations are regarded as a small sample out of the diverse populations from their indigenous origin in North America. It is expected that many outcomes will be produced from those diverse populations that remain unexplored.

Secondly, the population dynamics of *B. xylophilus* in culture and within a single tree were discussed. A population may begin with a few individuals, rapidly spread throughout a tree, increase in numbers and kill the tree. Then, the distribution may become clumped as its population density decreases, and only a small portion of the individuals that have aggregated around the pupal chamber of *M. alternatus* may be carried from the tree by the insect vector to be transmitted to other host pines. This cycle inevitably causes a strong bottlenecking effect on *B. xylophilus* populations so that the populations in dead pine trees are genetically rather uniform. Genetic diversity seems to be maintained mainly among trees and localities, not within trees.

Finally, the evolution of virulence is discussed. If we consider that a field populations of *B. xylophilus* should be composed of individual nematodes with different levels of virulence (at least virulent and avirulent), and that the virulence of the population is determined by the proportion of virulent individuals therein, a variety of populations varying in their virulence should evolve from one moderately virulent population as a result of genetic drift. The virulence of *B. xylophilus* tends to increase among its populations as the consequence of competition for colonizing new host pine trees; however, highly virulent populations may be rapidly removed when the population of the host pine becomes sparse, and only weakly virulent populations may survive by chance without causing symptoms until the host population flourishes again. These analytical results predict the fluctuation of the virulence of *B. xylophilus* populations synchronized to the dynamics of the host population. Apparently, the fluctuation indicates that the relationship between the pathogen and the host has not matured, that is, the analysis may show only a transient status in the process of the coevolution between *B. xylophilus* and a host pine. In fact, *B. xylophilus* in its native habitat in North America does not cause a disaster on the pine forests, and the relationship between this nematode species and the pine species seems stable. A more profound understanding is needed regarding the process of the coevolution between *B. xylophilus* and pines, which may provide a novel way managing this disastrous forest pest.

13
Transmission Biology of *Bursaphelenchus xylophilus* in Relation to Its Insect Vector

Takuya Aikawa

13.1 Introduction

The pine wood nematode (PWN) *Bursaphelenchus xylophilus* is transmitted by cerambycid beetles in the genus *Monochamus* (Coleoptera: Cerambycidae) with *M. alternatus* being its principal vector in Japan (Mamiya and Enda 1972; Morimoto and Iwasaki 1972), China and Korea (Kishi 1995), *M. carolinensis* in the USA (Linit 1988) and *M. galloprovincialis* in Portugal (Sousa et al. 2001). The nematode is believed to have originated in North America (Tarés et al. 1992a) and was introduced into Japan in the early 1900s (Mamiya 1984a). Then, the nematode spread to China, Korea in the 1980s (Kishi 1995) and Portugal in the late 1990s (Mota et al. 1999).

Pine wilt disease is characterized by a close relationship between the PWN and its vector beetle. As such, fully understanding the relationship between the two organisms is necessary to understand the epidemic mechanism of pine wilt disease. The relationship between the PWN and *M. alternatus* is as follows: the PWN has two developmental forms in its life cycle, the propagative form (egg, first-, second-, third-, fourth-stage propagative juveniles and adult males and females) to multiply in pine trees, and the dispersal form (third- and fourth-stage dispersal juveniles) which is transferred to new host trees (Mamiya 1975a) (Fig. II.37). During summer, the fourth-stage dispersal juvenile (J_{IV}), a special stage of individuals for transfer, is carried by the vector beetles from dead to healthy host trees. J_{IV}s that successfully invade the trees molt to adults and mate. The female adult then initiates oviposition.

Tohoku Research Center, Forestry and Forest Products Research Institute, 92-25 Nabeyashiki, Shimo-Kuriyagawa, Morioka 020-0123, Japan

Tel.: +81-19-648-3961, Fax: +81-19-641-6747, e-mail: taikawa@affrc.go.jp

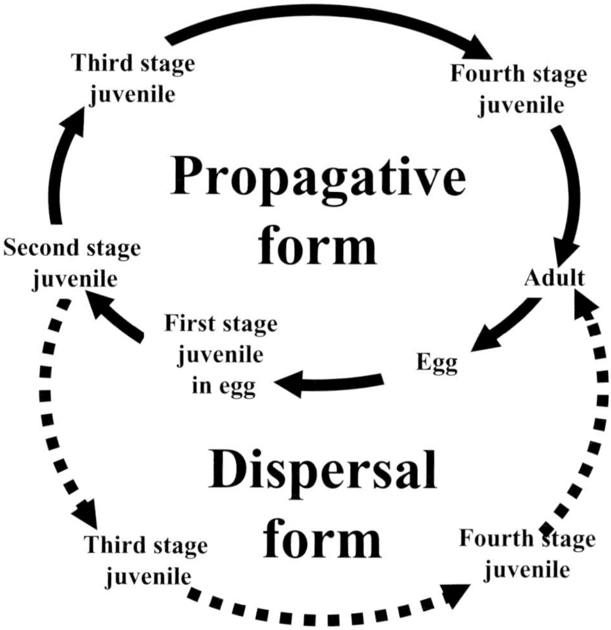

Fig. II.37 Life cycle of *Bursaphelenchus xylophilus*. The *solid arrow* and *dashed arrow* show the propagation cycle in pine trees and that for transmission to new host trees, respectively

Reproduction occurs through the propagative forms, which feed on parenchymatous cells of the host's resin canals and wood-infesting fungi, such as blue-stain fungi (Kobayashi et al. 1974, 1975), resulting in tree mortality. *M. alternatus* females deposit eggs under the bark of trees recently killed by the PWN. Generally, the larvae inhabit the inner bark in the first, second and third instars, begin to bore tunnels into the xylem in the fourth instar, and then makes pupal chambers in the xylem (Fig. II.38) (Togashi 1989a, b, 1991a). The beetles overwinter as larvae and pupate between late spring and early summer the following year. As for the nematode, when the population increases in the pine trees, the third-stage dispersal juvenile (J_{III}) appears (Kiyohara and Suzuki 1975). During winter, J_{III}s aggregate around the pupal chambers made by beetle larvae. They molt to J_{IV}s following adult eclosion of the beetle in the pupal chamber and then invade the beetle body. After emerging from host trees, beetles move to young branch of healthy trees, and make feeding wounds (maturation feeding) on them, and then J_{IV}s leave the beetles' body and invade new host trees. This is the epidemiological cycle of pine wilt disease resulting from the collaborative operation of the PWN and its vector beetle. The beetle serves as a carrier of the nematode and the nematode works as a producer of oviposition resources for the beetle. Namely, there is a close

Fig. II.38 *Monochamus alternatus* larva in the pupal chamber. A pine log infested with beetle larvae was chopped in two with a hatchet to observe the pupal chamber. Larva bores a tunnel into the xylem while plugging it with frass to make a pupal chamber. The *arrow* shows the entrance of the tunnel (*scale bar* = 2 cm) (see Color Plates)

mutualistic relationship between the PWN and its vector beetle at the population level.

After the roles of *B. xylophilus* and *Monochamus* beetles in the epidemic development of pine wilt disease were clarified, the biology and interaction between the two organisms has been intensively investigated. Especially in the last decade, study on the interaction between the PWN and its beetle vector has been rapidly advanced by discoveries such as the: (1) development of artificial pupal chambers (described in detail in the "Concluding Remarks"), (2) analysis of factors affecting nematode transfer to and departure from vector beetles, and (3) discovery of new transmission pathways of nematodes to pine trees. In this chapter, the transmission biology of PWN is dealt with in association with the vector beetle by reviewing the latest information and discussing future work on this topic.

In this chapter, the transmission biology of nematodes is divided into four steps: (1) transmission of nematodes to vector beetles in dead pine trees, (2) movement of nematodes in the tracheal system of vector beetles, (3) departure of nematodes from the tracheal system of vector beetles and (4) transmission of nematodes into pine trees.

13.2 Transmission of Nematodes to Vector Beetles in Dead Pine Trees

13.2.1 Development from the Second-stage Propagative Juvenile to the Third-Stage Dispersal Juvenile

As a prerequisite for transmission to the body of vector beetles, the nematode has to switch from the propagative form to the dispersal form, namely, molt from the second-stage propagative juvenile to J_{III} (Fig. II.37). One of the traits of J_{III}s is that they possess a large quantity of neutral storage lipid in their body (Kondo and Ishibashi 1978). A high density of nematodes is a prerequisite for the appearance of the J_{III}s (Fukushige 1991; Maehara and Futai 2000) and the J_{III} larvae first appear when nematode density starts to rise (Kiyohara and Suzuki 1975). Then, the proportion of J_{III} increases with time and peaks when or just after the nematode density peaks in trees (Fukushige and Futai 1987; Fukushige 1991; Maehara and Futai 2000). In pine tissues where the nematodes live, desiccation and degradation progress with time after the tree dies. J_{III} is an adaptable stage to such adverse ambient conditions (Mamiya 1983). During winter, J_{III}s aggregate around the pupal chambers made by beetle larvae. The walls of the pupal chamber are contaminated by unsaturated fatty acids such as palmitoleic, oleic and linoleic acids deposited by insect larvae. These acids are thought to stimulate J_{III}s to aggregate around pupal chambers (Miyazaki et al. 1977a, b).

13.2.2 Development of the Third-Stage Dispersal Juvenile to the Fourth-Stage Dispersal Juvenile

When pupae of vector beetles in the pupal chambers eclose between late spring and early summer, J_{III} nematodes around the pupal chamber develop into J_{IV}s. As is the case of J_{III}s, J_{IV}s also possess a large quantity of neutral storage lipid in their body, but J_{IV}s can be distinguished from J_{III}s by some morphological traits; for instance, J_{IV}s have a domed head, an invisible stylet and a pointed tail terminus (Fig. II.39). The existence of beetle pupae or callow adults in the pupal chambers is necessary for J_{III}s to molt to J_{IV}s (Morimoto and Iwasaki 1973; Maehara and Futai 1996). An in vitro experiment also showed the same results (Ogura and Nakashima 2002). In *M. carolinensis*, the molting from J_{III} to J_{IV} occurred when pulverized late pupae or callow adults were placed in an artificial hole drilled in pine logs infested with nematodes (Necibi and Linit 1998). This suggests that a genus-specific substance(s) associated with *Monochamus* adult eclosion induces the formation of J_{IV}. However, an experiment using an artificial pupal chamber showed that the formation of J_{IV} occurred even when *Psacothea hilaris*, a longicorn beetle, which attacks mulberry and fig trees, was placed in the artificial pupal chamber, although

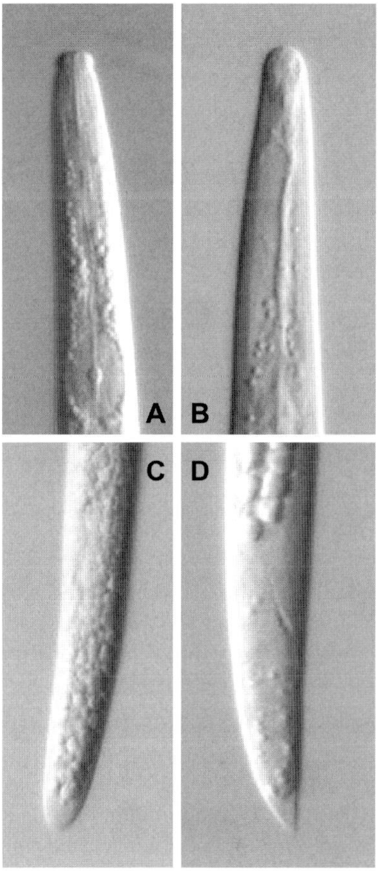

Fig. II.39 Anterior and terminal portions of third-stage dispersal juvenile (J_{III}) and fourth-stage dispersal juvenile (J_{IV}) of *Bursaphelenchus xylophilus*. **A** Anterior portion of J_{III}. **B** Anterior portion of J_{IV}. **C** Terminal portion of J_{III}. **D** Terminal portion of J_{IV}. Modified after Aikawa (2006, J Jpn For Soc), with permission

the rate was extremely low (Maehara and Futai 2001). Presumably, not only *Monochamus*-specific substances induce the development from J_{III} to J_{IV}.

13.2.3 Migration of the Fourth-Stage Dispersal Juvenile to the Body of Vector Beetle

J_{IV}s appear on the wall surface of the pupal chamber, and climb up the long perithecial neck of blue-stain fungi, which have grown abundantly on the wall of the pupal chamber. At the tip of the neck, they attach to spore masses of blue stain fungi with sticky substances. Then, they move onto the body of callow beetle adults, and finally J_{IV}s invade the tracheal system from spiracles of the beetle. The volume of carbon dioxide (CO_2) released by the respiration of beetle reaches a maximum level at eclosion, and CO_2 apparently stimulates J_{IV} to migrate to callow beetles (Miyazaki et al. 1978a, b). Most J_{IV}s are found in the tracheal system in the thorax, with a few occurring in the abdomen of *M. alternatus* and *M. carolinensis* (Linit et al. 1983; Kobayashi et al. 1984; Kondo 1986). Similar J_{IV} distribution occurs within the body of the beetle *M. galloprovincialis* (Naves et al. 2006a).

The number of J_{IV}s carried by a beetle from dead pine trees (initial nematode load) is one of the most important factors influencing the damage level of pine wilt disease (Togashi 1985, 1991b). The initial nematode load greatly varies among beetles and these variations are explained by several biological or non-biological factors. *M. alternatus* beetles that emerged from pine logs where the water content was extremely high or low, had very few J_{IV}s (Morimoto and Iwasaki 1973; Terashita 1975; Togashi 1989g) and the initial nematode load increases with beetle body size (Linit et al. 1983; Humphry and Linit 1989; Aikawa and Togashi 1998). In an experiment with artificial pupal chambers, the number of J_{IV}s that invaded the beetle body rapidly increased from the day of beetle eclosion to the third day thereafter, then remained stable until the eighth day (Maehara and Futai 2001). Generally, *M. alternatus* adults stay in the pupal chamber for 4–8 days after eclosion in the field (Ido and Takeda 1972; Kishi 1976; Takizawa 1979; Enda 1980); therefore, the beetles may emerge from dead pine trees after they received as many J_{IV}s as possible. The kind of fungi infesting dead pine trees is important in affecting the initial nematode load. If the fungus is favorable for the nematodes to feed on, the nematodes increase rapidly. The high density accelerates development from the propagative to the dispersal form, consequently, many J_{IV}s invade the beetle's body. By contrast, if the fungus infesting dead trees is unfavorable for the nematodes, development from the propagative to dispersal form is low and consequently, the initial nematode load is very low (Maehara and Futai 1996, 1997). In addition, the virulence (the ability to kill pine trees) of *B. xylophilus* is also thought to be a factor affecting the initial nematode load. There is great variation in the virulence among nematode populations in the field (Kiyohara 1989; Kiyohara and Bolla 1990). A virulent population has a greater capacity to reproduce, and its developmental capacity (from the second-stage propagative juvenile to J_{III} and from J_{III} to J_{IV}) is greater than an avirulent population. As a result, the initial nematode load of beetles is much greater for a virulent than an avirulent population (Aikawa et al. 2003).

13.3 Movement of Nematodes in the Tracheal System of Vector Beetles

Many reports have described the movement of J_{IV}s on adult beetles from diseased trees, but little is known about the movement of J_{IV}s in the tracheal system of beetles (Fig. II.40). Tracheae are the respiratory organ that transports air from outside the beetle's body to tissues within the body and metabolic gases outward. Spiracles (external opens of the respiratory system) occur on the thorax and abdomen of

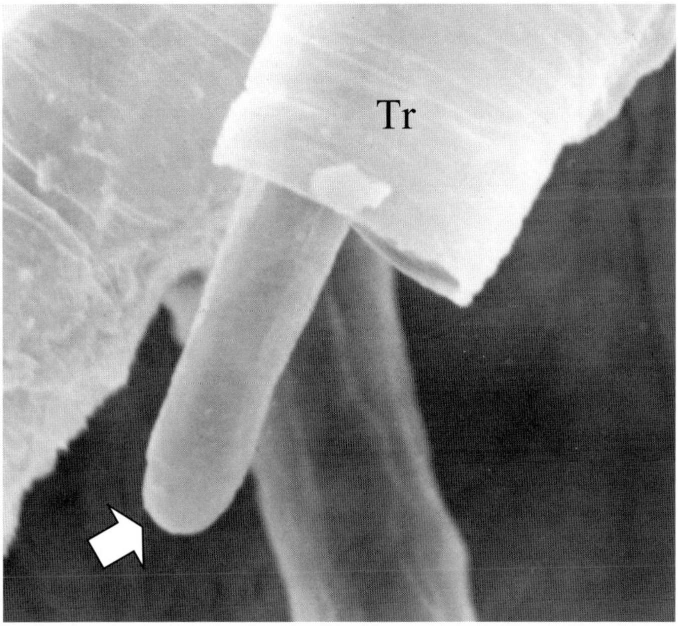

Fig. II.40 Fourth-stage dispersal juvenile (J_{IV}) of *Bursaphelenchus xylophilus* in the trachea (Tr) of *Monochamus alternatus* adult. *Arrow* shows the head of J_{IV}

beetles (Fig. II.41) and each spiracle has an atrium. The atrium has relatively large tracheal tubes behind it and the tracheal tubes are distributed throughout the body with branching. The first abdominal spiracles (Fig. II.41), located on both sides of the metathorax, are the largest of the spiracles and a "ball" of nematodes forms within the atrium of the spiracle when the initial nematode load is heavy (Zhang et al. 1995; Stamps and Linit 1998a; Fig. II.42). When a beetle contains many J_{IV}s, they occupy not only the tracheae near the spiracles but also those in the antennae and legs (Fig. II.43). Based on the number of J_{IV}s within those atria and tracheae the initial nematode load has been estimated (Kobayashi and Hosoda 1978; Enda and Makihara 1982; Zhang et al. 1995).

To be successfully transmitted to a new host tree, J_{IV}s in the trachea must first leave the spiracles and then the beetle's body. The movement of J_{IV}s in the tracheal system has been determined as the result of work on the orientation of J_{IV}s in the trachea of different-aged adult beetles. In newly emerged *M. carolinensis* adults most J_{IV}s are oriented in the same direction in the trachea (Kondo 1986). The same occurs in *M. alternatus* and the J_{IV}s are headed inward toward the distal end of the trachea (Aikawa and Togashi 2000). When 0-, 15-, 30-, 45- and 60-day-old beetles were dissected and the J_{IV}s heading inwards to the distal end of the spiracles and

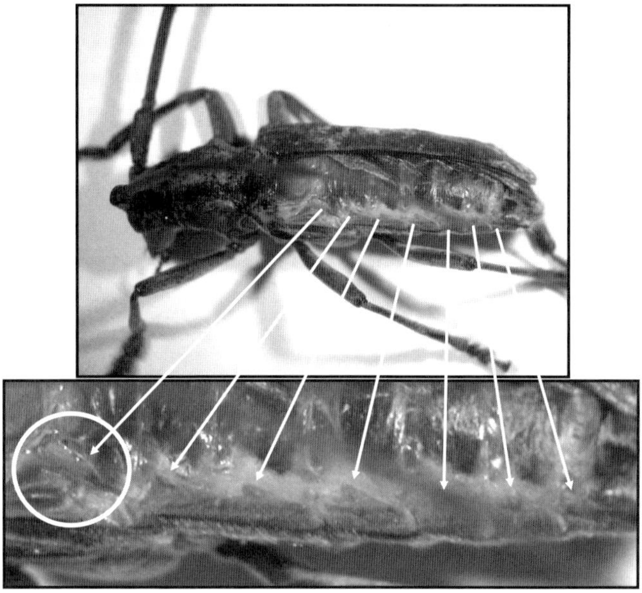

Fig. II.41 Abdominal spiracles of *Monochamus alternatus* adult. The left elytron was removed to observe the spiracles. Seven pairs of spiracles are arranged in pairs on seven abdominal segments. The first abdominal spiracle (*circle*) is the largest among the spiracles and a great number of fourth-stage dispersal juveniles enter the trachea through here (see Color Plates)

those outwards to spiracles in the trachea were counted, the proportion of J_{IV}s pointing outwards tended to increase with beetle age and was highest at 45 days after beetle emergence. Also, the proportion of the trachea where J_{IV}s were headed outwards was higher for 15 to 45-day-old beetles than for newly emerged beetles. As many U-shaped J_{IV}s, which are considered to be turning, also have been observed in trachea (Kondo 1986; Aikawa and Togashi 2000; Mamiya 2003a), these results suggest that J_{IV}s move forward in the tracheal system (from the spiracles) and then exit from spiracles after turning around in the trachea.

13.4 Departure of Nematodes from the Tracheal System of Vector Beetles

After J_{IV}s exit from the beetle spiracles, they move to the abdominal posterior end on the surface of the beetle body (Fig. II.44) and then drop from the body onto pine trees (Kobayashi et al. 1984). Many studies have reported a certain tendency

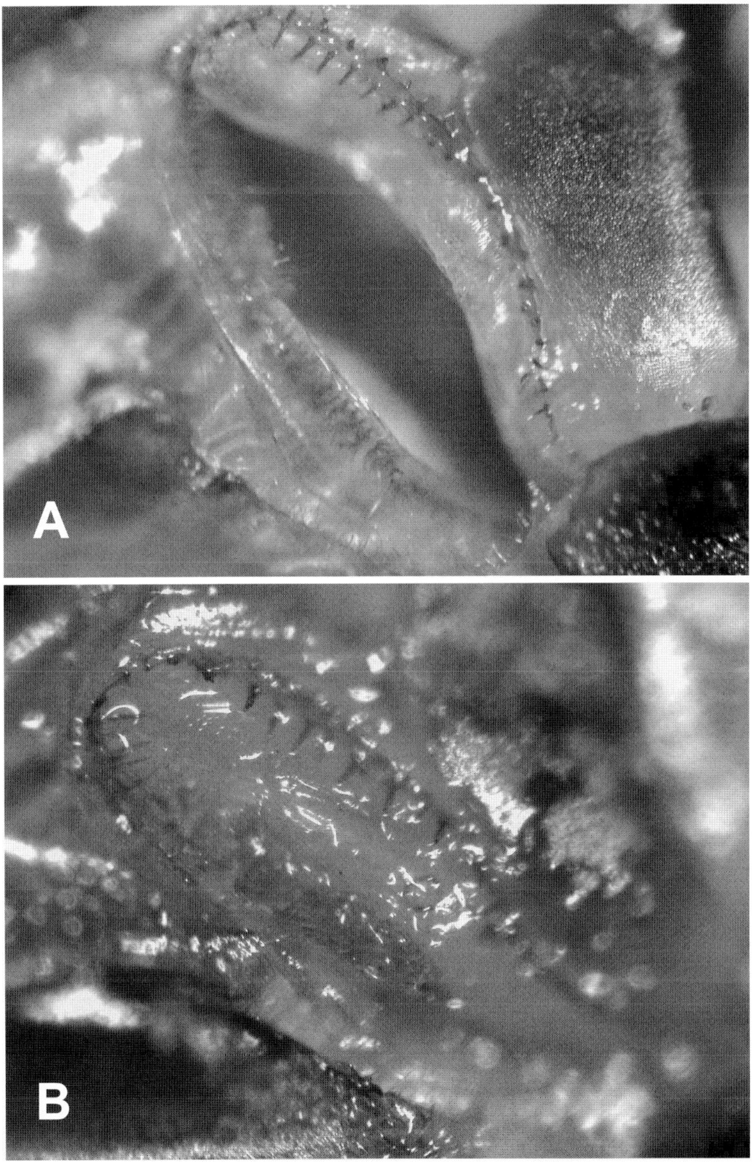

Fig. II.42 Atrium of the first abdominal spiracle of *Monochamus alternatus* adult. **A** The atrium of a beetle infested with no fourth-stage dispersal juvenile (J_{IV}). The cavity of the atrium can be clearly observed. **B** The atrium of a beetle infested with numerous J_{IV}s. The cavity of the atrium cannot be observed as the result of being packed with many J_{IV}s. Modified after Aikawa (2006), with permission (see Color Plates)

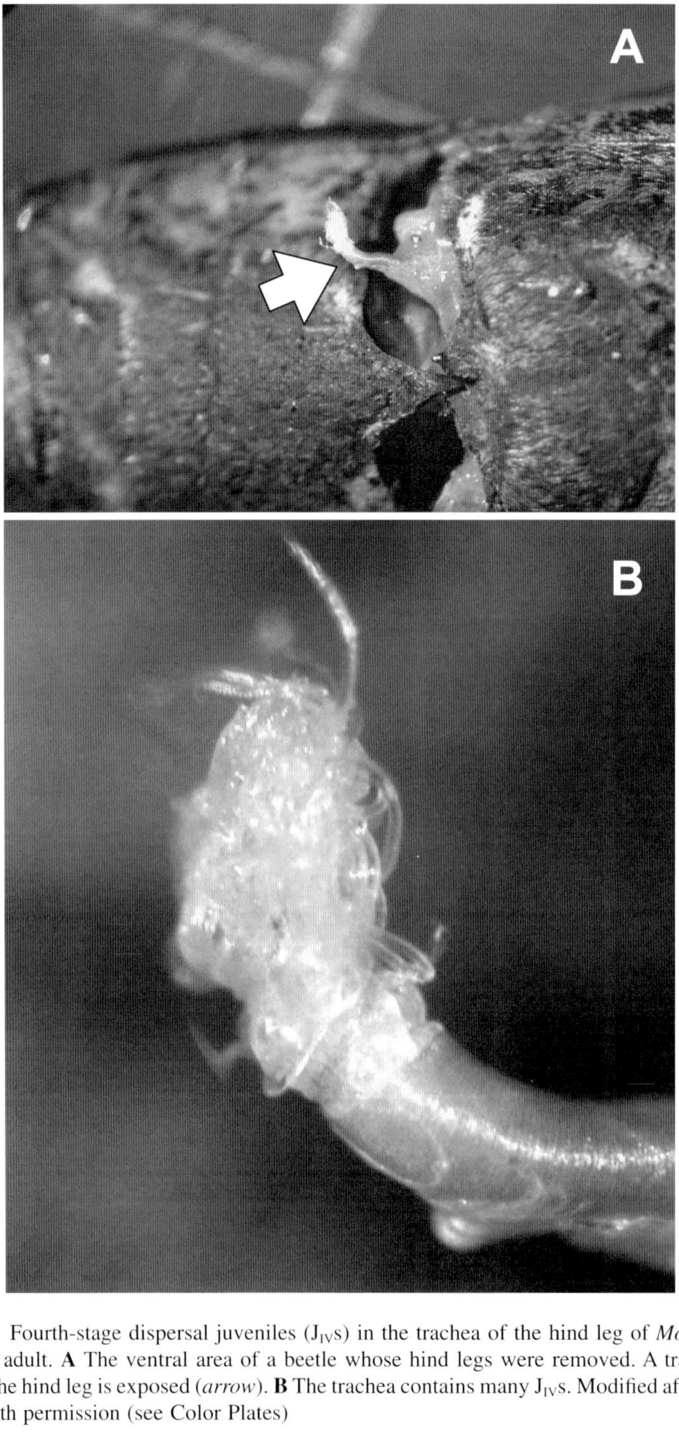

Fig. II.43 Fourth-stage dispersal juveniles (J_{IV}s) in the trachea of the hind leg of *Monochamus alternatus* adult. **A** The ventral area of a beetle whose hind legs were removed. A trachea connected to the hind leg is exposed (*arrow*). **B** The trachea contains many J_{IV}s. Modified after Aikawa (2006), with permission (see Color Plates)

Fig. II.44 Fourth-stage dispersal juveniles (J_{IV}s) crawling on the body surface of *Monochamus alternatus* adult. After J_{IV}s exited from the spiracles, they move to the abdominal terminal to depart from the beetle's body (see Color Plates)

regarding the temporal pattern of J_{IV} departure in relation to beetle age (nematode departure curve). The tendency is that J_{IV} departure hardly occurs just after beetle emergence and then occurs primarily between 10 and 40 days after emergence, and the highest peak in the departure curve usually occurs at 2–3 weeks after beetle emergence (Morimoto and Iwasaki 1972; Togashi and Sekizuka 1982; Togashi 1985; Aikawa and Togashi 1998). Since in North America many J_{IV}s depart from *M. carolinensis* adults between 2 and 3 weeks after emergence (Linit 1989), it is supposed that the nematode already had such a characteristic before being introduced into Japan.

The life span of adult beetles is inferred to affect the departure of J_{IV}s because it takes more than 1 week for J_{IV}s to start leaving the beetle. The proportion of J_{IV}s, which successfully departed from beetles is higher in long-lived *M. alternatus* adults than short-lived ones (Togashi 1985); however, it has been demonstrated that the life span of adult beetles is negatively correlated with the initial nematode load (Togashi and Sekizuka 1982). As a result, the proportion of nematodes remaining in the beetle body (= failed to depart from beetle) increases with the increase in initial nematode load (Togashi 1985). When the tracheal system is packed with many J_{IV}s, gas transfer between the exterior of the beetle and the beetles tissues decreases. A heavy nematode load would therefore have a detrimental effect on the beetle. Such a detrimental effect was shown in an experiment on flight performance of *M. carolinensis* adults. Beetles that carried >10,000 J_{IV}s had a significantly shorter flight distance and duration than beetles carrying <10,000 nematodes (Akbulut and Linit 1999). Namely, the heavy load of J_{IV}s may shorten the life span of beetles, and consequently, they may lose the opportunity to be transferred to new hosts.

Another factor affecting the transfer to the host tree is temperature. When *M. alternatus* adults were reared at 16, 20 or 25°C, the life span of beetles at 16 or 20°C was shorter than that at 25°C (Jikumaru and Togashi 2000). Thus temperature influences the life span of beetles; thereby determines the opportunity of J_{IV}s to be transferred. Also, the lower temperature delayed the timing of J_{IV} departure from the beetles (Jikumaru and Togashi 2000).

J_{IV}s must depart from the beetle while the beetle is on pine trees. The nematodes seem to perceive exogenous factors such as pine volatiles in the tracheal system of beetles as a means for determining the timing of departure. When J_{IV} departure was compared between *M. carolinensis* adults placed over pine chips and those placed over distilled water, J_{IV} departed in greater numbers and at a higher frequency from beetles over pine chips than from beetles over distilled water (Edwards and Linit 1992). Among the volatile components, β-myrcene showed a strong attraction for J_{IV}s on an agar plate and accelerated nematode departure from beetles (Enda and Ikeda 1983; Ishikawa et al. 1986; Stamps and Linit 1998b). This suggests that volatiles such as β-myrcene play an important role in the departure of J_{IV}s from beetles; however, J_{IV} departure from beetles is not controlled only by pine volatiles. When adults of *M. alternatus* were reared on autoclaved pine twigs that emitted no volatiles, 84% of the initial nematode load departed from the beetles by the day 51 after emergence (Aikawa and Togashi 1998). This suggests that the nematode has a trait of spontaneous departure from beetles. This spontaneous trait is thought to be related to the stored neutral lipid of J_{IV}s, as those remaining within *M. carolinensis* adults have a larger amount of neutral storage lipid than those that exited from beetles, and the lipid content of J_{IV}s remaining in beetles decreased with beetle age (Stamps and Linit 1998a). The neutral lipid is consumed as an energy source by nematodes and the decrease in the amount of lipid is inferred to be associated with behaviors leading to nematode departure from the beetle. In addition, J_{IV}s with the lowest neutral storage lipid content were attracted to β-myrcene, while J_{IV}s with the highest lipid content were attracted to toluene, a beetle cuticular hydrocarbon (Stamps and Linit 2001). The behavior of J_{IV} departure can be explained as follows: as J_{IV}s in the tracheal system of newly emerged beetles are filled with neutral storage lipid and are attracted to toluene as a beetle cuticular component, they remain in the trachea for a while. The quantity of neutral storage lipid in J_{IV}s decreases with time and this intrinsic cue stimulates J_{IV}s to leave the tracheal system of beetles. As J_{IV}s that have almost consumed the lipid content respond strongly to β-myrcene emitted from pine trees, J_{IV} departure from beetles is accelerated further. In this way, the departure behavior of J_{IV} from beetles was found to be related to compound actions, including intrinsic (neutral storage lipid) and extrinsic (beetle cuticle hydrocarbon and pine volatiles) cues.

13.5 Transmission of Nematodes into Pine Trees

13.5.1 Transmission of the Fourth-Stage Dispersal Juveniles Through Feeding and Oviposition Wounds Made by Vector Beetles

There are two pathways for J_{IV}s departing from beetles to invade pine trees. One is the invasion of healthy pine trees via wounds made by beetle feeding (Mamiya and Enda 1972; Morimoto and Iwasaki 1972) and the other is the invasion of recently killed trees via wounds by beetle oviposition (Wingfield 1983; Wingfield and Blanchette 1983; Linit 1988). J_{IV}s succeeding in invading pine trees are part of the J_{IV}s that departed from adult beetles. Accordingly, the temporal pattern of J_{IV} transmission into pine trees in relation to the age of beetles (nematode transmission curve) can be regarded as same as the nematode departure curve. For instance, the highest peak in the nematode transmission curve via feeding wounds by *M. alternatus* adult is observed in the second week after beetle emergence (Shibata 1985), between days 15 and 35 after emergence (Togashi 1985) or between the 10th and 30th day after emergence (Jikumaru and Togashi 2000), and these transmission curves tend to be unimodal. Also, in an investigation of *M. carolinensis* and *M. galloprovincialis*, the highest peak in the nematode transmission curve through feeding wounds is occurs in the third to fourth week and second to sixth week after emergence, respectively (Linit 1990; Naves et al. 2007). It is also reported that the greatest number of J_{IV}s are transmitted during the first 5 days after beetle emergence (Kishi 1978; Togashi 1985). The initial nematode load of beetles and the ambient temperature affect the transmission of J_{IV}s *via* feeding wounds. The peak of the nematode transmission curve becomes higher as the number of J_{IV}s carried by *M. alternatus* becomes greater (Togashi 1985; Jikumaru and Togashi 2000). The number of J_{IV}s transmitted by *M. carolinensis* increases with its initial nematode load (Linit 1990). As the ambient temperature decreases from 25 to 16°C, J_{IV} transmission efficiency (the number of J_{IV}s successfully invading pine twigs/that departed from beetles) decreases, and the peaking period of J_{IV} transmission is delayed and its peak height decreases (Jikumaru and Togashi 2000). In contrast, for *M. carolinensis* the number of J_{IV}s transmitted via oviposition wounds is constant irrespective of the age of female beetles, and J_{IV}s tend to be transmitted more frequently and in slightly greater numbers by beetles carrying more J_{IV}s (Edwards and Linit 1992). In *M. alternatus*, J_{IV}s are believed to be transmitted by female beetles via oviposition wounds, although this has not been confirmed.

13.5.2 Other Transmission Pathways

In recent years, two transmission pathways of J_{IV}s have been newly discovered. One is the transmission of J_{IV}s by males of *M. alternatus* via oviposition wounds.

When nematode-infested males, with mandibles fixed experimentally to prevent feeding, were placed for 48 h with pine bolts containing oviposition wounds that had been made by nematode-free females, nematode populations reproduced were obtained from the pine bolts 1-month later (Arakawa and Togashi 2002). Another pathway is the invasion of pine trees via horizontal transmission of J_{IV}s between sexes of *M. alternatus*. When nematode-infested beetles of one sex and nematode-free beetles of the opposite sex were paired in a container for 48 or 72 h, J_{IV}s migrated from nematode-infested to nematode-free beetles and then the J_{IV}s acquired by nematode-free beetles could be transmitted to pine trees (Togashi and Arakawa 2003). The number of J_{IV}s carried by nematode-free beetles tends to increase with increase in the number of J_{IV}s carried by nematode-infested beetles or the duration of sexual mounting and, in addition, the transmission of J_{IV}s from males to the spermatheca of females is also known (Arakawa and Togashi 2004). Transmission efficiency does not differ between male-to-female transmission and female-to-male transmission. The number of J_{IV}s that transfer between sexes of beetles during one sexual mounting is not so great (less than 100 J_{IV}s) (Togashi and Arakawa 2003); however, nematode-free beetles may receive many J_{IV}s from nematode-infested beetles by multiple copulations. Therefore, the transmission of J_{IV}s between beetle individuals may have possibly caused an increase of the PWN-vectors.

13.6 Concluding Remarks

In the case of *M. carolinensis*, adults infested with numerous J_{IV}s are easily obtained in the laboratory by allowing them to oviposit on pine logs that have been inoculated with a blue-stain fungus *Ophiostoma minus*, and PWNs (Warren and Linit 1993; Warren et al. 1995). Presumably, this is possible because of the short insect life cycle of ca. 2 months at 27°C. In contrast, it has been difficult to obtain *M. alternatus* infested with J_{IV}s by a method similar to *M. carolinensis* due to a decline in the nematode population in pine logs because in the laboratory *M. alternatus* requires 7–9 months to complete its life cycle due to larval diapause. However, Maehara and Futai (1996) developed an artificial pupal chamber to rapidly load adult *M. alternatus* with J_{IV}s using wood blocks of pine inoculated with *O. minus*, nematodes of the propagative form and a sterilized final instar larva of *M. alternatus*. This method made it possible to investigate the relationships between the nematode and *M. alternatus*, irrespective of season. Furthermore, the method was improved to allow the procurement of *M. alternatus* adults infested with many J_{IV}s (Fig. II.45) (Aikawa and Togashi 1997; Aikawa et al. 1997). By these methods, the effect of fungal species, beetle species, or virulence of nematodes inoculated into artificial pupal chambers on the initial nematode load of *M. alternatus* was elucidated (Maehara and Futai 1997, 2001; Aikawa et al. 2003b). In addition, the numerical comparison of boarding abilities on *M. alternatus* between *B. xylophilus* and *B. mucronatus*, which is the most closely related species to *B. xylophilus*, became possible using this method (Jikumaru and Togashi 2003, 2004). Recently,

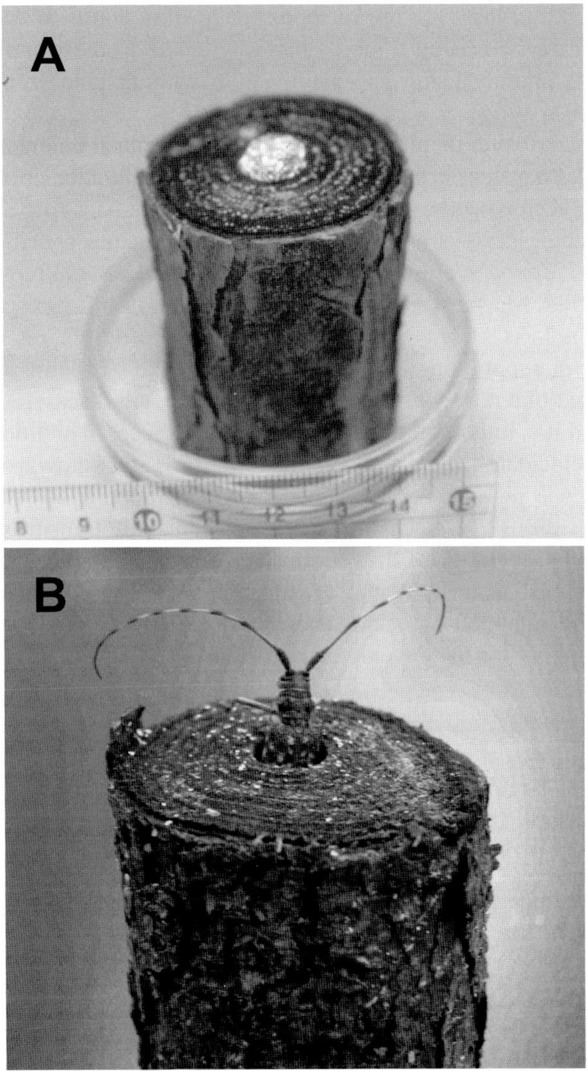

Fig. II.45 Pine bolt for loading *Monochamus alternatus* adult with fourth-stage dispersal juveniles (J_{IV}s) of *Bursaphelenchus xylophilus*. A hole 1.05 cm across and 5 cm deep was drilled in the center of the xylem at the cut end of the pine bolt (7.5-cm long and about 4-cm diameter) as an artificial pupal chamber for *M. alternatus* larva. The bolt was set upright in a layer of quartz sand in a polycarbonate container. The container with the pine bolt was autoclaved and then the bolt was inoculated with *Ophiostoma minus* under sterilized conditions. Two weeks later, 5,000 propagative nematodes suspended in 0.5 ml water and a fourth-instar *M. alternatus* larva were put into the hole at the same time and the hole was plugged with aluminum foil (**A**). During 30–40 days after inoculation of the nematodes and larva, adult beetle infested with a great number of J_{IV}s gnawed through the aluminum plug and emerged from the bolt (**B**). A beetle infested with more than 150,000 J_{IV}s has so far been obtained by this method (T. Aikawa, unpublished data) (see Color Plates)

more simplified and labor saving methods for loading adults *M. alternatus* with J_{IV}s have been devised using a mixed barley grain and pine wood chip medium (Togashi 2004) or only using an agar medium without pine wood (Ogura and Nakashima 2002).

As mentioned above, the development of artificial pupal chambers for loading beetles with J_{IV}s has accelerated the recent studies on the interaction between the PWN and the vector beetle, though many questions remain to be resolved. For example, although it has already been reported that a genus-specific substance(s) associated with *Monochamus* adult eclosion stimulates J_{III} to molt to J_{IV} (Morimoto and Iwasaki 1973; Maehara and Futai 1996; Necibi and Linit 1998), the substance is still unknown. J_{IV} is the only stage carried by vector beetles; therefore, if a method for preventing or suppressing the formation of J_{IV} could be found, it could lead to the development of a new control techniques for pine wilt disease. As the first step toward this end it is important to determine this substance. In addition, it would be interesting to elucidate the roles of two new transmission pathways of J_{IV}s in the spread of pine wilt disease, the maintenance of PWN populations or the evolution of virulence in the PWN. Such findings about nematode transmission would be valuable not only for clarifying the transmission mechanism of the pathogen in pine wilt disease but also as an epidemiological study on worldwide forest diseases. In the future, more detailed associations between the two organisms needs to be determined to allow for the development of studies and new control techniques.

14 Concluding Remarks

Current research topics on the nematode, *Bursaphelenchus xylophilus*, and its relatives, were introduced in this chapter. Research targets on the PWN are now shifting from "disease control" to "scientific interest", probably because research related to "control" is coming to an end. Except for several of the most difficult problems the groundwork research has been completed. However, there are many interesting and important questions that still need to be answered. The taxonomy and systematics of PWN and its related species and genera were described in Chap. 9. The taxonomic status of PWN has already been clarified, that is, PWN, a species of mycophagus or entomophilic nematode, belongs to the "xylophilus" group of the genus *Bursaphelenchus*; however, the definition and the limits of the genus are still unclear, and the framework of the genus needs to be defined based on proper systematic and evolutional evidence. Genetic analysis of PWN and its applications are described in Chap. 10. Molecular biology methods are now indispensable tools for doing research on these detailed physiology and pathological problems. In this context, EST data could be a breakthrough in the study of pine wilt disease. The EST database is now on line, and is available to researchers around the world. The information found during the project could be indispensable in solving the disease mechanisms and for physiological interactions between PWNs and their vectors. In Chap. 11, the embryology and cytology of PWN are introduced. The developmental biology and cytology of PWNs have been studied in detail taking advantage of the knowledge base obtained from a model organism, *Caenorhabditis elegans*; because not much is known these subjects as they relate to the PWN. The development and cytology of the PWN are basically the same as those of *C. elegans*, though there are several interesting differences. The PWN could be a possible model in the embryology and cytology for mycophagus/plant parasitic nematodes. From a practical viewpoint, these detailed observations could help develop effective pine wilt disease control methods. Current disease–control procedures targeting the vector beetle, and/or the pathogen, PWN, are not adequate to allow for us to obtain good results. Based on detailed observations and analyses, new effective control procedures must be developed. In Chap. 12, the population structure of the PWN in host trees and short-term evolution of pathogenicity are discussed from the viewpoint of population genetics. It is almost impossible to eradicate PWN from heavily affected forests, for example, the Japanese islands, where the PWN has spread over almost all of the country. Thus, in the future, disease control may become more locality-specific, that is, each geographical region has to be protected using methods suitable for each reagion under specified conditions. The modeling and prediction of the pathogenicity of each PWN population are important not only for the scientific interest, but also for a practical purpose, because information on the pathogenicity of each PWN population can be essential for establishing effective control strategies for each location. The basic life history and vector association of PWN are discussed in Chap. 13. Actually, the life history of the nematode has already been determined, and the information has already been utilized in devising control procedures, while the physiological traits of PWN are still unclear, that is, there are only a few studies about PWN physiology such as host (feeding resource) preferences and auxotrophy of the PWN. The controlling factors of PWN behavior and its beetle vectors, that is, chemical cues for dauer induction and vector identification (selection) and chemical substances for their pheromones still need to be clarified. Also, the largest and most important subject related to the host, specifically, how PWNs kill host trees, has not been determined (the details of current research on this topic are presented in Part IV).

Up to the present, the basic biological characteristics of the PWN have been only clarified as regards it being the pathogen of pine wilt disease, and detailed physiological knowledge is still lacking. To examine these details more precisely, molecular methods and genetically pure materials are needed. Currently, the PWN, host pines and vector beetles used for experiments are all genetically diverse, that is, they are heterologous populations derived from field populations. The establishment of genetically purified "experimental strains" of these materials and genetic engineering methods should provide solutions to the many unsolved questions.

Part III
The Vector Beetle

15
Introduction

Soon after *Bursaphelenchus xylophilus* (pine wood nematode, PWN) was shown to be the causative agent of pine wilt in Japan a pine sawyer beetle, *Monochamus alternatus,* was identified as its vector (Mamiya and Enda 1972; Morimoto and Iwasaki 1972). Later, in the USA, *M. carolinensis* was recognized as the primary vector of the PWN (Kondo et al. 1982; Wingfield and Blanchette 1983). Following these discoveries biological studies began on *M. alternatus* and *M. carolinensis*.

Chapter 16 mainly reviews the biology of *M. alternatus*, which occurs in East Asia from Japan to northern Vietnam and Laos. Thus, there is a great variation in the life cycle; a 1- or 2-year life cycle in Japan and two or three generations a year cycle in subtropical China. Different life cycles are controlled by the insect's obligate or facultative larval diapause. After terminating diapause in winter, the larvae resume their development at temperatures above a lower limit (developmental zero). Chapter 16 summarizes the latitudinal cline in developmental zero of post-diapause development and the difference in the thermal constant among local populations. The author discusses the temporal relationship between seasonal occurrence of newly killed (from pine wilt disease) host trees and oviposition of *M. alternatus* in different localities to understand the differences in pine wilt disease epidemiology.

This part also reviews life-history traits of *M. alternatus* such as adult longevity, fecundity, flight performance, spatial distribution of immature stages, larval cannibalism, sound production, and key stages responsible for the survival rate of the immature stage. Chemicals affect adult behavior; for example, (1) acetone extracts from 1-year-old twigs of *Pinus thunbergii* elicit a strong feeding response from adult beetles, (2) ethane and saturated hydrocarbons with straight C_5–C_{10} chains that are contained in *P. densiflora* foliage show a strong feeding repellent activity for adult beetles, (3) monoterpenes and ethanol emitted from dying and newly killed pine trees attract reproductively mature beetles, and (4) secretion from the spermathecal gland of beetle female deters the oviposition. Some of chemicals associated with the beetle's life history are introduced in Chap. 16. The prevalence of recently killed pine trees as larval food resources is considered to be one of the most important factors in determining the population growth of *M. alternatus*. Chapter 16 discusses the relationship between the life-history traits of *M. alternatus* and the different, seasonal occurrence patterns of newly killed pine trees before and after the PWN was introduced into Japan.

More knowledge is needed about PWN transmission biology before a complete understanding can be obtained regarding the pine wilt epidemic. Chapter 17 introduces the effect of PWNs on the longevity and flight performance of adult vector beetles. *Monochamus alternatus* adults with a heavy load of PWNs live for a shorter time than adults with a light load. A heavy PWN load reduces the flight performance of *M. carolinensis*; however, vector beetles with a heavy PWN load can induce pine wilt incidence on their own. Chapter 17 deals with the temporal pattern of PWNs transmitted per unit time from a beetle to pine twigs and the effects of temperature on PWN transmission. The PWN transmission curve of insect vectors changes from being L-shaped to unimodal in the early stages of a disease outbreak in a *P. densiflora* stand. This change is discussed in relation to the virulence-transmission association. This chapter does not deal with chemicals that are emitted from beetles and pine trees and which partly control the PWN's entry into and departure from beetles. This is reviewed in Chap. 13 of this book.

Chaper 17 also deals with the epidemiology of pine wilt disease. The spread pattern of PWN has so far been analyzed by modeling and analytical approaches. The spread pattern of the PWN within a pine stand is determined mainly by short-distance flight of the vector beetles relating to reproductive maturity, the effect of PWN load on the transmission pattern and beetle performance, and a positive, spatial correlation of wilt-diseased trees in two consecutive years. The spread pattern of PWN over pine stands is determined by the long-distance flight of beetles. The spread rate of PWN increases from zero with the increasing proportion of long-distance dispersers and abruptly reduces beyond a certain proportion due to the Allee effect. The spread of PWN over prefectures, Japanese administrative districts with a mean area of 8,038 km^2, is caused by humans transporting pine logs infested with PWNs and insect vectors. The mechanisms and rates of PWN spreading at various levels of spatial scales may be indicative for the control of pine wilt disease.

Insect vectors of the PWN are considered to have increased their range on a geological scale by flight. Now they can move by flight and transportation related to human activities. The genetic structure of a beetle population is made up by gene flow, mutation, genetic drift, and natural selection. The interactions among the PWN, beetle vectors, and trees in a pine stand are likely to change the respective ecological traits, resulting in the evolution of a pine wilt system. Thus, it is necessary to determine the genetical relationships between vector populations and those between beetles within a population. DNA markers are helpful tools to elucidate the genetic structure. Chapter 18 introduces molecular-ecology studies using two different resolution levels of DNA markers, mitochondrial DNA and microsatellite regions of nuclear DNA, to successfully represent the phylogenic relationship among *M. alternatus* populations in East Asia and the northward expansion pathways in the Tohoku district of Japan. Chapter 18 also describes a method for developing microsatellite markers.

16
Vector–Host Tree Relationships and the Abiotic Environment

Katsunori Nakamura-Matori

16.1 Introduction

Some cerambycid beetles of the genus *Monochamus* (Coleoptera: Cerambycidae) are known vectors of the pine wood nematode (PWN). These include *M. alternatus* in East Asia (Mamiya and Enda 1972; Morimoto and Iwasaki 1972; Lee et al. 1990; Yang 2004), *M. saltuarius* in Japan (Sato et al. 1987), *M. carolinensis* in North America (Linit et al. 1983), and *M. galloprovincialis* in Portugal (Sousa et al. 2001). Among these vectors, *M. alternatus* has been most intensively investigated, especially in Japan, because of its relatively long history and importance as a vector of the PWN. The purpose here is to review the literature focusing on the biology of *M. alternatus*.

Monochamus beetles are secondary insects that can attack only weakened or dying trees (Cesari et al. 2005). *Monochamus alternatus* is thought to have been a rare insect in Japan until the introduction of the PWN about 100 years ago. The life-history traits of *M. alternatus* acquired before the invasion of the PWN are, however, considered to be helpful in clarifying the outbreak and PWN epidemics (Togashi 2002, 2006b). Here an outline of the geographical distribution and life history of *M. alternatus* will be presented, followed by a discussion on the features of its food resources, weakened or dying host trees that strongly affect the development of the insect's life-history traits. As well, the effects of abiotic conditions on *M. alternatus* resource availability will be discussed in relation to the bionomics of the insect and pine wilt disease symptom development in host trees.

Tohoku Research Center, Forestry and Forest Products Research Institute, 92-25 Nabeyashiki, Shimo-Kuriyagawa, Morioka 020-0123, Japan

Tel.: +81-19-648-3962, Fax: +81-19-641-6747, e-mail: knakam@ffpri.affrc.go.jp

16.2 Geographical Distribution of *Monochamus alternatus* and Other Vectors

Monochamus alternatus was originally described by Hope (1982) based on specimens from the Zhoushan islands in Eastern China (Zhejiang Province) (Makihara 1997). *Monochamus tesserula* described by White (1858) is a synonym of *M. alternatus*. It occurs in Japan, Korea, Taiwan, continental China, northern Laos and northern Vietnam (Makihara 2004; Fig. III.1). In Japan, *M. alternatus* is present on three of the major islands, Honshu, Shikoku, and Kyushu, and small nearby islands, but the presence of *M. alternatus* in the northernmost area of Honshu before the pine wilt disease epidemic is questionable because the cool temperatures in those areas are unfavorable for the insects' oviposition and embryonic development (Takizawa 1982). In some of the isolated islands, such as Ogasawara, *M. alternatus*, accompanied by the PWN, is thought to have been introduced by human activity, that is, the transportation of contaminated wood material (Makihara and Enda 2005). Although PWN was introduced to Okinawa Island in southwestern Japan in 1973 (Kuniyoshi 1974), collection records of the 1940s indicate that *M. alternatus* is native to Okinawa (Makihara and Enda 2005). In Korea, *M. alternatus* was introduced to Pusan City probably from Japan and its range is expanding in the Korean Peninsula, although the insect is indigenous to Jeju-do Island (Makihara and Enda 2005). Though Makihara (2004) proposed dividing *M. alternatus* into two subspecies, the continental China-Taiwanese group and the Japan-Korean

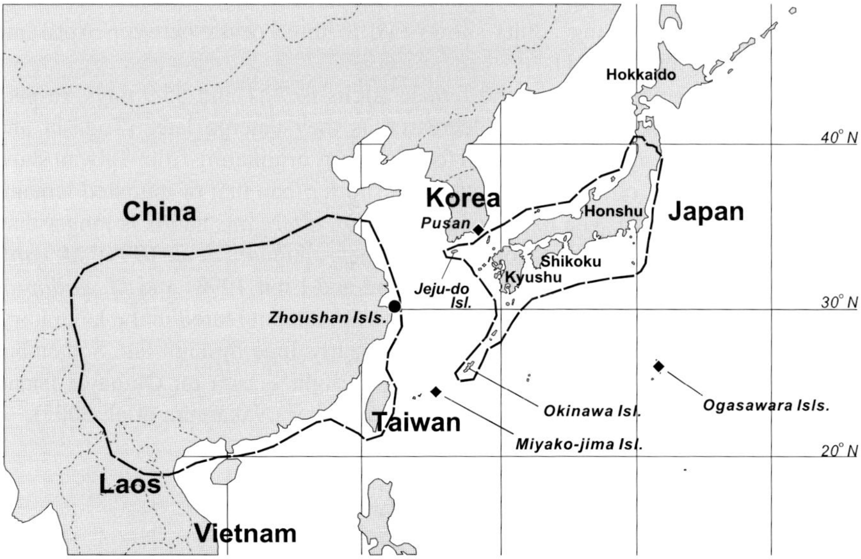

Fig. III.1 Geographical distribution of *Monochamus alternatus* (after Makihara 2004). Areas enclosed by *dotted lines* show the estimated natural distribution of *M. alternatus*. *Solid squares* indicate population spread by human activities

group, based on different morphological features, the two groups are not genetically distinct from one another (Kawai et al. 2006).

Monochamus saltuarius has been found in Finland, northern Italy, Russia, northeastern China, Korea, Sakhalin, and Japan, suggesting a wide distribution across Eurasia. In Japan, its distribution is considered to be rather restricted to the cool summer areas in Honshu, Shikoku and a part of Kyushu, but not in the northernmost part of Honshu and Hokkaido (Makihara 1997). *Monochamus galloprovincialis* is widely distributed in Eurasia through North Africa, but not in Japan (Makihara and Enda 2005). The distribution of *M. carolinensis* is restricted to North America. It has been collected from southeastern Canada through Mexico, including the east coast and mid-western USA.

16.3 Life History of *Monochamus alternatus*

16.3.1 Adult Emergence and Flight Season

Monochamus alternatus has 1- and 2-year life cycles in Japan (Kishi 1995) where adults emerge from dead host trees once a year, in late spring through summer. The emergence period differs among locations; early April through August in Okinawa in the southwestern islands (Irei et al. 2004), May through July (Togashi and Magira 1981) or May through August (Kishi 1995) in central Japan, and late June to August in northern Honshu (Chida and Sato 1981; Hoshizaki et al. 2005). In southern China, *M. alternatus* has two to three generations per year, thus, the seasonal change in the number of emerging adults shows two to three peaks between April and November (Song et al. 1991).

The mean longevity of male and female adults is 70.1 and 65.9 days, respectively, in outdoor cages, and this decreases as they emerge later (Togashi and Magira 1981). In the laboratory, mated female adults originating from Taiwan show an average longevity of 179.7 days, which is shorter than that of unmated females (Zhang and Linit 1998). The mean longevity of *M. alternatus* adults is longer than that of *M. saltuarius* (57.3 days for fertile females; Jikumaru et al. 1994), *M. carolinensis* (103.4 days for mated females; Zhang and Linit 1998) and *M. galloprovincialis* (64.0 days for mated females; Naves et al. 2006c) reared in the laboratory. In central Japan, *M. alternatus* adults occur in early June through late September in the field (Shibata 1981; Togashi 1988). The flight season on Okinawa Island starts in April and lasts through October or November (Nakamura et al. 2005).

16.3.2 Adult Activities and Fecundity

Monochamus alternatus adults are reproductively immature at emergence. They feed on the bark of pine twigs or other conifers for survival and sexual maturation (Fig. III.2). This is often referred to as "maturation feeding". The pre-oviposition

Fig. III.2 Life cycle of *Monochamus alternatus*. **A** female adult coming out of an emergence hole (*arrow*), **B** feeding wounds by adult beetles on a *Pinus thunbergii* twig, **C** oviposition scars (indicated by the *arrows*), **D** an egg laid in the inner bark, **E** a feeding larva and frass under the bark, **F** a mature larva in a pupal chamber (*arrows* show the entrance hole) (see Color Plates)

period is 16–30 days (Enda and Nobuchi 1970), although the first egg deposition has been recorded as early as 6 days after emergence in a warm area (Ido and Takeda 1974) and as late as 61 days in a cool area (Takizawa 1982). The development rate of the female's ovary is affected by the quality of food: Ovary development is faster for females provided with current-year twigs of *Pinus densiflora* than those provided with 1- or 2-year-old twigs (Katsuyama et al. 1989). Males need about 5 days after emergence to inseminate females (Nobuchi 1976). Feeding response is elicited by extract of pine twig bark (Miyazaki et al. 1974) and is inhibited by ethane and other chemical components of pine needles (Sumimoto et al. 1975).

Immature adults randomly disperse by flying (Togashi 1990c) whereas mature adults are strongly attracted to volatiles emitted from dying or newly killed trees (Ikeda and Oda 1980; Ikeda et al. 1980b, 1981), and as a result are concentrated around such trees (Shibata 1986; Togashi 1989b). On those trees, they mate, and oviposit. A tethered flight experiment indicated that reproductively immature adults have higher flight activity than mature adults (Ito 1982), and the laboratory result was supported by a mark-and-recapture study in the field (Togashi 1990b). Adults move by walking or short-range flights in the pine canopy, where the distance traversed is estimated to be 7–40 m week^{-1} (Togashi 1990c) or 10.6–12.3 m during their lifetime (Shibata 1986). The beetles sometimes disperse over a long distance. A laboratory experiment with a flight mill recorded a maximum distance of 3.3 km or a maximum duration of 58 min of continuous flight (Enda 1985). Under field conditions, a released adult was recaptured 2.4 km away from the release point 4 days later (Ido et al. 1975). Wild adults flew over the sea to bait pine logs set on pine-free islands, showing a flight of at least 3.3 km from the nearest pine forest infested with *M. alternatus* (Kawabata 1979). These results demonstrate that *M. alternatus* adults can disperse over long distances of 2–3 km by flight, though such dispersal may be infrequent. A mathematical model indicated that long-distance dispersal of *M. alternatus* contributes greatly to the spread of pine wilt disease (Takasu et al. 2000).

Both males and females are attracted to the volatiles of the weakened host trees (Ikeda et al. 1980a,b), which likely helps them find a mating partner. Males have a volatile pheromone to attract females and both sexes have a contact pheromone on the body surface that elicits copulatory behavior in males (Kim et al. 1992). Thus, in mating, males are initially passive, just waiting for a female, and when a female touches his body, he dashes forward and mounts her (Fauziah et al. 1987). They form a long period of pair bond (23–390 min) and copulate repeatedly even after the female starts oviposition (Fauziah et al. 1987). Adults of both sexes copulate with several mates throughout their lifespan, suggesting that *M. alternatus* has a polygynous mating system (Fauziah et al. 1987). In addition, a female adult can fertilize all eggs laid in her lifespan when inseminated once (Nobuchi 1976). Sperm removal behavior like short-time penis insertion (Yokoi 1989) was not observed for congeneric *M. saltuarius* (Kobayashi et al. 2003). Kishi (1995) suggested that pheromones emitted from *M. alternatus* adults of both sexes are dispersed over a wide range in the field, which has not been proven.

Larvae of *M. alternatus* can grow only in dying or newly killed pine trees, thus female adults try to find and lay their eggs on such trees. This is why adults are attracted to volatiles emitted from weakened trees. When ovipositing, the female adult makes a slit-like wound on the bark surface with her mandibles, then turns 180° to insert the ovipositor through the center of the wound into the inner bark (Nishimura 1973; Fig. III.2). When the outer bark is thick enough the oviposition site may appear as a cone-shaped pit. In most cases females deposit a single egg or no eggs in the inner bark through a wound; less frequently they lay two to three eggs. The mean number of eggs per oviposition scar (oviposition ratio) is about 0.5 (Ochi and Katagiri 1979; Togashi and Magira 1981). Oviposition is stimulated by some compounds present in the inner bark of pine trees (Islam et al. 1997; Sato et al. 1999a,b), and is inhibited by a jellylike secretion deposited by female adults on oviposition scars as well as larval frass (Anbutsu and Togashi 2001, 2002).

An age-specific fecundity curve (m_x curve) of female adults reared under outdoor conditions is unimodal (Togashi and Magira 1981). The mean lifetime fecundity of early emerged females (157.3 eggs) was greater than that of mid- and late-emerged females (78.0 and 23.5 eggs, respectively) (Togashi and Magira 1981). The mean lifetime fecundity of mated *M. alternatus* originating from Taiwan under laboratory conditions was 581.0 eggs (Zhang and Linit 1998), greater than that of

M. saltuarius (69.7; Jikumaru et al. 1994), *M. carolinensis* (451.3; Zhang and Linit 1998) and *M. galloprovincialis* (67.0; Naves et al. 2006c). Lifetime fecundity differs greatly among *M. alternatus* female adults and correlates positively with body size (Togashi 1997).

Oviposition scars show a clumped distribution among trees (Shibata 1984), because the number of oviposition scars per tree varies depending on the time when the trees are weakened (Togashi 1989b). A clumped distribution pattern of oviposition scars is also observed among parts of a tree trunk (Shibata 1984). Female adults tend to avoid parts of tree trunks with thick outer bark because it requires much time and energy to make an oviposition site (Nakamura et al. 1995a,b). As a result, more oviposition scars tend to be in the middle and upper parts of trunks (Yoshikawa 1987; Nakamura et al. 1995b). The number of oviposition scars in a unit area of the bark surface, however, shows a uniform distribution, probably resulting from deterrent oviposition at already-occupied sites by other eggs (Shibata 1984). The inhibitory effect of the jellylike secretion deposited in oviposition scars and larval frass on oviposition behavior may be responsible for this phenomenon (Anbutsu and Togashi 2001, 2002).

Adults are nocturnal (Nishimura 1973). Feeding is observed in the daytime as well as at night, while movement (flight and walking) and reproductive behavior (mating and oviposition) occur mostly at night (Nishimura 1973; Kichiya and Makihara 1991), although the beetles tend to be inactive from midnight through dawn, probably because of the low temperatures (Fauziah et al. 1987; Kichiya and Makihara 1991).

16.3.3 Development of Immatures

It takes 5 to 6 days for eggs to hatch (Ochi 1969). Hatchability decreases sharply when the daily average temperature is below 15°C (Takizawa 1982). Eggs are often killed by oleoresin from host trees (Togashi 1990a) that are weakened but probably maintain a low oleoresin exudation. Larvae grow by feeding on the inner bark and later partly on the sapwood. The larval gallery under the bark is packed with frass, which is a mixture of excreta and wooden fibrous shreds (Fig. III.2). Larvae under the bark make a clicking sound (Izumi and Okamoto 1990; Izumi et al. 1990), but the mechanism and function of this sound are unknown. *Monochamus sutor* larvae seem to make a sound by scratching the bark with their mandibles, and the sound could serve in protecting food resources from conspecific competitors (Victorsson and Wikars 1996). *Monochamus alternatus* larvae develop through four instars (Morimoto and Iwasaki 1974a; Yamane 1974a). Most fourth instar and some third instar larvae make tunnels in the xylem in late summer through autumn (Togashi 1989c), which have an oval-shaped entrance hole on the xylem surface and an elongated tunnel upward in the trunk (Fig. III.2). The terminal end of the tunnel usually curves toward the bark surface, thus the longitudinal section of the tunnel has a U-shape. Fully grown fourth-instar larvae make a pupal chamber at the terminal end by packing wooden fibrous shreds toward the entrance hole, and overwinter there. Some larvae living under thick bark excavate depressions on the xylem surface and pupate there (Togashi 1980). The depth of the pupal chamber from the bark surface is mostly less than 10 mm, and tends to be deeper in cool areas (Kishi et al. 1982).

Pupation occurs after overwintering. The pupal stage lasts for 12–13 days at 25°C (Yamane 1974b). Newly eclosed adults require about 1 week for sclerotization. Sclerotized adults gnaw a round hole (ca. 1.0 mm in diameter) from the pupal chamber to the bark surface and exit to the outside (Fig. III.2).

16.3.4 Survival Rate of Immatures

The survival rate of *M. alternatus* in dead host trees, from egg to adult emergence, is as high as 28.9% (Ochi and Katagiri 1979) or 24.9% (Togashi 1990a; Table III.1). Key factor analysis shows that the third and fourth larval instars in the pupal chamber are the stages responsible for the fluctuation in the survival rate (Togashi 1990a). When the density of oviposition scars per unit area (approximately proportional to the egg density) of the bark surface increases, the density of emerging adults first increases and then saturates at a certain level (Morimoto and Mamiya 1977; Togashi 1986; Fig. III.3). This density-dependent process is materialized by deadly interference among the larvae and exploitative competition for food resources (Togashi 1986). Typically, the upper limit of the number of emerging adults is estimated to be about 20 per 1 m^2 bark surface (Morimoto and Mamiya 1977). *Monochamus alternatus* has many natural enemies such as entomogenous microorganisms, parasitoids, predatory insects, and woodpeckers (Table III.2). Natural enemies such as insect predators and parasitoids are density-independent mortality factors (Togashi 1986). Some natural enemies including the insect pathogenic

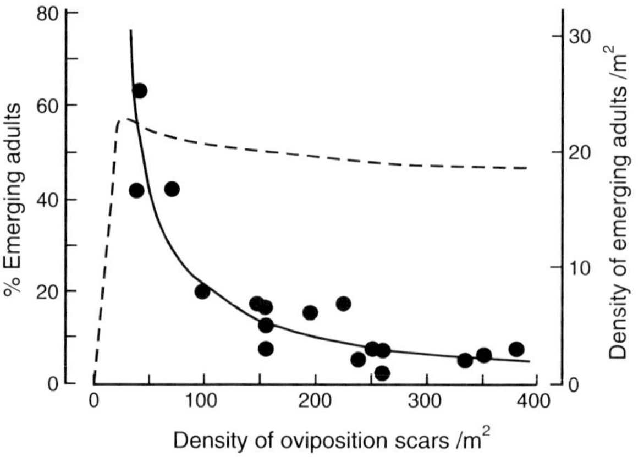

Fig. III.3 Relationship between densities of oviposition scars and emerging adults of *Monochamus alternatus* per 1 m^2 on the surface of the bark on pine logs (reproduced from Morimoto and Mamiya 1977, with permission). *Solid* and *dotted line* show the trends in the percentage of emerging adults to the number of oviposition scars and number of emerging adults, respectively

Table III.1 Life tables of *Monochamus alternatus* within dead *Pinus thunbergii* trees in Ishikawa, central Japan (Togashi 1990a)

Age class (x)	Causes of mortality (d_xF)	1980–1981 generation					1981–1982 generation					1982–1983 generation					1983–1984 generation					1980–1984 generations				
		l_x	$100q_x$	d_x	l_A	l_B	l_x	$100q_x$	d_x	l_A	l_B	l_x	$100q_x$	d_x	l_A	l_B	l_x	$100q_x$	d_x	l_A	l_B	l_x	$100q_x$	d_x	l_A	l_B
Egg		1,000.0					1,000.0					1,000.0					1,000.0					1,000.0				
	Resin		5.2	6				0.5	1				3.1	3				0.0	0				2.2	10		
	Unknown		28.7	33				17.9	34				18.8	18				31.0	18				22.4	103		
	Total		33.9	39	76	6		18.4	35	155	0		21.9	21	75	0		31.0	18	40	0		24.6	113	346	6
$L_{1,2}$		660.9					815.8					781.2					689.7					753.8				
	Bird predation		0.0	0				0.0	0				0.0	0				2.5	1				0.3	1		
	Unknown		7.8	4				7.9	10				16.9	13				17.5	7				11.5	34		
	Total		7.8	4	47	18		7.9	10	117	28		16.9	13	64	7		20.0	8	32	0		11.9	35	260	53
$L_{3,4}$ (Before completion of PC)		609.0					751.6					649.4					551.7					664.4				
	Conspecific bites		4.2	3				0.8	1				1.3	1				5.6	1				2.0	6		
	A. initiator		0.0	0				1.6	2				2.6	2				0.0	0				1.4	4		
	Insect predation		0.0	0				0.0	0				1.3	1				11.1	2				1.0	3		
	Bird predation		0.0	0				1.6	2				0.0	0				5.6	1				1.0	3		
	Unknown		11.1	8				15.9	20				31.2	24				11.1	2				18.4	54		
	Total		15.3	11	61	17		19.8	25	101	68		36.4	28	49	11		33.3	6	12	4		23.9	70	223	100
$L_{3,4}$ (in PC)		516.0					602.4					413.2					367.8					505.7				
	Failure to pupate		1.2	1				1.0	4				0.8	1				0.0	0				0.9	6		
	Conspecific bites		0.0	0				0.3	1				0.0	0				0.0	0				0.1	1		
	Insect predation		9.4	8				3.8	15				7.5	10				3.7	3				5.2	36		
	Unknown		42.4	36				10.6	42				7.5	10				23.7	19				15.4	107		
	Total		52.9	45	40	5		15.7	62	334	27		15.8	21	112	6		27.5	22	58	9		21.6	150	544	47
Pupa (in PC)		242.8					508.1					348.0					266.7					396.4				
	Insect predation		0.0	0				0.5	1				0.0	0				2.6	1				0.6	2		
	Unknown		9.4	3				4.0	8				4.2	3				12.6	5				5.6	19		
	Total		9.4	3	29	0		4.5	9	189	0		4.2	3	69	0		15.4	6	33	0		6.2	21	320	0
Adult (in PC)		220.1					485.0					333.5					225.6					372.0				
	Failure to emerge		0.0	0				1.6	3				1.4	1				0.0	0				1.2	4		
	Insect predation		3.4	1				0.0	0				0.0	0				0.0	0				0.3	1		
	Unknown		37.9	11				37.0	70				20.3	14				18.2	6				31.6	101	214	0
	Total		41.4	12	17	0		38.6	73	116	0		21.7	15	54	0		18.2	6	27	0		33.1	106		
Adult (in flight)		129.0					297.7					216.0					184.6					248.7				

l_A shows the number of insects in more advanced age classes than the age class for estimation of mortality rate. l_B and d_x shows the numbers of live and dead insects in the age class concerned. Mortality rate ($100q_x$) was calculated as $\{d_x/(l_A + d_x)\} \times 100$. PC in age class column represents pupal chamber in xylem. In causes of mortality, insect predation includes predation by *Trogassita japonica*, *Stenogostus umbratilis* and *Gonolabis marginalis*

Table III.2 List of natural enemies of the pine sawyer beetle *Monochamus alternatus*

Category	Order: Family	Species	Stage(s) attacked	Literature
Entomogenous microorganisms				
		Serratia marcescense	L	e
		Serratia spp.	L	d, f
		Beauveria bassiana	L, P, A	d, e, f
Parasitoides				
	Hymenoptera			
	Braconidae	*Atanycolus initiator*	L	f
		Spathius sp.	L	f
		Doryctus sp.	L	e
		Ecphylus hattorii	L	e
		Iphiaulax impostor	L	e
	Bethylidae	*Sclerodermus nipponicus*	L	f
		Sclerodermus guani	L	g
	Pteromarlidae	*Cleonymus* sp.	L	f
	Ichneumonidae	*Dolichomitus* sp.	L	f
		Megarhysa sp.	L	e
	Coleoptera			
	Bothrideridae	*Dastarcus helophoroides*	L, P, A	e
		Dastarcus kurosawai	L	h
	Diptera			
	Tachinidae	*Billaea* sp.	L	f
Insect predators				
	Hymenoptera			
	Formicidae	*Monomorium intrudens*	E	a, e
	Coleoptera			
	Elateridae	*Stenagostus umbratilis*	L	f
		Paracalais berus	L	e
		Paracalais larvatus	L	h
	Histeridae	*Platysoma lineicollis*	L	e
	Rhizophagidae	*Mimemodes emmerichi*	E	e
		Rhizophagus sp.	E	f
	Trogossitidae	*Trogossita japonica*	L, P, A	a, e
	Cleridae	*Thanasimus lewisi*	L	a, e
	Hemiptera			
	Reduviidae	*Velinus nodipes*	A	e
	Neuroptera			
	Inoceliidae	*Inocellia japonica*	L	e
	Dermaptera			
	Anisolabididae	*Anisolabella marginalis*	L	f
		unidentified	L	e
Chilopoda				
	Scolopendromorpha	unidentified	L	e

(continued)

Table III.2 (Continued)

Category	Order: Family	Species	Stage(s) attacked	Literature
Arachnida				
	(Spiders)		A	b
Woodpecker				
	Piciformes		L	c
	Picidae	*Dendrocopos major*		

E eggs, *L* larvae, *P* pupae, *A* adults
a: Ochi and Katagiri (1979)
b: Morimoto and Mamiya (1977)
c: Igarashi (1980)
d: Katagiri and Shimazu (1980)
e: Taketsune (1983)
f: Togashi (1989c)
g: Zhang and Song (1991)
h: C. Kiyuuna (personal communication)

fungus, *Beauveria bassiana* (Okitsu et al. 2000; Shimazu and Sato 2003), parasitoid insects such as *Sclerodermus guani* (Zhang and Song 1991) and *Dastarcus helophoroides* (Urano 2004), and the Great Spotted Woodpecker, *Picoides major* (Yui et al. 1993) have been intensely studied for the use in controlling *M. alternatus.*

Monochamus alternatus immatures can be reared aseptically on an artificial diet containing sliced or minced host tree tissue (Kosaka and Ogura 1990).

16.4 Predictability of Food Resources for *Monochamus alternatus*

Togashi (2002, 2006a) pointed out the spatio-temporal unpredictability of food resources for *M. alternatus* larvae before the invasion of Japan by PWN unlike the adults. He also pointed out that the life-history traits of *M. alternatus*, such as a long adult flight season due to long adult longevity and 1- and 2-year life cycles, are adapted to the unpredictable larval food resources. He then related the life-history traits to the pine wilt disease epidemic. Here his views are outlined along with additional novel information.

16.4.1 Before PWN Introduction

Monochamus alternatus adults fed on the bark of twigs of healthy or weakened pine trees. In general, pine twigs in a pine stands are abundant and considered not to be limited food resources regardless of their long life span. In addition, the adults can feed on various species of the genus *Pinus* (Furuno and Uenaka 1979) or non-pine conifers such as trees of the family Pinaceae, Japanese cedar (*Cryptomeria japonica*) and cypress (*Chamaecyparis obtusa*) (Yamane and Akimoto 1974; Nakamura and Okochi 2002). This flexibility of food selection may be helpful for adults lacking pine trees to survive until they find a new safe site after wandering around. Consequently, food shortage rarely occurs for *M. alternatus* adults.

On the other hand, *M. alternatus* larvae feed on the inner bark of weakened or newly killed trees in the family Pinaceae, especially the genus *Pinus* (Table III.3). They can not survive in live host trees because of oleoresin, the main defense agent

Table III.3 List of trees species on which *Monochamus alternatus* larvae live

Genera	Species	Literature
Pinus	*armandii* var. *amamiana*	f
	banksiana	a, b
	densiflora	a, b
	elliottii	a, c
	engelmannii	c
	greggii	c
	khasya	c
	koraiensis	a
	leiophylla	c
	luchuensis	a, b
	massoniana	a, c
	nigra	c
	palustris	c
	palustris	e
	pinaster	d
	ponderosa	c
	pungens	b
	radiata	c
	strobus	a, b
	sylvestris	c
	taeda	b, c
	thunbergii	a, b
Abies	*firma*	a, b
Cedrus	*deodara*	a, b
Larix	*kaempfer*	a
Picea	*abies*	a, b
	jezoensis var. *hondoensis*	a, b
	morinda	a, b

a: Kojima and Okabe (1960)
b: Kojima and Nakamura (1986)
c: Furuno (1972)
d: Mineo and Kontani (1973)
e: Kishi (1980)
f: Akiba et al. (2000)

in conifers. Trees that passed several months or more after death are also unsuitable for *M. alternatus* larvae because their inner bark is already occupied by other subcortical insects such as weevils and scolytids (Yoshikawa et al. 1986) or because the inner bark is completely exploited. Also, those trees may not satisfy the nutrition requirements for *M. alternatus* larval development. The occurrence of weakened or newly dead host trees are supposed to be relatively rare in pine forests unless pine wilt disease has spread. Although stressed pine trees resulting from shading by other trees constantly occurs in forest settings, the total amount is limited and not all trees are weakened during the flight season (Togashi 2006a). Many dead trees can be supplied by disasters such as forest fires, drought, wind and snow damage. Those events are, however, rare and quite unpredictable. We have pests and diseases of pine trees such as lepidopteran defoliators and *Armillaria* fungi, but these are usually not as devastating as pine wilt disease. Over all, we can regard food resources for *M. alternatus* larvae as rare or unpredictable, or both, in time and space, and this unpredictability of food resources should have worked as a strong determinant for *M. alternatus* to maintain the population (Togashi 2006a). In fact, it is said that *M. alternatus* was seldom captured before the pine wilt disease epidemic in Japan (Nawa 1937), suggesting that the unpredictability of food resource prevented their outbreak.

To utilize unpredictable food resources, *M. alternatus* seems to have refined its lifestyle. The long adult flight season (Shibata 1981; Togashi 1988) owing to long adult longevity (Togashi and Magira 1981; Zhang and Linit 1998) would increase the chance of encountering the few available weakened trees. The adults would reach host trees suitable for oviposition by their highly developed sensory system and flight ability. They have high reproductive potential (Ochi and Katagiri 1979; Togashi 1990a), so that they can surely produce offspring when they encounter scarce food resources. One- and 2-year life cycles within a population (mentioned later) lower the probability of extinction due to a lack of newly dead trees in a year (Togashi 2006b). Varying developmental times has a stabilizing effect on population density under unstable environmental conditions (Takahashi 1977).

16.4.2 After PWN Introduction

PWN is thought to have been introduced into Japan in the early 1900s (Mamiya 1983) from North America (de Guiran and Brugier 1989; Tarès et al. 1992a,b), whereas *M. alternatus* is indigenous to Japan. It can be inferred that the introduced nematode could easily establish a mutual relationship with *M. alternatus* once they co-occurred in a weakened tree, because they both originally had relationships with equivalent counterpart organisms of the same genus (i.e., North American *Monochamus* beetles such as *M. carolinensis* for the PWN, and the nematode *B. mucronatus* for *M. alternatus*). Unfortunately, the PWN is highly pathogenic to Japanese pine trees such as *P. densiflora* (Japanese red pine), *P. thunbergii*

(Japanese black pine) and *P. luchuensis* (Ryukyu pine). As a result, *M. alternatus* utilizes a massive number of weakened and dying trees killed by the PWN.

Epidemics of pine wilt disease caused by the PWN change the weakened pine trees from unpredictable to predictable resources. In central Japan, infection of the PWN on pine trees occurs only in the adult flight season of *M. alternatus*, primarily during summer, thus tree death resulting from nematode infection is mostly found in mid-summer through autumn (Togashi 1989b). In such a case, most newly killed trees are ready for oviposition of *M. alternatus*. The trees killed from pine wilt disease show contagious distribution probably because of the attractiveness of newly dead trees to reproductively mature *M. alternatus* adults (Togashi 1991b). Moreover, a positive, spatial correlation is observed between trees diseased the previous year and those diseased in the early season of the current year (Togashi 1991b). This may be caused by the latent infection of the PWN (Futai 2003a; see also Sect. 23.4) or transmission of the PWNs through grafted roots (Tanaka 2003). The occurrence of killed trees from pine wilt disease is predictable in time and space. In addition, the number of newly dead trees in PWN-infested pine stands is far greater than in uninfested forests. Yearly loss of pine trees often accounts for over 30% in PWN-infested stands and less than 10% in uninfested stands (Kishi 1995).

After the introduction of PWN, the occurrence pattern of newly dead pine trees that serve *M. alternatus* larvae as food resources changed completely. Life-history traits adapted to limited food resources for immature stages allow the prosperity of *M. alternatus* by using newly dead trees, and make it difficult for us to control pine wilt disease.

16.5 Temperature Effect on Development and Adult Prevalence of *Monochamus alternatus*

As an ectothermal organism, the development of *M. alternatus* is subject to ambient temperature. In Japan, *M. alternatus* has a 1-year life cycle (univoltine) when the larvae have adequate time for development at a suitable temperature. In areas with a short, cool summer, the larvae cannot complete their development within one season and thus have a 2-year life cycle (semivoltine) (Togashi 1986, 1989a,c). The two different life cycles are observed within a single *M. alternatus* population and the proportion of 1-year life cycles tends to decrease with an enhanced latitude or altitude with cooler summers (Kishi 1995). Japanese populations of *M. alternatus* have an obligate larval diapause (Togashi 1989a). Facultative diapause is observed in the southern part of the range (Enda and Kitajima 1990), resulting in two generations in a year (bivoltine) in Taiwan (Enda and Makihara 2006) and two or three generations in subtropical Guangdong Province, China (Song et al. 1991). In Okinawa, an island off southwestern Japan, the *M. alternatus* population is not bivoltine, and thus seems to have an obligate diapause, regardless of the subtropical climate (Kosaka et al. 2001a; Irei et al. 2004; Enda and Makihara 2006).

In winter, the first to fourth instar larvae of *M. alternatus* appear (Togashi 1989a). Some are in diapause and others are not. Diapause larvae are fourth instars with a yellowish white to yellow body and no food in the intestine. They enter diapause prior to winter and low temperatures in winter break the diapause (Togashi 1995). The diapause is terminated by mid-February in Ishikawa, central Japan (Togashi 1991c). Diapause larvae do not resume development even under suitable temperature conditions.

Pre-diapause third and fourth instar larvae skip the diapause when they experience low temperatures. They resume feeding in spring and become pupae and adults the following summer (Togashi 1989a,c, 1995). Larvae that overwinter as first and second instars resume feeding after winter and continue to grow. They enter the diapause as final instar before the second winter and terminate the diapause in the second winter; namely, those insects become semivoltine.

The development of post-diapause larvae to pupae and adults is subject to the thermal summation law. Thermal constant and developmental zero for post-diapause development has been estimated for *M. alternatus* populations originating from different locations in Japan as well as Taiwan and South China (Igarashi 1977; Kosaka et al. 2001a; Enda and Makihara 2006; Fig. III.4). Developmental zero shows a latitudinal cline; it tends to increase as the latitude of places where *M. alternatus* populations were sampled decreased. This may indicate the adaptation of *M. alternatus* populations to local climate. In contrast, the thermal constant varies greatly among *M. alternatus* populations, although it is constantly low in populations with facultative diapause. The thermal constant for post-diapause development also shows great fluctuation from year to year even for a population in specified place (Fig. III.5).

Adult emergence of *M. alternatus* has been regarded as one of the most important pieces of information needed for determining the timing of spraying of insecticides for controlling pine wilt disease (Nakamura and Yoshida 2004; Yoshida 2006). Many efforts have been made to develop a prediction system for adult emergence time on the basis of thermal summation law, but they have not been fruitful. The uncertainty of the values mentioned above could be one reason. In addition, thermal conditions for immatures in dead pine trees subject to factors relating to the micro-habitat, for example, location of pupal chamber in the trunk (depth, sunny or shaded, etc.). Estimating the developmental zero and thermal constant for *M. alternatus* local populations is significant to better understand their life history, but is not sufficient to predict adult emergence time, especially under field conditions.

16.6 Different Availability of Host Trees in Different Locations

The occurrence pattern of weakened and newly killed trees, caused by pine wilt disease, changes in different locations, as does the adult emergence time (Fig. III.6). In general, high temperatures and low water content of the soil promote

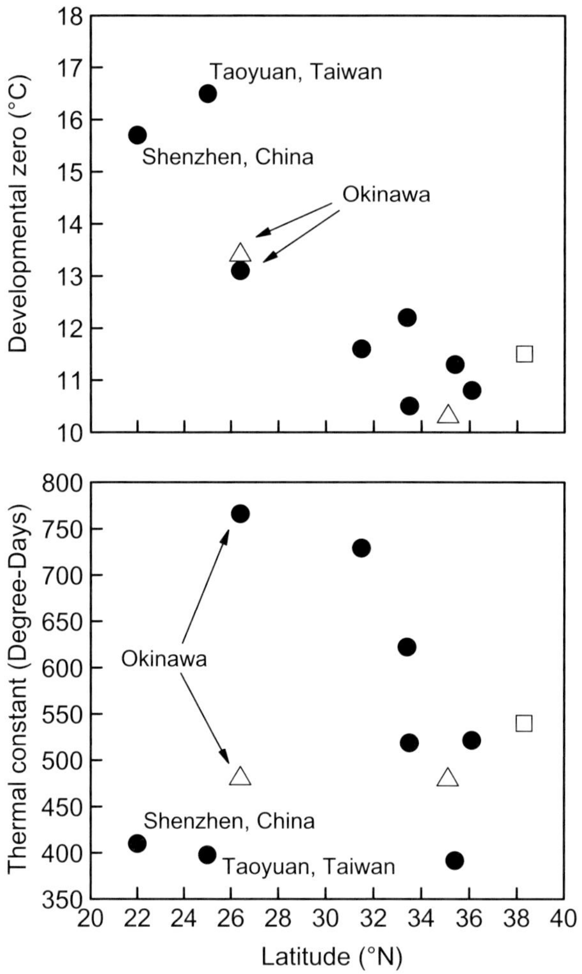

Fig. III.4 Cline of developmental zero (*upper*) and thermal constant of post-diapause development of *Monochamus alternatus*, which was calculated as the thermal sum required for 50% of adults to emerge. *Different symbols* represent data sources: *squares* Igarashi (1977), *triangles* Kosaka et al. (2001a), *solid circles* Enda and Makihara (2006)

disease development (Hotta et al. 1975; Suzuki and Kiyohara 1978). In central and south-western Japan, weakened and newly killed pine trees due to natural infection by the PWN occur mostly in early through mid-summer when hot temperatures and low precipitation prevail. The occurrence of weakened and newly dead trees coincides with the oviposition period of *M. alternatus*, resulting in a high proportion of dead trees infested with PWN and *M. alternatus* (Togashi 1989b).

In cool summer areas at high latitude or altitude, the emergence of *M. alternatus* adults starts in late June to July and ends in August (Chida and Sato 1981; Hoshizaki et al. 2005) resulting in a short flight season, thus, in such areas, PWN transmission is delayed. Moreover, disease development itself is slowed by cool summer temperatures. Consequently, a substantial proportion of diseased trees develop

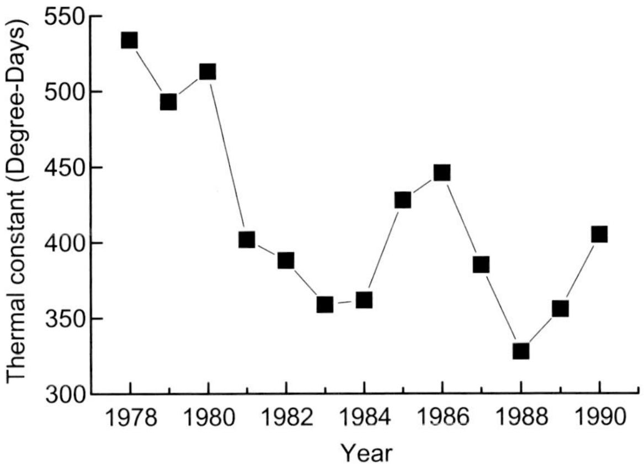

Fig. III.5 Yearly fluctuation in the thermal constant of post-diapause development of *Monochamus alternatus* at Forestry and Forest Products Research Institute at Tsukuba, Ibaraki (Data from Enda 2006). Adult emergence from trees killed from pine wilt disease in the field was recorded every year. The thermal constant was calculated as the thermal sum required for 50% of the adults to emerge using daily average temperatures from meteorological data assuming a developmental zero of 11°C

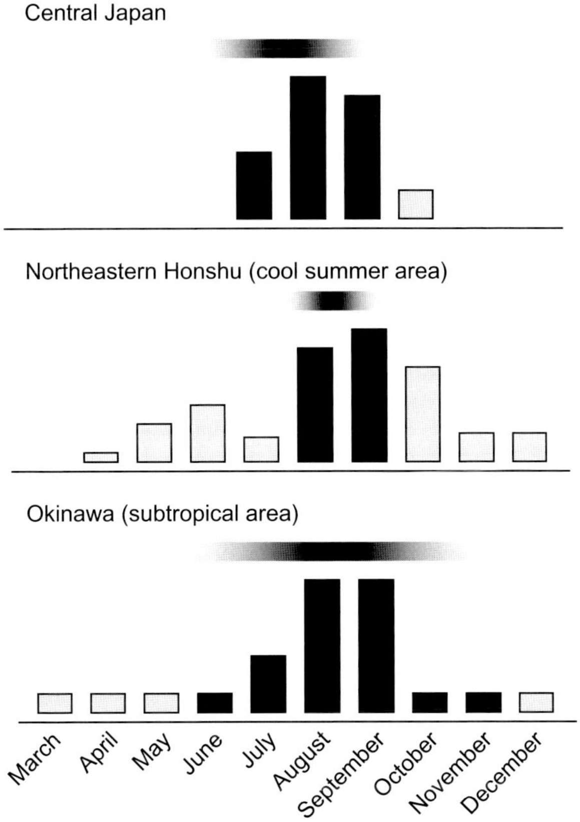

Fig. III.6 Temporal relationships between the prevalence of newly killed pine trees, caused by pine wilt disease, and the oviposition season of *Monochamus alternatus* in different localities. The *vertical bars* indicate the relative abundance of newly killed trees. The oviposition season of the insects in each location is represented by the *horizontal bar* on each graph, in which *dark shading* shows a high oviposition activity. Newly killed trees occurring during the oviposition season are infested with *M. alternatus* (*solid*) and the others are not infested (*gray*)

symptoms after autumn. Chlorosis does not progress in winter because of the suppression of symptom development under low temperatures; it resumes in spring when temperatures rise sufficiently. Asynchronism between the oviposition period and tree mortality lowers the proportion of dead trees infested with *M. alternatus*; however, some newly dead trees in the year after PWN infection are used by *M. alternatus* for oviposition.

In the subtropical climate of Okinawa, pine wilt symptom development progresses even in winter, though the occurrence of newly dead trees has a distinct peak in June to July (Nakamura et al. 2005). The flight season of *M. alternatus* in Okinawa is as long as April through November, but they still can not utilize newly dead trees occurring in winter and early spring. As a result, we can find a numerous dead pine trees without *M. alternatus*, unlike the situation in central Japan with a moderate climate.

Pine wilt disease provides food resources to *M. alternatus*, while the temporal pattern of the occurrence of diseased trees varies owing to different climates and the flight seasons of adult insects, thus, we can conclude that the availability of diseased trees to *M. alternatus* differs depending on the biotic and abiotic conditions relating to pine trees, PWN and vector insects. It seems that the conditions occurring in central and southwestern Japan facilitate the efficient propagation of *M. alternatus* and pine wilt disease epidemics.

16.7 Concluding Remarks

Monochamus alternatus is a secondary insect that attacks only weakened or dying host trees, like other *Monochamus* species (Cesari et al. 2005). To utilize such rare and unpredictable resources, secondary insects should have the life-history traits shown in *M. alternatus*, at least to some extent (Togashi 2006b). Most of them, however, could not cause serious damage such as the mass mortality of trees, merely provoking problems in the wood production process, such as deterioration of wood quality by boring holes or staining, when the trees are stressed or killed by natural disasters or forestry operations. Some scolytid species associated with pathogenic fungi change from secondary to primary pests by mass attack at high population levels (Paine et al. 1997). *Monochamus alternatus* does not need mass attack to kill a pine tree because of the extreme pathogenicity of PWN to susceptible host trees. They seem to have changed from secondary to primary pests just by establishing a relationship with an alien pathogen introduced by humans. The prosperity of *M. alternatus* accompanied by the PWN must be recognized as an exceptional case of biological interaction.

Monochamus alternatus is widely distributed from subtropical to cooltemperate areas (Fig. III.1) and PWN has already spread throughout much of this area. There is considerable difference in the levels of environmental factors among the areas, and such differences are reflected in the development of *M. alternatus*, symptom development in host trees and their interactions (Fig. III.6). Moreover, there are at least two geographically separate groups showing facultative and obligate diapause (Enda and Makihara 2006). We must take into account local differences in the bionomics of the vectors and host response to better understand the epidemic and make a control system against pine wilt disease.

17
Vector–Nematode Relationships and Epidemiology in Pine Wilt Disease

Katsumi Togashi

17.1 Introduction

Pine wilt is a serious, infectious disease of susceptible pines, occuring in North America, East Asia, and Portugal (Yang and Wang 1989; Kishi 1995; Mota et al. 1999). The disease is caused by the pine wood nematode (PWN), *Bursaphelenchus xylophilus*, and is vectored by cerambycid beetles in the genus *Monochamus* (Kiyohara and Tokushige 1971; Mamiya and Enda 1972; Linit 1988; Fig. III.7). *Monochamus* beetles deposit their eggs in PWN-killed trees. The PWN reproduces in such trees and adult beetles acquire the nematode when they emerge from such dead trees, beginning another cycle of the disease.

Pine wilt is thought to have been introduced to Japan from North America in the early 1900s (Mamiya 1988). Since then the disease has spread in Japan because Japanese native pine species, *Pinus thunbergii* and *P. densiflora*, are very susceptible to the disease (Kishi 1995). This fact leads to two interesting subjects: (1) an evolutionary change in the pine wilt system comprising PWN, its insect vectors and host trees during and after the spread in Japan, and (2) mechanisms of PWN spread within a pine stand, between pine stands, and over government administrative districts.

When considering the evolutionary change in the pine wilt system, a great variation in virulence (the ability to kill host trees) among PWN isolates (Kiyohara and Bolla 1990) is important. After PWNs invade a pine stand, they kill pine trees and reproduce therein, therefore virulent PWNs gaining a higher fitness than avirulent PWNs in newly killed trees. Actually, inoculation of a virulent and an avirulent isolate on identical *P. thunbergii* trees revealed a large proportion of virulent isolate in the PWN-killed seedlings (Aikawa et al. 2006). Mean number of PWNs carried by a beetle for a virulent isolate is greater than that for an avirulent isolate in

Laboratory of Forest Zoology, Graduate School of Agricultural and Life Science, The University of Tokyo, Yayoi, Bunkyo-ku, Tokyo 113-8657, Japan

Tel.: +81-3-5841-5217, Fax: +81-3-5841-7555, e-mail: togashi@fr.a.u-tokyo.ac.jp

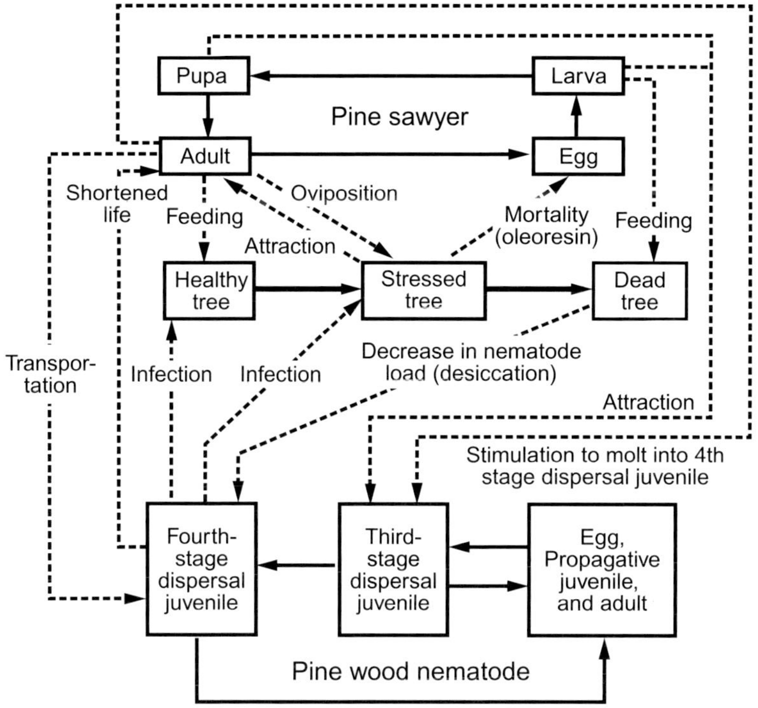

Fig. III.7 The relationships between *Bursaphelenchus xylophilus*, its insect vector, and pine trees in the pine wilt disease system. Solid and broken arrows represent the change in stages or states and the action, respectively (modified from Togashi 1989b)

laboratory (Aikawa et al. 2003). However, the proportion of PWNs transmitted successfully into pine twigs in the PWNs carried by the vector may differ between virulent and avirulent PWNs via their effects on longevity and activity of vector beetles. In addition, surviving pine trees may show a reduced susceptibility to the disease. These factors lead to a discussion on the evolutionary change in the interrelation between vector insects, PWN, and host trees including the virulence of the PWN in the course of pine wilt epidemic within a pine stand. Consequently, the early half of this chapter comprises brief biologies of the vectors, especially *Monochamus alternatus*, and PWN, the effects of PWN on adult longevity and flight performance of *M. alternatus* and *M. carolinensis*, and the relationship between transmission and virulence of the PWN.

Invasive PWN spreads the range by flying vector beetles and humans transporting pine logs infested with PWN and its vectors (Togashi and Shigesada 2006). The relative importance of both factors in the PWN spread may differ at the different scales of the space because of a variation and limitation in flight distance by beetles. Consequently, the late half of this chapter presents spatio-temporal pattern of PWN spreading within a stand, among stands, and among regions within Japan, and then concludes the mechanisms of PWN spread by analytical approaches and modeling.

To review the literature focusing primarily on the PWN–vector relationship and the epidemic of pine wilt in this chapter, I compiled our two recent reviews, because my colleagues and I had already completed these reviews about PWN spread and the evolutionary change in the PWN–insect vector relationship (Togashi and Shigesada 2006; Togashi and Jikumaru 2007).

17.2 PWN Vectors

Insects in the orders Coleoptera (families Cerambycidae, Curculionidae, Buprestidae, and Scolytidae) and Isoptera (family Termitidae) carry fourth-stage dispersal juveniles of the PWN (Table III.4). Among these insects species of the genus *Monochamus* carry many PWNs (Linit 1988). Species that have been confirmed to vector PWNs to host trees are *M. alternatus* and *M. saltuarius* in East Asia, *M. carolinensi*s, *M. mutator*, *M. scutellatus* and *M. titillator* in North America, and *M. galloprovincialis* in Portugal (Mamiya and Enda 1972; Morimoto and Iwasaki 1972; Sato et al. 1987; Linit 1988; Sousa et al. 2001). *Monochamus alternatus*, *M. carolinensis*, and *M. galloprovincialis* are most important in East Asia, North America, and Portugal, respectively. Their conifer hosts belong to the family Pinaceae.

17.3 Biology of *Monochamus alternatus*

Monochamus alternatus usually has a 1-year life cycle in Japan. It sometimes requires 2 years when the eggs are deposited late in the flight season (Togashi 1989a). The adult beetles emerge from dead pine trees in June and July regardless of the developmental time of immature stages (1 or 2 years). They feed on the bark of twigs on healthy host trees and reproductively mature 5–30 days after emergence. Mature beetles locate dying or recently dead pine trees for copulation and oviposition. In pine stands infested with the PWN, the beetles primarily use PWN-killed trees for oviposition (Togashi 1990a). At oviposition, adult females make wounds on the bark of the trunk (bole) and branches of pine trees into which they insert their ovipositor under the bark (through the wound), and deposit eggs (Anbutsu and Togashi 2000). The mean fecundity is 76–86 eggs per female in a normal summer and 41 eggs in a cool, rainy summer (Togashi 1989c). In central Japan the flight season is as long as 4 months from June through September (Togashi 1989e).

Larvae hatch from eggs a week after oviposition and feed on the inner bark. They usually molt three times and bore tunnels into the xylem at the third or fourth instar (Morimoto and Iwasaki 1974a; Togashi 1989a,b). Before winter,

Table III.4 Insect species known to carry fourth-stage dispersal juveniles of *Bursaphelenchus xylophilus*, with notes on geographic range, host tree genus, and documentation of transmission. Adapted from Linit (1988)

Family, species	Range	Host	Transmission	References
Cerambycidae				
Acaloptera fradatrix	Japan	Pi.	ND	3
Acanthocinus griseus	Japan	Pi	ND	10, 3
Acanthocinus gundaiensis	China	Pi	ND	16
Amniscus sexguttatus	NA	Pi, Pc	ND	8, 15
Arhopalus rusticus	Japan	Pi	ND	10
Arhopalus rusticus obsoletus	NA	Pi	ND	8
Aromia bungii	China	Pi	ND	16
Asemum striatum	NA	Pi, Pc	ND	8
Corymbia succedanea	Japan	Pi	ND	10
Monochamus alternatus	Japan	Pi, Pc, A, L, C	Yes	10, 11, 6
Monochamus carolinensis	NA	Pi	Yes	7, 15
Monochamus clamator	NA	ND	ND	1
Monochamus galloprovincialis	Portugal	Pi	Yes	13
Monochamus marmorator	NA	A, Pc	ND	15
Monochamus mutator	NA	Pi	Yes	15
Monochamus obtusus	NA	Pi, Ps, A	ND	4
Monochamus scutellatus	NA	Pi, A, Pc, L	Yes	4, 15
Monochamus titillator	NA	Pi, Pc, A	Yes	9, 14
Monochamus nitens	Japan	Pi	ND	3
Monochamus saltuarius	Japan	Pi	Yes	12
Neacanthocinus obsoletus	NA	Pi, A,	ND	2, 5
Neacanthocinus pusillus	NA	Pi, A, Pc	ND	15
Spondylis buprestoides	Japan	Pi	ND	11
Uraecha bimaculata	Japan	Pi	ND	3
Xylotrechus saggitatus	NA	Pi	ND	15
Buprestidae				
Chysobothris spp.	NA	Pi	ND	8, 15
Curculionidae				
Hylobius pales	NA	Pi	ND	8
Pissodes approximatus	NA	Pi, Pc	ND	8
Scolitidae				
Tomicus piniperda	China	Pi	ND	16
Termitidae				
Odontotermes formosanus	China	Pi	ND	16

Range: China, Japan, and Portugal reveal the countries where studies were done. NA = North America
Host: A = *Abies*, C = *Cedrus*, Pc = *Picea*, Pi = *Pinus*, L = *Larix*, ND = not documented
Transmission: ND = not determined, Yes = transmission documented
References: 1 = Boweres et al. (1992), 2 = Carling (1984), 3 = Enda (1972), 4 = Holdeman (1980), 5 = Kinn (1987), 6 = Kishi (1995), 7 = Kondo et al. (1982), 8 = Linit et al. (1983), 9 = Luzzi et al. (1984), 10 = Mamiya and Enda (1972), 11 = Morimoto and Iwasaki (1972), 12 = Sato et al. (1987), 13 = Sousa et al. (2001), 14 = Williams (1980), 15 = Wingfield and Blanchette (1983), 16 = Yang et al. (2003)

well-developed larvae plug the entrance of a U-shaped tunnel with fibrous, wood shreds, before occupying an enlarged portion of the tunnel called the pupal chamber at the terminal where they over-winter (Togashi 1989a, b). Poorly developed larvae do the same around the end of autumn or over-winter under the bark. The larvae pupate between late spring and mid-summer of the following year or 2 years later, and develop into adults. As the rate of survival from the egg to adult stage at emergence is as high as 0.249 for the 1-year life cycle (Togashi 1990a), the net reproductive rate is estimated to be 10.7. Density-dependent mortality due to cannibalism occurs at the larval stage, whereas natural enemies such as insect parasitoids and predators are density-independent mortality factors (Togashi 1986). Consequently, at high egg densities a definite number of adult beetles emerge from a unit area of the bark of dead trees (Morimoto and Iwasaki 1974b; Togashi 1986).

17.4 Dispersal of *Monochamus alternatus* Adults

Reproductively immature adults of *M. alternatus* likely have higher mobility than mature adults. Tethered, immature beetles are more active fliers than mature beetles (Ito 1982). When 1- to 5-day-old, marked beetles were released in a 15–19-year-old *P. thunbergii* stand, the recapture rate decreased up to the age of 17 days and then remained constant (Togashi 1990b). The recapture rate of wild beetles in a pine stand is lower (ca. 10%) in June and July than in August and September (ca. 20%; Togashi 1990b). The mean age of wild beetles increases as the season progresses because of the limited season of adult emergence between June and July. In the case of *M. carolinensis*, unmated females show significantly longer flight distance and duration than mated females (Akbulut and Linit 1999).

Short- and long-distance dispersals by beetles have been studied by a mark-recapture method in a pine stand and over pine stands, respectively. When immature beetles were released on healthy trees in a 15–19-year-old *P. thunbergii* stand, some of them did not leave the trees where they were released and others left them within a week (Togashi 1990c). The proportion of beetles that fly away from the pine stand where they have been released increases as stand density decreases (Togashi 1990c). The mean duration of stay on release trees was estimated to be 11.5 days in a closed canopy and 1.4–5.4 days in a sparse canopy. A similar value of 2.2–2.5 days was obtained in a 6-year-old *P. thunbergii* stand (Morimoto et al. 1975). An analysis of the distribution of distances between release and recapture points in a pine stand indicated random dispersal by immature beetles (Togashi 1990c). The mean distance traversed was estimated to be 7.1–37.8 m in a week in stands with a sparse canopy. The mean distance traversed by wild beetles was estimated to be 10–20 m week^{-1} through the first 3 weeks after release. In an 8-year-old *P. thunbergii* stand it was estimated that wild beetles traversed a mean distance of 12.3 m for females and 10.6 m for males between the first and final captures (Shibata 1986).

As for long-distance dispersal, Ido and Kobayashi (1977) showed that one of 756 beetles released was recaptured at a distance of 2.4 km from the release point although 75.5% of recaptures were within 100 m of the release point. Fujioka (1992) gave a distance distribution between release and recapture points up to 2 km and Takasu et al. (2000) estimated the mean traversed distance to be 1.82 km.

The beetles were distributed independently of pine-wilt diseased trees in June, the early flight season, while they overlapped spatially with diseased trees from July onward (Togashi 1989c). The degree of overlapping tended to become large in quadrat sizes of 25 m^2 or larger, indicating that the beetles were attracted and stopped in an average area of 25 m^2 containing diseased trees. After a diseased tree occurred in late July, beetles aggregated on healthy trees surrounding the diseased tree (Shibata 1986). The seasonal change in the spatial distribution of beetles is probably related to their reproductive maturation (Togashi 1991b).

17.5 Biology of PWN in Reference to Transmission

PWN reproduces in recently killed trees (Mamiya 1976a, 1983). Second-stage juveniles emerge from eggs and molt into third-stage propagative or dispersal juveniles (Mamiya 1975a). Third-stage propagative juveniles molt to fourth-stage propagative juveniles, which develop into adults. It takes 4–5 days for eggs to become adults at 25°C (Mamiya 1975a).

The proportion of third-stage dispersal juveniles increases in dead trees as time elapses (Mamiya 1976a). The juveniles concentrate around the pupal chambers occupied by *M. alternatus* larvae or pupae and molt to fourth-stage dispersal juveniles, a special stage for transportation, around the time of vector beetle eclosion (Warren and Linit 1993; Maehara and Futai 1996; Necibi and Linit 1998). Then they enter the beetle's tracheal system via openings in the vector beetle's exoskeleton called spiracles. Most of the nematodes enter through the first abdominal spiracles, during a week-long period prior to the emergence of the adult beetle from the pupal chamber.

After beetle emergence, the fourth-stage dispersal juveniles leave the tracheal system via the spiracles (for details, see Chap. 13), crawl to the beetle's abdominal terminal along sterna, and then leave the beetle (Enda 1977). The nematodes enter healthy trees via feeding wounds made by beetles and then molt into adults (Mamiya and Enda 1972; Morimoto and Iwasaki 1972). When the number of entering nematodes exceeds the threshold for the disease to develop under water stress, the nematodes deprive susceptible trees of the ability to exude oleoresin within several weeks (Hashimoto and Sanui 1974; Mamiya 1983; Fukuda 1997). After this stage of disease development, the foliage of most trees becomes brown or red brown (Togashi 1989f).

Reproductively mature beetles of both sexes are attracted to dying and recently dead trees including PWN-infected trees. As well, there are other transmission pathways such as when female beetles transmit PWNs to dying trees via oviposition wounds directly (Wingfield and Blanchette 1983; Edwards and Linit 1992). Beetle males searching for mates also transmit the nematodes to dying trees via already existing wounds on the bark caused by conspecific oviposition and other factors (Arakawa and Togashi 2002). The reason for this is closely related to movement by the nematodes from landing sites to beetle oviposition wounds. Before transferring from vector beetles to trees, PWNs sometimes move between beetles during the mating process (Edwards and Linit 1992; Togashi and Arakawa 2003). Half-day pairing of reproductively mature beetles, one sex harboring PWNs and the other free from PWNs, indicates that female beetles carry the fourth-stage dispersal juveniles of PWN in spermatheca, a sac for storing spermatozoa (Arakawa and Togashi 2004), although it is unknown if PWNs in spermatheca are transmitted to pine trees. The transmission pathways by mature beetles mentioned above might contribute to the persistence of PWN in pine forests resistant to PWN or forests where pine wilt disease can not develop due to cool summer temperatures.

17.6 Effects of PWN on Population Performance of *Monochamus alternatus*

There is a great difference among *M. alternatus* adults in the number of PWNs carried by a beetle at emergence from dead trees (initial nematode load). Most beetles carry fewer than 5,000 PWNs, however, a record 289,000 PWNs were obtained from a single *M. alternatus* (Kishi 1995).

A large proportion of PWNs leave the *M. alternatus* body between 10 and 40 days after the beetles emerge from dead trees (Togashi and Sekizuka 1982). The number of PWNs transmitted per unit time from a single beetle to pine twigs via feeding wounds changes depending on the initial load of PWNs and the age of the

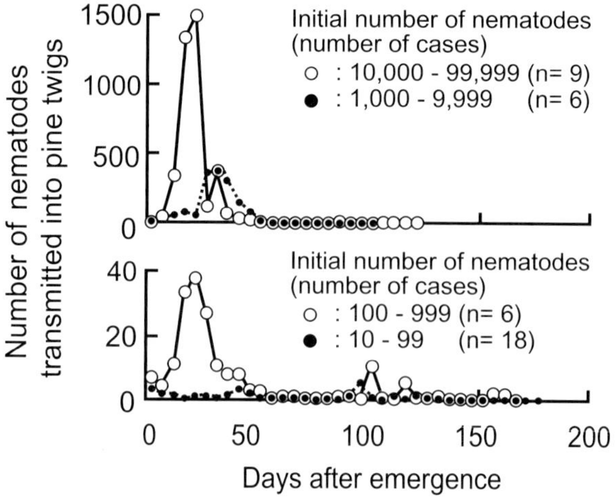

Fig. III.8 *Bursaphelenchus xylophilus* transmission curves for *Monochamus alternatus* adults. Beetle adults are grouped by the level of the initial number of nematodes carried (after Togashi 1985)

beetle (Fig. III.8; Togashi 1985). The transmission curve of PWN is defined for individual beetles as the age-specific change in the number of transmitted PWNs, and can be divided into L-shaped and unimodal types. In a unimodal type of transmission curve, PWN transmission from a beetle to pine twigs through feeding wounds peaks between 20 and 35 days after emergence (Togashi 1985). The ambient temperature has an effect on PWN transmission (Jikumaru and Togashi 2000). A laboratory experiment showed that as temperature decreases from 25 to 16°C, the mean longevity of *M. alternatus* adults decreases from 143.6 to 65.9 days, the mean efficiency of PWN transmission (efficiency of transmission = the total number of PWNs transmitted successfully into pine twigs/thc total number of PWNs leaving the vector) decreases from 0.40 to 0.13 for beetles with an initial load of 100 to 999. As well, the peak period of PWN transmission was delayed from 10–15 to 25–30 days, and its mean peak height decreased from 53 to 13 nematodes for beetles with an initial load of 100–999 nematodes, and from 276 to 23 nematodes for beetles with an initial load of 1,000 to 9,999 nematodes. Cool temperature inhibits the transmission process of PWNs (Jikumaru and Togashi 2000).

A heavy initial load of PWNs increases the number of PWNs transmitted from *M. alternatus* adults and reduces their longevity, resulting in a beetle population structured by PWN (Fig. III.9; Togashi and Sekizuka 1982; Togashi 1985, 1991b). Beetles with a heavy load of more than 10,000 nematodes transmit a mean of 1,500 nematodes during 5 days at the peak (Fig. III.8). As they stay on a healthy tree for an average time of 2–3 days (Togashi 1990c), each of them transmits on average 750 nematodes to the tree at the peak of transmission. Hashimoto and Sanui (1974) indicate a minimum inoculum level of 300 fourth-stage dispersal juveniles is sufficient to induce pine wilt. Thus, such beetles are considered to be producers of food resources for the larvae because females oviposit on the diseased trees; however, beetles with a heavy nematode load do not live long, resulting in a poor contribution to reproduction (Togashi and Sekizuka 1982). Beetles with a moderate load of 1,000–10,000 PWNs can transmit a sufficient number to induce the disease when they focus on a healthy tree. Beetles with a light load of less than 1,000 PWNs

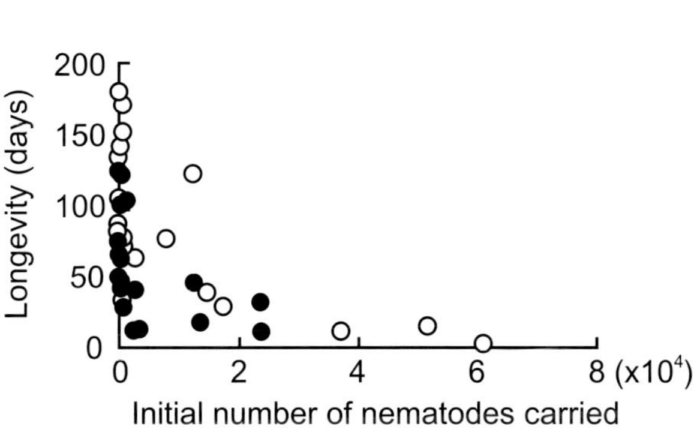

Fig. III.9 Relationship between the initial number of *Bursaphelenchus xylophilus* carried and longevity of *Monochamus alternatus* adults. *Open* and *solid circles* represent beetle females and males, respectively (after Togashi and Sekizuka 1982, with permission)

are not related to pine wilt incidence in the current year. Beetle longevity is scarcely affected by initial loads of less than 10,000 PWNs; thus, such beetles can contribute greatly to reproduction. Based on this information, Togashi (1985) suggested that nematode load divides a beetle population into three functionally different groups; the first group relating to the production of the progeny's food, the second relating to reproduction, and the third to both functions.

A study using a flight mill indicated that *M. carolinensis* adults carrying more than 10,000 PWNs had a significantly shorter flight distance and duration than those carrying less than 10,000 PWNs (Akbulut and Linit 1999).

17.7 Two Types of PWN Transmission Curves in Relation to Virulence

The proportion of *M. alternatus* adults with a unimodal nematode transmission curve varied greatly among populations; 100% for Mie and Nara populations (Shibata and Okuda 1989), 92% for an Ishikawa population (Togashi 1985), and a small percentage for an Ibaraki population (Kishi 1978).

There is a great variation in the virulence of PWN in the field in Japan (Kiyohara and Bolla 1990). When transmitted to healthy, susceptible pine trees, virulent PWNs kill them and reproduce therein soon after tree death. As such, virulent PWNs that are transmitted early can initiate reproduction earlier than those transmitted late, possibly resulting in greater fitness of early-transmitted PWNs than late-transmitted ones, so L-shaped transmission curves are expected to favor virulent PWNs. Alternatively, only when transmitted to dying and recently dead host trees can avirulent PWNs reproduce in them (Wingfield and Blanchette 1983). *Monochamus alternatus* adults mature reproductively a few weeks after emergence and visit dying and recently dead host trees to copulate and oviposit (Togashi 1989). It is considered that avirulent PWNs gain greater fitness when transmitted after the reproductive maturation of beetles than before their maturation. So, a unimodal nematode transmission curve favors avirulent PWNs, although there have been no studies on the relationship between virulence and the PWN transmission curve.

When less than 30% of the trees survive a pine wilt epidemic in *P. densiflora* and *P. thunbergii* stands, they contain a low proportion of trees having a substantially reduced susceptibility (Toda and Kurinobu 2002). Thus, the mean level of tree susceptibility to pine wilt in a pine stand is expected to decrease in the course of the epidemic. A reduced mean level of tree susceptibility likely reduces the fitness of early-transmitted PWNs.

Using *M. alternatus* adults that emerged from dead trees in a *P. densiflora* stand, Ibaraki Prefecture, Kishi (1995) showed that the averaged nematode transmission curve changed from L-shaped to unimodal type, and the beetle age at which the peak of the transmission curve was observed increased during 4 years in the early half of a pine wilt epidemic (Fig. III.10), although the transmission curve came back to an L-shaped type during the later 2 years. Assuming a close relation

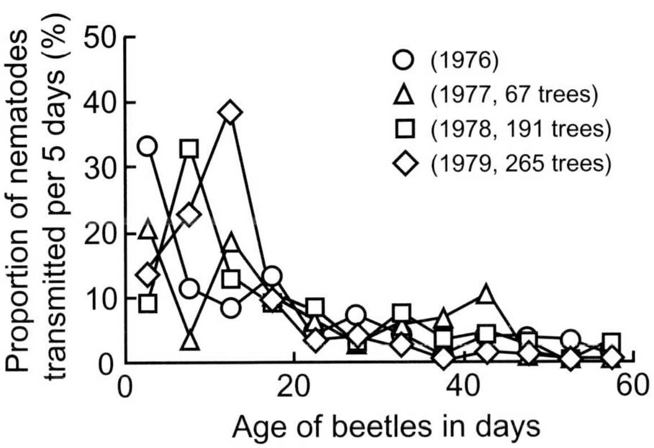

Fig. III.10 Yearly changes in the average transmission curves of *Bursaphelenchus xylophilus* into *Pinus densiflora* twigs by *Monochamus alternatus* from 1976 to 1979 (after Table 48 of Kishi 1995). Fifty *M. alternatus* adults, which emerged from dead trees in a *P. densiflora* stand in Naka, Ibaraki Prefecture, were examined each year. The stand contained 667 living trees before the 1976 *B. xylophilus* infection season. The ordinate represents the mean proportions of nematodes transmitted each 5 days to the total number of nematodes transmitted. The two numbers in *parentheses* indicate the year and the number of pine wilt-killed trees in the pine stand in that year (after Table 54 of Kishi 1995)

between the virulence and PWN transmission, the change in the average nematode transmission curve during the early half of the epidemic suggests decreasing virulence of PWN to reach the maximum basic reproductive rate (Anderson and May 1982). The progress of a pine wilt epidemic reduces the mean susceptibility level of surviving trees. Reduced susceptibility acts as a selective pressure on PWNs, tending to select the higher virulent PWNs. This may induce a change in the averaged nematode transmission curve during the later half of the epidemic.

The transmission curves of the avirulent congener, *Bursaphelenchus mucronatus*, were unimodal for all 43 *M. saltuarius* vectors examined (Jikumaru and Togashi 2001). This supports a close relationship between virulence and the PWN transmission curve.

17.8 Temporal Change in the Number of Wilt-Diseased Trees in a Pine Stand

Seasonal changes in the number of newly diseased trees that are unable to exude oleoresin, an initial symptom of pine wilt, may differ between areas. In Chiba Prefecture facing the Pacific Ocean, 87% of newly diseased trees occurred between June and August in a *P. thunbergii* stand (Mamiya 1976a), whereas in Ishikawa Prefecture facing the Sea of Japan, the proportion of newly diseased trees increased from June through August and then remained constant up to October (Togashi 1989d). After the invasion of a pine stand by the PWN, the annual number of killed trees accelerated, reached a peak, and then decreased without the application of any control measures (Kishi 1995).

A mark-recapture study showed that the adult density of *M. alternatus* per tree increased from early June, peaked in early July, and then remained almost constant for a month (Togashi 1989e). It began to decrease in mid- or late-August and the adults had disappeared by October. The mean density of adult beetles between June and August varied yearly in a pine stand. There was a positive correlation between the mean density of adult beetles and the incidence of pine wilt (Togashi 1989e).

The annual number of disease-killed trees (tree mortality) has been recorded for some pine stands. A simple mathematical model was constructed using the following assumptions to evaluate the susceptibility to pine wilt of a pine stand. The healthy tree density per unit area in a pine stand is N before invasion by PWN. After invasion, a definite number of beetles emerge from each dead tree the year following PWN infection. There is no difference in the initial PWN load among beetles. The transmission rate (β) of disease between trees remains constant. Let x_t and y_{t-1} be the densities of healthy and disease-killed, infective trees just before the infection season of year t, respectively. As pine seeds are difficult to germinate during a pine wilt epidemic due to a shaded floor covered with undergrowth, we can ignore the reproduction of pine trees. Thus, we obtain the following equations (Togashi et al. 1992),

$$N = x_t + \sum_{i=1}^{t-1} y_i \tag{1}$$

$$y_t = \beta x_t y_{t-1} \tag{2}$$

Arranging Eq. 2 and substituting x_t in Eq. 1 into Eq. 2,

$$y_t / y_{t-1} = \beta x_t = \beta \left(N - \sum_{i-1}^{t-1} y_i \right) \tag{3}$$

This model indicates increasing tree mortality ($y_2/y_1 > 1$), which leads to a pine wilt epidemic, at a healthy tree density above a threshold determined by the transmission rate ($x_2 > 1/\beta$), whereas it shows no epidemic at $x_2 < 1/\beta$, which indicates that the initial tree density is important in the disease epidemic incidence because of $x_2 \approx N$ due to an extremely small rate of initial tree mortality. When an epidemic occurs, this model gives a peak of yearly change in tree mortality. The simulation indicates that the density of trees that survive a pine wilt epidemic increases with decreasing initial density of healthy trees N and with decreasing transmission rate β. It also indicates that the period of coexistence of PWN and pine trees is longest at a certain transmission rate, which decreases with increasing initial healthy tree density.

Using Eq. 3, β values are estimated to be 0.00062–0.00242 for four pine stands with no control measures compared to 0.00037–0.00837 for 17 stands with control measures (Togashi et al. 1992). Aged pine stands exhibit large β values of more than 0.0024 irrespective of the control practice, indicating severe infestation of the infective disease.

Yoshimura et al. (1999) developed a deterministic model for the population dynamics of the host-vector association between pine trees and beetles carrying PWNs. Let H_t and P_t be the healthy tree density and vector density before PWN infection (nematode transmission) at year t, respectively. Assuming a constant transmission rate for each vector, a proportion of healthy trees, $\exp(-\alpha P_t)$, escapes PWN infection induced by feeding of vectors. Consequently, the density of healthy trees in the following year H_{t+1} and density of diseased trees $\tilde{H}_t$ are:

$$H_{t+1} = \exp(-\alpha P_t) H_t \tag{4}$$

$$\tilde{H}_t = \{1 - \exp(-\alpha P_t)\} H_t \tag{5}$$

As vector density at year $t + 1$ is determined by the densities of vector and diseased trees at year t,

$$P_{t+1} = (1-\theta) F\left(P_t, \tilde{H}_t\right) \tilde{H}_t, \tag{6}$$

where θ is the eradication rate of beetles and $F(P_t, \tilde{H}_t)$ is the mean number of adults emerging from a diseased tree. Yoshimura et al. (1999) provided the following equation based on empirically measured parameters, which involves density-dependent mortality for the immature stages:

$$F\left(P_t, \tilde{H}_t\right) = \frac{0.98\sigma S K P_t}{S\left\{a + \tilde{H}_t\right\} + 0.065\sigma K P_t}, \tag{7}$$

where σ is the sex ratio of the beetle, K is the mean number of eggs that a female beetle can deposit maximally, S is the mean surface area of a pine tree, and $1/a$ is the efficiency of oviposition. The parameters are estimated as $\alpha = 7.7\ \text{m}^{-2}$, $\sigma = 0.48$, $K = 80$, $S = 2.4\ \text{m}^2$, and $a = 0.022\ \text{m}^{-2}$, based on data obtained from a pine stand on the northwest coast of Japan (Togashi and Magira 1981; Togashi 1989c, e).

The results of Yoshimura et al. (1999) are indicative. There is a minimum pine density below which the disease always fails to establish itself (Fig. III.11); however, even if the pine density exceeds the threshold, the disease fails to establish due to the Allee effect when the density of beetles is extremely low. The Allee effect is caused by the low production of wilt-diseased trees and poor reproduction of vector beetles at low beetle density. The minimum pine density increases disproportionately with an increase in the eradication rate. The probability that a healthy tree escapes infection until the epidemic dies out decreases sharply with an increase in initial pine density or initial beetle density.

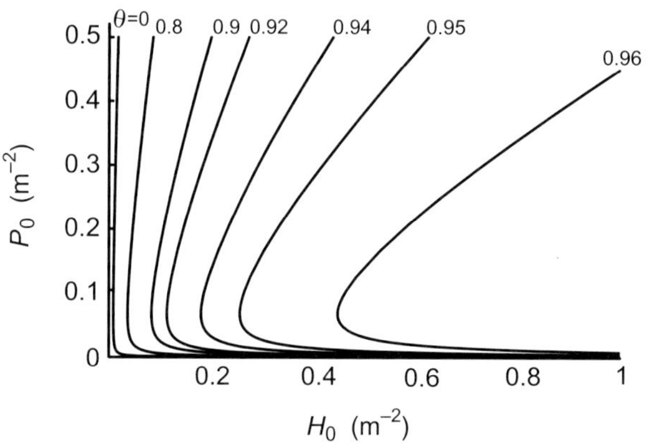

Fig. III.11 Boundary curves for the successful establishment of *Monochamus alternatus* in the (H_0, P_0) plane for various eradication rate values, θ. H_0 and P_0 are the initial densities of healthy pine trees and adult beetles, respectively. On the right side of each curve, the beetle density increases the following year, while on the *left side*, it decreases (after Fig. 6 of Yoshimura et al. 1999)

17.9 Temporal Change in the Spatial Distribution of Wilt-Diseased Trees in a Pine Stand

The spatial distribution pattern of wilt-diseased trees within a pine stand varies depending on the season and number of years after the PWN invades. In the first few years after invasion, diseased trees occur at limited sites within a pine stand in the early half of the beetle flight season and then around the early diseased trees in the late half (Fig. III.12; Togashi 1991b). Reproductively immature, single beetles with a heavy PWN load are considered to induce the early occurrence of diseased trees and the subsequent occurrence might be due to mature beetles with a moderate initial PWN load that concentrate on early diseased and neighboring healthy trees. Thus, diseased trees show a clumped distribution. The degree of aggregation is high in the early season and then low in the late season (Togashi 1991b).

After a few years after PWN invasion, diseased trees occur throughout the stand in the early half of the beetle flight season and subsequently around the early diseased trees, showing a regular to random distribution at first and then a clumped distribution. The occurrence of diseased trees in the early season is results from both beetles with a heavy PWN load and a positive, spatial correlation between trees diseased early in the current year and those diseased in the previous year (the aftereffect of the previous year's disease), which occurs even if all trunks and branches of dead trees are removed completely before beetle emergence (Togashi 1991b). Plausible causes for the positive, spatial correlation are small PWN populations that over-winter in dead branches on healthy trees (Futai 2003a) and the movement of PWNs via root grafts between infected trees and healthy trees (Tamura 1983a). The seasonal increase in the degree of aggregation of diseased trees is explained by the concentration of sexually mature beetles on diseased trees and neighboring, healthy trees.

A simulation model for evaluating various techniques and the combinations thereof to control PWN spread was made following systems analysis (Togashi

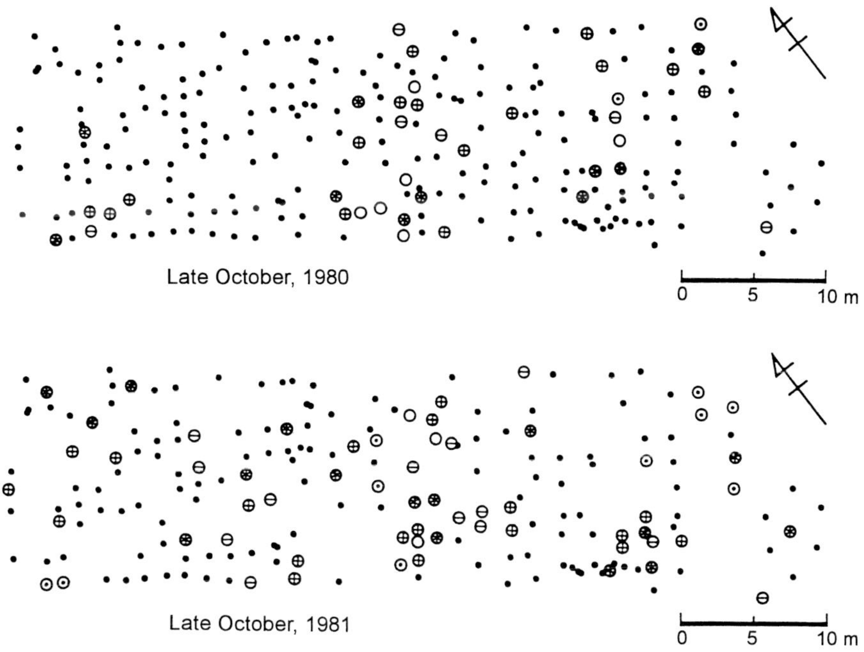

Fig. III.12 Distribution of pine wilt diseased and healthy trees within a *Pinus thunbergii* stand. Diseased trees are designated by the month during which they began to die (*open circle* for June, *circle with a dot* for July, *circle with a horizontal bar* for August, *circle with a cross* for September, and *circle with three bars* for October). Healthy trees are shown by *small solid circles* (after Togashi 1991b, with permission)

1989c). In the simulation, individual adult beetles fly, transmit nematodes, oviposit, and die in a pine stand with a definite number of trees planted in a lattice fashion. The model used empirical data such as adult emergence, survival, short-distance dispersal, fecundity, and PWN transmission that are related to beetle age and initial PWN load as well as the density-dependent mortality of beetle immature stages, residual effect of insecticide, and the positive, spatial correlation of pine wilt incidence (Togashi and Magira 1981; Togashi and Sekizuka 1982; Togashi 1985, 1986, 1989e, 1990c, 1991b). The model involves parameters representing the level of tree susceptibility, eradication rate for immature stages, and number of insecticide sprayings for adult beetles. The results are suggestive. Assuming a definite frequency distribution of initial PWN load on beetles, a small number of beetles with low dispersal ability cause an epidemic of pine wilt in a stand, whereas those with high dispersal ability can not do so even when no control measures are applied. There is only a difference of less than twice the time required for cumulative tree mortality to exceed 80% even when the initial number of emerging beetles increases from 25 to 1,000 in a stand with 900 trees or when the number of healthy trees in a stand with 100 beetles emerging increases from 400 to 2,500 at a constant density of 0.25 m^{-2}. Diseased trees show a clumped distribution irrespective of the level of susceptibility of trees to pine wilt. Intriguingly, the degree of aggregation increases seasonally when trees are highly susceptible, whereas it decreases seasonally for trees with incomplete resistance.

17.10 PWN Spread over Pine Stands

Long-distance dispersal of PWN by beetle flight and the human transportation of pine logs infested with PWN and the insect vector both accelerate the spread of pine wilt. Local spread of the disease from infested pine stands to surrounding, un-infested pine stands is likely caused by long-distance dispersal by beetles.

The rate of spread of PWN's range over pine stands has been determined in several areas in Japan (Table III.5) by mapping the expansion front of pine wilt disease incidence over 1–9 years. The observed values are between 2 and 15 km $year^{-1}$, resulting in a mean of 6 km $year^{-1}$. In the case of Ibaraki Prefecture with an area of 6,096 km^2, Kishi (1995) recorded the occurrence of pine wilt on 2 × 2 km^2 resolution. The disease occurred in the center of the prefecture and spread concentrically year by year. The relationship between year and the diameter of the range of disease incidence in Ibaraki Prefecture gives a spread rate of 4.2 km $year^{-1}$ (Takasu et al. 2000).

Mathematical models incorporating the reproduction of insect vectors within pine stands and long-distance dispersal of adult vectors estimated the mean spread rate of PWN's range to be several kilometers per year (Takasu et al. 2000; Yamamoto et al. 2000). Takasu et al. (2000) postulated that adult beetles disperse soon after emergence and then infect healthy trees with PWNs to reproduce on wilt-killed trees. Let $H_t(x)$, $P_t(x)$, and $P'_t(x)$ be the densities of healthy trees, pre-dispersal beetles, and post-dispersal beetles at year t at site x in a one-dimensional space, respectively. When $f(|x - y|)$ is the possibility that a beetle emerging at site y reaches site x, the relation between $P'_t(x)$ and $P_t(x)$ can be expressed as follows:

Table III.5 Spread of *Bursaphelenchus xylophilus* in Japan

Speed (km $year^{-1}$)	Study period (years)	Prefecture	Reference
3–15	4	Chiba	Matsubara (1976)
4–5	5	Shizuoka (east)	Fujishita (1978)
9–10	5	Shizuoka (west)	Fujishita (1978)
4	3	Aichi	Kato and Okudaira (1977)
2–3	1	Fukuoka	Hagiwara et al. (1975)
4.2	9	Ibaraki	Takasu et al. (2000)

After Table 1 of Togashi et al. 2004

$$P'_t(x) = \int_{-\infty}^{\infty} f(|x-y|)P_t(y)\,dy = \int_{-\infty}^{\infty} f(z)P_t(y)\,dy, \tag{8}$$

where travel distance, z, is used as $z = |x - y|$ for simplicity. By substituting P_t in Eqs. 4–7 with $P'_t(x)$ in Eq. 8, the disease incidence and the reproduction of the beetle population at site x can be described after dispersal. Some beetles disperse over a long distance and others over a short distance. Thus,

$$f(z) = (1-\sigma_L)f_S(z) + \sigma_L f_L(z), \tag{9}$$

where σ_L, f_S, and f_L are the proportion of long-distance dispersers, and the distribution of traveling distance for short-distance and long-distance dispersers, respectively. The following functions were used as f_S and f_L.

$$f_S(z) = \sigma_S \frac{u}{2\upsilon\Gamma(1/u)} \exp\left\{-\left(\frac{z}{\upsilon}\right)^u\right\} + (1-\sigma_S)\delta(x), \tag{10}$$

where σ_S, Γ, and δ represent the proportion of short-distance dispersers, and gamma and delta functions, respectively. The parameter values were estimated as $u = 2.554$ and $\upsilon = 35.69$ (m) using data obtained by Shibata (1986). For long-distance dispersal,

$$f_L(z) = \frac{\mu}{2}\exp(-\mu z), \tag{11}$$

where the μ value was estimated to be 5.5×10^{-4} m^{-1} from the data of Fujioka (1992), resulting in a mean dispersal distance of 1,820 m.

Takasu et al. (2000) simulated the effect of short-distance dispersal on the range expansion of PWN using Eqs. 4–10. They found that when the initial tree density (H_0) is low, no expansion of range occurs because the beetle can not establish itself and becomes extinct. As H_0 increases beyond a threshold, the beetle can establish itself and the rate of range expansion is transformed into a traveling wave. The range expansion rate of PWN is sensitive to changes in initial tree density around the threshold value. After an initial sharp rise, this rate gradually reaches a point where it can not increase further because the beetle's reproduction has an upper limit due to the effect of density, as included in Eq. 7. The rate of range expansion decreases with an increasing eradication rate and reaches zero at a certain eradication rate. In every case, the range expansion rate is at most 50 m $year^{-1}$.

When beetles disperse long distances in a given proportion, the rate of range expansion attained is several kilometers per year (Fig. III.13a,b; Takasu et al. 2000). The expansion rate increases from zero with the increasing proportion of long-distance dispersers and suddenly drops to zero beyond a certain proportion (Fig. III.13a). The threshold proportion decreases with increasing eradication rate (Fig. III.13a) and with decreasing initial tree density (Fig. III.13b).

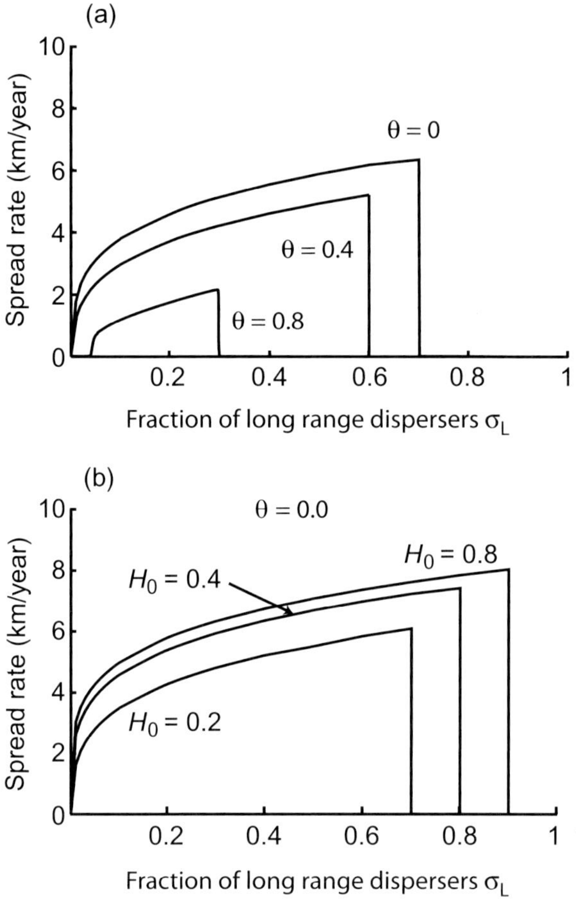

Fig. III.13 Dependence of the range expansion rate of *Bursaphelenchus xylophilus* on the fraction of long-distance dispersers, σ_L. **a** The initial pine density is fixed as $H_0 = 0.263\ m^{-2}$, while θ varies as indicated, **b** the eradication rate is fixed as $\theta = 0$, while H_0 varies as indicated (after Fig. 7 of Takasu et al. 2000, with permission)

17.11 PWN Spread Among Government Jurisdictions

Long-distance transportation of infested pine logs by humans is considered to be the primary cause of the regional spread of PWN as observed in Japan and South Korea (Togashi et al. 2004). As Japan is divided into 47 prefectures of administrative districts, the mean area of which is 8,038 km^2 (SD = 11,700 km^2) ranging from 1,876 to 83,453 km^2, the spread of PWN over prefectures was analyzed by Togashi et al. (2004).

Though PWN was determined to be the causative agent of pine wilt in 1971, pine tree mortality presumably resulting from pine wilt first occurred in Nagasaki City, Nagasaki Prefecture in 1905, followed by Aioi Town, Hyogo Prefecture, 730 km east of Nagasaki City, in 1921 (Fig. III.14; Kishi 1995). According to Kishi (1995), the first nematode population in Nagasaki was supposedly eradicated

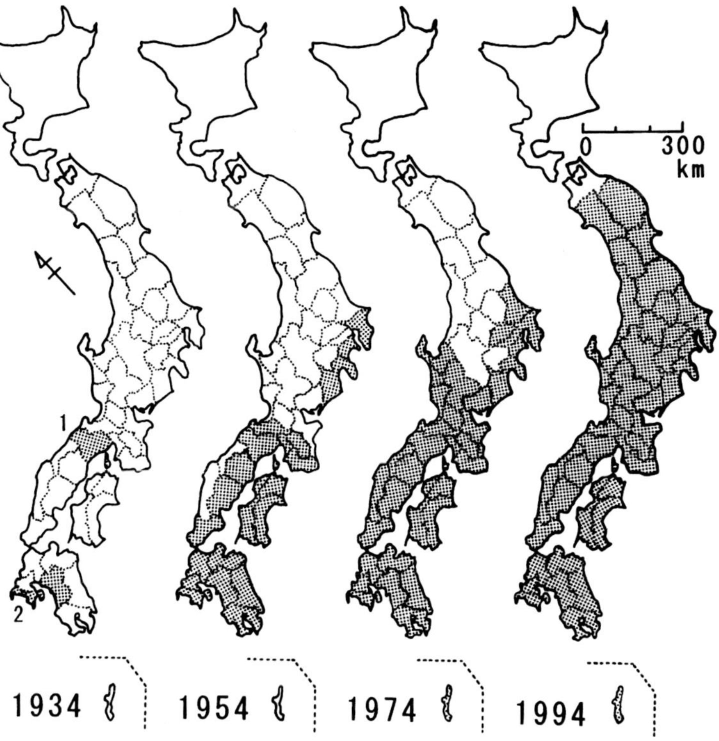

Fig. III.14 Spread of pine wilt in Japan. *Shaded parts* represent prefectures invaded by *Bursaphelenchus xylophilus*. The two numbers in the leftmost figure show the prefectures the nematode invaded early; 1 and 2 for Hyogo and Nagasaki Prefectures, respectively (after Fig. 3 of Togashi et al. 2004)

shortly after 1912. Togashi et al. (2004) determined the year of invasion of individual prefectures by PWN based on Kishi (1995) and others (Akasofu 1974; Matsueda 1975). The number of prefectures that had been invaded since the occurrence of pine wilt in Hyogo Prefecture increased in an accelerated manner between 1936 and 1947, during which PWN spread over prefectures in western Japan and those facing the Pacific Ocean (Figs. III.14, III.15). Between 1948 and 1958, the spread was stopped because the US army, which occupied Japan after World War II, recommended the intensive control of pine wilt using felling and burning (Kishi 1995). After 1959, the number of newly invaded prefectures increased again and reached 45 in 1982 then stopped increasing (Fig. III.15). Currently, only the two northernmost prefectures, Aomori and Hokkaido, with a cool summer climate, have not been invaded.

It is difficult to determine the origin of PWN that induced the first occurrence of pine wilt in each prefecture. Thus, assuming that PWN invades each prefecture from the nearest already-invaded prefecture and that pine wilt is found in the year when the infection (effective invasion) occurs, the time required for invasion, the

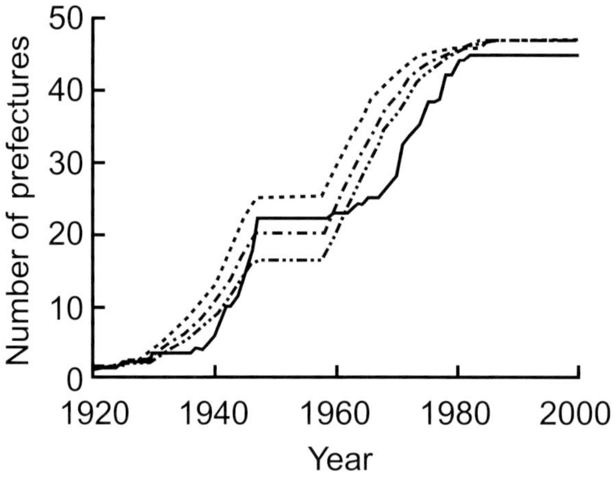

Fig. III.15 Observed and simulated changes in the number of prefectures invaded by *Bursaphelenchus xylophilus* in Japan. The *solid line* represents the observed data. *Broken* and *chain lines with one and two dots* indicate the results of simulation with $p = 0.0034$ and $\tau = 1$, $p = 0.0035$ and $\tau = 2$, and $p = 0.0037$ and $\tau = 3$, respectively (after Fig. 4 of Togashi et al. 2004)

distance transported, and the year of invasion are determined for each newly invaded prefecture. There is a great difference in the time required for PWN to invade a prefecture from the nearest already-invaded prefecture, the mean of which is 9.6 years (SD = 9.4 years, range of 1–34 years). In 44% of the invaded prefectures, the time required for PWN invasion is 1–4 years, suggesting that the transportation of infested logs is important in the PWN's spread considering a mean diameter of prefectures of 92.5 km and a mean rate of nematode range expansion of 6 km year^{-1} (Togashi and Shigesada 2006). The distance between a newly invaded prefecture and the nearest already-invaded prefecture averaged 125.4 km (SD = 161.0 km) with a maximum of 924 km between Hyogo and Nagasaki Prefectures, and a minimum of 33 km between Saitama and Tokyo Prefectures. Long-distance dispersal of more than 200 km occurred four times, sporadically between 1925 and 1982. There is no clear relationship between the time required for PWN invasion and the distance between two prefectures (Togashi and Shigesada 2006).

Togashi et al. (2004) analyzed the spread of PWN over prefectures using a simple model. Let $I(t)$ be the number of prefectures where pine wilt is newly found at year t after the natural infection period. Then $I(t)$ decreases with the decreasing number of un-invaded prefectures, $U(t)$, prior to natural infection. $I(t)$ also increases with the increasing number of prefectures which can distribute pine logs infested with PWN and its vectors to other prefectures. Thus, let τ be the time delay between the PWN invasion of a prefecture and the distribution. Consequently,

$$I(t) = pU(t)\left\{\sum_{i=1}^{t-\tau} I(i)\right\} \tag{12}$$

$$U(t) = N - \sum_{i=1}^{t-1} I(i), \tag{13}$$

where N is the total number of prefectures (N = 47) in Japan and p is the rate of invasion of un-invaded prefectures by PWN per year per already-occupied prefecture. Substituting Eq. 13 into Eq. 12 and rearranging it, we obtain the following,

$$I(t)\Big/\left\{N-\sum_{i=1}^{t-1} I(i)\right\}=p\left\{\sum_{i=1}^{t-\tau} I(i)\right\}. \tag{14}$$

The left side of Eq. 14 represents the rate of invasion per un-invaded prefecture and the right side represents the rate of transport from already-occupied prefectures. The value of p is estimated by fitting the data to a line passing through the origin. Simulation with the estimated values of p and τ showed that it took at least 2 years for a prefecture to distribute PWN to un-invaded prefectures after being invaded (Fig. III.15).

To determine the spatial characteristics in the regional spread of the PWN, the diameter of individual prefectures and the focal distance between a newly invaded prefecture and the nearest already-invaded prefecture were determined following Suarez et al. (2001) (Togashi et al. 2004). The diameter of prefectural area was estimated as two times the positive square root of area divided by π. It averaged 92.5 km (SD = 41.4 km) with a maximum diameter of 326.0 km for Hokkaido and a minimum of 48.9 km for Kagawa Prefecture. As stated earlier, the distance between a newly invaded prefecture and the nearest invaded prefecture averaged 125.4 km (SD = 161.0 km). When we considered the already-occupied, nearest prefectures as those that were invaded at least 2 years earlier, the distance averaged 127.8 km (SD = 160.4 km) with a maximum of 924 km and minimum of 33 km (Togashi et al. 2004). Consequently, there was no difference in frequency distribution between the three distances. The result suggests that PWN invaded many prefectures from neighboring already-occupied prefectures.

Re-distribution of PWN two or more years after the spread to a prefecture may be due to the great rate of initial increase in tree mortality and the extensive availability of wilt-killed pine trees. Tree mortality from the PWN increased from 6 to 67 during the first two consecutive years in a *P. densiflora* stand of 858 healthy trees without any control measures in Ibaraki Prefecture (Kishi 1995), suggesting that transportation of a substantial number of infested logs could begin a few years after PWN invasion of a pine stand. Pine wilt-killed trees were harvested and kept around houses as fuel up to the 1950s. A portion of wilt-killed trees have been harvested and hauled to lumber mills together with healthy trees to be distributed to mines, roads, and ships for use in construction (e.g., Kuniyoshi 1974). Wood chips of PWN-infested pine trees have been used in the pulp mills. The extensive amount of available wilt-killed pine trees might have decreased the time required for a newly invaded prefecture to export the nematode to un-invaded prefectures.

17.12 Concluding Remarks

The PWN is an invasive species in East Asia including Japan. Low proportions of *P. densiflora* and *P. thunbergii* trees that survive an epidemic of pine wilt show substantially reduced susceptibility (Toda and Kurinobu 2002); thus, the mean level of tree susceptibility to pine wilt in a pine stand is expected to decrease in the course of an epidemic. A reduced mean level of tree susceptibility likely reduces the fitness of early-transmitted PWNs. A change in the shape of the PWN transmission curve in relation to the progress of the epidemic (Kishi 1995), suggests a relationship between the peak time of PWN transmission and tree susceptibility. A great difference in virulence among Japanese PWN isolates (Kiyohara and Bolla 1990) indicates that there may be a possibility of reducing the mean virulence of PWN. Relationships between the virulence of PWN, PWN transmission by vector, and tree susceptibility are very important in order to better predict the future pattern of the epidemic of pine wilt.

A negative correlation between adult longevity and PWN transmission of *M. alternatus* through an initial PWN load divides a vector population into three groups contributing to reproduction, the production of larval food resources, and both. A heavy load of PWN also reduces the flight performance of *M. carolinensis*. Deleterious effects of PWN on the performance of vector insects influence the seasonal pattern of pine wilt incidence within a pine stand.

On the scale of a pine stand, the spatial spread pattern of PWN is determined mainly by short-distance flight by beetles relating to reproductive maturity, the effect of PWN load on the transmission pattern and beetle performance, and a positive, spatial correlation of wilt-diseased trees in two consecutive years. (Togashi 1989c, 1991b). A simulation based on systems analysis suggested the level of tree susceptibility to the disease to be one important component (Togashi 1989c). Another population dynamics model of pine trees and adult beetles indicated reduced reproduction at low densities of adult beetles (Allee effect) explicitly using empirical data in pine wilt systems (Yoshimura et al. 1999). This effect is caused by nematode infection and oviposition by beetles on trees and not by reduced mating opportunities. The model also shows that the minimum host density above which the nematode can successfully establish itself in a pine stand depends on the rate of eradication of beetle populations (Yoshimura et al. 1999).

The local and regional spread of invading organisms has been modeled by using reaction-diffusion equations or integro-difference equations incorporating reproduction and dispersal processes when the spread occurs almost exclusively by the movement of the organism or its vector (Andow et al. 1990; Shigesada et al. 1995; Veit and Lewis 1996; Takasu et al. 2000; Yamamoto et al. 2000). In the spread of PWN over pine stands, Takasu et al. (2000) elucidated the generation of a traveling wave of beetle density in a one-dimensional space using integro-difference equations, which involves the actual population dynamics of the beetle vector within a pine stand and the observed frequency distribution of the long distance traveled by marked beetles. They also indicated that the expansion rate of the PWN range is influenced by the initial tree density and eradication rate. Interestingly, the expansion rate decreases to zero due to the Allee effect when a large proportion of beetles are engaged in long-distance dispersal. The maximum rate of expansion obtained by the modeling is included in the rates observed in the field.

Suarez et al. (2001) stated that the spread dependent on accidental or intended transportation of the organism remains to be analyzed, although stratified dispersal models indicate the importance of artificial transportation for an invasive moth, the horse chestnut leafminer (Gilbert et al. 2004). Analysis of PWN spread over prefectures in Japan indicates that PWN invaded many prefectures from neighboring already-occupied prefectures and was re-distributed to un-invaded prefectures by man in two or more years. Inhibition of artificial transportation of infested pine logs is most important to control the spread of PWN on a regional scale.

Acknowledgments The author greatly appreciates the kind permission of American Society of Ecology, Ecological Society of Japan, Japanese Society of Applied Entomology and Zoology, Society of Population Ecology, and the Springer Science and Business Media for the use of figures and tables. He is also grateful to Dr. S. Jikumaru for drawing some figures. This study was supported in part by a Grant-in-Aid for scientific research from JSPS (no. 18208013).

18
Molecular Ecology of Vectors

Etsuko Shoda-Kagaya

18.1 What Has Molecular Ecology Brought to Forest Entomology Study?

Molecular ecology is a recently developed discipline in biology. During the mid-twentieth century, ecology and molecular biology occupied almost opposite positions in biological studies. Molecular biology is the study of organisms at the molecular level mainly for DNA, RNA and proteins. The disconnection between ecology and molecular biology may have been due to the differences in the size of objects being studied and investigative procedures. Molecular biologists seldom studied individual organisms in the field and most ecologists were not aware of DNA in their studies; however, the improvement of molecular techniques and advances in theoretical studies in evolutionary ecology and population genetics enable both disciplines to be combined.

What is molecular ecology? This is briefly stated in the journal information of "MOLECULAR ECOLOGY", which pioneered this study area. There are two statements: (1) utilizing molecular genetic techniques to address consequential questions in ecology, evolution, behaviour and conservation, and (2) studies may employ neutral markers for inference about ecological and evolutionary processes or examine ecologically important genes and their products directly (Molecular Ecology).

Since the 1980s, DNA has been isolated from non-model organisms for making templates of polymerase chain reaction (PCR). PCR is a method used to make replicate copies of a specific region of DNA by enzyme reaction. The first challenge for adopting a molecular technique into the study of forest insects was done with gypsy moths *Lymantria dispar* (Harrison et al. 1993). The gypsy moth is an invasive

Insect Ecology Laboratory, Forestry and Forest Products Research Institute, 1 Matsunosato, Tsukuba 305-8687, Japan

Tel.: +81-29-829-8251, Fax: +81-29-873-1543, e-mail: eteshoda@affrc.go.jp

alien species in North America, and the American populations were shown to have lower genetic diversity than native populations in Europe and Asia, and were suggested to be bottlenecked populations (Harrison et al. 1993; Bogdanowicz et al. 1997). There were four major lineages in the gypsy moth and the American individuals were on the European lineage (Bogdanowicz et al. 2000). Their invasion into North America from Europe could be confirmed from these data. Studies on the pine shoot beetle *Tomicus piniperda*, which bores into the trunk of weakened or dead pines, revealed intra- and inter-specific genetic structures (Kerdelhué et al. 2002; Duan et al. 2004). *Tomicus piniperda* in the Mediterranean area and in Yunnan Province (China) differs markedly from *T. piniperda* in France and has been shown to be new species. Including other *Tomicus* species, the evolutionary histories of this genus could be inferred. The mountain pine beetle *Dendroctonus ponderosae* is a native pine pest in North America, and the genetic components differs between eastern and western populations isolated by the Mojave Desert (Mock et al. 2007). Thus, study of the genetic structure suggests no dispersal across the desert.

These earlier studies suggest two ways to utilize molecular biology in forest insect studies. First, the new technique makes it possible to infer their dispersal history on both evolutionary and ecological time scales. Ongoing gene flow can also be revealed. For controlling forest pest damage, dispersal ability is one of the most important pieces of information; however, it is difficult to observe small insect dispersal directly in the field. Mark-recapture methods are laborious and sometimes unsuccessful. If molecular markers can be used, they are powerful tools for estimating their dispersal ability. Second, they can clarify the taxonomic status of the focal species and detect cryptic species (a different species with a similar morphology). It often shows different ecological traits, for example, host preference and life cycles. It is important to distinguish cryptic species because a protection plan against forests pests should be decided separately for each species with a different ecology.

The dispersal ability of the adult pine sawyer *Monochamus alternatus* has been determined by mark-recapture methods for sawyer adults and the rate of range expansion of pine wilt disease. Molecular ecological techniques can address information about the long-distance dispersal of *M. alternatus* and ongoing gene flow between populations. If there is some cryptic species within *M. alternatus*, its biology might be different from that of *M. alternatus*. To assess *M. alternatus* dispersal and genetic relationships within Northeast Asia, molecular ecological studies were done.

18.2 Genetic Structure of *Monochamus alternatus* in Northeast Asia

Changes in the genetic structure of a population are formed by balances among gene flow, genetic drift and mutation. If the individuals disperse widely or expand their range rapidly with substantial immigrants, populations reveal similar genetic components. A few immigrants might cause genetic drift and change the frequency of some alleles. The isolation of populations for a long time makes each population genetically different because of the accumulation of genetic mutations and/or genetic drift brought about by demographic events. Population histories are reflected in genetic components and we can infer the history of the population by studying the genetic structure.

Monochamus alternatus is distributed across Japan, Taiwan, Korea, China, Tibet, Laos, and Vietnam (Makihara 2004). In Japan, pine wilt disease was first recorded about 100 years ago. Since the 1980s, the disease has spread to other Asian countries (for details see Part I). Before the invasion of the pine wood nematode (PWN), the pine sawyer was a rare species in the pine forest community in Japan; however after the invasion, the sawyer's populations increased (Makihara 2004) and its migration rate might have been enhanced (Morris et al. 2004). The movement of pine sawyer populations associated with the PWN in the past should provide useful information on the future range expansion of pine wilt disease. Its past colonization and dispersal process of the PWN were studied using DNA markers, mitochondrial DNA (mtDNA) and microsatellite (simple sequence repeat, SSR) regions (Kawai et al. 2006). Each type of molecular marker has a specific level of genetic resolution. The mtDNA is used for phylogenetic studies because it is transmitted predominantly through maternal lines and its evolution rate is high compared to nuclear DNA. The SSR has very high-resolution Mendelian markers. Each SSR locus consists of reiterated short sequences, for example, GAGA-GAGA—that are tandemly arrayed at a particular genomic location. As the rate of mutation in the numbers of repeats is often high, it is useful for detecting genetic diversification among populations.

After collecting sawyers from 24 sites in Japan, Taiwan and China (Fig. III.16), all the sawyers were stored in 99.5% ethanol. Mitochondrial DNA analysis was conducted on 2–4 individuals for all populations, and 18–30 individuals from Japanese populations were subject to SSR analysis. Genomic DNA was isolated using the Chelex method, one of the easiest methods to prepare template DNA for PCR. One of the legs of an adult was dissected from the body, and a piece of leg muscle (about 2 mm × 1 mm) was used for the analysis. The muscle was mixed with proteinase K and incubated in 5% Chelex solution (SIGMA) for several hours. After incubation, the samples were heated at 95°C for 10 min and then used as template DNA for PCR. Using primers COII-Croz (5′-CCACAAATTTCTGAACATTGACC -3′, Roehrdanz 1993) and tRNALeu-F (5′-GTGCAATGGATTTAAACCCC-3′), a ca. 700 bp fragment of mtDNA that included portions of cytochrome *c* oxidase II (COII) and $tRNA^{Leu\text{-}UUR}$, were amplified. The PCR products were sequenced directly.

Seven haplotypes were detected in the Asian populations. The nucleotide sequence designated haplotype A, which had the highest frequency among all samples, was deposited in the DDBJ/EMBL/GenBank nucleotide sequence databases (accession number AB191523). The mtDNA sequence data divided the pine sawyer populations into two major groups based on the phylogenetic tree

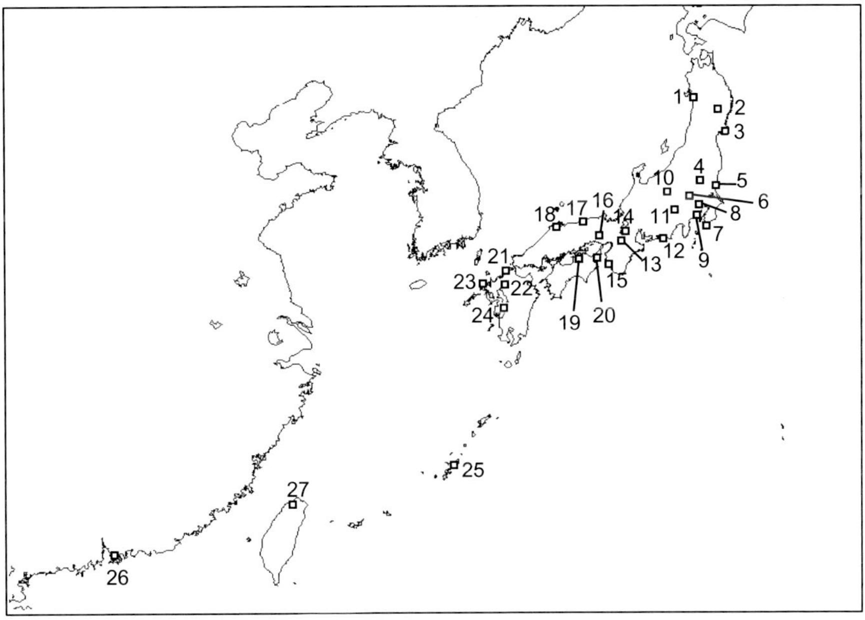

Fig. III.16 Collection sites in northeast Asia where *Monochamus alternatus* population structure was studied, from Kawai et al. (2006), with permission

(Fig. III.17): Clade A (haplotypes A–F) and Clade B (haplotype G). Clade B is monotypic, and occurs only in some populations of central Japan, while Clade A is widely distributed across Northeast Asia. The "clade" is a monophyletic group of taxa sharing a close common ancestry. If the rate of nucleotide substitution in the mtDNA studied sites can be assumed to be approximately constant over evolutionary time the proportion of accumulated mutations to total genes should relate to divergence time between clades. This is known as the molecular clock hypothesis. The molecular clock for coleopteran mitochondrial genes for the genus *Tetraopes* (Coleoptera: Cerambycidae) is inferred to be 1.5% per million years (Farrell 2001). Based on this information, the divergence of the two clades is calculated to have begun 1.45 million years ago, which corresponds to the Lower Pleistocene.

This haplotypic distribution can be considered to have been formed by two alternative scenarios. In the first, the sawyer colonized Japan at least twice from the Eurasian continent and descendent populations have since fused and dispersed together. Clade B would have colonized Japan during the Lower Pleistocene period because it is monotypic and was likely formed through a genetic bottleneck in a small region, maybe in Japan. Then Clade A would have come to Japan. The second

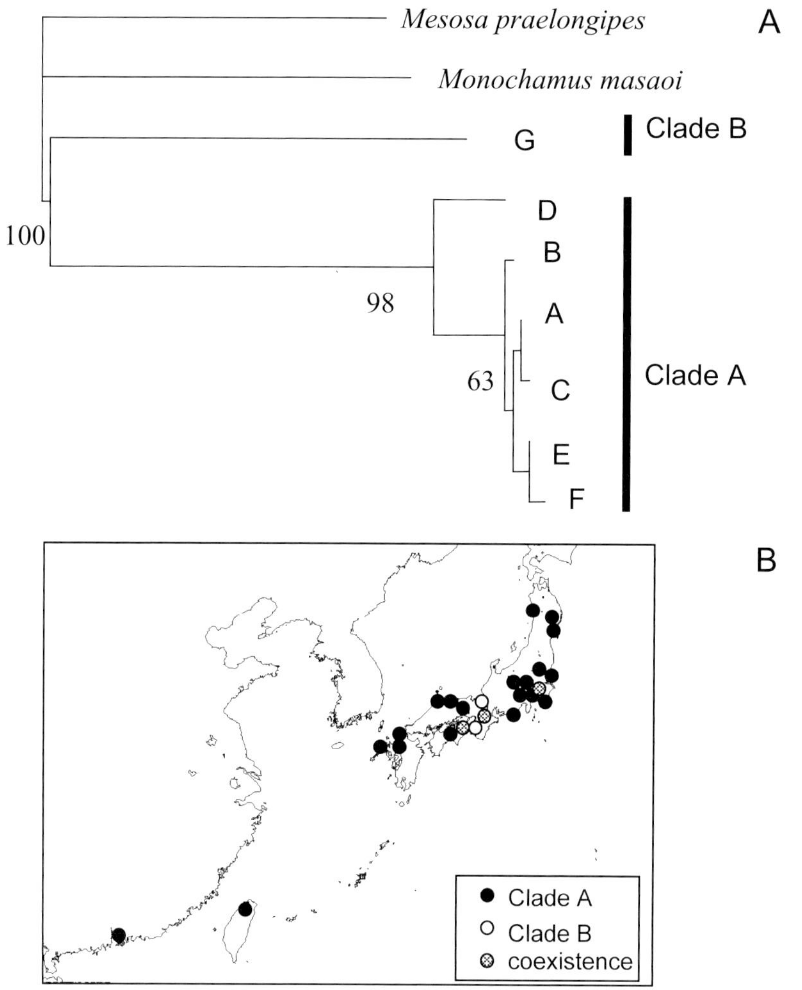

Fig. III.17 The neighbor-joining tree of *Monochamus alternatus* populations based on COII sequence variation using Kimura's two-parameter method. Bootstrap percentages of 1,000 replicates are shown for each branch; only bootstrap values >50% are shown for both trees (**A**), from Kawai et al. (2006), with permission. Distribution of each clade (**B**)

possibility is that the coalescence of genes, that is. tracking the ancestry of the genes of individuals between the two lineages, has been extended by population subdivision. Discordance between the mitochondrial lineage and phylogenetic relationship arose because mitochondrial separations predate the phylogenetic split and remain sympatrically. If this is the case, it is likely to have occurred by demographic chance, i.e. random loss of genetic variants in different lineages derived from a polymorphic common ancestry, so-called "lineage sorting".

Do two mitochondrial lineages represent cryptic species in the pine sawyer? Clade A and Clade B are probably the same species, because the population which includes rare mitochondrial lineage shows no special characteristics in SSR analysis (mentioned below) of genetic structures. As populations including Clade B are not differentiated from populations composed of Clade A, Clade B should have mixed completely with Clade A in nuclear DNA. There might not been reproductive isolation between Clade A and B, and ecological traits may not have differentiated between the two different lineages of mtDNA; however, further study is needed to determine whether genetic variations in ecological traits exist or not using rearing experiments and/or observation of population dynamics. Rearing in a constant environment and breeding experiments should reveal the extent of genetic variations and heritability of life-history traits. If there are some differences in population dynamics among populations in a similar environment, geographical genetic variation of ecological traits might affect the dynamics.

While mtDNA is useful for determining phylogenetic relationships because of its maternal inheritance and rare recombination, we should be aware of a fundamental distinction between mitochondrial gene trees and organismal trees, which are the historical pathway of the whole genome. The coalescence of genes might be extended by population subdivision, and mtDNA tends to introgress more easily than nuclear DNA. The population structure revealed by SSR analysis of nuclear DNA more clearly illustrates gene flow and demographic history. Population differentiations of *M. alternatus* are also investigated using SSR regions. Five SSR loci (N. Maehara et al., unpublished) developed with the dual-suppression PCR technique (Lian and Hogetsu 2002) were used. Genetic differentiation can be measured using the F_{ST} index, the standardized variance of allele frequencies among populations. The high overall F_{ST} of 0.0920 estimated from SSR data indicates differentiation among populations. There was significant differentiation among the populations of *M. alternatus*, even between the populations collected from nearby prefectures. Thus, there seems to be little ongoing gene flow between the *M. alternatus* populations of nearby prefectures.

Whether or not the genetic drift and gene flow in a population are in equilibrium can be determined by the "isolation by distance" test, regression analysis of geographic distances and pairwise population differentiation, $F_{ST}/(1 - F_{ST})$ estimated from SSR data (Hutchinson and Templeton 1999). If there is a correlation between geographic and genetic distances among populations, gene flow and genetic drift are in regional equilibrium and gene flow is reduced due to spatial separation of one population from another with low dispersal ability. Although significant differentiations were detected among *M. alternatus* populations, there was no correlation between geographical distances and genetic differentiation between populations (Fig. III.18), indicating that the populations throughout Japan were not in demographic equilibrium between genetic drift and gene flow, and were subject to genetic drift.

Analysis of the genetic structure from SSR data makes it possible to infer the dispersal and colonization processes of the populations. Strong genetic drift suggests intense population fluctuation. Thus, colonization into new habitats seems to

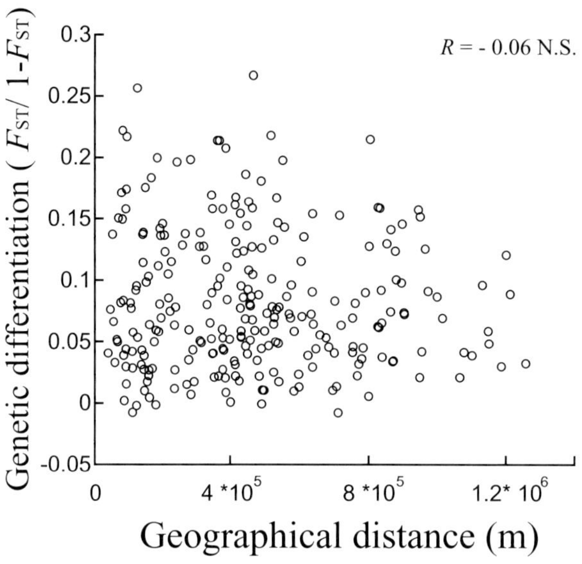

Fig. III.18 The relationship between genetic differences [$F_{ST}/(1 - F_{ST})$] and geographical distances in *Monochamus alternatus* populations sampled in Japan

be attained by a small number of adults. The sawyer population should rise rapidly due to the large number of PWN-killed trees. Range expansion of the sawyer population may have occurred by natural dispersal of sawyer adults not only on a small spatial scale but also by long-distance dispersal. If there were dispersal only on a small spatial scale, there would have been isolation by distance pattern; however, the populations of *M. alternatus* were revealed to have no geographical tendency in their genetic structure. Relocation of infected and damaged wood has enhanced the dissemination of pine wilt disease (Makihara 1998). Artificial introduction of sawyers might have aided in the long-distance dispersal of pine sawyer beetles. Repeated invasions of small numbers of individuals and rapid population growth may have formed the genetic structure of the sawyers.

18.3 Dispersal of *Monochamus alternatus* in the Northern Frontier of Pine Wilt Disease in Japan

Prevention of the invasion of uninfected areas by the PWN is one of the most important purposes of control. The move northward of pine wilt disease is a serious problem. The Tohoku (northeast) district of Honshu Island, Japan is the frontier area affected by pine wilt disease. In 1975, the first damage was reported in the southeastern part of Tohoku district and it has since spread (Fig. III.19; after Kamata 1996). The Ohu Mountain Range runs from north to south through the

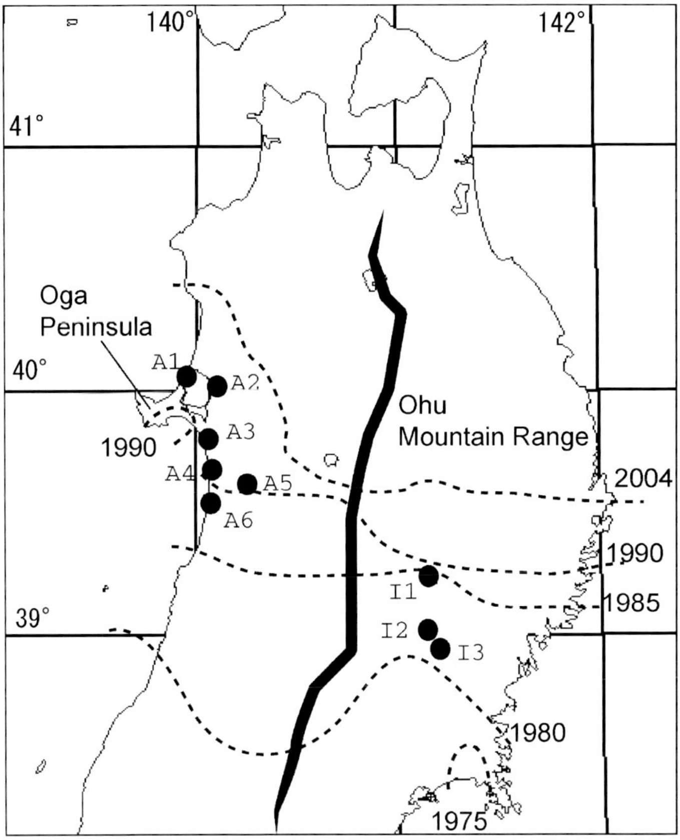

Fig. III.19 Sampling locations for *Monochamus alternatus* in the Tohoku district and pine wilt disease spread from 1975 to 2004, from Shoda-Kagaya (2007), with permission

district. Today, Akita Prefecture, on the western side of the Ohu Mountain Range, and Iwate Prefecture, on the eastern side, are the northern limits of the disease. To stop the spread of pine wilt disease, it is important to elucidate the sawyer's spatio-temporal process of range expansion associated with PWN in the Tohoku district. Knowledge of the dispersal of the pine sawyer is essential when developing an intensive protection program against pine wilt disease. Although the introduction of PWN was shown to influence the genetic structure of the sawyers beyond the regional scale (see Sect. 18.2), we have not been able to identify their dispersal route or the origin of introduced populations in a region.

According to the reports of the occurrence of damage, there seems to be two main trajectories of invasion of the PWN. The dispersal of PWN has occurred on both routes, the east and west sides of the Ohu Mountain Range, and has expanded its range in synchronism between the east and west sides. Thus, it is estimated that dispersal of the sawyer over the mountain range could have enhanced the PWN range expansion. In Akita, there seems to be another route from the southern part of the prefecture. Pine wilt disease occurred suddenly in the Oga Peninsula in 1988, maybe the result of artificial introduction of PWN-infected pine logs to a wood yard. Hence, there may be two invasion routes in Akita, from the southern front and from the peninsula.

Shoda-Kagaya (2007) showed the genetic population structure of *M. alternatus* in Tohoku district using SSR markers. The present study had three goals. The first was to map the population structure of *M. alternatus* near the frontier of the pine wilt disease-infested area using SSR markers. The second was to clarify the dispersal of *M. alternatus* populations and individuals to answer the following questions: (1) Do pine sawyers disperse between Iwate and Akita Prefectures across the Ohu Mountain Range? (2) Have the two populations from the south front and from the peninsula merged in Akita? The final goal was to relate genetic structural patterns on local and regional scales and to infer the sawyer's gene flow and associated stochastic processes on both scales.

The sawyers used in the present survey were collected near the frontier populations, and 9–41 individuals were collected per site in 2003 and 2004. Six populations from Akita Prefecture (A1–A6) and three from Iwate (I1–I3) were examined (Fig. III.19). Within each prefecture, the geomorphology is flat and few obvious geographical barriers exist. Adults were trapped using lures or collected from damaged logs. When sawyers were collected using traps, propylene glycol was added to the collection bottle. This is a non-volatile solution that prevents the DNA of sawyers from degrading. As propylene glycol is generally recognized as safe and biodegradable, it is a useful solution to collect samples for DNA analysis in the field. After collection, the legs were dissected and stored in 99.5% ethanol before use. Each individual was genotyped using the five SSR loci used in the previous study.

The genetic relationship among populations is shown using Multidimensional scaling (MDS) (Fig. III.20). MDS is a class of ordination techniques that displays complex relationships among populations in a small number of dimensions. These two axes could explain 98.2% of the variance. As can be seen, there are two groups on the first axis. There is an analysis for genotype data similar to analysis of variance (ANOVA), called analysis of molecular variance (AMOVA). Approximately 95% of genetic variability within populations could be explained by AMOVA. The remaining variation could be explained by the significant genetic heterogeneity between Iwate and Akita Prefectures; however, variations among populations within prefectures were not significant. While the sawyers diverged between the prefectures, their genetic components remained similar within the prefectures. It is interesting that significant isolation by distance was detected among populations within Akita Prefecture (Fig. III.21). This suggests that the population from the Oga Peninsula merges with the population from the south front, and/or restricted gene flow between populations occurs in a stepping stone manner, which is formed by

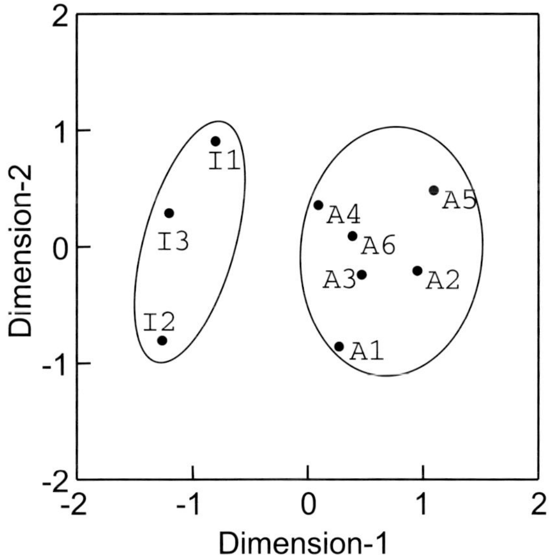

Fig. III.20 Multidimensional scaling analysis of frontier populations of *Monochamus alternatus* in Japan based on genetic distances, from Shoda-Kagaya (2007), with permission

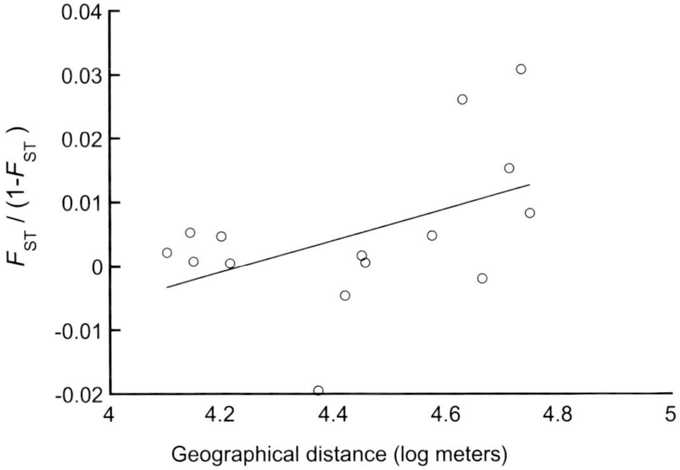

Fig. III.21 The relationship between genetic differences $[F_{ST}/(1 - F_{ST})]$ and geographical distances in *Monochamus alternatus* populations within Akita Prefecture, from Shoda-Kagaya (2007), with permission

migration restricted to adjacent populations. The latter assumes population equilibriums between gene flow and genetic drift. In case of the sawyer's recent colonization of this area, it seems difficult to propose that sufficient time has passed for the establishment of regional equilibrium, and the isolation by distance structure might have been formed by two invasion events from the peninsula and the south front.

The populations throughout Akita Prefecture are not genetically homogenous from the analysis of isolation by distance, that is, they have not dispersed freely among collection sites. The Akita populations occurred within a 28-km radius, and it is too large an area to disperse in one generation. The sawyer's ability of both short- and long-distance dispersal was examined in earlier studies. Studies using the mark-recapture method revealed that most adult dispersal takes place over a short distance (Shibata et al. 1986; Togashi 1990; Fujioka 1993) and that the dispersal distance of an adult throughout its life is 50–260 m (Togashi 1989). The longest measured dispersal of a marked adult captured by bait logs was 2.4 km (Ido et al. 1975). Takasu et al. (2000) simulated its mean distance for both short- and long-distance dispersal, which was calculated to be 14.2 m and 1.82 km, respectively. Based on pine wilt disease tracking, the largest dispersal of an adult sawyer was estimated to be 2–20 km (Kamata 1996). The propagation of pine wilt disease, however, can also be caused by human transportation of infested wood. Although it is difficult to infer the average or maximum dispersal distance from a genetic standpoint, the dispersal distance of 20 km of substantial individuals appears to be an overestimate. Further study for estimating dispersal distance using molecular data is needed. For correct estimation, we should use sophisticated statistical models and collect additional data of other loci.

These results lead us to the following conclusions: pine sawyers rarely spread over the mountain range, and the synchronized expansion on both the east and west sides of the range is most likely accidental. The Ohu Mountain Range is composed of mountains with altitudes of 1,000–2,000 m above sea level. Mountains of this altitude should work as geographical barriers to pine wilt disease spread. In Akita Pref., two invasion routes, from the Oga Peninsula and the southern part of the prefecture, have probably already merged. The dispersal ability of the sawyers is not so high and they cannot disperse over 20 km in one generation.

18.4 Development of Microsatellite (SSR) Markers

If a molecular technique is needed for vector species of pine wilt disease other than *M. alternatus*, you must find primer sets for target region DNA amplification. As universal primers of mitochondrial and nuclear DNA amplification for insects are commonly used, and mtDNA of *Monochamus* beetles is confirmed to be amplified by the primers used in Sect. 18.2, it is easy to start experiments for phylogenetic study; however, SSR primers are often species specific. In many fields on insects, it is necessary to develop SSR markers for the species. To develop such markers, the SSR region in the genome should be isolated, and this is suspected to be time consuming and to require a higher technique (e.g., DNA recombination and cloning technique using competent cells). It is true that developing SSR markers needs various equipment for carrying out DNA recombination and a P1 level laboratory according to Cartagena Protocol on Biosafety, but it has become easier than the traditional colony hybridization method because of advances in techniques for enriching SSR loci. It is not difficult for researchers familiar with DNA sequencing to develop new SSR markers. If the experiment proceeds without problems, these markers can be developed within 3 or 4 weeks. Here some efficient methods for isolating the SSR DNA region in sawyers are introduced and the methods are compared (Table III.6).

Enrichment of the SSR region using magnetic beads (Fig. III.22) is very popular in isolating SSR. Hamaguchi et al. (2007) showed that ants have a high rate of concentrating SSR loci for insects. First, the genomic DNA is isolated and digested using restriction enzyme *Sau*3AI, and cassette DNA (short DNA of known sequences, e.g., TaKaRa Cassette, *Sau*3AI) is ligated to the digested fragments. The

Table III.6 Comparison of isolating methods for simple sequence repeats (SSR)

	Hamaguchi (2007)	Lian and Hogetsu (2002)	Lian et al. (2007)	Traditional colony hybridization method
Duration of SSR isolation	Short	Medium	Very short	Long
Requirements for experimental equipment	Medium	Low	Low	High
Null allele	Low	Low	Medium	Low

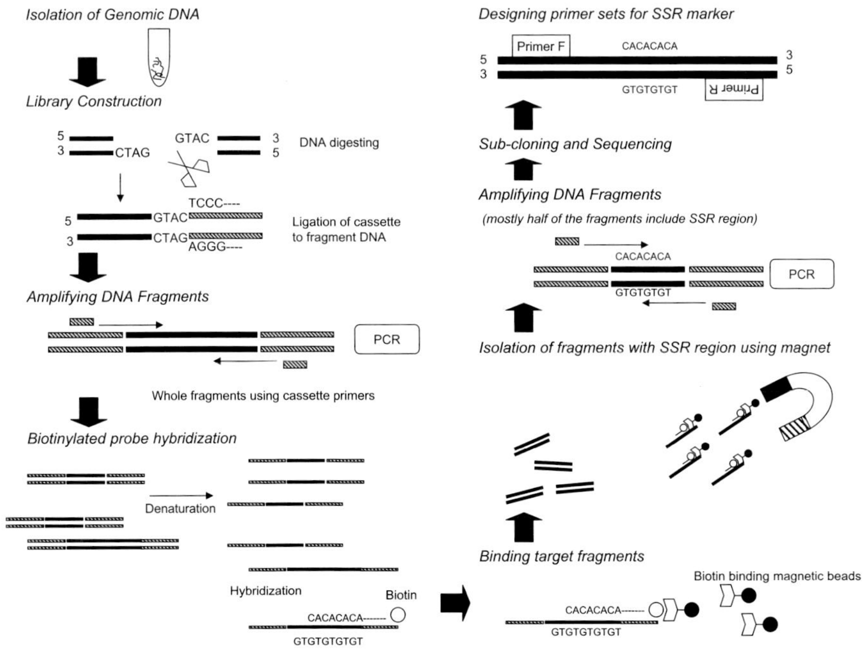

Fig. III.22 Diagram showing the steps to isolate SSR using the magnetic beads method (Hamaguchi et al. 2007)

fragments with cassette DNA at both ends are then amplified using cassette PCR primers complementary to the cassette DNA. Second, the SSR region is hybridized with a biotinylated SSR probe and fragments enriched with SSR regions. GT or CA repeats are suitable for SSR probes. Fragments including such a region are hybridized with these probes after denaturation. Then, biotinylated DNA fragments are captured to magnetic beads (e.g., Dynabeads M-280 Streptavidin, Dynal Biotech) and isolated using a magnet. After rinsing the beads, target DNAs are recovered by eluting them in boiling water. The enriched library is amplified again by cassette PCR, and the products are ligated into pGEM-T vector (Promega) and transformed into Epicurian Coli XL1-Blue MRF' supercompetent cells (Stratagene). The positive colonies are directly sequenced and SSR regions are screened. After designing primer sets over the SSR region, frequencies of polymorphisms can be checked. Primers complimentary to the flanking regions of the repeats can be easily designed using PRIMER 3 software (http://frodo.wi.mit.edu/primer3/input.htm; Rozen and Skaletsky 2000).

If it is necessary to develop SSR markers with the minimum experimental equipment, Lian and Hogetsu (2002) and Lian et al. (2006) have the best solution (Fig. III.23). The former is called a dual-suppression PCR technique and the latter is a compound microsatellite method. In the dual-suppression PCR technique, six kinds of adapter-ligated libraries should first be constructed using restriction enzymes, *Eco*RV, *Ssp*I, *Alu*I, *Afa*I, *Hinc*II and *Hae*III. The restricted fragments are

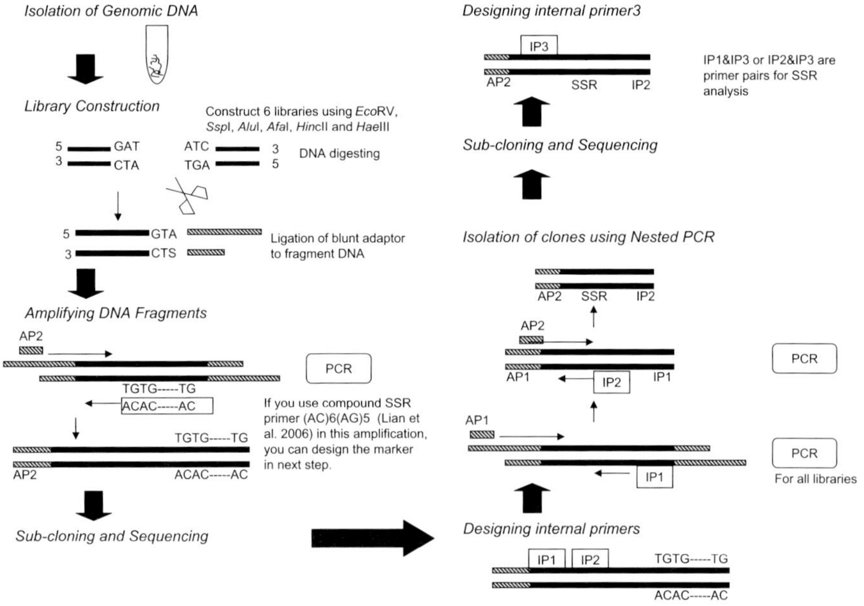

Fig. III.23 Diagram showing the steps to isolate SSR using the dual-suppression-PCR technique (Lian and Hogetsu 2002)

ligated with a specific blunt adaptor consisting of a 48-mer: 5′-GTAATACGACT-CACTCACTATAGGGCACGCGTGGTCGACGGCCCGGGCTGGT-3′ and an 8-mer with the 3′-end capped by amino residue: 5′-ACCAGCCC-NH_2–3′. Using the adapter primer, AP2 (5′-CTATAGGGCACGCGTGGT-3′) and SSR primer [e.g., $(AC)_{10}$], fragments which have SSR at one end are amplified. The products should be subcloned and to be sequenced. Between the adapter primer and SSR sequences, two internal primers (IP1 for outer and IP2 for inner) should be designed. These primers direct this 3′-end to SSR. Using an adapter primer, AP1 (5′-CCATCGTA-ATACGACTCACTATAGGGC-3′) and IP1, products including IP1 sequences are amplified for all libraries. Nested PCR for all of the former PCR products are carried out using IP2 and AP2. Mostly, some of the libraries produce a single and adequately sized band. The PCR product is again sub-cloned and sequenced. The other primer is designed between the adapter primer and SSR sequences with 3′-end to SSR. "IP1 and IP3" or "IP2 and IP3" can be used as primer sets of SSR markers.

The compound SSR method is the easiest way to develop SSR markers and its cost is maybe lowest during development and analysis. The target region of this method is compound SSR, for example, $(AC)_n(AG)_n$. The construction method of the library is the same as in dual-suppression PCR, but it needs only one kind of restriction enzyme, *Eco*RV. Using a compound SSR primer $(AC)_6(AG)_5$ or $(TC)_6(AC)_5$ and AP2, fragments are amplified from the library. After the product is subcloned and sequences of the fragment are determined, a specific primer is designed from a sequence flanking the compound SSR. The compound SSR primer and the designed primer can be used as a primer set. With this method, it is very convenient to use only one or two kinds of fluorescent primers in polymorphism analysis using sequencers, because the compound SSR primer is common to all loci; however, if there are short tandem variations (less than five repeats) in each compound SSR, genotyping of the allele should be inaccurate or the allele will become null.

For all methods, it is important to avoid contamination with other species, especially for the PWN when isolating genomic DNA of vector species. If you rear *Monochamus* species larvae, you can obtain nematode-free samples to take the larvae before making pupal chambers in wood. As only adults were used here, legs were dissected from the pine sawyers and the muscle removed to avoid contamination from the PWN genome. Ten mg of muscle could be harvested from the legs of one male pine sawyer and it yielded enough genomic DNA to construct a DNA library. However, a vector with a heavy load of PWNs will also carry the nematode in its legs; thus, this method is not recommended. QIAGEN DNeasy® Blood and Tissue Kit could isolate high-quality DNA for a precise digestion reaction in the following steps.

18.5 Conclusions

Molecular biology has brought new tools to forest pest studies which are especially useful for investigating past dispersal. The genetic population structure of *M. alternatus* in Asia suggested their dispersal and colonization process. Isolation by distance test showed strong genetic drift on the populations. As strong genetic drift suggested intense population fluctuation, colonization into new habitats seems to be performed by a small number of adults of the beetle and then the population should rise rapidly. Ongoing gene flow between prefectures seems not to be high between prefectures in Japan because of high genetic differentiations among populations. The genetic structure of the frontier population in Japan revealed their ongoing dispersal area. They could not go over mountain ranges with an altitude of 1,000–2,000 m above sea level and they could not disperse over 20 km in one generation in a flat landscape. At the end of this article, some efficient methods to isolate SSR regions were introduced, which can be used as the most powerful markers to research the dispersal of *Monochamus* vectors. I strongly hope to develop SSR markers for other and potential vectors of PWN, which can be utilized to plan how to protect pine forests from pine wilt disease all over the world.

Acknowledgments I thank K. Togashi for his valuable comments and K. Futai for suggesting this article. I am grateful to K. Kobayashi, M. Kawai, T. Maehara, R. Iwata, A. Yamane, C. Lien, Z. Zhou, T. Hogetsu, K. Fujita, T. Goto, K. Nakamura, and H. Kinuura for their cooperation with this study.

19
Concluding Remarks

In this part, we dealt primarily with *Monochamus alternatus* because there is a great deal of information available on this vector of the PWN; however, this information has been obtained from a relatively limited area, Japan, compared to the PWNs range in East Asia. This presents some problem in understanding the complete biology of *M. alternatus* in relation to pine wilt disease. Most likely future research will effectively clarify the differences in morphology, physiology, behavior, and genetic relationship among *M. alternatus* populations inhabiting different localities. However, the population performance of *M. alternatus* in pine stands is essential for understanding the pine wilt epidemic in East Asia. Studies such as morphology and molecular ecology may help us choose beetle populations to be targeted.

Modeling is also a helpful tool in understanding the spread of PWN within and over pine stands. Molecular ecology studies may support the predictions by models or may point out the lack of information essential to construct a model that will represent pine wilt. Modeling, analytical, and molecular ecology approaches complement each other.

Part IV
Host Responses and Wilting Mechanisms

20
Introduction

Following the discovery that the pine wood nematode (PWN), *Bursaphelenchus xylophilus*, is the pathogen of pine wilt disease (Kiyohara and Tokushige 1971), researchers focused on identifying the wilting mechanism. In recent years that effort has been aided by dramatic improvements in the various techniques and other technological advancements that have been employed in these studies. The purpose of the authors here is to discuss some of those techniques and advancements as related to wilting of host pines. The artificial propagation of PWN and an inoculation technique were first established by Mamiya (1980, 1984b), and then physiological measurements were conducted (Tamura et al. 1987, 1988). During the 1980s, highly virulent PWNs were selected from many cultures obtained from dead pines (Kiyohara and Dozono 1986; Kiyohara and Bolla 1990). It is very important to conduct inoculation tests with virulent races of PWN that successfully kill pine trees, as observed during natural infection. Early studies used trees 10–20 cm in diameter later while smaller and younger saplings less than 5 years old were used (Mamiya 1985). Supported by these fundamental achievements, detailed investigations to clarify the wilting mechanism have been conducted in Japan since the 1980s (Tamura and Dropkin 1984; Tamura et al. 1988). By the end of the 1990s hundreds of reports had been published on the phenomena related to trees inoculated with the PWN (Fukuda 1997; Yamada 2006) as well as reports on incidents associated with natural infection and the behavior of vector beetles (Kishi 1995). Unfortunately, most research papers, especially in the two decades after the discovery of the pathogen, went unnoticed outside Japan because they were written in Japanese, and too they lacked an English summary (Hashimoto 1980; Kiyohara and Dozono 1986). Kishi (1995) compiled a vast amount of historical and research data obtained from over 1,700 publications on pine wilt in Japan into a book (translated into English).

In this part, due to space limitations, only the highlights of the reports are cited. Recently, the most significant progress has been made in studies on the distinction of primary and secondary changes, or the main events and byproducts occurring in infected trees (Fukuda 1997). Today, many of the events that occur during disease development are theoretically interpreted from a physiology or biochemistry

Table IV.1 External symptoms and internal changes observed in pine wilt-susceptible saplings inoculated with the pine wood nematode, *Bursaphelenchus xylophilus*

		Early phase		Developing phase	
	Stage	**1**	**2**	**3**	**4**
External	Symptom	None	→	Discoloration of old needles	Discoloration of young needles → Death
	Oleo-resinosis	Normal →	Decreasing →	None	
Internal	Cells	Change in secondary metabolism	→	Partial necrosis →	Necrosis in wide area
	Sap ascent	Normal →	Blockage start	Low conductivity →	Completely stop
Pine wood nematode		Low population	→	Propagation	Extensive propagation
Time (weeks) Example of *Pinus thunbergii*		1	2	3	4 and beyond

viewpoint. Here we first describe chronologically the various phenomena occurring in pine trees as related to the activities of PWNs and describe how infected trees are killed. Disease development is summarized in Table IV.1, and is based the ideas of Suzuki (1992), Fukuda (1997), and Yamada (2006). We focus on the internal phenomena that occur, especially, in the early phase of infection before symptom development. Based on the available data, we discuss the causal factors to explain the wilting mechanism, the fate of infected trees, and their susceptibility and resistance. In the last chapter of this part, a hypothesis about a possible symbiotic relationship between the PWN and bacteria is presented in relation to pine wilt development.

21
Physiological Incidences Related to Symptom Development and Wilting Mechanism

Keiko Kuroda

21.1 Introduction

There are only a few wilt diseases of trees, and most of these are caused by fungi, for example Dutch elm disease, oak wilt, and blue stain disease of conifers (Sinclair et al. 1987). In contrast, pine wilt disease and red ring disease of coconut palm (Nowell 1919) are caused by nematodes. Despite a long research history, more than 100 years in some cases, there is no tree disease in which the wilting mechanism has been completely defined. In the case of pine wilt, the wilting mechanism has been relatively well defined. It would not be entirely true to say that the wilting mechanism of pine wilt is unexplained because some of the information remains hypothetical. This misunderstanding probably comes from the fact that the specific toxins directly inducing wilt such as cerato-ulmin in Dutch elm disease (Takai 1974) have not been identified in pine wilt disease. In this section a description is given of what is currently known about the wilting mechanism in pine wilt disease.

To understand the host reaction and disease development (Fig. IV.1), knowledge of the anatomy and physiology of the host tree is essential. Here, the functional anatomy of pine trees is briefly introduced in relation to the behavior of the pine wood nematode (*Bursaphelenchus xylophilus*, PWN). More information on tree anatomy and physiology is available in Evert (2006), and Tyree and Zimmermann (2002). Next the reaction of tree cells against the pathogen is described. The wilting mechanism is explained with a focus on the blockage of sap flow, and on the cytological aspects. Finally, the resistance or tolerance against the PWN observed in some species and resistant cultivars (clones) that were obtained from selective breeding are compared to the phenomena in susceptible trees. The PWN inoculation

Kansai Research Center, Forestry and Forest Products Research Institute, Momoyama, Fushimi, Kyoto 612-0855, Japan

Tel.: +81-75-611-1201, Fax: +81-75-611-1207, e-mail: keiko@affrc.go.jp

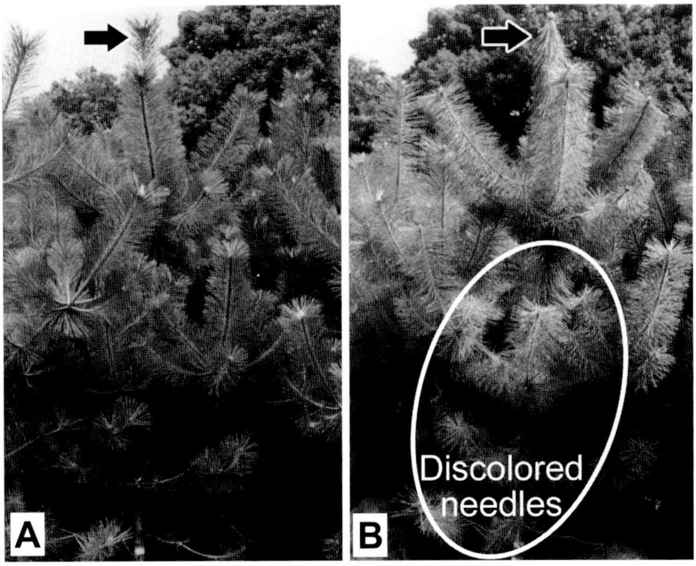

Fig. IV.1 Symptom development of *Pinus thunbergii* sapling inoculated with pine wood nematode, *Bursaphelenchus xylophilus*: **A** healthy 5-year-old sapling; **B** start of old-needle discoloration and drooping of apical needles 3 weeks after inoculation. Trees indicated by *arrows* are the same tree of different dates (see Color Plates)

technique is described in the last of this chapter (Column) for those interested in conducting further experiments on wilting mechanisms.

21.2 Structure of Pine Tissue and Cell Functions Related to Infection and Defense

In the stems of conifers, most xylem tissue consists of a fistulous water conduit called a "tracheid" (ca. 1-mm long and 30 μm in diameter). Ray tissue crosses orthogonally to tracheids. In the case of *Pinus* species, the ray tissue consists of ray tracheids and ray parenchyma cells. Ray parenchyma cells survive for many years, sometimes in excess of 10 years, and play important roles such as the accumulation of assimilates as starch grains, secondary metabolism, and the accumulation of toxic waste products (Hillis 1987). The function of ray cells, however, has not been fully defined. *Pinus* species contain vertical and horizontal resin canals (ducts) in the xylem (Figs. IV.2, IV.3). The horizontal resin canals pierce the center of some ray tissues. The cortex contains vertical resin canals (Fig. IV.2). Between the phloem and xylem, cambial initials produce new cells and young immature cells exist on each side of the cambium.

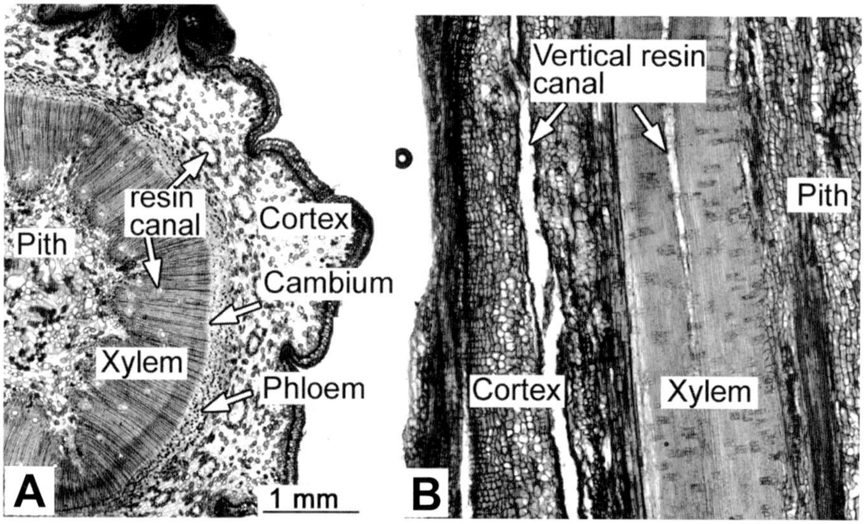

Fig. IV.2 Structure of a current year *Pinus densiflora* shoot. **A** Cross section, **B** radial section. Stained with nile blue (see Color Plates)

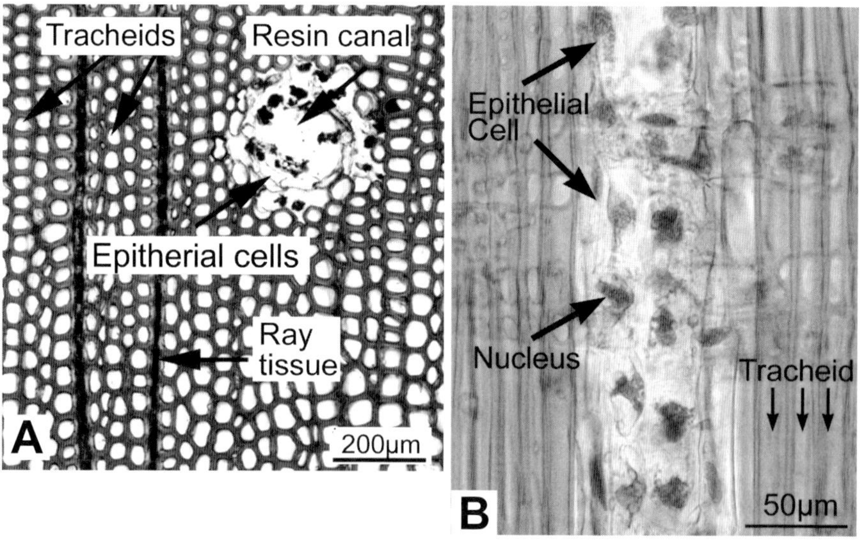

Fig. IV.3 Xylem tissue of *Pinus* species. **A** Cross section of *Pinus thunbergii*. **B** Radial section of a vertical resin canal in the xylem of *P. densiflora*. The diameter of the resin canal in *P. densiflora* is thinner than that in *P. thunbergii*. Stained with nile blue (see Color Plates)

When pine sawyers bearing the PWN feed on current and 1-year-old pine shoots and branches, PWNs invade the resin canals from the broken end of the canals in the cortex and xylem caused by pine sawyer feeding. The PWN moves down the shoots of susceptible pine trees through the vertical resin canals and invades the main stem of the tree. The PWN spontaneously pokes into some tracheids but cannot move a long distance through the tracheids because it is much thicker than the bordered pits (pit aperture is about 2–3 μm; see Fig. IV.8) that link the tracheids.

Disease development progresses in two phases and four stages (Table IV.1). In the early phase of the disease (Stages 1 and 2 in Table IV.1), the PWN moves exclusively in the resin canals and does not feed on or kill the cambial cells. In the final stages of pine wilt disease, as the tree is approaching death, the PWN population increases (Stage 4 in Table IV.1), the nematode is present in the cambial zone, and the immature cells around the cambium appear degraded (Kusunoki 1987), probably by hydrolytic enzymes.

In branches and stems younger than 2 or 3 years old, the epidermis, cortex and thin secondary phloem are outside the cambium (Fig. IV.2). As the branches and stems grow thicker, the periderm, which is dead tissue, forms in the cortex (Evert 2006). Finally, pine stems and branches consist of the outer bark (periderm), inner bark (secondary phloem), cambium and xylem as the dead cortex is sloughed off the surface of stems and branches (Fig. IV.4). The PWN has to use the resin canals in the xylem for vertical migration in stems and branches older than 3 or 4 years of *Pinus thunbergii* and *P. densiflora* because the secondary phloem does not have vertical resin canals (Ichihara et al. 2000a,b).

The resin canals arise as intercellular spaces formed by separation of parenchyma cells recently derived from the cambium (Evert 2006). Epithelial cells (epithelium) that produce resin surround the space (Fig. IV.3). The inner diameter of the resin canal in the xylem is 50–100 μm and that in the cortex is much larger, sometimes over 500 μm in *P. thunbergii* and *P. densiflora*, although the size varies among *Pinus* species. The length of the vertical resin canals has not been determined. Vertical resin canals are in contact with horizontal resin canals and have an opening in the contact area.

21.3 Behavior of the PWN in Tree Tissue and the Reaction of Tree Cells

When pine tissue is injured mechanically or by insect feeding, epithelial cells immediately synthesize and exude resin that contains resin acid and volatile terpenes (Hillis 1987). Most of the PWNs on the wounded surface of pine tissue are trapped in the sticky resin and therefore cannot invade the resin canals. Large numbers, from 3,000 to 10,000, of the PWNs are used in inoculation tests because only about 10% of the inoculum successfully invades the tissue (unpublished data). Although the resin effectively prevents the initial invasion of nematodes, it does

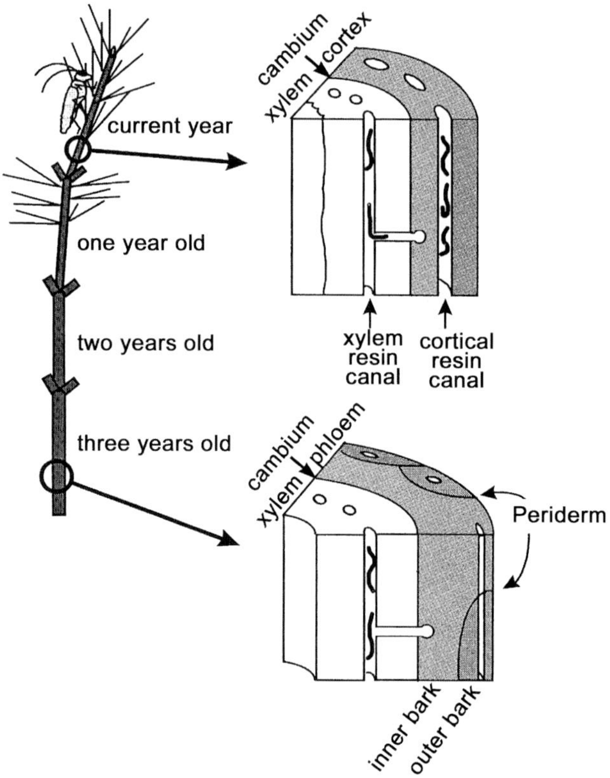

Fig. IV.4 Three-dimensional structure of a pine shoot (modified from Ichihara 2000, with permission)

not prevent the migration of PWNs in the tissue since nematodes easily move through the resin canals of pine wilt-susceptible pine trees. PWNs that invade resin canals presumably feed on the contents of living cells during migration (Tamura and Mamiya 1979; Kuroda and Mamiya 1986). PWNs exude hydrolytic enzymes, such as cellulase and pectinase (Odani et al. 1985; Kuroda 1987; Kikuchi et al. 2004), and absorb the contents of host cells. The effect of hydrolytic enzymes, the break down of cell walls, is easily observed on callus tissue used to culture PWNs (Tamura and Mamiya 1979).

The first anatomical change in the host can be observed under the light microscope by staining of the thin sections (about 20–30 μm in thickness) of host tissue with safranin and fast green (Mamiya 1980). A few living cells change their stainability along the migration route of the PWN and in the adjacent ray tissue at the early phase (first stage) of infection (Table IV.1). Necrosis and swelling of epithelial cells (tylosoid formation) occur spontaneously in such resin canals, except where

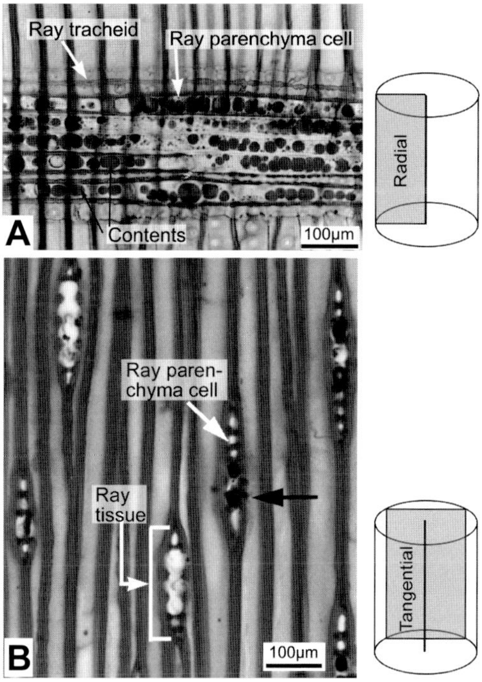

Fig. IV.5 Ray parenchyma cells in the xylem of *Pinus thunbergii* stained with nile blue. **A** Contents of healthy cells (*radial section*). **B** Exudates (*arrow*) and changes in the cell contents in shape and stainability are observed at the developing phase of the disease (*tangential section*) (see Color Plates)

cells are directly killed by the many PWNs at the inoculation sites. In addition, changes in cell contents, such as enlargement and destruction of vacuoles, have been reported (Nobuchi et al. 1984; Fukuda et al. 1992a). Changes in the stainability of affected ray parenchyma cells were also detected using Sudan III and nile blue, dyes for staining lipids (Fig. IV.5; Mamiya 1985; Ichihara et al. 2000a; Hara and Futai 2001). The production of secondary metabolites by the defense reaction (Hillis 1987) and subsequent decline and death were observed in those cells. The secondary metabolism is the biochemical process that does not relate to the maintenance of host life. In trees, materials such as terpenoids, phenolic compounds, and stilbenoids, which are formed at heartwood formation and as a reaction to wounding or infection, are called secondary metabolites. Here such reaction of trees is referred to as defensive. Dyes for light microscopy such as nile blue or Sudan III, do not stain specific substances; therefore, the materials produced in parenchyma cells cannot be identified by using these dyes.

By using gas chromatography, an increase in volatile terpenes, such as α-pinene and β-pinene in the xylem of *P. thunbergii* and *P. densiflora*, was detected as early as 3 days after PWN inoculation, and the concentration peaked approximately 2 weeks in the second stage of disease development (Table IV.1; Fig. IV.6; see

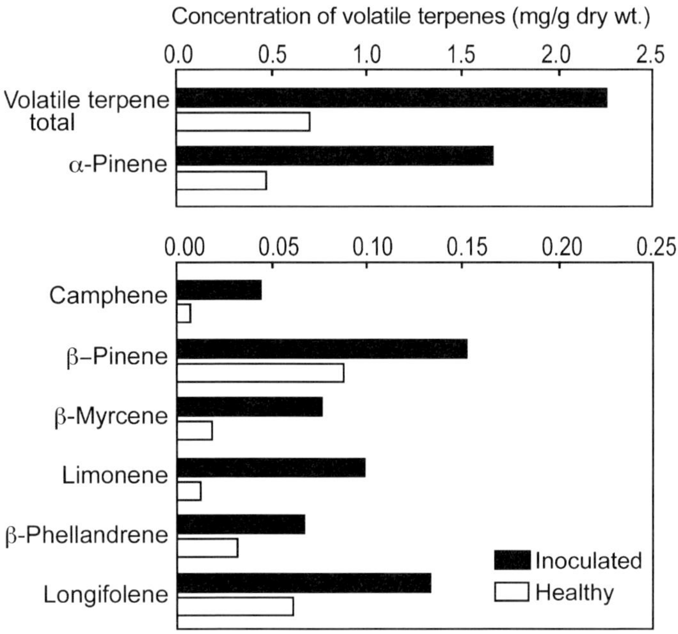

Fig. IV.6 Increase in volatile terpenes in the xylem of *Pinus densiflora* 1 week after pine wood nematode inoculation, at the initiation of dehydration (Kuroda et al. 1991b)

Chap. 23; Kuroda 1989, 1991). These substances are assumed to be synthesized in the epithelium and ray parenchyma cells as a defense reaction and then released into the lumina of neighboring tracheids (Fig. IV.5B). Cells contributing to the secondary metabolism degrade and then die as occurs during heartwood formation (Hillis 1987). In the second and third stages of the disease (Table IV.1), oily droplets are observed in the tracheids around resin canals and along the ray tissues, adhering to the inner wall of tracheids. During this time, oleoresin exudation from the wound decreases and stops (Table IV.1). Therefore, this cessation has been used to diagnose pine wilt disease just before symptom initiation. Biochemical incidences related to defense reaction are described in detail in Chap. 22.

PWNs migrate very rapidly in the shoot and branches, 150 cm day^{-1} at maximum, and invade the main stem (Kuroda and Ito 1992). Since the horizontal and vertical resin canals are distributed densely (Fig. IV.2) in the xylem tissue, the PWNs can spread throughout a 10-year-old sapling (height: 2.5–4.0 m) within approximately a week (Hashimoto and Kiyohara 1973; Kuroda and Ito 1992); however, not all PWNs reach the roots. Some stay in the branches or upper part of the stem, and sometimes they congregate in restricted areas of the stem in older and big trees. In seedlings less than 1-year-old or 30 cm in height, PWNs quickly spread throughout the seedlings, and symptoms may develop quickly (Ichihara et al. 2000a; Kuroda et al. 2007).

Activation of the secondary metabolism both in the parenchyma cells and the epithelium do not occur synchronously throughout a tree. Ichihara et al. (2000a) confirmed that changes in cell content occur in connection with the distribution of the PWN. The host reactions sometimes appears to occur with no PWN contact because the PWN moves very fast and is seldom found near affected cells. Cytological changes, aging, and necrosis of parenchyma cells observed under a microscope are local in the early phase, and various stages of cell degradation are observed within a few millimeters. Therefore, microscopic incidences observed in a small tissue sample from an infected tree cannot be used to assess disease development in the whole tree.

This pathogenic nematode mates and reproduces in pine tissue. Even in susceptible hosts, the population of the PWN in the tree tissue is very low, less than 10 per g dw (oven-dried weight) during the early phase of infection (Table IV.1; Kuroda et al. 1988). After the needles start to discolor at about 3 weeks (Fig. IV.1), the PWN population starts to increase and exceeds over 100 per g dw. During this period, necrotic cells increase in ray tissue and the cambium, and the water content in the xylem decreases dramatically (Kuroda and Ito 1992). This suggests that at this time the environment in the tree stem becomes suitable for nematode propagation.

21.4 Wilting Mechanism: Disturbance and Blockage of Sap Flow Resulting from the Dehydration of Tracheids

The first symptom observed in *P. thunbergii* and *P. densiflora* is the sudden reddish discoloration of old needles (formed in previous years; Table IV.1; Fig. IV.1). Symptom development in these pine species varies depending upon tree age, weather, and soil moisture. In an inoculation experiment with 5–12-year-old *P. thunbergii* and *P. densiflora* saplings, this period was shown to be around 3 weeks when inoculation was done at the end of July. The disease then develops to the discoloration and wilt of younger needles and results in host death 1–2 months after inoculation (Kuroda et al. 1988). In naturally infected trees, wilt and death usually occur following summer drought, September to October in Japan. In cooler areas, some trees die the following spring (Kishi 1995). Seedlings less than 1-year-old or below 30 cm in height develop symptoms more rapidly than older saplings and older trees (Kuroda et al. 2007).

In the early phase before the initiation of visible symptoms, 2–3 weeks after inoculation of saplings, the PWN population in pine stems is still low. At this time, a small number of PWNs are spreading in the resin canals, however, another significant change has already started in the pine tissue: the disturbance and partial blockage of the sap ascent occurs in the xylem as some tracheids that transport

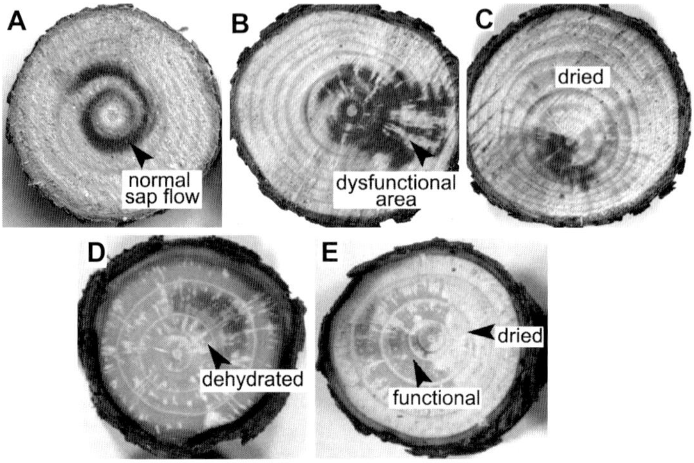

Fig. IV.7 Development of the disturbance and decrease of sap flow after inoculation of *Pinus thunbergii* with the pine wood nematode. An acid fuchsine solution was injected in the base of trunks **A** to **C** before the tree was cut down. **A** Normal spiral sap flow in healthy condition before inoculation of the pine wood nematode. **B** Three weeks after inoculation. Sap flow is disturbed by the partial blockage of sap ascent. **C** Four weeks after inoculation. Sap flow decreases extensively, and necrosis is occurring in the cambium. **D** *White patches* of dehydrated areas are observed in the early phase (2 weeks after inoculation). **E** Desiccation of the xylem is progressing at the start of old-needle discoloration in the developing phase (3 weeks after inoculation) (see Color Plates)

water from the roots to the shoot become dysfunctional (Fig. IV.7). In a cross section of the main stem, small white patches can be observed with the naked eye as shown in Fig. IV.7D (Kuroda et al. 1988, 1991b). The whitish part of the xylem is filled with gas or air (Fig. IV.8). A dye solution, acid fuchsine (0.1% aqueous solution), injected from a hole on the lower trunk, indicates the route of xylem sap ascent (Tyree and Zimmermann 2002). The white patches remain unstained by the dye (Fig. IV.7A–C); therefore, the tracheids in those patches have been determined to be dysfunctional (Tamura et al. 1987) by the dehydration of the tracheids (Kuroda 1991; Fukuda et al. 1992a).

The dehydration of water conduits such as vessels and tracheids by gas is known as embolism. This occurs every day even in healthy plants; however, the empty conduits are usually refilled with water, and water transport resumes. Therefore, plants do not wilt easily (Sperry and Tyree 1988). In contrast, in PWN-infected pines, the dehydrated areas shown in Fig. IV.7 do not refill with water even when the trees are well watered. This type of dehydration in sapwood is known as a dry zone, which occurs following infection by pathogenic fungi (Hillis 1987). The formation of a dry zone is a part of the defense reaction to infection by microorganisms, and sometimes prevents the wide distribution of fungi in the host tissue. The dehydrated area increases in the case of pine wilt, whereas the dry zone

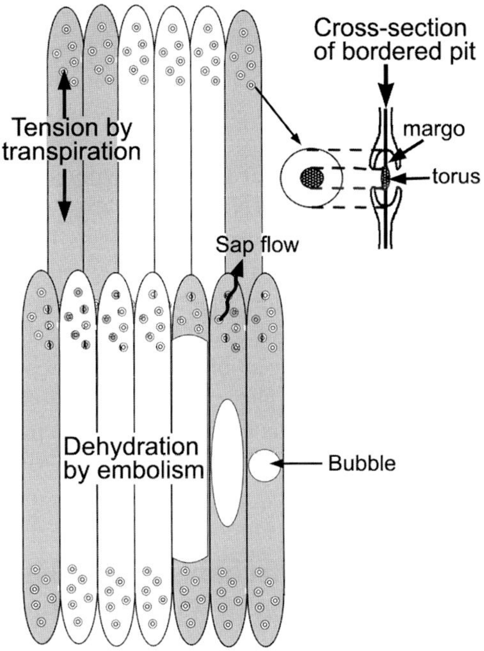

Fig. IV.8 Embolism occurring in the tracheids of pine sapwood infected with the pine wood nematode and structure of a bordered pit. Bubble formed in the tracheids immediately enlarges and dehydrates the tracheid by strong tension induced by transpiration. The dehydrated area spreads to adjacent tracheids in a chain reaction (see Color Plates)

formed by fungal infection usually remains smaller. As the dehydrated area increases from 2 to 3 weeks after inoculation with PWNs (Kuroda 1991), water conductivity significantly decreases (Ikeda and Suzaki 1984). During the time when the xylem water content decreases and the tissue is drying (Fig. IV.7C, E), needles begin to discolor. Photosynthesis decreases in the developing phase for the water deficit in leaves (Melakeberhan 1990; Fukuda et al. 1992b; Fukuda 1997).

As the desiccation of the xylem progresses, the cambium and phloem become necrotic (Fig. IV.7C), and the tree dies (Kuroda et al. 1988, 1991b). In 5–10-year-old saplings of *P. thunbergii* and *P. densiflora*, xylem sap ascent stops in the stem at 4–6 weeks after inoculation. This period varies depending upon the tree's age and size. The process of xylem desiccation by the dehydration of tracheids progresses rapidly compared to the cuttings of pine shoots that are exposed to the atmosphere for a few weeks.

In wilt diseases of trees, the period from infection to death is long, a couple of months or more (Christiansen and Solheim 1990), in contrast to that of herbaceous plants. Regarding the mechanism of symptom development in tree diseases, various physiological and biochemical processes are taking place in tree stems in the early period preceding the first appearance of symptoms. Some anatomical incidences and changes in metabolism detected in the late stage of disease development do not cause the symptoms, rather, they are the result of either byproducts or disease development. To detect the actual cause of these symptoms, research must be conducted using tree samples obtained at a much earlier stage of disease development. When pine trees infected with PWN are dying (developing phase: stage 4 in Table IV.1), various kinds of fungi, bacteria, and other microorganisms are detected throughout the tree. It has been hypothesized that the real pathogen of pine wilt disease is not the PWN but a fungus (Morooka et al. 1983). However, it was later demonstrated that the microorganisms in pine trees inoculated with PWNs were the same in healthy trees before symptom initiation and that only a few weakly pathogenic fungi and bacteria were present (Kuroda and Ito 1992). Blue-stain fungi, species of *Ceratocystis* and *Ophiostoma*, brought into pine stems by bark beetles are not the causal agent of pine wilt either because infection with these fungi occurs in the developing phase of the disease (Table IV.1) when sap flow has almost stopped. The population dynamics of the pathogen must be also considered in relation to internal host changes rather than visible symptoms.

21.5 Wilting Mechanism: Cytological and Physiological Elucidation of Xylem Dysfunction

For the discussion about what causes the dehydration of tracheids in infected trees and why sap flow does not recover, understanding sap flow in healthy plants is essential. Research on sap ascent in healthy trees has progressed in the last few decades (Canny 2001; Tyree and Zimmermann 2002). Although the detailed mechanism of sap flow is still in dispute, the following explanation of xylem dysfunction in pine trees infected with the PWN can be given on the basis of the "cohesion-tension theory", which was well established during the 1980s.

The cohesion-tension theory supports the view that healthy conifer trees can absorb water from the roots and transport the sap to the top of tall trees because the water in the xylem conduits is pulled up by the tension caused by transpiration from leaves and the cohesive strength of water molecules. When water molecules cannot endure the strong tension, the water column in the conduits (tracheids in pine trees) is broken, and bubbles appear in the xylem sap. This phenomenon resembles the boiling of water under low pressure. Very tiny bubbles generated in the sap are assumed to expand suddenly. The water in the conduits is drained immediately and water vapor fills the lumina (Fig. IV.8; Sperry and Tyree 1988; Tyree and Zimmermann 2002). The blockage of sap flow caused by such dehydration is called embolism. In healthy plants, the tension induced by transpiration becomes weaker after sunset or rainfall. Under such conditions, dehydrated tracheids are refilled with water and the function of water transport recovers (Sperry and Tyree 1988); therefore, plants do not wilt as easily when soil moisture is adequate.

Xylem sap is thought to be almost pure water (Tyree and Zimmermann 2002). Sperry and Tyree (1988) reported that some substances with surface tension lower than xylem sap, such as butyl alcohol, are blended in the xylem sap, the tensile strength of the sap becomes weaker, and embolism is thus promoted in the tracheids. A similar process is assumed to occur in pine stems infected with PWNs. In the xylem of *P. thunbergii* and *P. densiflora* inoculated with PWNs, an increase of monoterpenes is detected in the early phase (Fig. IV.6; Kuroda 1989, 1991). The surface tension of volatile terpenes (α-pinene: ca. 25 dyn cm^{-1}) is much lower than that of water (ca. 72 dyn cm^{-1}). Those substances weaken the tensile strength of sap and can induce embolism when they are released from epithelial and ray parenchyma cells into the sap. The injection of α-pinene into the stem of *P. thunbergii* artificially induced dehydration (Kuroda 1991). Other substances with lower surface tension may be involved in this event besides monoterpenes, but these substances remain unknown.

When an embolism occurs in healthy plants, ultrasonic acoustic emission (AE) is detected from the stems (Tyree and Sperry 1989). Ikeda and Ohtsu (1992) successfully monitored the extensive increase of AE events from the stems of pine trees inoculated with PWNs. AE monitoring showed that abnormally frequent embolisms occurred with two peaks during the period from the infection to the death of a pine tree. At the first increase of AE frequency, white patches of dehydrated area appeared (Fig. IV.7D; Kuroda 1995). At the second AE increase, desiccation of the xylem progressed dramatically and completely stops the sap flow (Ikeda and Ohtsu 1992). The results indicate that changes in the physical characteristics of the xylem sap contribute to the dehydration of tracheids in pine wilt-susceptible trees (Kuroda 1991).

Light microscope observations show that globules exuded both from epithelial cells and ray cells are present in tracheids, mostly along the ray tissues and around resin canals from 1 or 2 weeks after nematode inoculation (Fig. IV.5B; Nobuchi et al. 1984; Kuroda 1989; Hara and Futai 2001). The stainability of these substances with nile blue and Sudan III has demonstrated the presence of oily components (e.g., lipids or terpenoids) in them (Conn 1977). Water-repellent substances effectively block the refilling of once-dehydrated tracheids with water if they adhere to the pit membrane (Fig. IV.8). Oily droplets (globules) are rare during the early phase of the disease, when a dehydrated area consisting of small, white patches has just appeared (Fig. IV.7B,D). Occlusion of the tracheids progresses in the developing phase after the symptom of needle discoloration has started (Table IV.1). When drying of the xylem is extensive (Fig. IV.7C,E), a vast amount of oily exudates is observed in tracheids. The complete cessation of sap ascent might be promoted by such a physical occlusion. Various investigations have concluded that the cause of xylem dysfunction and the symptoms of needle discoloration could be attributed to extensive dehydration by embolism (Ikeda and Suzaki 1984). From the findings on sap ascent, AE monitoring, and those on histological changes, pine wilt disease is thought to develop in two steps, early phase and developing phase. In the early phase, before the initiation of symptoms, metabolism changes in epithelium and parenchyma cells, water blockage is induced by embolism, and the dysfunctional area remains narrow in a cross section of the stem. In the developing phase, dehydration of tracheids progresses extensively, probably by occlusion with oily and sticky substances, and xylem desiccation and cambial necrosis occur. Sap ascent was long thought to be a pure physical phenomenon occurring in dead structures such as tracheids or vessels; however, recently, it has been suggested that ray parenchyma cells play an important role in controlling xylem sap flow (Canny 2001). In pine trees infected with the PWN, necrosis of ray parenchyma cells may contribute to the dysfunction of the tracheid, although this is unproven.

Other physiological events, such as the decreases in xylem water potential, transpiration, and assimilation, occur during the developing phase in association with the progress of water deficit in the stem. These events are not the cause of wilt, but rather, secondary phenomena (Fukuda et al. 1992b), and are thus omitted from this discussion of the wilting mechanism.

21.6 Defense Reaction of Pine Tissue

A specific characteristic of pine wilt is the very rapid spread of the pathogen throughout the tree, up to 150 cm day^{-1} (Kuroda and Ito 1992). The defense system of pine trees is effective in preventing some fungal infections and insect feeding activities when the infection or attack is limited to a restricted area; however, in the case of pine wilt-susceptible trees, defense against the PWN is not effective in spite of the presence of defense metabolites (secondary metabolites) such as terpenoids (Kuroda 1991; Kuroda et al. 1991b; see Chap. 22) or tannin-like substances (Futai 2003) synthesized in the tree from the early period of infection. Xylem extracts of *P. thunbergii* effectively immobilize PWNs (Bentley et al. 1985). Judging from this, susceptible species do not synthesize metabolites sufficiently to prevent nematode activity in the tree stems nor do not have time to accumulate sufficient amounts of defense metabolites to affect the nematodes, which spread rapidly. The defense reaction actually occurs in the epithelium and parenchyma cells throughout the entire tree along the PWN migration route (Ichihara et al. 2000a).

In the case of fungal infection in an incompatible host tree, a dry zone, i.e., an area that is dehydrated and filled with gas (air), forms in the sapwood around the infected site, and hypha elongation is blocked (Coutts 1977; Hillis 1987). This is one defense mechanism of trees. In pine wilt-susceptible species, tracheid dehydration, that is, the same as dry-zone formation occurs widely as a reaction to the rapid distribution of PWNs throughout the tree; therefore, the sapwood becomes dysfunctional, and sap ascent stops. The extensive defense reaction throughout a wide area of a tree is common to tree wilt diseases. For instance, the wilting of *Picea* spp. by the blue-stain fungus (*Ceratocystis* and *Ophiostoma* species) occurs when vector beetles inoculate pathogenic fungi at many points on the stems and induce complete stoppage of sap flow (Yamaoka et al. 2000; Kuroda 2005). Infection at a few points does not induce wilt because sap flow does not stop completely. In pine wilt, the wide and rapid spread of PWNs induces complete stoppage of sap flow, although there are fewer infection sites than those observed in other fungus-caused wilts.

The developing phase of the pine wilt disease (Table IV.1) can be summarized as follows. As the tracheids dehydrate and are filled with gas or air, water supply to the shoots decreases, and during this time the water content of the xylem decreases to 20% that of healthy pine trees (Kuroda et al. 1989). Increasing needle discoloration, drooping, and color fading are observed. When the desiccated area enlarges in the current annual ring (Fig. IV.7E), cambium necrosis starts. There may be a mechanism that promotes extensive desiccation of the xylem and necrosis of the cambium in the developing phase (Fukuda 1997). Judging from the sudden population growth of the PWNs and enlargement of the dehydrated area on the current annual ring, the affected tree must be at a physiological turning point at this time, that is, the termination of the defense reaction due to the mass necrosis of parenchyma cells.

21.7 Resistant Species and Selective Breeding of Resistant Trees in Susceptible Pine Species

Native Japanese pines, *P. thunbergii* and *P. densiflora*, are highly susceptible to pine wilt. More than 70% of the saplings of these species die when inoculated with virulent PWN strains (Kiyohara and Bolla 1990). Conversely, *P. taeda* and *P. strobus*, native to North America, are resistant. The mortality rate of these species is not zero but it is far lower than that of the Japanese species (Futai and Furuno 1979; see Chap. 23). In *P. taeda* saplings, nematode distribution and propagation are strongly suppressed (Fig. IV.9) (Hashimoto and Dozono 1975; Kuroda et al. 1991a). The PWN population in the main stems of *P. taeda* is very low immediately after inoculation and then decreases to an undetectable level, although the inoculated branch sometimes dies. Perhaps some chemicals suppress PWN activities in the tissues of resistant species (Futai 1979). Bentley et al. (1985) reported that extracts from *P. taeda* xylem displayed high activity in immobilizing the PWN; however, no specific substance has been discovered. In saplings of *P. strobus*, the initial migration of PWNs is sometimes rapid and slight drooping of the needles is simultaneously observed (Kuroda et al. 1991a), but the mortality rate is lower than that in native Japanese species (Tamura et al. 1988; Yamada and Ito 1993). The following is a detailed explanation of host resistance, focusing on the highly resistant *P. taeda*.

Inoculated PWNs can invade the resin canals of *P. taeda* and move in the stem. White patches of dehydrated tracheids are also formed in the xylem of *P. taeda*; however, they are scarce, and do not enlarge (Fig. IV.10; Kuroda et al. 1991a); therefore, water deficit does not become serious in infected trees. The increase of volatile terpenes is observed after infection, as in the case of susceptible pines, but active synthesis terminates earlier, about 2 weeks after inoculation. The defense reaction (secondary metabolism) in *P. taeda* is restricted to a narrow area in the

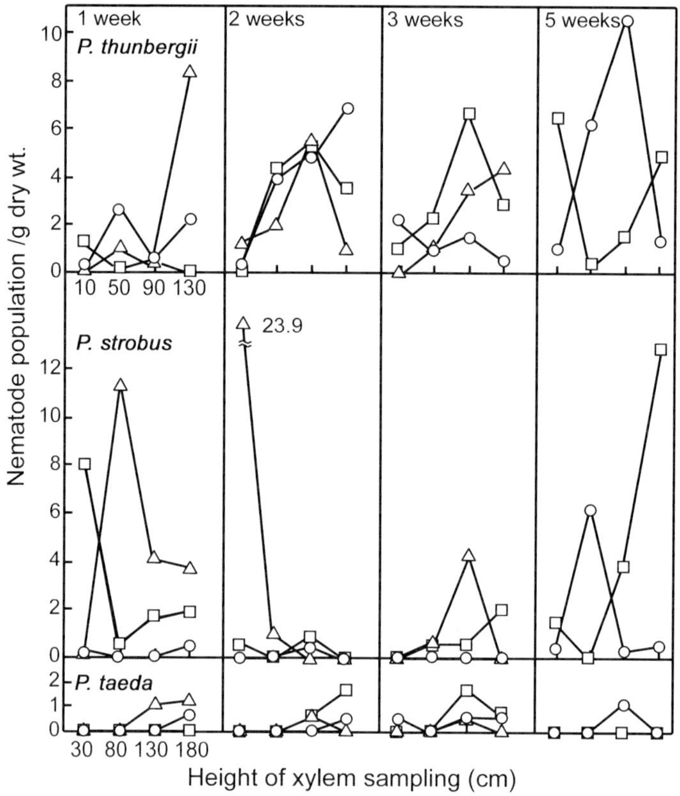

Fig. IV.9 Population growth of the pine wood nematode (PWN) observed in the stems of *Pinus thunbergii*, *P. strobus*, and *P. taeda*. Ten thousand PWNs were inoculated into branches at a height of ca. 130 cm (Kuroda et al. 1991a, with permission)

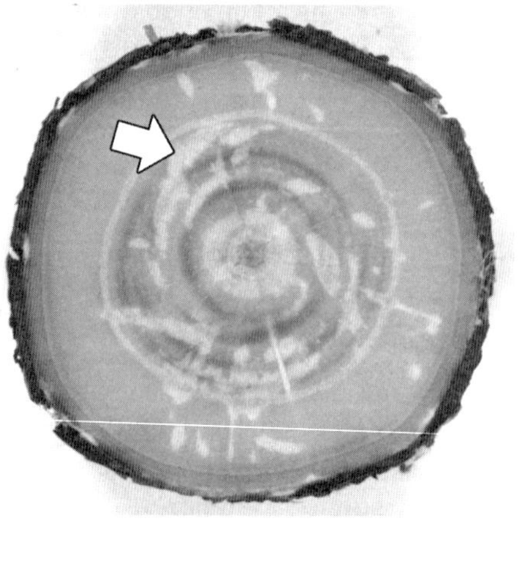

Fig. IV.10 Cross section of a *Pinus taeda* stem 5 weeks after inoculation with pine wood nematode. White patches of the dehydrated area (*arrow*) are narrower than those in *P. thunbergii* (Fig. IV.7) (see Color Plates)

xylem and seems to finish early. Generally surviving trees of resistant species are characterized by the termination of disease development in the early phase (Table IV.1) without reaching the developing phase. It must be noted, however, that juvenile seedlings (less than 1-year old) of resistant species are easily killed by inoculation with PWNs (Tamura and Dropkin 1984).

In Japan, the breeding of resistant trees has been practiced by selecting trees surviving in heavily damaged stands of susceptible *P. thunbergii* and *P. densiflora* (Toda 1999; Toda et al. 2002). Cuttings obtained from surviving trees are grafted onto seedlings from non-selected trees, and inoculation tests are done. Trees that survived the inoculation test are designated as resistant. Such resistant trees are propagated by grafting and the clones are outplanted into seed orchards. Seedlings produced by open pollination of these clones are called "resistant families" and are used for reforestation (Toda 1997). At present, more than 100 *P. densiflora* trees and 16 *P. thunbergii* are maintained as resistant mother trees. A high proportion of the seedlings from those mother trees survive after inoculation with PWN (Toda et al. 2002). The suppression of nematode spread and population growth in the host tissue was observed in such seedlings (Kuroda 2004; Kuroda et al. 2007), although the suppressive ability is weaker than that in resistant species, such as *P. taeda*. Seedlings about 1-year old or less than 30 cm in height do not always indicate clear resistance (Kuroda et al. 2007). This suggests that the PWN can easily become wide spread in short seedlings before the defense reaction occurs in host tissue. At present, the production of resistant seedlings does not meet the great demand. Investigation into technical improvements to produce many resistant seedlings is underway. This subject is addressed in Part VI.

The defense or protective reaction against infection is commonly placed into either of two possibilities: one is a defense system that exists before infection, and the other is one that develops after infection stimulated by the pathogen's activities. The active synthesis of monoterpenes detected 2–3 days after infection in both susceptible and resistant species belongs to the latter category. Even in a short twig segment 5 cm or cuttings 20 cm in length, PWN spread is retarded in resistant clones. However, in cuttings or boiled material, the PWN population increases occurred earlier than in intact seedlings of both susceptible and resistant clones (Togashi and Matsunaga 2003; Kuroda et al. 2007; Kuroda 2008). These results suggest that the defense system occurs to a certain extent, even in susceptible seedlings. A possible resistant mechanism, as suggested by anatomical investigations, is the contribution of the structural characteristics of resistant trees (Kawaguchi 2006; Kuroda and Kawaguchi 2007), besides chemical compounds, as the most likely candidate for providing resistance. The diameter or distribution density, or both, of resin canals may contribute to the impediment of PWN spread in pine tissue. This is the defense system that exists before infection.

Regarding the defense mechanism, it is still unclear what kinds of metabolites affect the PWN in resistant species and seedlings from resistant cultivars (clones). Genetic studies are now being done to find specific substances that strongly suppress nematode activities, or genes related to this process (Kodan et al. 2001; Kuroda 1999).

21.8 Concluding Remarks

The process of disease development in the bole of pine trees can be summarized as follows (Fig. IV.11): (1) stimuli by nematode invasion, and migration and feeding in resin canals, (2) activation of the secondary metabolism in epithelium and parenchyma cells, (3) release of oily metabolites into tracheids, (4) enhancement of dehydration of tracheids by embolism and partial blockage of sap flow, (5) enlargement of dysfunctional xylem, (6) complete stoppage of sap ascent, and (7) cambial necrosis and death. One to four represents the early phase, and 5–7, the developing phase of the disease (Table IV.1). A low proportion of infected trees in the early phase may survive, but host survival from the developing phase is impossible.

As already described, recent determination of the main events related to disease development from secondary incidents or byproducts is a significant achievement. For instance, necrosis of cambial cells is not the direct cause of tree mortality, but it is a second event associated with or following serious water deficit. Anatomical observations and physiological investigations from immediately after PWN inoculation have revealed various internal symptoms preceding the appearance of visible symptoms. To plan experiments on wilt disease, it is very important to identify the

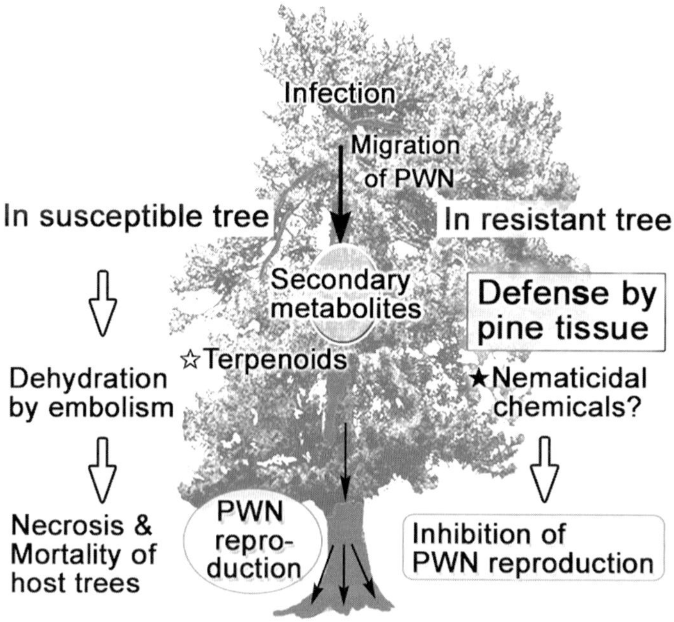

Fig. IV.11 Concept of defense in a tree trunk and disease development. Regarding the process of disease in a susceptible tree, refer to Table IV.1. In resistant trees, pine wood nematode reproduction may be inhibited by the effect of some unidentified secondary metabolites

target period relating to the fate of affected trees. The information described here will help with detailed investigations to develop techniques for decreasing pine wilt in forests.

Column

Propagation and Inoculation Protocols for the PWN

PWN Propagation

The PWN has not yet been cultured on a synthetic medium, consequently it is propagated on a fungus, usually a non-sporulating strain of *Botrytis cinerea*. For growing *B. cinerea*, potato dextrose agar (PDA) or boiled and sterilized rice or barley (with or without hull) is used. The medium is placed into flat-bottomed flasks so that vast numbers of PWNs can be propagated for the inoculation test. PDA, in test tubes, is used to keep PWNs over a longer period. The culture of *B. cinerea* is done at 25°C for a few weeks to a month. When sufficient hyphae have grown, a small amount of PWNs are transferred from the old medium and cultured at 25°C for 2 weeks to a month until the hyphae are consumed. Several virulent PWN strains have been subcultured for many years. Some occasions, however, pathogenicity decreases during long subculture.

Inoculation Method

The inoculation test is a fundamental method used to investigate wilting or resistant mechanisms in pine trees. To compare the results of researchers, standardization of experimental techniques is desirable. In this section, an outline of the common techniques that can induce symptoms and mortality in healthy pine trees is given. Original studies by Kiyohara and Tokushige (1971) and Mamiya (1980) are available for reference. Before the inoculation test, PWNs in the culture medium are mixed with water and extracted by a common technique (Thorne 1961). The PWN concentration in the water is counted and adjusted for inoculation by dilution with water. Usually, 3,000–10,000 PWNs in 50–100 μl of a suspension are used for inoculation into each seedling or sapling. The concentration of the PWNs in the suspension must be changed depending upon the size of specimens. The population should be determined by reference to reliable research studies.

Incision wounds for the inoculation are made on the shoot, branch, or main stem of the seedling. Then, the PWN suspension is quickly pipetted onto the wound. If this process is slow, the PWNs cannot invade the pine tissue because of the exudation of resin, and the inoculation will not succeed. Some training is necessary to conduct this type of inoculation. Inoculation at the base of juvenile or thin seedlings is not recommended because it sometimes induces unusually rapid wilt.

The mortality of susceptible pine trees has been induced by the inoculation with as few as five pairs of PWN (Mineo 1986); therefore, inoculation with 3,000 PWN is sometimes thought to be unnatural. The numbers of PWN carried by pine sawyers varies depending on the beetles, and consequently the number of nematodes invading the pine twigs is assumed to vary too. To investigate the wilting mechanism or resistance using many pine seedlings, it is desirable that symptoms develop simultaneously among inoculated specimens, and the mortality rate is stable. As such it is common to use inoculum with large numbers of PWNs. Symptom development is observed in the seedlings and saplings after inoculation with several thousands of PWN is not unnatural but is confirmed to be the same process as that occurring during natural infection.

22 Biochemical Responses in Pine Trees Affected by Pine Wilt Disease

Toshihiro Yamada

22.1 Introduction

A series of pathological changes occur in pine trees affected by the pine wood nematode (PWN, *Bursaphelenchus xylophilus*) through the interaction between the nematode and pines, that is, the action of the PWN and host responses. Included in such changes are: the action of the pathogen, host responses which induce and accelerate disease development leading to host death, by-products of the wilting process, and host defense responses. Substances toxic to pines and defense-related substances have also been reported, and the role of such substances has been investigated in different pine species and in various environments. Responses, especially biochemical ones, occurring in pine trees during the development of pine wilt disease and during the development of resistance are described here.

22.2 Responses Related to Disease Development

Disease development of pine wilt has been divided into the early and the advanced stages (Fukuda 1997; Chap. 21; Table IV.1). The first internal histological changes occur when a part of the epithelial cells in the cortex and xylem near infection sites are destroyed soon after nematode infection. Then, temporal and partial reduction of oleoresin exudation and degeneration of parenchyma cells begins (Hashimoto 1980a; Mamiya 1975d, 1976b, 1980, 1985). These changes spread throughout the

University Forest in Chiba, The University Forests, The University of Tokyo, 770 Amatsu, Kamogawa 299-5503, Japan

Tel.: +81-4-7094-0621, Fax: +81-4-7094-2321, e-mail: yamari@uf.a.u-tokyo.ac.jp

entire tree with the rapid spread of the PWNs. Enlargement of vacuoles in the parenchyma cells and leakage from parenchyma cells into tracheid lumina are also observed (Futai 1984; Nobuchi et al. 1984; Hara and Futai 2001). Inoculation of an avirulent nematode species *Bursaphelenchus mucronatus*, a close relative of the PWN, also induces the enlargement of vacuoles, but in a limited area (Futai 1984; Nobuchi et al. 1984). Death of parenchyma cells and destruction of cortex, phloem and cambium spread throughout the tree in the advanced stage of disease development together with cessation of transpiration and expanded water blockage (Chap. 21), finally leading to the death of the entire tree.

The production of ethylene and tannins, lipid peroxidation, leakage of electrolytes, and accumulation of phytotoxic substances are known biochemical changes that occur during pine wilt development. In the very initial stage of infection, before the reduction of oleoresin exudation, ethylene release increases (Mori and Inoue 1986; Fukuda et al. 1994). This first release of ethylene only occurs in resistant pines (resistant clones of *Pinus thunbergii* and *P. densiflora*, resistant species *P. taeda* and *P. jeffreyi*) and in susceptible pines, which survive infection. In susceptible pines, ethylene production decreases once, then considerable ethylene production occurs and the pine trees die (Mori and Inoue 1986). Inoculation with a less virulent isolate of *B. xylophilus* did not induce ethylene production (Fukuda et al. 1994). Ethylene was released more rapidly and more abundantly from water-stressed trees (Mori and Inoue 1986), suggesting that ethylene production reflects the damage level of pine cells. Ethylene began to be released from the inoculation site several hours after inoculation. This was slower than with mechanical damage such as excision, suggesting that the invasion and migration of the PWN itself does not induce ethylene production (Mori and Inoue 1986). Nematode secretions such as cellulase were thought to play an important role in ethylene production during disease development, as cellulase or PWN secretion induced ethylene production (Mori and Inoue 1986).

Tannins accumulate in parencyma cells afer PWN infection (Futai 1984, 2003b). In the case of *P. thunbergii* seedlings, catechin-like substances, precursors of condensed tannins, accumulate. Then, with the enlargement of vacuoles in the parenchyma cells, condensed tannin content increases rapidly prior to PWN multiplication. In the case of inoculation with the non-pathogenic *B. mucronatus*, however, no obvious changes were recognized. Most tannins disappeared at the stage of dramatic nematode multiplication and the appearance of wilting symptoms. In the bark (cortex and phloem) of infected *P. densiflora* and *P. thunbergii*, active oxygen species are produced prior to lipid peroxidation and the leakage of parenchyma cell contents. Tannins are produced in plants as scavengers of active oxygen species. So, the hypothesis has been proposed that hypersensitive reaction-like responses are induced by PWN infection, but the responses expand and cause blockage of sap ascent throughout the whole tree. The result is that the tree wilts as the result of its own metabolites. Several researchers interpret the destruction and death of pine tissue in this disease as dynamic host defense responses, that is, hypersensitive reaction/cell death (Futai 1984; Myers 1984; Iwahori and Futai 1993). Further research is necessary to reveal whether this hypothesis is appropriate.

The level of lipid peroxidation, indicating damage to the cell membrane, usually increases in both the resistant and susceptible responses. Lipid peroxidation progresses when necrosis in the xylem parenchyma and partial destruction of cortex and phloem are observed, and the level increases parallel to the extent of cell destruction or necrosis and nematode reproduction (Yamada 1987; Fig. IV.12). Abnormal electrolyte leakage as the result of dysfunction of the cell membrane also increases in the early phase of nematode infection, and an increase in parenchyma cell death follows this increased leakage (Zheng et al. 1993). These can be considered as the development of damage caused by the nematode infection.

Changes in host protein content and enzyme activity are also detected before histological changes, other than the destruction of epithelial cells, are observed. The protein content and activity of peroxidase and polyphenol oxidase increase conspicuously during disease development, and the increments parallel the level of lipid peroxidation (Yamada et al. 1984; Yamada 1987; Fig. IV.12). The carbohydrate content of pines decreased rapidly after PWN infection (Bolla et al. 1987). Responses such as the production of secondary metabolites and increased activity of several enzymes could occur by this consumption of carbohydrates. Decrease of carbohydrate content seems to derive from increased respiration after infection (Hashimoto 1979; Ohyama and Kaminaka 1975a,b; Mori and Inoue 1983). Such respiration is uncoupled from energy generation, and so cell division at the cambium decreases dramatically and as such host growth is reduced after PWN infection (Hashimoto 1980b). These responses, that is, ethylene production, tannin accumulation, lipid peroxidation, increases in peroxidase activity and respiration, and so forth, do not occur unsystematically, but instead in a sequence of responses linked to each other.

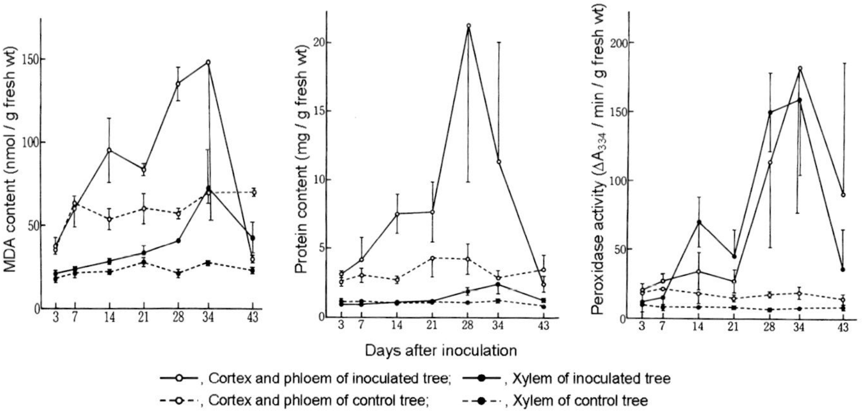

Fig. IV.12 Lipid peroxidation, protein content and peroxidase activity in pine wood nematode-infected *Pinus thunbergii*. MDA: malondialdehyde, an idex of lipid peroxidation. Vertical bars indicate standard deviations (Yamada 1987, with permission)

22.3 Phytotoxins

PWN infection leads to the death of pine trees. So, what substances cause damage to host cells? Early on, Oku et al. (1979) found pine toxins from PWN cultures, PWN-inoculated *P. densiflora* and *P. thunbergii* trees, and from pine trees naturally infected with the PWN, but later they found that the toxins originated from bacteria (Oku et al. 1980). Subsequently, benzoic acid, catechol, dihydroconiferyl alcohol (DCA), and 8-hydroxycarvotanacetone (8-HCA) were detected as toxins in inoculated *P. thunbergii* and *P. densiflora* (Oku 1984; Oku et al. 1985; Ueda et al. 1984; Fig. IV.13). Besides the above instances, several toxins were detected in pine trees.

When 2–3-year-old *P. densiflora* seedlings were treated with benzoic acid and 8-HCA, browning of the needles occurred (Ueda et al. 1984). It is not clear, however, whether or not the initial characteristic symptom of pine wilt disease, that is, reddish browning of 1-year-old needles was reproduced. The susceptibility of pine seedlings to the above-mentioned toxins is higher in susceptible *P. thunbergii* than in resistant *P. taeda* and *P. rigida*. Synergistic action of the toxins was also reported (Oku et al. 1985).

Although L-8-HCA and DCA suppress nematode reproduction to some extent, benzoic acid, catechol and D-8-HCA, which are highly toxic to pines, cannot suppress nematode reproduction even at a concentration of 100 ppm (Oku et al. 1985). This suggests the diverse action of each toxin. Toxins, 10-hydroxyverbenone (10-HV) and 8-HCA, were detected from naturally-infected *P. sylvestris* (Bolla et al. 1982a, 1984a,b; Shaheen et al. 1984; Fig. IV.13). They were produced only when

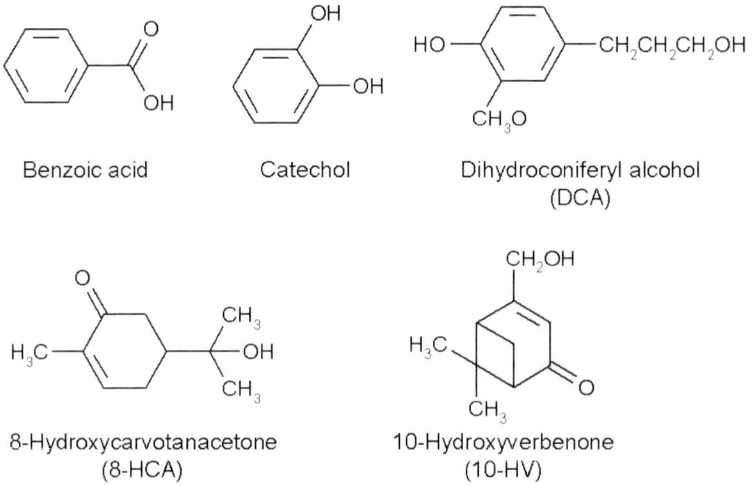

Fig. IV.13 Toxins isolated from PWN-infected pines

susceptible hosts were inoculated with pathotypes of PWN (Bolla et al. 1986). These data support the contribution of phytotoxins in disease development. Further research, however, is necessary to determine their actual significance in wilting and their relationship to pathogenesis, resistance and susceptibility at each stage of disease development, or whether or not they are just byproducts.

Among toxins, the sequence of benzoic acid accumulation in wilting was shown to be as follows (Kawazu 1990). Although benzoic acid is produced as a host response soon after PWN infection, it is detoxified by conversion to glucose ester (glucose 2-benzoate), and glucose ester concentration gradually increases. Subsequently, benzoic acid production increases rapidly with the nematode population growth, and overwhelms the detoxifying capacity, resulting in massive accumulation of benzoic acid and the death of pine trees (Kawazu 1990), however artificial blockage of water conduction, treatment with cellulase and cycloheximide do not induce the accumulation of benzoic acid and its glucose ester (Kawazu 1990). This suggests a difference in the wilting mechanism between such artificial treatments and nematode infection.

These phytotoxins are produced mostly in the last stage of disease development, suggesting that metabolites are produced secondarily in the dying process of pines. The responses leading to toxin production, however, are important as the wilting mechanisms.

The above-mentioned host origin toxins are, of course, produced as a result of the action of pathogens. Pathogen-produced cues that induce toxin production of pine trees and symptom development have been searched for. Although PWN itself produces lytic enzymes as toxic substances (Chap. 10), no other clue has been found.

Although several strains of PWN-accompanying bacteria toxic to pines have been found, it is still controversial to assign a role for bacteria in pine wilt disease. There is a hypothesis that the accompanying bacteria are necessary for the pathogenicity of the PWN, based on the result that axenized PWNs loose their pathogenicity (Kawazu and Kaneko 1997; Zhao et al. 2003). In contrast, several reports have rejected the role of bacteria based on the result that axenic seedlings are killed by axenic PWN (Tamura 1983b; Kuroda and Mamiya 1986). More information is needed on the role of bacteria and nematodes in the wilting mechanisms.

Kawazu et al. (1996b) identified a major bacterial toxin as phenylacetic acid, and considered that its elicitor action induced benzoic acid production in pines and led to tree death. Non-axenic *P. thunbergii* (3 years old) seedlings were killed by axenic PWNs (Tamura 1983b), which is explained by bacteria inhabiting pine trees in the context of bacterial toxin. Many kinds of bacteria isolated from naturally PWN-infected pine trees produced substances toxic to pines (Han et al. 2003; Zhao et al. 2003). It is suggested that related bacteria are not specific and do not always accompany nematodes, but bacteria inhabiting pine trees can be the pathogens. In all cases, the participation of the PWN is required, because bacteria alone cannot kill pines (Tamura 1983b; Kawazu 1990). The topic of associated bacteria and their toxins is covered in Chap. 24.

22.4 Contribution of Dynamic Responses to Resistance

Resistance expression against *Bursaphelenchus* spp. is negatively related to nematode reproduction in pine trees. Although the PWN cannot reproduce in standing, resistant *P. taeda* trees (Hashimoto and Dozono 1975), no difference in PWN reproduction rate, in excised stems and branches, is found between *P. taeda* and susceptible pine species, that is, the PWNs reproduce vigorously in all species (Hashimoto and Dozono 1973; Mamiya 1982a). Moreover, nematodes can multiply in dead tissue even in resistance species (Tamura and Dropkin 1984; Yamada and Ito 1993a), suggesting that the resistance of pines against the PWN is dynamic. In other words, constitutive resistance factors observed in healthy pines such as *P. taeda* contribute to suppressing nematode migration and reproduction, but it is not enough. As additional nutrients and energy are not supplied to excised stems and branches, defense responses will be overwhelmed over time.

Even callus of resistant pines is not resistant. All virulent *B. xylophilus*, less virulent *B. xylophilus* and avirulent *B. mucronatus* can reproduce well on the callus of resistant *P. taeda* and *P. massoniana* (Iwahori and Futai 1990, 1993) as well as that of the susceptible *P. densiflora* and *P. thunbergii* (Tamura and Mamiya 1976, 1979). Callus cells are killed where the PWNs multiplied. Similarly, the nematode can reproduce on the callus of non-host plants, such as alfalfa and tobacco, and the callus dies there too (Tamura and Mamiya 1975; Iwahori and Futai 1990). These results suggest that tissue structure or special substances specific to differentiated cells are necessary for the expression of resistance.

PWNs spread more slowly in resistant *P. taeda* (Hashimoto and Dozono 1975; Suzuki 1984), but move widely throughout entire trees of both resistant and susceptible pines. More nematodes invade and migrate rapidly in resistant *P. strobus* than in susceptible pines (Tamura et al. 1988; Kuroda et al. 1991a; Yamada and Ito 1993a), suggesting diverse resistance mechanisms. Resistance is maintained in the seedlings of resistant clones of *P. densiflora* even when nematode spread is not restricted (Kuroda 2004). PWN spread seems to be severely restricted in highly resistant pine species. When PWNs were inoculated into excised stems of highly resistant hybrid *Pinus rigitaeda* seedlings, their spread ceased within 5 cm of the-inoculation site (Oku et al. 1989). These results are interpreted to mean that dynamic host responses were induced during nematode spread, because such spread was not reduced in 2.5 cm long stems or in heat-treated stems. The number of nematodes passing through excised branches of *P. thunbergii* is high for virulent *B. xylophilus* isolates, but low in less virulent *B. xylophilus* and avirulent *B. mucronatus*, suggesting that virulent nematodes have a greater ability to invade the cortex (Ishida et al. 1993), or that host dynamic resistance may be involved. Although the relationship between the virulence of PWNs and their axial migration in pines was not observed, dynamic responses are suggested to be that the longer the excised branch, the fewer nematodes passing through (Togashi and Matsunaga 2003).

Needle removal and photosynthesis inhibitor treatment enabled even less virulent *B. xylophilus* and *B. mucronatus* to reproduce in pine trees, and kill *P. thunbergii* seedlings (Fukuda 1993). Lowering the light intensity also accelerates disease development and increases host mortality (Kaneko and Zinno 1986; Kaneko 1989). *P. thunbergii* branch tissue held in the dark loses its resistance against spread and invasion by the PWN (Hirai et al. 1994). Water stress also accelerates disease development (Suzuki 1984; Suzuki and Kiyohara 1978). These results indicate that dynamic host responses are necessary for resistance to occur to PWN infection.

22.5 Biochemical Factors in Dynamic Resistance Response

An aqueous extract of healthy pines immobilizes PWNs, and this activity is higher for *P. taeda* than for *P. thunbergii* (Mamiya 1982a; Bentley et al. 1985; Table IV.2). This activity probably contributes to the resistance of *P. taeda*, because it further increases by nematode inoculation with virulent and less virulent isolates of *B. xylophilus* and *B. mucronatus* (Iwahori and Futai 1996; Table IV.2), however the host factor has not yet been identified. Oku et al. (1989) felt the induced factor that prevents nematode spread in *P. rigitaeda* to be a water soluble, nematode-immobilizing substance. In addition, a nematicide, dehydroabietic acid, increased after the inoculation of *P. sylvestris* with incompatible nematodes, whereas abietic acid, which is abundant in healthy pines, had no nematicidal activity (Bolla et al. 1989). Such substances are classified as inhibitin or phytoanticipin, as they preexist in healthy trees and increase after PWN infection.

Inhibitory substances accumulate in the cortex, phloem and xylem of PWN-inoculated *P. strobus* (Table IV.2). The substances in the xylem are stilbenes, pinosylvin (PS) and pinosylvin monomethyl ether (PSME) (Yamada and Ito 1993b; Figs. IV.14, IV.15), which are known as phytoalexins of pine xylem. Flavanones, pinobanksin and pinocembrin, also accumulate as inhibitory substances in the xylem. The components induced in the cortex and phloem are a stilbene, 3-*O*-methyldihydropinosylvin (dihydropinosylvin monomethyl ether, MDPS), and

Table IV.2 Comparison of resistance responses among pine species to the pine wood nematode (PWN)

	Spread of PWN	Chemical defenses	Anatomical changes
Susceptible species			
Pinus thunbergii	Rapid	—	—
Resistant species			
P. taeda	Slow	Water soluble compounds, Increase after infection	Resin canal occlusion, partial necrosis, wound periderm
P. strobus	Rapid	Phytoalexin accumulation	Resin canal occlusion, partial necrosis, wound periderm
Induced resistance (*P. thunbergii* and *P. densiflora*)	?	?	Resin canal occlusion, partial necrosis, wound periderm

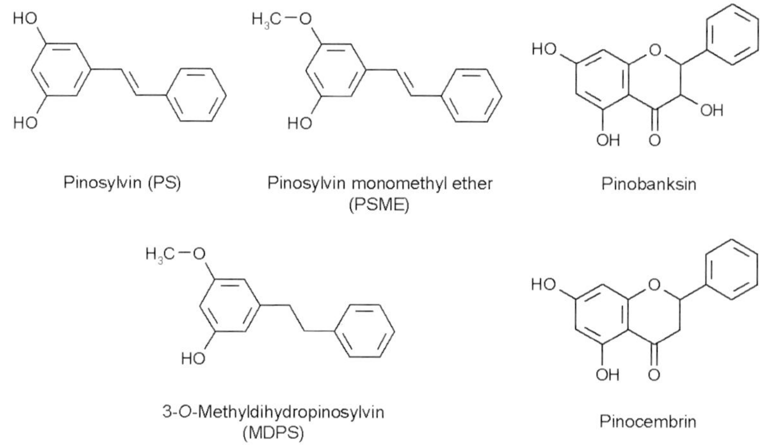

Fig. IV.14 Phytoalexins produced in *Pinus strobus* infected with the pine wood nematode

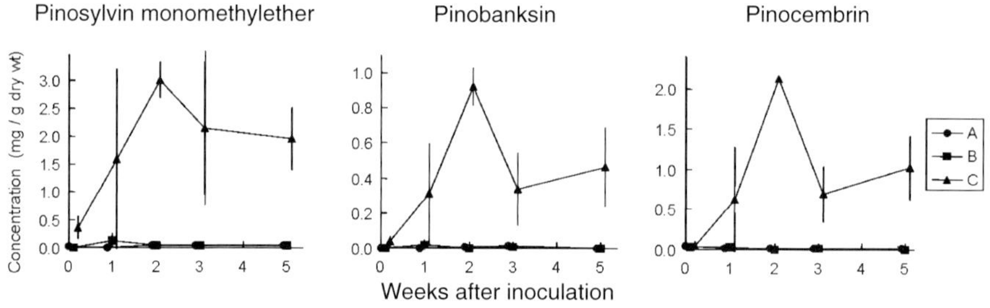

Fig. IV.15 Phytoalexins accumulation in the branch xylem of *Pinus strobus* inoculated with PWN: 6–11 cm (*A*), 16–21 cm (*B*), 26–31 cm (*C*, branch base) from inoculation site. *Vertical bars* indicate standard deviations (Yamada and Ito 1993b, with permission)

pinocembrin (Hanawa et al. 2001; Figs. IV.14, IV.16). MDPS was induced also in the xylem. Large amounts of all these accumulated within a week after inoculation, and the concentration of MDPS reached the level that completely kills nematodes (Table IV.3). These inhibitory substances probably contribute to the resistance of *P. strobus*, because PS and PSME have strong nematicidal activity (Suga et al. 1993) as does MDPS. These inhibitory substances only exist in the heartwood of healthy pines, and their production is induced by infection, and as such, they are classified as phytoalexins. These phytoalexins were originally known as antifungal substances. Plants do not produce highly pathogen-specific and highly active substances like antibiotics, but it is common for them to produce substances with some broad spectrum. Accumulation of these inhibitory substances, however, was not observed in resistant *P. taeda* whose sapwood produces PS and PSME (Shain 1967). The fact that nematode infection and the resulting damage does not induce their

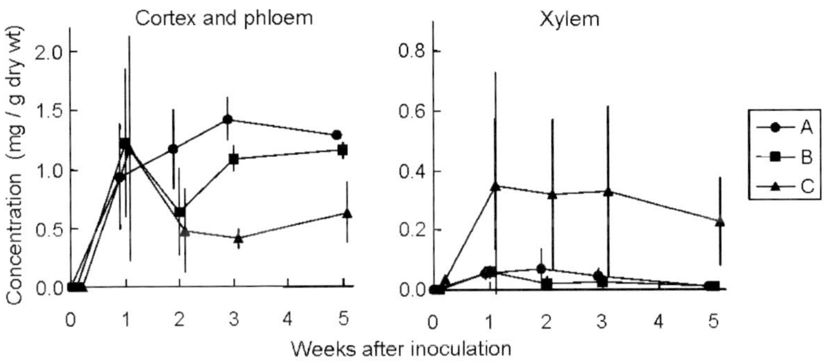

Fig. IV.16 Accumulation of 3-*O*-methyldihydropinosylvin (MDPS) in the branch of *Pinus strobus* inoculated with PWN: 6–11 cm (*A*), 16–21 cm (*B*), 26–31 cm (*C*, branch base) from inoculation site. *Vertical bars* indicate standard deviations (Hanawa et al. 2001, with permission)

Table IV.3 Effect of 3-*O*-methyldihydropinosylvin on the mobility of the pine wood nematode (Hanawa et al. 2001)

Dose (ppm)	Immobilized nematode (%)					
	One day treatment			Two days treatment		
	Days after transfer			Days after transfer		
	0	1	3	0	1	3
250	100 ± 0	100 ± 0	100 ± 0	100 ± 0	100 ± 0	100 ± 0
100	54 ± 1	27 ± 8	25 ± 6	46 ± 7	22 ± 10	18 ± 6
50	7 ± 2	4 ± 1	12 ± 2	7 ± 3	7 ± 0	13 ± 6
25	2 ± 1	6 ± 2	12 ± 5	4 ± 1	4 ± 2	8 ± 3
0	2 ± 1	3 ± 2	11 ± 2	4 ± 2	5 ± 1	8 ± 2

Means ± SD

accumulation suggests that their production is suppressed, and as such other resistance mechanisms act against PWN infection. Accumulation of these inhibitory substances is not observed in *P. thunbergii* doomed to wilting. As phytoalexin usually accumulates in necrotic tissue of susceptible plants, some mechanisms that suppress phytoalexins accumulation may also function in *P. thunbergii*.

Diverse modes exist in the relationship between terpene production and resistance among resistant pine species. Terpenes accumulate temporarily and partial cavitation occurs in *P. taeda* after PWN infection, whereas almost no cavitated areas were observed in *P. strobus* when terpenes increased (Kuroda et al. 1991a). This suggests that at least two types of resistance exist in pine species, that is, no terpene induction and tolerance to terpene production and its activities.

It is difficult for the PWN to invade tissue of excised branches of resistant *P. taeda*, while they can easily invade resistant *P. strobus* tissue as well as PWN-susceptible *P. densiflora* and *P. thunbergii* (Futai 1985a, b). No difference was detected in the initial infection and subsequent spread of PWNs in resistant *P. strobus* or susceptible *P. densiflora* (Arihara 1997a, b). These results are consistent

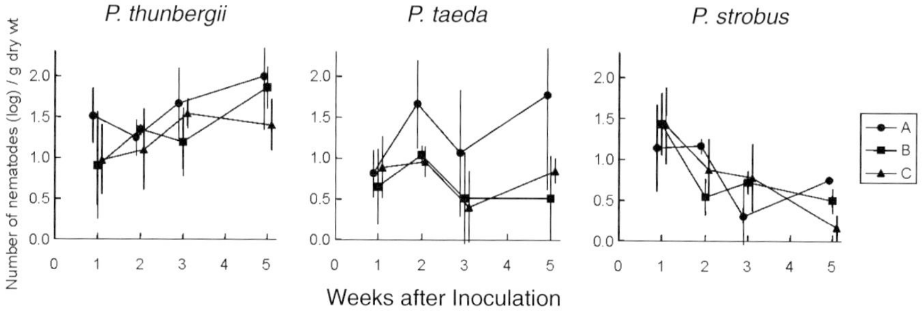

Fig. IV.17 Population dynamics of pine wood nematode (PWN) in the branch of PWN-inoculated pines: 6–11 cm (*A*), 16–21 cm (*B*), 26–31 cm (*C*, branch base) from inoculation site. *Vertical bars* indicate standard deviations (Yamada and Ito 1993a, with permission)

with the results that dynamic responses after nematode infection are important in the case of *P. strobus* (Yamada and Ito 1993a, b). PWNs move quickly in *P. strobus* trees just after invasion while the spread and distribution of PWNs is somewhat slow in *P. taeda*, which has pre-existing inhibitory substances. Conversely, phytoalexins are induced in *P. strobus* after nematode invasion, and so the activity of nematodes is expected to be suppressed, thus the population decreases and eventually disappears (Fig. IV.17).

The remarkable phytoalexin accumulation that was observed in *P. strobus* was not detected in resistance-induced *P. thunbergii* and *P. densiflora*, and not in either species when they were subsequently challenged with a virulent PWN isolate (Table IV.2). Phytoalexins appear not to be included in induced resistance to PWN. These results indicate that the resistance mechanisms and major active substances concerning resistance are quite different among *Pinus* spp., as shown in *P. strobus* and *P. taeda* (Table IV.2). When such differences in the patterns of terpene production and phytoalexin-associated resistance are known, the relationship between the resistance mechanisms of pine species and their phylogenetic relationship will be clarified (Futai and Furuno 1979).

22.6 Constitutive Resistance Factors

Constitutive resistance factors are recognized in all stages of pine wilt development, that is, nematode attraction to pines, invasion of pines and PWN activity in pines. Among the monoterpenes of pines β-myrcene is most attractive to PWNs (Ishikawa et al. 1986), and speeds up PWN reproduction (Hinode et al. 1987). Ishikawa et al. (1987) reported the β-myrcene content as a determinant in the susceptibility in *Pinus* species with various resistance levels, that is, the higher the β-myrcene content, the more susceptible the tree. However, it has also been reported that no correlation between β-myrcene content and the level of resistance was found for susceptible *P. thunbergii* and resistant *P. taeda* and *P. strobus* (Kuroda et al. 1991a).

Sap (aqueous extracts) of resistant pines such as *P. taeda* also attracts PWNs, but no correlation has been observed between the level of attraction and susceptibility (Futai 1980b). No difference was found in the attraction to sap even between *B. xylophilus* and *B. mucronatus*. A similar result was obtained when conifers other than pines such as western larch (*Larix occidentalis*) and black spruce (*Picea mariana*) were included (Futai and Sutherland 1989).

However, in an experiment using 17 pine species a relationship was found between the resistance of pine species and the difficulty of PWN spread in pine tissue, especially in bark (cortex and phloem) (Futai 1985a, b). Also, factors controlling the nematode's invasion were detected in water extracts of pines. An extract from resistant *P. taeda* was more repellent than that from *P. thunbergii* (Futai 1979). Healthy trees of resistant clones of *P. thunbergii*, *P. densiflora* and their hybrids have several characteristics such as high catechol tannin content in the bark (cortex and phloem), low pH of the bark (less than 4.1), and a high transpiration rate (Ohyama et al. 1986). This suggests that PWN cellulase, at optimum pH 6.5 (Odani et al. 1985), activity is suppressed in such resistant pine clones because of their low pH. Activity of *B. xylophilus* pectinase appears to become higher in susceptible clones, because their optimum pH is 4.5–5.0 (Kuroda 1987). In addition, higher nematicidal activity of methanol extract from healthy trees of resistant species (*P. massoniana*, *P. strobus* and *P. palustris*), when compared to susceptible species (*P. thunbergii* and *P. sylvestris*), suggests its involvement in resistance (Suga et al. 1993). *P. massoniana* contains a repellent, α-humulene and the nematicides, pinosylvin monometyl ether (PSME) and (–)-nortrachelogenin, in the heartwood, and contains nematicides, methyl ferulate and (+)-pinoresinol, in the phloem. PSME is present in the heartwood of *P. strobus*, and the heartwood and phloem of *P. palustris*. Inhibitory substances in the heartwood cannot contribute to resistance, because the PWN do not occur in the heartwood. As mentioned later, however, these substances may be produced in the sapwood where they are active against nematode infection and thereby contribute to resistance.

In the relationship of healthy tree constituents to resistance, a correlation was also found between the level of resistance and diterpenes, resin acids and amino acid content in *P. massoniana* of different age classes and provenances (Xu et al. 1994b, 1998). As the comparison of such components tends to be insufficient for reproducibility, there is a need for more information, especially in various environments and regions. The role of such constituents in actual resistance mechanisms should also be studied.

22.7 Concluding Remarks

Although biochemical responses that occur during pine wilt disease development and during resistance expression have been discovered from many aspects of wilt development, research on the mutual relationship among such responses is not sufficient, and our knowledge is far from comprehensive. We should note that just a partial understanding can be applied to practical purposes such as breeding for resistance and other control measures.

Recently, new attempts began in the field on the expression of resistance (Kuroda 2004). In addition, details of lytic enzymes of the PWN and bacterial toxins are being discovered (Chaps. 10, 24). Recent advances in molecular biology should shed new insight on the pathological changes and resistance mechanisms of pine wilt disease.

23 Host Fate Following Infection by the Pine Wood Nematode

Yuko Takeuchi

23.1 Introduction

Although the pine wood nematode (*Bursaphelenchus xylophilus*, PWN) seriously damages the physiology of its host pine trees, as described in Chaps. 21 and 22, it does not always kill host trees even when it successfully invades them. The main reasons for this are that both the virulence of the PWN and the susceptibility of pine trees to the nematode vary. Consequently, whether or not the disease progresses largely depends upon the host-pathogen combination (see Sect. 23.2). Furthermore, host fate after infection by the PWN can also be affected by other biotic or abiotic conditions, or both, such as precipitation and temperature plus other factors including the number of nematodes in the inoculum, and the season when PWN infection occurs (reviewed by Kishi 1995).

So what happens inside the surviving host tree? Some avirulent isolates of the PWN can induce host resistance against subsequent infection by virulent PWN isolates, that is, so-called "induced resistance" occurs (reviewed by Kosaka et al. 2001b). In contrast, there have been several reports on asymptomatic carrier trees that show few, if any, symptoms even when infected with PWNs (e.g., Bergdahl and Halik 1999; Futai 2003a). In this chapter these two phenomena will be discussed to show the current status of research in this area.

23.2 Compatibility Between Host Trees and the PWN

The fate of host trees after infection by the PWN largely depends upon the compatibility between the host and the nematode. When the host tree is susceptible and the PWN is pathogenic/virulent, their relationship is termed "compatible". However,

Laboratory of Environmental Mycoscience, Graduate School of Agriculture, Kyoto University, Kitashirakawa Oiwake-cho, Sakyo-ku, Kyoto 606-8502, Japan

Tel.: +81-75-753-6060, Fax: +81-75-753-2266, e-mail: yuuko@kais.kyoto-u.ac.jp

when the host tree is resistant or the PWN is not pathogenic/virulent, or both, they are incompatible and the host will not die. Here, the factors that determine the compatibility of host-PWN relation will be discussed.

23.2.1 Susceptibility of Pinaceae Trees to the PWN

Several species within the genus *Pinus* are susceptible to the PWN, including Japanese native *Pinus thunbergii*, *P. densiflora*, *P. luchuensis*, *P. parviflora* var. *parviflora*, and *P. koraiensis*, and some exotic species such as *P. pinaster* (reviewed by Kishi 1995; Akiba 2006; Table IV.4). They can be the hosts that harbor PWNs after either natural or artificial infection. Several researchers have tried to rank pine species according to their degree of susceptibility to the PWN based on the mortality of trees after artificial inoculation, but such ranking is sometimes inconsistent with natural incidence. For example, Furuno (1982) ranked *P. flexilis* and *P. radiata* as rather susceptible, which is in contrast to Ohba (1982) who ranked both of these species as resistant. In addition, even *P. taeda* and *P. elliottii* stands, both of which are considered as highly resistant (Futai and Furuno 1979; Furuno 1982; Ohba 1982), were observed to naturally suffer losses resulting from the PWN (Ogawa and Hagiwara 1975). Focusing on the origin of pine species, Northeast American species, such as *P. strobus* and *P. taeda* are generally resistant and many Eurasian species including Japanese native species and *P. sylvestris* are susceptible to the PWN (Adams and Morehart 1982; Wingfield et al. 1986; Linit and Tamura 1987; Bedker and Blanchette 1988). Perhaps this is because the PWN originated in North America (Tares et al. 1992a). In North America, the PWN seldom causes pine wilt disease on native pine trees, but instead it usually inhabits and completes its life cycle in dead pines or trees weakened by other pests or stressed by environmental factors such as drought and high temperature (Dropkin et al. 1981; Robbins 1982; Wingfield et al. 1982). This is also true for the nematode *Bursaphelenchus mucronatus*, a taxonomically closely related species to the PWN, in Japan. A relationship between the susceptibility to the PWN and taxonomic position of each pine species has been proposed (Tanaka 1973; Futai and Furuno 1979; Furuno 1982); species belonging to the subsection *Australes* and *Contortae* are highly and moderately resistant, respectively, while those in the *Ponderosae* and *Oocarpae* are susceptible, although susceptibility can vary even within the same subsection. The taxonomy of *Pinus* spp. is now being rearranged based on molecular phylogeny at the species and subsection levels (Price et al. 1998; Liston et al. 1999; Gernandt et al. 2005). Future studies are needed to determine whether or not the taxonomic position of the host is actually related to susceptibility to the PWN, as indicated by Akiba (2006).

Other Pinaceae trees besides those in the genus *Pinus* can also be hosts of the PWN (Table IV.5), and in nature *Larix kaempferi*, *Picea abies*, and *Cedrus deodara* are known hosts. Such species have resin canals or form traumatic resin canals after injury, through which the PWN can spread; however, most cases involve other

Table IV.4 Species of the genus *Pinus* reported as hosts of the pine wood nematode, *Bursaphelenchus xylophilus*

Subgenus	Section	Subsection	Species[a]	References	
				Natural infection	Artificial infection
Pinus	*Pinus*	*Pinus (Sylvestres)*	*densiflora*	Many	Kiyohara and Tokushige (1971), Futai and Furuno (1979)
			heldreichii (syn. *leucodermis*)		Ishii et al. (1981)
			kesiya (syn. *insularis, khasya*)	Furuno (1982)	
			luchuensis	Many	Kiyohara and Tokushige (1971)
			massoniana	Furuno (1980), Yang and Wang (1989)	Futai and Furuno (1979)
			mugo	Robbins (1982)	Futai and Furuno (1979), Dropkin et al. (1981)
			nigra		Futai and Furuno (1979)
			pinaster	Mineo and Kontani (1973), Mota et al. (1999)	Futai and Furuno (1979)
			resinosa	Robbins (1982)	Dropkin et al. (1981), Futai and Furuno (1979)
			sylvestris	Furuno (1982), Robbins (1982)	Futai and Furuno (1979)
			tabulaeformis		Futai and Furuno (1979)
			taiwanensis	Yang and Wang (1989)	Futai and Furuno (1979)
			thunbergii	Many	Kiyohara and Tokushige (1971), Futai and Furuno (1979)
			yunnanensis	Yang and Wang (1989)	Furuno et al. (1984)
		Halepenses	*brutia*		Dwinell (1993)
			halepensis	Robbins (1982)	Dropkin and Linit (1982)

(continued)

Table IV.4 (Continued)

Subgenus	Section	Subsection	Species[a]	References	
				Natural infection	Artificial infection
		Contortae	*banksiana*	Robbins (1982)	Dropkin et al. (1981), Wingfield et al. (1986)
			clausa	Robbins (1982)	Jones et al. (1982), Mamiya (1984)
			contorta	Robbins (1982)	Futai and Furuno (1979), Dropkin et al. (1981)
			virginiana	Robbins (1982)	Dropkin et al. (1981), Dwinell (1985a)
		Australes	*caribaea*		Ohba et al. (1984)
			echinata	Ogawa and Hagiwara (1975), Robbins (1982)	Dropkin et al. (1981)
			elliotti	Ogawa and Hagiwara (1975)	Kiyohara and Tokushige (1971)
			glabra		Jones et al. (1982)
			palustris	Steiner and Buhrer (1934)	Dropkin et al. (1981)
			pungens		Mineo (1974)
			rigida	Robbins (1982)	Kiyohara and Tokushige (1971)
			serotina		Dwinell (1985a)
			taeda	Ogawa and Hagiwara (1975), Robbins (1982)	Futai and Furuno (1979)
		Ponderosae	*engelmannii*	Furuno (1982)	Futai and Furuno (1979)
			jeffreyi		Tamura and Dropkin (1984)
			ponderosa	Furuno (1982), Robbins (1982)	Futai and Furuno (1979), Dropkin et al. (1981)
			devoniana (syn. *michoacana*)	Furuno (1982)	
			hartwegii (syn. *rudis*)	Furuno (1982)	Futai and Furuno (1979)
			pseudostrobus (syn. *estevezii*)	Furuno (1982)	Ohba et al. (1984)

Table IV.4 (Continued)

		Attenuatae	*muricata*	Furuno (1982)	Futai and Furuno (1979)
			radiata	Furuno (1982), Robbins (1982)	Dropkin et al. (1981), Futai and Furuno (1979)
		Oocarpae	*greggii*	Furuno (1982)	
			oocarpa	Furuno (1982)	Futai and Furuno (1979)
		Leiophyllae	*leiophylla*	Furuno (1982)	Futai and Furuno (1979)
Strobus	*Parrya*	*Gerardianae*	*bungeana*	Furuno (1982), Furuno et al. (1977), Ogawa and Hagiwara (1975)	Ogawa et al. (1973)
	Strobus	*Strobi*	*armandii*	Yang and Wang (1989)	Akiba and Nakamura (2005)
			ayachahuite (syn. *strobiformis*)		Futai and Furuno (1979), Dropkin et al. (1981)
			fenzeliana (syn. *kwangtugensis*)	Yang and Wang (1989)	
			flexilis		Dropkin et al. (1981)
			lambertiana		Dropkin et al. (1981)
			monticola		Futai and Furuno (1979), Dropkin et al. (1981)
			morrisonicola		Lee (1986)
			parviflora	Sakura et al. (1978), Kishi (1980), Robbins (1982)	Futai and Furuno (1979)
			strobus	Robbins (1982), Dropkin and Foudin (1979)	Futai and Furuno (1979)
			wallichiana (syn. *excelsa, griffithii*)	Yang and Wang (1989)	Futai and Furuno (1979)
		Cambra	*cembrae*		Dropkin et al. (1981)
			koraiensis	Futai (2003)	Futai and Furuno (1979)

Modified from Akiba (2006). Species names according to Price et al. (1998)
[a]The synonyms given are those used in the references

Table IV.5 Non-*Pinus* species, in Pinaceae family reported as hosts of the pine wood nematode, *Bursaphelenchus xylophilus*

Genus	Species[a]	References	
		Natural infection	Artificial infection
Abies	*amabilis*		Sutherland et al. (1991)
	balsamea	Robbins (1982), Dropkin et al. (1981), Wingfield et al. (1983)	
	firma	Agriculture, Forestry and Fisheries Research Council (1987)	
	grandis		Sutherland et al. (1991)
	homolepsis		Ogura et al. (1983)
	sachalinensis		Ogura et al. (1983)
Tsuga	*mertensiana*		Sutherland et al. (1991)
Cedrus	*atlantica*	Robbins (1982), Dropkin et al. (1981)	
	deodara	Kishi (1980), Robbins (1982)	Ogura et al. (1983)
Pseudotsuga	*memziesii*		Sutherland et al. (1991)
Larix	*decidua* (syn. *europaea*)	Robbins (1982), Dropkin et al. (1981)	
	kaempferi (syn. *leptolepis*)	Mamiya and Shoji (1985)	Ogura et al. (1983)
	laricina (syn. *americana*)	Robbins (1982), Dropkin et al. (1981)	
	occidentalis		Futai and Sutherland (1989)
Picea	*abies* (syn. *excelsa*)	Ebine (1981), Koiwa et al. (2004)	Ogura et al. (1983), Koiwa et al. (2004)
	engelmannii		Sutherland et al. (1991)
	glauca (syn. *canadensis*)	Robbins (1982)	
	mariana		Futai and Sutherland (1989)
	pungens	Robbins (1982)	Myers (1982)
	rubens		Sutherland et al. (1991)
	sitchensis		Sutherland et al. (1991)

Modified from Akiba (2006). Species names conform to Taxonomy of Farjon (1990)
[a]The synonyms are those used in the references

factors that enhanced disease development, such as drought and insect defoliation (Kishi 1980b; Ebine 1980, 1981), and it remains unclear if the PWN is solely responsible for the wilting of these non-*Pinus* species.

The susceptibility of each species should be related to their morphological and physiological properties, and *Pinus* species that are resistant to the PWN often have active or constitutive defense reactions (reviewed by Hara and Takeuchi 2006; Yamada 2006). For details see Chaps. 21 and 22.

23.2.2 Pathogenicity and Virulence of PWN

When considering the PWN as a pathogen it is necessary to qualify and quantify the nematodes' ability to kill the host and to determine its "pathogenicity" and "virulence" (Akiba 2006). Besides using several Pinaceae species as hosts, the PWN also kills some of them (see above), indicating that it is pathogenic to them. Several reports suggest that the pathogenicity of the PWN is differentiated within species in North America. Some PWN races isolated from dead conifers in Missouri, MN, USA, and Western Canada varied in their pathogenicity, with some being unable to cause disease in species of the genus *Pinus*, but only in species of other tree genera such as *Abies*, *Picea* and *Larix* (Dropkin 1982; Kondo et al. 1982; Wingfield et al. 1983; Futai and Sutherland 1989). Such intraspecific differences in pathogenicity were associated with morphological (Wingfield et al. 1983) and genetic differences (Bolla et al. 1988) in the nematodes. In Japan, no variation in pathogenicity has been found (see also Sect. 12.2).

So, how much disease does PWN cause in susceptible host trees when it possesses pathogenicity to such trees? The virulence of the nematode varies widely and some isolates have only slight or no virulence to even very susceptible host pines. Such isolates are termed "avirulent isolates" (Kiyohara and Bolla 1990). Different geographical origins, different *Monochamus* beetle individuals and different host trees, from which PWN isolates originated, all enter into the virulence picture (Kiyohara et al. 1983). The virulence of each PWN isolate is usually based on the mortality rate of host trees or seedlings after inoculation with a PWN isolate, although the results can differ depending upon host age (Bergdahl 1982; Mamiya 1983; Wingfield et al. 1983, 1986; Bedker et al. 1987).

Several differences have been reported between the characteristics of virulent and avirulent isolates. For example, virulent and avirulent isolates differ somewhat in their base sequences of the ribosomal DNA ITS region (Iwahori et al. 1998; Mota et al. 2007); however, it remains unclear if such genetic differences are directly related to virulence. Physiologically, ecologically and developmentally avirulent isolates are at a disadvantage because in vitro both their reproduction rate is lower (Kiyohara and Bolla 1990) and they move more slowly than virulent isolates (Iwahori and Futai 1995; Kaneko et al. 1998). Such traits have been shown to exist as well in vivo (Kiyohara and Bolla 1990; Ichihara et al. 2000a), in their lower ability to invade the cortex tissues of host shoots compared to virulent isolates (Asai and Futai 2006). As to the relationship to their vector beetle *Monochamus alternatus*, avirulent PWNs have less potential for being carried by the vector (Aikawa et al. 2003b) and they are less likely to encounter beetles and compared to virulent PWN isolates they make much fewer trees available for oviposition (for details see Chaps. 12, 13). As such, avirulent PWNs are inferior to virulent PWNs in the infection process in the field. However, a recent study showed that an avirulent PWN isolate (C14-5) could survive for a long time inside *P. thunbergii* seedlings without causing any symptoms (Takeuchi and Futai 2007b). Such an avirulent PWN should be able to survive in the field without being "pitched out" because living trees would not be removed from forest stands when infected (see also Sect. 23.3). So does such asymptomatic infection by the PWN have any positive or negative effects on the host trees? Two such examples are given in the following sections.

23.3 Induced Resistance

In general, plant species have a sophisticated system for overcoming infection by a wide range of pathogens, which can be compared to the immune system of animals, even though plants do not possess antibodies. When previously infected with avirulent bacteria, fungi, or viruses, plants can immediately induce defense responses systemically or locally after infection with a virulent pathogen (systemic or localized acquired resistance, respectively; e.g., Hunt et al. 1996; Heil and Bostock 2002). This is reminiscent of antigen vaccinations for activating animal immunity, but induced resistance in plants does not involve an antibody–antigen reaction. Induced resistance is also found in the pine-PWN relation. A host pine tree, which has been previously inoculated with an avirulent isolate of the PWN systemically induces resistance against subsequent infection by a virulent PWN isolate, although the detailed mechanisms are not fully known (Kiyohara 1984; Kosaka et al. 2001b). A series of studies by Kiyohara and his colleagues showed several characteristics of the induced resistance concerning pine wilt disease. Firstly, the degree of resistance induced in the host pine increases with the number of avirulent PWNs pre-inoculated (Kiyohara 1997), secondly the interval between pre- and post-inoculation, that is, inoculation with avirulent nematodes and that with virulent nematodes affects the efficiency of resistance induction. For example, the highest survival rate of host *P. thunbergii* saplings (90–95%) occurred when pre-inoculation with an avirulent PWN (K-48) was done 3–4 weeks prior to inoculation with the virulent PWN (S6-1) isolate, while shorter or longer intervals caused lower survival rates (as low as 60%; Kiyohara 1997; Fig. IV.18B). Next, the studies showed that multiple inoculations (repeated inoculations or simultaneous inoculations) with avirulent PWNs at different sites resulted in more efficient induction of host resistance against subsequent inoculation with virulent PWNs than a single inoculation (Kiyohara and Kusunoki 1987). Also, it was found that induced resistance is not host species specific since it occurs widely in *P. thunbergii*, *P. densiflora*, *P. taeda*, and *P. massoniana* (Kiyohara 1984). The results also showed that induced resistance occurs regardless of host age, although the degree may differ (Kiyohara et al. 1999). It is interesting that the effect of induced resistance also was found in *P. taeda*, an innately resistant species, which may be useful for comparing innate species-specific resistance and induced resistance.

The effective duration of the induced resistance, however, is not constant or predictable, and some trees become symptomatic, and finally die after 1 or more years (Fig. IV.18; Kiyohara 1997), especially mature trees (Kosaka et al. 1998, 2001b; Kiyohara et al. 1999). This suggests that induced resistance resulting

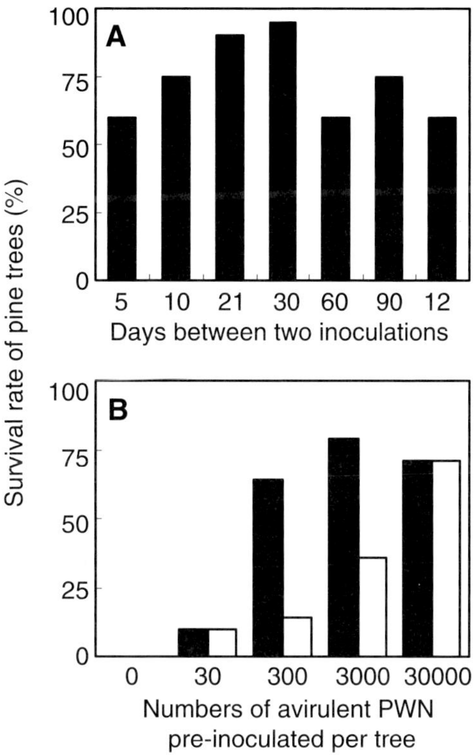

Fig. IV.18 Effect of (**A**) time and (**B**) pre-inoculation dosage with an avirulent isolate of *Bursaphelenchus xylophilus* (K-48) on the survival of *Pinus thunbergii* trees subsequently inoculated with a virulent, *B. xylophilus* isolate (S6-1). **A** The resistance induced in trees increased with the interval between the two inoculations, peaked when pre-inoculation was done around one month before the second inoculation, and gradually decreased. **B** Resistance induced in the trees increased with nematode pre-inoculation dosage. Survival of trees was observed the same year (*closed column*) and again the subsequent year (*open column*). The two inoculations were conducted at an interval of 60 days. Data from Kiyohara (1997)

from pre-inoculation with avirulent PWNs is effective only in delaying symptom development, not in preventing trees from being killed. So what happens inside avirulent PWN pre-inoculated host trees? Although the avirulent isolate OKD-1 interbred with the virulent Ka-4 isolate in vitro (propagated on a fungus), these two isolates seldom produced hybrid offspring in vivo, within a 6-year-old *P. thunbergii* tree (Aikawa et al. 2006); thus, avirulent PWN pre-inoculation may not contribute to propagation of post-inoculated virulent PWNs even when they encounter one another inside the host. However, even avirulent PWNs that survived inside a host pine for a long time (Takeuchi and Futai 2007b) may kill the tree under stress conditions (Kiyohara et al. 1999; Asai 2002). Thus the protective effect induced by avirulent PWNs is far from clear, other possibilities need to be examined for inducing host resistance.

Maybe some fungal or bacterial species plus avirulent isolates of the PWN can induce host resistance against virulent PWNs; for example, microbial inoculations using several bacteria and fungi such as species of *Cladosporium*, *Lophodermium*, and *Septoria* (Kiyohara et al. 1990), and *Botrytis cinerea* (Takeuchi et al. 2006a) can induce host resistance although the effect is quite short-lived. Such microbial inoculation might be safe and efficient for pine wilt disease control if the protection can be effectively maintained by using some technique such as repeated inoculations. As for some *Bursaphelenchus* species, including *B. mucronatus*, the results were highly unpredictable; they sometimes succeeded (Kiyohara et al. 1989, 1990) but also failed to induce resistance (Kiyohara 1982).

After inoculation with avirulent PWNs, susceptible pines exhibit partial necrosis in the cortex and the formation of wound periderm, as is often the case with resistant pines such as *P. taeda* and *P. strobus* inoculated with virulent PWNs (Yamada and Ito 1993a; for details see Chaps. 21, 22). Perhaps these reactions are involved in inhibiting nematode spread and contribute to the short-term effect of induced resistance to infection. The key "on/off" determinants of induced resistance in host pines are being investigated. Takeuchi et al. (2006a) suggested that the viability of the microbes used in pre-inoculation inside the host plant and consecutive stimuli may be required to keep resistance induction switched on, by comparing the results of pre-inoculation with avirulent PWNs (C14-5) and *B. cinerea*. This confirms the fact that both heat-killed and crushed nematodes, and also culture filtrates of the avirulent PWN (K-48) cannot induce resistance in young *P. thunbergii* (Kiyohara 1984). The substances that trigger induced resistance and the enzymes or genes involved in the induction pathway need to be clarified before it is possible to fully understand the host defense reaction against PWN infection and the pathogenicity.

How then does natural infection with avirulent PWNs come about in the field? A certain proportion of the PWN population is avirulent in the field (see Sect. 23.2.2) and Takemoto et al. (2005) suggested that this may be very prevalent especially in regions where the epidemic damage caused by PWN has already stabilized, based on a simulated model (for details see Chap. 12). Therefore, field-grown pines may be exposed to both virulent and avirulent PWNs. This poses the questions as to whether or not host resistance is naturally induced when the tree is infected with avirulent PWNs and if so, whether this resistance protects the trees. Based on past reports on induced resistance to pine wilt disease it may be that natural infection with avirulent PWNs induces host resistance and delays symptom development. Such resistance could result from subsequent infection by virulent PWNs but the effect would be short-lived which would mean that duplicate infection with avirulent and virulent PWNs might be a factor causing off-season wilt, including delayed wilt appearance in the year following inoculation in the field.

23.4 Asymptomatic Carriers

When susceptible pines are infected by PWNs, most of the trees, within a month of artificial inoculation, stop oleoresin exudation (Kiyohara and Tokushige 1971), decrease sap flow (Ikeda and Suzaki 1984), and begin to show external wilting symptoms such as needles discoloration (Table IV.1; see also Chap. 21). Some of the trees, however, can, depending upon the environmental or physiological conditions survive PWN infection even when the host-PWN relationship is compatible.

For example, daily watering dramatically reduced the mortality of potted, 3-year-old *P. thunbergii* seedlings inoculated with 50,000 PWNs, although low numbers of nematodes remained in the seedlings for 2 months (Suzuki and Kiyohara 1978). In such cases, do the surviving trees defeat PWN or do they just temporarily remain asymptomatic?

Most of the surviving pine trees often show a kind of stigma or a disorder. For example, susceptible pine trees, which survived inoculation with PWNs showed a dramatic reduction in height growth in the year following inoculation (Furuno and Futai 1983). Partial wilt and death of branches is often observed in trees, especially in resistant pines after inoculation with PWNs, but this phenomenon is rare in mature and naturally-infected trees (Kishi 1995). It is known for wilt which occurs in the year following infection that the delayed wilt symptoms can be attributed to late (in the season) inoculation or infection (Shoji and Jinno 1985), or in the areas where the climate is cool (Halik and Bergdahl 1994). Inoculation experiments also revealed that symptom development was delayed when the inoculum dose was lower or when inoculation was done in September or later (Kiyohara and Tokushige 1971). Kishi (1995) found that the first distinct symptoms such as dysfunction of oleoresin exudation and needle discoloration appeared during any season, and that the number of pine trees in which symptom development was delayed until the following year increased with a decrease in ambient temperatures.

On the other hand, Bergdahl and Halik (1999) inoculated 20-year-old trees of *P. sylvestris* with 30,000 PWN per tree and reported asymptomatic ("symptomless" or "latent" carrier) trees serving as hosts for PWNs for as long as a decade, indicating the epidemic dangers involved in transportation of logs, that is, pine logs derived from asymptomatic trees may carry and disseminate the PWN. As well, such asymptomatic trees are of ecological importance as pointed by other studies (Kanzaki and Futai 1996; Futai 2003a; Takeuchi and Futai 2007a). In the eradication of dying or dead trees, PWN infection is generally diagnosed by the amount of oleoresin exuded after artificially wounding trees exhibiting external symptoms, such as yellowing or browning of the needles (Futai 1999). The danger here is that trees without any symptoms can be overlooked and easily evade eradication because of the difficulty of detecting them by conventional diagnostic techniques. To detect asymptomatic carrier trees, Takeuchi and Futai (2007a) conducted intensive diagnoses with molecular techniques in two pine stands naturally suffering from the disease. Briefly, samples of woody tissues were taken from the experimental trees and DNA samples from them served as templates for use in nested PCR-based PWN detection, targeting the ribosomal DNA ITS region of the PWN with specific primers (Takeuchi et al. 2005; Takeuchi and Futai 2007a; Fig. IV.19). This technique successfully distinguished the PWN from three, closely related *Bursaphelenchus* species, which share morphological characteristics with the PWN, and showed its usefulness in detecting even a single PWN in 8 g of sample tissue. The results demonstrated that many *P. thunbergii* and *P. densiflora* trees harbored PWN without any evidence of infection, that is, external symptoms and dysfunction in oleoresin secretion (12 out of 54 trees and 2 out of 39 trees, respectively; Fig. IV.20). These data strongly suggest that both visual symptoms and the amount of oleoresin

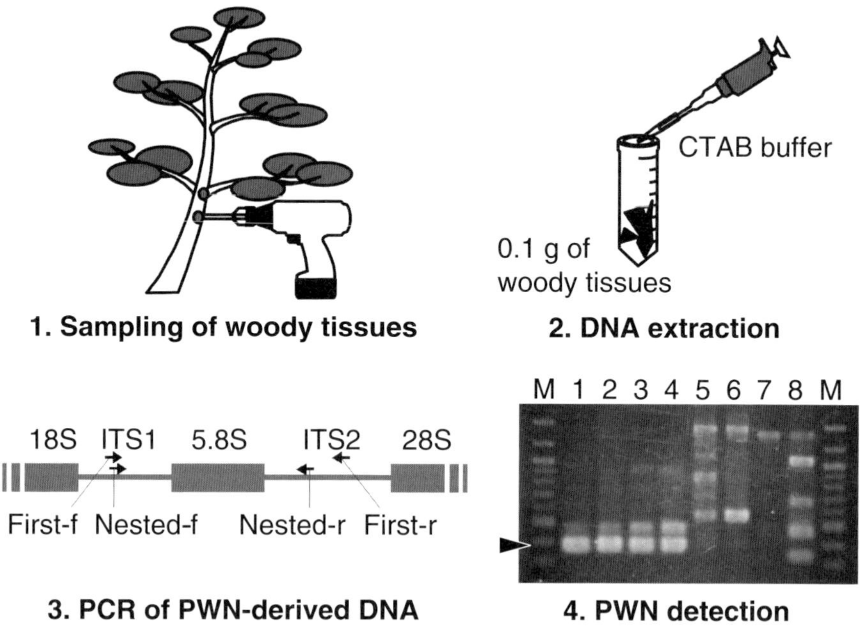

Fig. IV.19 Schemes showing the steps in the molecular diagnosis of pine wilt disease. (1) Samples of woody tissues are taken from test pine trees, and (2) about 0.1 g of tissue is used for DNA extraction using the CTAB method. (3) To detect *Bursaphelenchus xylophilus* the ribosomal DNA ITS region of the nematode is amplified by double PCR with two sets of specific primers as follows: TAC GTG CTG TTG TTG AGT TGG and CAC GGA CAA ACA GTG CGT AG for first PCR; GTG CTG TTG TTG AGT TGG CG and AGA CGA CTG TCA CAA CGT GC for nested PCR. (4) PCR products are separated by electrophoresis and the samples that yield a *B. xylophilus*-specific band are regarded as infested. M, size marker (100-bp ladder); *1* S10; *2* T4; *3* OKD-1; *4* C14-5 isolates of *B. xylophilus*; *5* intact *Pinus thunbergii*; *6* intact *P. densiflora*; *7 B. conicaudatus*; *8 B. luxuriosae*. Modified from Takeuchi et al. (2005), Takeuchi and Futai (2007a), with permission

exudation are not sufficient indicators to detect "hidden" PWN infection. Some of the asymptomatic trees, especially those of *P. thunbergii*, survived for at least 2 years and others quickly became symptomatic, and died quickly. This indicates that asymptomatic trees can be a reservoir for PWN and play a role in the dissemination of the disease in the field where they serve as oviposition sites for the PWN vector, *Monochamus* beetles, when the trees die.

When naturally or artificially infected with PWN, susceptible host pine saplings (Ikeda and Oda 1980; Kuroda 1989, 1991; Fukuda et al. 1994) and trees (Takeuchi et al. 2006b; Fig. IV.21) emit a characteristic bouquet of volatile compounds. For example, ethylene, ethanol and terpenoids including monoterpenoids such as α- and β-pinene and sesquiterpenoids such as junipene in a different manner from both trees, which have been mechanically injured or remain intact. Some of the volatile compounds are bioactive to the vector beetles of the PWN. It was found that monoterpenoids, including α-pinene mixed with ethanol at a specific ratio, are stronger

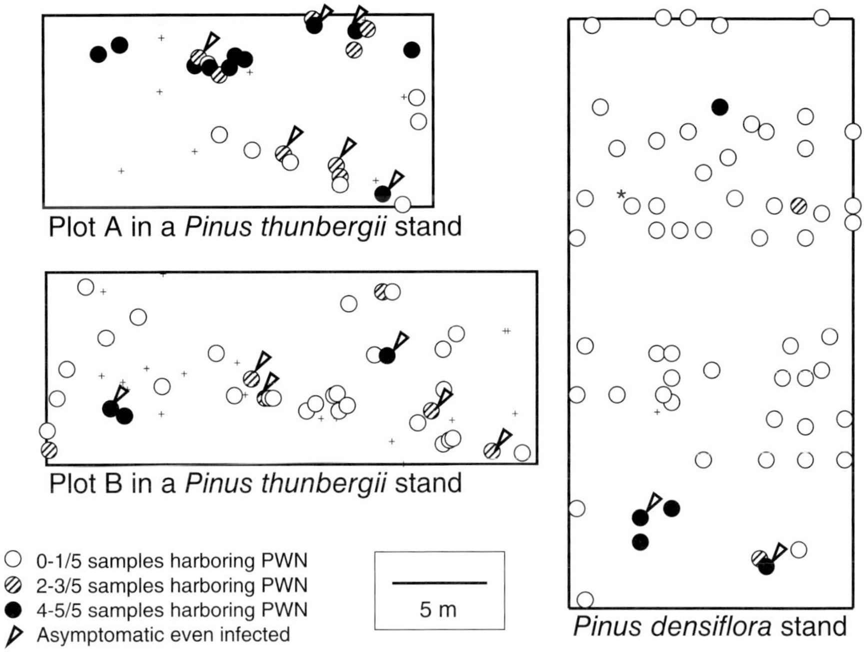

Fig. IV.20 Infection with *Bursaphelenchus xylophilus* (PWN) in natural stands of *Pinus thunbergii* at a sand dune in Tottori and *P. densiflora* on a mountainside in Ishikawa. Sample tissues were taken from five points on each tree before the pine wilt infection season and were used for PWN detection with molecular techniques. Oleoresin exudation and external symptoms were also checked at the same time. To avoid possible cross-contamination between samples, the pine trees in which PWNs were detected in less than one of the samples were considered as unaffected. Among infected trees, 14 lacked both discoloration of needles and dysfunction of oleoresin exudation (indicated by open arrows). Data from Takeuchi and Futai (2007a), with permission

attractants for *Monochamus* beetles than monoterpenoids alone (Ikeda et al. 1980a). Such volatile emissions were also observed in a *P. thunbergii*, forest tree that was naturally infected with PWN but remained asymptomatic until the following infection season (Takeuchi et al. 2006b).

The occurrence of dead pine trees within a stand often overlaps spatially, with recently diseased trees being near trees or stumps or trees killed the previous year (Togashi 1989c; Futai and Okamoto 1989). Initially the migration of PWNs from dead to healthy trees was attributed to their movement via root grafts, but subsequent work showed that there were few natural root grafts (Terada 1986). Futai (2003a) showed that asymptomatic 3-year-old *P. thunbergii* seedlings played a significant role as attractants for the newly emerged vector beetles that could then transmit PWNs to neighboring trees. He also found some possible asymptomatic *P. koraiensis* trees in a natural stand, following which he proposed a "chain infection model" taking asymptomatic trees into consideration to show how they are involved in the infection cycle and epidemic spread of pine wilt (Fig. I.4). The

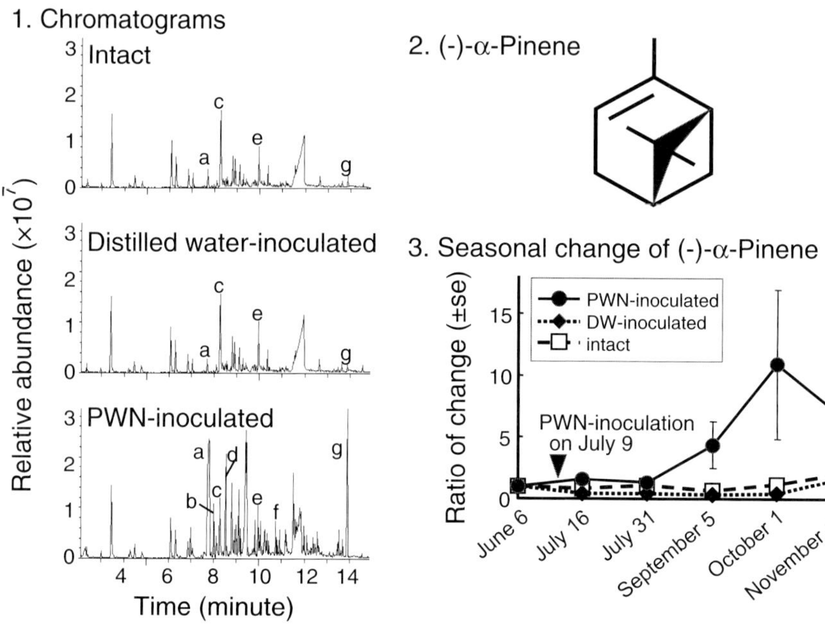

Fig. IV.21 Volatile emission from the trunk of *Pinus thunbergii* trees inoculated with a virulent isolate (S10) of *Bursaphelenchus xylophilus* (PWN). Inoculation with 20,000 PWNs or distilled water (DW) was done on 9 July. (1) Total ion chromatogram of volatiles emitted from PWN- or DW-inoculated and intact trees in October (*a* (−)-α-pinene; *b* camphene; *c* benzaldehye; *d* β-pinene; *e* 1-phenylethanone; *f* α-campholene aldeyde; *g* junipene). (2) Chemical structure and (3) seasonal fluctuation of (−)-α-pinene, a major component of pine volatiles, emitted from PWN- or DW-inoculated and intact trees. Emissions from PWN-inoculated trees peaked in late summer to autumn. Data from Takeuchi et al. (2006b), with permission

model is based on the fact that a few pine trees infected with PWNs remain asymptomatic and evade eradication, and *Monochamus* beetles attracted by the volatile compounds, which they release. As such the PWN is introduced into the surrounding trees. Aggregative distribution of *Monochamus* beetles around dying trees in the field (Shibata 1986) supports this hypothesis. The volatile compounds from infected pine trees, however, only attract mated beetles; while unmated beetles, which are beetles carrying PWNs, preferred volatiles from healthy trees (Hao et al. 2006). Volatile compounds released from infected, asymptomatic trees need to be examined for their attractiveness to unmated beetles.

The importance of asymptomatic trees has only recently been recognized in pine wilt disease epidemiology and also in a bacterial disease of willows (Hauben et al. 1998) and a fungal disease of oak (Luchi et al. 2005), suggesting the ecological and economical significance of such trees. The recurrence of pine wilt disease within the same stand even after thorough eradication of dead pine trees has been a serious problem, hampering disease control efforts. Asymptomatic trees may be partially responsible for pine wilt recurrence and as such they deserve intensive attention as a "third alternative" in the host's fate, coming after "survival" and "death".

23.5 Concluding Remarks

The host responses against infection by the PWN have been intensively investigated from the chemical, physiological, and histochemical aspects (see Chaps. 21, 22) since the first report of the PWN being the cause of pine wilt disease (Kiyohara and Tokushige 1971). However, the principal factor that decides host fate between "to die and to survive", "symptomatic and asymptomatic", and "susceptible and resistant" remains unclear, though phenomena and processes involved have been mostly elucidated. What holds the key, the PWN or the host tree? In the tomato-root knot nematode relation, the resistance gene *Mi* in tomato codes the receptor specific to the root knot nematode *Meloidogyne incognita* (Hwang and Williamson 2003), which is widely conserved in organisms and acts as a starting point for defense responses in encounters with the pathogen. There have been few reports on genes involved in host responses concerning pine wilt disease (Kuroda and Kuroda 1998). Further studies are needed to fully determine the pathogenic mechanisms at the molecular level, and to shed light on the hidden turning-point that determines the fate of PWN-infected host trees.

24 The Role of Bacteria Associated with the Pine Wood Nematode in Pathogenicity and Toxin-Production Related to Pine Wilt

Bo Guang Zhao[1] and Rong Gui Li[2]

24.1 Introduction

The role of bacteria associated with the pine wood nematode (PWN) *Bursaphelenchus xylophilus*, in pathogenicity is a controversial topic. Some scientists think that bacteria associated with PWN are chance contaminants, since bacteria exist both inside and outside the host tree, and they are not pathogenic (Yang 2000). For a long time it was thought that the PWN was the only pathogenic agent causing the disease (Mamiya 1975d, 1983; Nickle et al. 1981; Nobuchi et al. 1984; Fukuda et al. 1992a). However, some other scientists hold the alternate opinion that bacteria associated with the PWN may play some role in the pathogenicity of the disease (Oku et al. 1979; Kawazu 1998; Zhao et al. 2003; Han et al. 2003) and certain bacteria carried by the nematode may be symbiotically associated with the PWN (Zhao et al. 2005).

A number of scientists working on the disease attributed the quick wilting of host trees to toxins in the host; however, the origin of these toxins was controversial. Some scientists thought they came from the PWN or the response of the host to the nematode (Tanaka 1974; Oku 1979, 1988; Bolla et al. 1982b, 1984b; Ueda et al. 1984) and some believed that the toxins were products of bacteria associated with the PWN (Oku et al. 1979; Kawazu et al. 1996b; Kawazu 1998; Han et al. 2003; Zhao et al. 2003, 2005; Tan et al. 2004; Guo et al. 2006, 2007a; Yin et al. 2007).

As experimental equipment and protocols are continuously being improved, especially in molecular plant pathology, and with more scientific efforts and investments on the disease, controversial ideas will be looked at and then mechanisms behind the disease symptoms will be fully understood.

[1]Department of Forest Protection, Nanjing Forestry University, Nanjing 210037, People's Republic of China

[2]Department of Biology, Qingdao University, Qingdao 266071, People's Republic of China

[1]Tel.: +86-25-85427302, Fax: +86-25-85423922, e-mail: zhbg596@126.com, boguangzhao@yahoo.com

24.2 A Brief History of Research on Bacteria Associated with the PWN and its Pathogenicity

For a long time it was thought that *B. xylophilus* was the only pathogenic agent causing the disease (Mamiya 1975d, 1980; Nickle et al. 1981; Nobuchi et al. 1984). This viewpoint was supported by many studies such as those showing that inoculation with the PWN led to pine tree wilt and that nematodes from different pine trees differ in their pathogenicity (Kiyohara and Bolla 1990; Fukuda et al. 1992a; Kojima et al. 1994; Hu et al. 1995). However, certain scientists thought that toxins play an important role in the wilting process (Oku 1988, 1990; Zhang et al. 1997). Some researchers reported that the PWN itself did not produce these toxins (Cao and Shen 1996). Oku et al. (1979) suggested that the production of toxins was associated with bacteria. Kawazu (1998) isolated three strains of a bacterium that were toxic to both the callus and the seedlings of Japanese black pine (*Pinus thunbergii*). He also isolated the toxic substance and identified it as phenylacetic acid (Kawazu et al. 1996b; Kawazu 1998). Zhao et al. (2000b) observed, using electron microscopy, that many bacteria were attached to the surface of the PWN. The average number of bacteria carried by nematodes isolated from diseased trees was approximately 290 per adult nematode (Guo et al. 2002). Han et al. (2003) isolated two bacterial species in the genus *Pseudomonas* from PWN. To determine pathogenicity, callus and aseptic black pine seedlings were inoculated using the following inoculum: aseptic PWNs, the bacterium only (each of the two bacterial strains) or aseptic nematodes plus the bacterium (each of the two bacterial strains). The results showed that inoculation with aseptic PWNs did not lead to browning of the callus or wilt of the aseptic black pine seedlings, but those inoculated with aseptic nematodes plus either one of the two bacterial strains showed severe symptoms. The bacteria carried by the PWN played an important role in pathogenicity. In addition, bacteria were cultured in liquid, shake-culture media. The filtered liquid was directly applied to the callus of Japanese black pine and the ability of culture filtrate of each bacterium to induce browning was determined. The results showed that the culture filtrate of either strain caused severe browning. The authors deduced from these results that wilting was due to toxins in the bacterial culture filtrate (Han et al. 2003).

Zhao and his colleagues did a national survey of bacteria carried by the PWN in the main provinces in People's Republic of China where pine wilt is epidemic. They found that 17 of the 24 identified strains produced phytotoxins in a bioassay, in which 60-day-old aseptic seedlings of *P. thunbergii* with the root cut off were put into an ampoule containing the cell-free culture filtrate from an identified bacterial strain. Each bottle with three seedlings was sealed and kept at ±27°C with 14 h light and 10 h dark. The seedlings were observed and their appearance was recorded daily for 15 days. The time taken before a seedling to wilt was recorded and the mean of the six replicates was used to indicate the phytotoxicity of each bacterium. If the average period was 12 days or less, the strain was considered to produce phytotoxin(s). Eleven of these 17 phytotoxin producers belonged to the genus *Pseudomonas* (Zhao et al. 2003).

In field tests, 3-year old black pine seedlings inoculated with 19,000 axenic PWNs per seedling. Nine trees were killed 27–95 days after inoculation (Tamura 1983b). Kawazu and Kaneko (1997) thought this was because the PWNs became contaminated with bacteria producing phenylacetic acid during inoculation under field conditions and thereby regained pathogenicity. Kuroda and Ito (1992) excluded the possibility that microorganisms were associated with the pine wilt, because in their experiments, xylem water-blockage was observed much earlier than the presence of fungi and bacteria in inoculated Japanese black pine (*P. thunbergii*); however, Xie et al. (2005) who examined bacterial species and their population dynamics in PWN-inoculated Japanese black pine found that in the inoculated branches, bacteria could be detected on the 11th day after inoculation. This was before a few pine needles lost their color, while most needles become yellow or brown in the last stages of the disease when both the populations PWNs and bacteria had increased. When the test trees were dying and needles were totally wilted, and the populations of both nematodes and bacteria had increased dramatically. According to the identification of the bacteria isolated from the test trees, the first bacterial species appearing in the branches were pathogenic, the same as those Zhao et al. (2003) found in a China-wide survey.

24.3 Inoculations to Determine the Role of Bacteria in Pine Wilt Disease

To determine the role of bacteria in pine wilt disease, inoculations have been made with non-axenic and axenic PWNs. Oku et al. (1980) inoculated 3-year-old seedlings with a suspension of a bacterium of the genus *Pseudomonas* isolated from pathogenic PWNs. The results showed that three out of five of the 3-year-old seedlings inoculated subsequently wilted. They concluded that the bacterium associated with PWN could play a significant role in the rapid wilting of pine trees. Kawazu and Kaneko (1997) inoculated aseptic *P. densiflora* seedlings with aseptic PWNs (OKD-3 isolate) under aseptic conditions and the results showed that the inoculated seedlings did not wilt. They concluded that aseptic PWNs were not pathogenic.

Tan et al. (2004) reported that when 1- or 2-year-old branches of *P. massoniana* were excised, and the cut stems kept in water, and then inoculated with aseptic *B. xylophilus* (Bx), the accompanying bacterium *Bacillus firmus* (Bf) of Bx, or the mixture of Bx and Bf, respectively, that the branches inoculated with the mixture of Bx and Bf became diseased, while control treatment branches inoculated with Bx or Bf only, showed no disease symptoms.

Chi et al. (2006) reported inoculating 10-year-old Japanese black pines with aseptic PWN. After 11 months, all three of the control trees inoculated with wild (not aseptic) PWNs died from pine wilt, while five of eight trees inoculated with aseptic PWN remained healthy. At the end of the field test, only aseptic PWNs were recovered from the xylem of the healthy-looking trees inoculated with the aseptic PWN. The results showed that the re-isolated nematodes were still aseptic. This indicated that aseptic PWNs could not induce wilting of the inoculated pine trees, even though they survived inside the xylem of the pine trees. Among the three trees showing wilting symptoms, one tree died and the other two trees showed wilting in part of their crown. Recovery tests showed that bacteria were carried by the PWNs from all three trees. The authors thought that the bacteria were acquired from the environment, possibly through the wounds created by inoculation, mechanical injury or insect feeding scars.

Guo et al. (2006) inoculated aseptic seedlings and found that PWNs within *P. thunbergii* affected the reproduction and pathogenicity of *Burkholderia cepacia* (formerly *Pseudomonas cepacia*) B619 strain carried by the PWN. Both the secretions from live PWNs and dead nematodes promoted the reproduction and pathogenicity of the bacterium by providing essential metabolites or nutrients, and that the promotion effect of living nematodes was stronger than that of dead ones. The promotion effect of the nematode secretions decreased after high temperature (over 100°C) treatment.

Zhao et al. (2003) studied the role of bacteria associated with the PWN by inoculating greenhouse-grown, 4-month-old aseptic black pine seedlings and 3-year-old Japanese black pine, with axenic PWNs, axenic PWNs + a pathogenic bacterium or axenic PWNs + a non-pathogenic bacterium. The results from both inoculated seedlings and 3-year-old trees (Tables IV.6, IV.7) showed that inoculation with the bacterium could not induce the disease, but the disease was induced only when inoculated with axenic PWNs + pathogenic or wild PWNs, while seedlings exhibited no or weak symptoms when inoculated with axenic PWNs or axenic PWNs + the non-pathogenic bacterium. The results indicated that the disease was induced by the synergic effect of axenic PWNs and the pathogenic bacterium. This led the authors to put forward a new hypothesis that pine wilt disease is a complex disease induced by both PWNs and its associated pathogenic bacteria.

The re-isolations showed that the number of nematodes per gram of wood in the pine trees inoculated with axenic PWNs was much less than that from trees inoculated with axenic PWNs + the pathogenic bacterium or with wild PWNs in the control. The phenomena led the authors to conclude that a mutualistic symbiosis relationship exists between the PWN and its associated bacteria (Zhao and Lin 2005).

Discovery of the phenomena of mutualistic symbiosis between the PWN and its associated bacteria provides evidence to reveal the role of bacteria in the disease.

24.4 Toxins Produced by a Strain of *Pseudomonas fluorescens*

Oku (1979, 1988, 1990), Ueda et al. (1984), and Bolla et al. (1982b, 1984b) reported that toxic substances, including benzoic acid and 8-hydroxycarvotanacetone, which were induced by invading PWNs, killed pine cells. Some reports also suggested a direct attack of needle cells by the nematodes (Tanaka 1974). It was observed that nematodes could excrete cellulase (Odani et al. 1984, 1985), which

Table IV.6 Effects of various combinations of pine wood nematodes (PWN: bacterial-free and wild isolates) and strains of bacteria isolated from the PWNs 5 days after inoculation of 4-month-old axenic Japanese black pine seedlings, and upon recovery of the nematodes and bacteria from stem segments at 5 days (means of 6 seedlings)

Inoculum			Fifth day after inoculation					Recovery	
			Symptom level[2]				Incidence (%)	bacteria	Nematode
Nematode	Bacterium	Strain	0	I	II	III			
Control			6	0	0	0	0	–	–
None	*Pseudomonas fluorescens*	GcM5-1A	6	0	0	0	0	–	–
None	*Pseudomoas cepacia*	JnB619	6	0	0	0	0	–	–
None	*Pseudomoas putida*	ZpB1-2A	6	0	0	0	0	–	–
None	*Pantoea* sp.	ZpB1-1A	6	0	0	0	0	–	–
Axenic PWN	*Peptostreptococcus asaccharolyticus*	AcB1C	6	0	0	0	0	–	–
Axenic PWN	None	None	4	2	0	0	33	–	+
Axenic PWN	*Pseudomonas fluorescens*	GcM5-1A	0	0	3	3	100	+	+
Axenic PWN	*Pseudomoas cepacia*	JnB619	0	3	1	2	100	+	+
Axenic PWN	*Pseudomoas putida*	ZpB1-2A	0	2	2	2	100	+	+
Axenic PWN	*Pantoea* sp.	ZpB1-1A	0	0	4	2	100	+	+
Axenic PWN	*Peptostreptococcus asaccharolyticus*	AcB1C	4	2	0	0	33	+	+
Wild PWN	None	None	0	1	2	3	100	+	+

Modified from Zhao et al. 2003
"+" and "–" indicates positive or negative recovery (isolation) of the bacteria or nematodes

Table IV.7 Greenhouse test of pathogenicity of axenic, pine wood nematode and a phytotoxin producing strain of a bacterium carried by the nematode using 3-year-old black pine trees (Modified from Zhao et al. 2003)

Inoculum		N	Symptom	Recovery of pathogen[1]		
Nematode	Bacterium			Bacteria	Nematodes g^{-1}	Fungus
Axenic PWN	None	16/20	–	–	1.2	–
		4/20	++++	+	264.1	–
Axenic PWN	*Pseudomonas fluorescens*	20	++++	+	201.6	–
None	*Pseudomonas fluorescens*	20	–	–	0	–
Wild PWN	–	20	++++	+	220.4	–
Control		20	–	–	0	–

"+" and "–" indicate positive or negative recovery (isolation) of the bacteria or nematodes

damages or kills pine cells. Ishida and Hogetsu (1997) reported that after PWN inoculation, nematodes did not kill cortical cells immediately, but killed them after living on the cortical cells for some time. They pointed out that this might result from the growth of fungi or bacteria in pine tissue. Cao and Shen (1996) studied the toxicity of the extraction of PWNs cultured on an artificial medium and found that nematode extracts were not toxic to the seedlings when the extracts were bioassayed against 30-day-old seedlings of *P. thunbergii* and *P. massoniana*, They concluded that wilt toxins were not produced by PWNs under artificial culture conditions. Jiang et al. (2005) tested the toxins in the cell-free filtration of the culture of a bacterial strain of *P. fluorescens* with non-host plants. The results showed that the toxins in the filtration were toxic to the non-host test plants. This indicated that the toxins were non-selective.

Guo et al. (2007b) isolated two compounds, which showed obvious toxicity to both suspension cells and seedlings of *P. thunbergii*, from the culture of a strain of *P. fluorescens* (*P. fluorescens* GcM5-1A) carried by PWNs (Fig. IV.22). The two compounds were identified as two cyclic dipeptides, cyclo (-Pro-Val-) and cyclo(-Pro-Tyr-), by MS, ^{1}H NMR, ^{13}C NMR, ^{1}H-^{1}H COSY, ^{1}H-^{13}C COSY spectra. In addition to the small molecular toxins, one protein with a molecular weight of 50 kDa was also purified from the culture of *P. fluorescens* GcM5-1A (Fig. IV.23) (Guo et al. 2007a). The N-terminal sequence of the protein was ALSVNTNITS, which indicated that the protein was flagellin. Bioassay results showed that the protein was also toxic to suspension cells and seedlings of *P. thunbergii*.

Three-year-old black pines were co-inoculated with aseptic PWNs and *P. fluorescens* GcM5-1A, or wild (non-axenic) PWNs alone isolated from diseased black pines. One week later, symptoms of wilt appeared. Proteins in the stem tissues of both inoculated and healthy branches were extracted and analyzed by Western blotting using antiserum against flagellin. The results revealed that no obvious proteins could be detected in healthy branches, while flagellins could be detected in the samples of inoculated branches. Only one positive protein with a relative

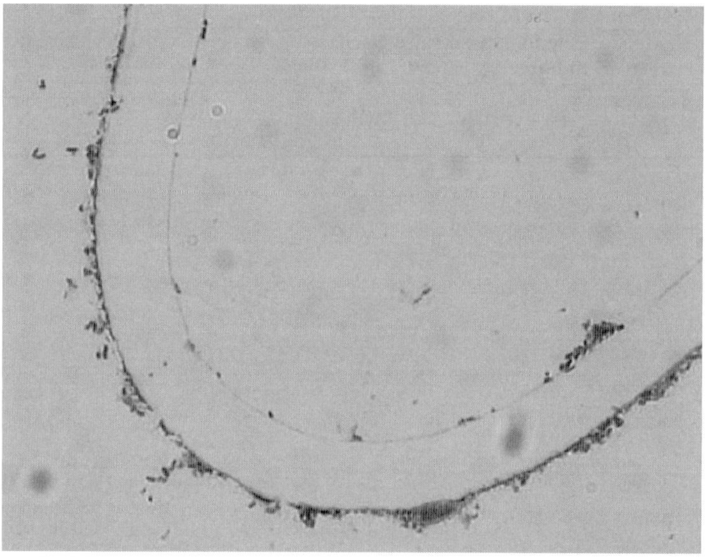

Fig. IV.22 Bacteria carried by pine wood nematode under a light microscope. The nematode was isolated from a naturally diseased Japanese black pine (stained with crystal violet) (see Color Plates)

molecular weight of 50 kDa was detected in seedlings co-inoculated with aseptic PWNs and *P. fluorescens* GcM5-1A, which was consistent with the molecular weight of flagellin purified from the culture of *P. fluorescens* GcM5-1A. Two main positive proteins with molecular weights of 50 and 35 kDa, were detected in branches inoculated alone with wild PWNs isolated from diseased Japanese black pines, and the 35 kDa protein was much more prevalent than the 50 kDa protein (Fig. IV.24). The 50 kDa protein was probably the flagellin of *P. fluorescens* GcM5-1A, while the 35 kDa protein was flagellin of other bacteria carried by the PWNs.

To further elucidate the roles of flagellin in pine wood wilt, recombinant flagellin of another strain of *P. fluorescens*, strain Pf-5, was overexpressed in *E. coli* and purified. Bioassays indicated that the protein showed a similar toxicity to suspension cells and seedlings of *P. thunbergii* to that of strain GcM5-1A. Fukuda (1997) reported that PWNs inoculations of Japanese black pine stem tissue caused cell membrane damage and resulted in electrolyte leakage. Treatment of suspension cells with flagellin also lead to the leakage of soluble carbohydrates and free amino acids (Yin et al. 2007). The morphology of suspension cells of *P. thunbergii* treated with flagellin was quite different from that of the control. The cytoplasm became condensed after suspension cells were treated with flagellin for 3 days and the cell wall became thick (Fig. IV.25). Both intercellular and intracellular peroxidase activities increased dramatically the day after flagellin was applied and a new peroxidase isozyme appeared in the treated cells.

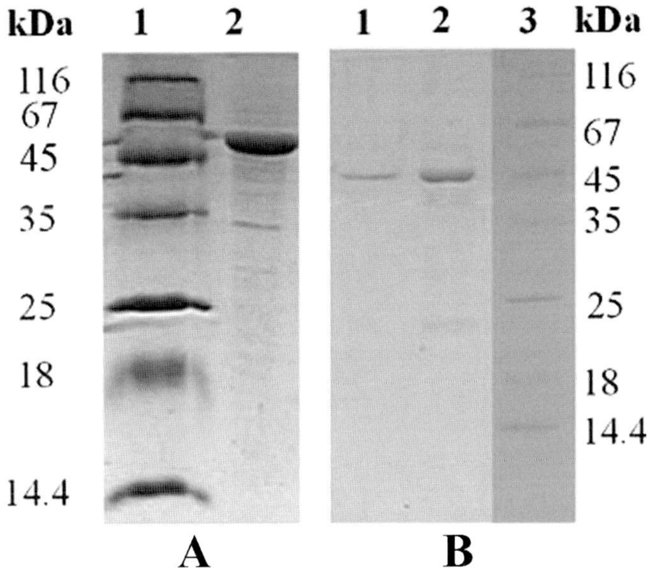

Fig. IV.23 Purification of flagellin from culture of *Pseudomonas fluorescens* GcM5-1A. Proteins in cell-free culture were precipitated using $(NH4)_2SO_4$ of 200 g l^{-1} dissolved in TE buffer (**A**), the flagellin was further purified through chromatography on DEAE-Sepharose FF column followed by Superdex 75 column (**B**). Fractions containing flagellin were analyzed by SDS-PAGE and the gel was stained with Coomassie Brilliant Blue R250. **A** *Lane 1* protein marker; *lane 2* proteins precipitated by ammonium sulfate. **B** *Lane 1* flagellin eluted from Superdex 75 column; *lane 2* proteins eluted from DEAE-Sepharose FF column; *lane 3* protein marker

Tanaka et al. (2003) reported that flagellin from an incompatible strain of *Acidovorax avenae* induced rapid H_2O_2 generation accompanying hypersensitive cell death and the expression of PAL, Cht-1 (encoding chitinase), and PBZ1 (a probenzanol (3-allyoxy-1,2-benzisothiazole-1,1-dioxide)-inducible gene homologous to *PR*-10) in rice, and the flagellin-deficient incompatible strain lost the ability. The flagellins purified from *P. syringae* pv. *tomato* and *glycinea*, incompatible pathogens of tobacco plants, induced fragmentation of chromosomal DNA and oxidative burst accompanied by programmed cell death in tobacco cells, but the flagellin from *P. syringae* pv. *tabaci*, a compatible pathogens did not (Taguchi et al. 2003). Another study demonstrated that basal resistance induced by flagellins in the leaves of *Nicotiana benthamiana* was accompanied by reduced vascular flow into minor veins (Oh et al. 2005). Considering this phenomenon as well as the toxicity of flagellin to suspension cells and seedlings of *P. thunbergii*, we proposed that flagellins of bacteria carried by the PWN play some roles in pine wilt disease. These roles may includes two aspects; firstly, flagellins as effector proteins probably induced a basal immune response to resist the invasion of plant pathogens, or secondly, flagellins might also induce cell death and reduce water flow into leaves, resulting in wilting. These studies provide some clue regarding the mechanism of pine wilt.

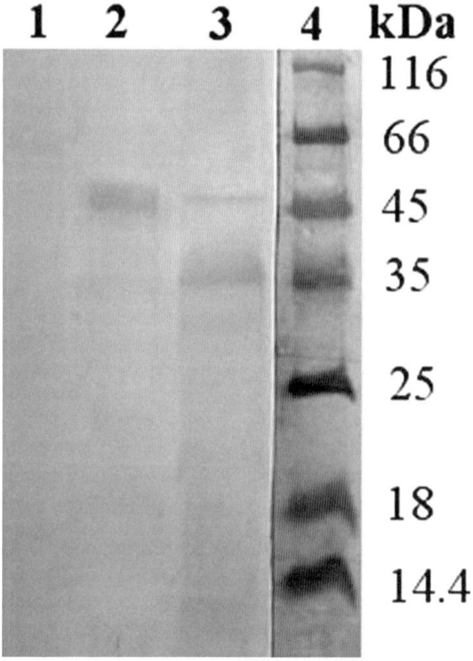

Fig. IV.24 Western blotting analysis of flagellins in branches of pine wood nematode (PWN)-inoculated *Pinus thunbergii*. Three-year-old *P. thunbergii* branches were inoculated with wild PWNs or a mixture of aseptic PWNs and *Pseudomonas fluorescens* GcM5-1A. The proteins from both healthy and diseased branches were analyzed by Western blotting. *Lane 1* healthy branches; *lane 2* diseased branches inoculated with a mixture of aseptic PWN and *P. fluorescens* GcM5-1A; *lane 3* diseased branches inoculated with wild PWN; *lane 4* protein marker

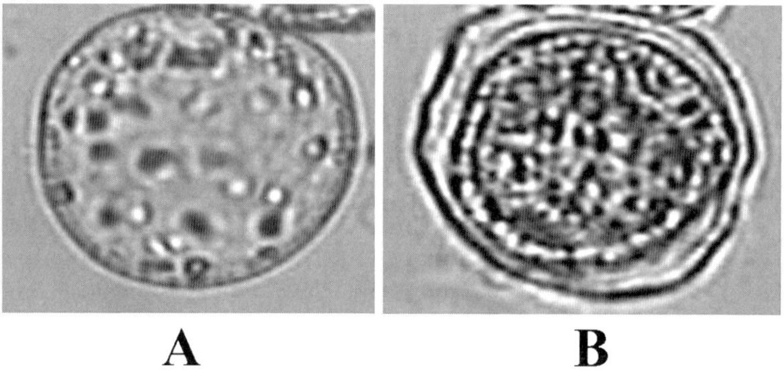

Fig. IV.25 Morphology of suspension cells of *Pinus thunbergii* under a light microscope. **A** Healthy cell; **B** cell treated with flagellin

24.5 Concluding Remarks

Pine wilt disease is a very complex disease, which has been studied from several viewpoints such as insects, fungi, bacteria and human activities. The role that bacteria play in the disease is controversial. The main difference in opinion is how to explain the results of inoculation tests with PWNs, axenic PWNs and the bacteria associated with PWNs in the field. The different results of field inoculations may result from axenic PWNs being re-contaminated with bacteria during inoculation in the field or from wounds created by the inoculation protocol. Another possible cause is that axenic PWNs used in field inoculation are not really axenic, since some methods to obtain axenic PWN are not effective for surface sterilizing PWNs (Lin 2005). New methods should be developed so that axenic PWNs used in field inoculation are not contaminated with bacteria.

It has been shown that toxins are produced by bacteria associated with the PWN and that those toxins exist in host trees inoculated with wild PWNs. As well, experiments showed that primarily purified flagellin induced leakage of soluble carbohydrates and free amino acids from suspension cells of Japanese black pine (Yin et al. 2007). This indicated that flagellin brought about changes in the cell membrane and may partly explain the cause of rapid wilting of pine wilt disease affected trees (Guo et al. 2007a, b). These results provide further evidence for this new hypothesis (Zhao et al. 2003).

Further research on the characterization of gene(s) related to the toxins and the symbiosis between PWNs and its associated bacteria are needed. The next step is to check the new hypothesis by making transgenic bacteria (*E. coli* or *P. fluorescens* which are avirulent or only weakly pathogenicity) with the toxin gene(s) to test their pathogenicity by making co-inoculations using the new bacterium and axenic PWNs. Further research on the role that bacteria play in the disease will provide new knowledge about the etiology of pine wilt disease.

25
Concluding Remarks

The process of disease development, possible factors causing wilting, and expression of resistance are described in this part. More detailed and historical reviews are available in the following references: Mamiya (1983), Fukuda (1997) and Yamada (2006). Investigation of anatomical, biochemical and physiological aspects of pine wilt disease have provided considerable insight into the processes and mechanisms of wilting through xylem dysfunction. Thus, many reactions and metabolites have been found that may induce or promote symptom development. Some links between the various aspects of our knowledge, however, are lacking, for example, between toxic metabolites of the pathogens, induced metabolisms, cell death of pine trees, and xylem dysfunction. Specific reactions and metabolites related to xylem dysfunction are important topics where knowledge is still lacking. Symptom development specific to pine wilt also needs to be considered.

Several researchers have hypothesized about the contribution of PWN-associated bacteria and their toxins in the wilting of pines. Recently, another possibility is that of the bacteria contributing a macromolecular toxin, which plays a role in the disease. Although we have conflicting results about the possible role of bacteria in the wilting process, there can be no doubt as to the major role of the PWN in pine wit disease.

Breeding of resistant pine trees is necessary if pines are to be used for afforestation. Although many pine wilt resistant *Pinus densiflora* and *P. thunbergii* trees have been selected in Japan (Part VI), detailed knowledge about the resistance factors and mechanisms is most important for the development of more resistant pines. Anatomical and physiological features and genes related to the resistance have been studied in recent years. Studies to determine the mechanisms of wilting and resistance have just begun, and future research should provide new insights into these mechanisms.

Part V
Related Microbes

26
Introduction

Organisms involved in pine wilt disease include pine trees (the hosts), the pine wood nematodes (PWN, *Bursaphelenchus xylophilus*) (the pathogen), and cerambycid beetles of the genus *Monochamus* (the vectors), however many other organisms are also associated with the disease; for example, numerous beetles of various genera in the families Cerambycidae, Curculionidae and Scolytidae, all of which colonize wilt-killed pines (Kishi 1995). Before Kiyohara and Tokushige (1971) showed that the PWN is the pine wilt pathogen, it was thought that these wood borers killed pine trees and they were commonly called "Matsukuimushi" in Japanese, meaning insect pests devouring pine wood.

Other *Bursaphelenchus* nematodes besides *B. xylophilus* also exist in pine wilt-killed trees, for example, *Bursaphelenchus mucronatus* and recently, new species of *Bursaphelenchus* have been found in several countries (See Chap. 9). As well, Futai et al. (1986) and Fukushige and Futai (1987) reported the presence of free-living nematodes and their seasonal population changes in wilt-killed trees. What's more, Sriwati et al. (2006) detected 15 species of nematodes belonging to several taxonomic orders, that is, (1) five mycophagous species in the orders Aphelenchida and Tylenchida, (2) nine saprophagous species in Monhisterida and Rhabditida, and (3) one predacious species in Mononchida. In Chap. 28, Sriwati reviews the succession of nematode fauna and the population dynamic relationships between nematode fauna and the PWN in wilt-affected pine trees.

Numerous microorganisms such as bacteria and fungi also exist in dead pine trees. Mamiya (1980) and Kusunoki (1987) observed bacteria in PWN-inoculated pine seedlings with a light microscope and electron microscopes, respectively. Kondo (1986) also reported the presence of bacteria on mites inhabiting the body of *Monochamus carolinensis* based on scanning electron microscopic observations. Kawazu and Kaneko (1997) and Zhao et al. (2003) reported that bacteria participated with the PWN in killing pine trees; however, Tamura (1983b) and Kuroda and Ito (1992) showed that bacteria did not participate in pine wilt symptom development (See Part IV). In Chap. 27, Zhao reviews the symbiotic relationships between the PWN and its associated bacteria.

Kobayashi et al. (1974, 1975) isolated various fungi from healthy and wilt-killed pine trees and the body of *Monochamus alternatus*. Kobayashi et al. (1975), Fukushige and Futai (1987), Kuroda and Ito (1992), and Sriwati et al. (2007) examined seasonal changes in fungal species and their detection frequency. In Chap. 28, Sriwati reports the succession of fungal flora and the relationship in the distribution between fungal flora and PWN in PWN-infected pines.

As well as being a plant pathogen the PWN is also a mycophagous nematode. Several researchers have examined the suitability of various fungi for PWN reproduction and showed that some fungi are suitable for reproduction while other fungi are unsuitable (Kobayashi et al. 1974, 1975; Fukushige 1991; Maehara and Futai 2000; Sriwati et al. 2007). These studies contribute to our knowledge on the important role of fungi in pine wilt disease. In Chap. 29, Maehara reviews the interactions among the PWN, wood-inhabiting fungi, and vector beetles, and discusses the role of fungi.

It is well known that besides wood-inhabiting fungi that are found in wilt-affected pine trees ectomycorrhizal fungi are associated with the roots of pines. Ectomycorrhizae also play an important role in pine wilt disease. In Chap. 29, Maehara briefly reports the effect of mycorrhizae on the disease.

27
Bacteria Carried by the Pine Wood Nematode and Their Symbiotic Relationship with the Nematode

Bo Guang Zhao

27.1 Introduction

Although pine wilt disease was found in Japan in 1905 (Yano 1913), and has been studied for more than half a century, all the factors associated with the disease have not been clearly defined. For a long time it was thought that the pine wood nematode (PWN), *Bursaphelenchus xylophilus* was the only pathogenic agent causing the disease (Mamiya 1975d, 1983; Yang 2002). Subsequent studies found that toxins play an important role in the wilting process (Mamiya 1980; Bolla et al. 1982b; Oku 1988, 1990; Zhang et al. 1997) and that the PWN itself does not produce toxins (Cao 1997; Kawazu and Kaneko 1997). Oku et al. (1979) suggested that the production of toxins was associated with bacteria. It was also reported that bacteria exist in PWN-inoculated pine seedlings (Kusunoki 1987) and they were associated with the PWNs (Oku et al. 1979; Higgins et al. 1999). Using an electron microscope Zhao et al. (2000b) observed many bacteria adhering to the surface of the PWN's body. The average number of bacteria carried by nematodes isolated from diseased trees was approximately 290 per adult nematode (Guo et al. 2002). Kawazu (1998) isolated three strains of bacteria that were toxic to both the callus and the seedlings of black pine. He also isolated the toxic substance and identified it as phenylacetic acid (Kawazu et al. 1996b; Kawazu 1998). Han isolated two species of bacteria from the nematode. Bioassay showed that the two species were strongly pathogenicity and possibly produced substances toxic to black pine seedlings and their callus (Han et al. 2003). Zhao and his colleagues made a national survey of bacteria carried by the PWN in the main epidemic provinces in the People's Republic of China. They found that 17 of the 24 identified strains produced phytotoxins in bioassays. Eleven of these 17 belonged to the genus *Pseudomonas* (Zhao et al. 2003). Based on their studies Zhao and Lin (2005) proposed that a

Department of Protection, Nanjing Forestry University, Nanjing 210037, People's Republic of China

Tel.: +86-25-85427302, Fax: +86-25-85423922, e-mail: zhbg596@126.com, boguangzhao@yahoo.com

mutualistic symbiosis exists between the PWN and its associated bacteria in the genus *Pseudomonas*.

This discussion focuses on the bacteria species associated with the PWN and their interactions with the nematode.

27.2 Bacteria in the Pine Forest Ecosystem

Some questions related to the possibility of the PWN carrying bacteria universal are: (1) where do PWNs carry bacteria, (2) does the vector insect, *Monochamus alternatus*, carry bacteria (3) do bacterial species exist in pine wilt-affected trees (4) if so, what are these species (5) do healthy pine trees contain bacteria (6) how do the populations of bacteria change in pine trees inoculated with PWNs and (7) could the bacterial species that PWN carries change when they are transported from one forest to another under different climate conditions or different geographic regions? All of these questions are important for understanding the role that bacteria may play in the disease. This chapter reviews the literature dealing with these questions.

Zhao and colleagues found that from all the diseased pine samples collected from five main provinces where pine wilt was epidemic in China, bacteria colonies appeared on the tracks left by nematodes on the surface of NB (Nutrient Broth) medium. Isolation experiments demonstrated that it was a universal for PWNs to always carry bacteria in their natural environment in China. No bacteria were found in any samples collected from healthy trees (Zhao et al. 2003). Zhang et al. (2004) isolated fungi and bacteria from trees and found bacteria in samples from pine wilt-killed trees, but not from healthy pine trees.

Using a transmission electron microscope (TEM) and a scanning electron microscope (SEM) Zhao et al. (2000b) observed the bacteria carried by the PWN and found many bacteria on the surface of the PWN's tegument and no bacteria within the body of living PWNs. With a light microscope it was also observed that bacterial cells existed on the surface of PWNs (Xie 2003).

The surface coat or glycocalyx lies external to the epicuticle of nematodes (Spiegel and McClure 1995). The surface coats of nematodes are polyanionic (probably due to sulfate or phosphate groups) and contain carbohydrate and mucin-like proteins (Page et al. 1992). Many carbohydrate and mucin-like proteins avoid eliciting host defense or wound responses even during the invasion and migratory phases, or when resident in the host. It is possible that the surface coat mimics host self-identity, as antibodies raised to the *Meloidogyne* surface specifically cross-react with host phloem cells (Blaxter et al. 1992). Sharon et al. (2002) characterized the surface coat of *Meloidogyne javanica* using antibodies and studied their effect on nematode behavior. The results showed that behavior was affected by all the antibodies that bound to the surface coat and it was demonstrated that the movement pattern of *M. javanica* was affected by these antibodies. Continuous binding of the antibodies to the surface of *M. javanica* inhibited infection of *Arabidopsis thaliana* roots on agar plates. Surface coats are a common feature of both parasitic and free-living nematodes (Blaxter et al. 1992), so those very direct means of immune evasion may be a simple adaptation to parasitism.

Zhao et al. observed the surface of PWNs using SEM, TEM and a light microscope (unpublished data) and found a thin surface layer on the cuticle of the PWN. Light microscope and TEM observations showed bacteria in the surface coat (Xie 2003).

Xie et al. (2005) examined bacterial species on the vector insect, *M. alternatus*, and found that most of the bacteria that the insect carried were taxonomy similar to those found on the PWN. In a national survey in China, Zhao et al. (2003) found 24 bacterial strains on PWNs that had been extracted from diseased pine trees (Table V.1). Xie et al. (2004) examined the bacterial species and their population dynamics in Japanese black pine inoculated with the PWN. They found that in the inoculated branches, bacteria could be detected when just a few pine needles had become yellowish; while when most needles changed to yellow or brown the abundance of both PWNs and bacteria increased. Finally, when the test tree was dying and the needles were totally wilted, the populations of both nematodes and bacteria had increased dramatically. Based on the identification of the bacteria isolated from the inoculated tree, the first bacteria to appear in the branches were pathogenic, the same as Zhao et al. (2003) found in the national survey. As the disease was developing, bacteria increased not only in quantity, but also in the number of species; however, the dominant species remained the same (Xie et al. 2004). The main species within the infected pine host are given below (Table V.2).

Could the bacterial species carried by the PWN change when the nematodes invade another forest in a different climate type or geographic region from that at its origin? According to existing data, bacteria carried by the PWN in different countries are mainly species of *Pseudomonas* and *Bacillus* (Oku 1979; Kawazu et al. 1996a,b; Kawazu 1998; Zhao et al. 2003; Tan and Feng 2004); however, the dominant pathogenic bacterial species carried by the PWN seems to be different. The dominant species in China belonged to the genus *Pseudomonas*, but were *Bacillus* spp. in Japan, with both genera in Korea (personal communication with Dr. Sung Chang). A field inoculation experiment in China showed that PWNs from Japan caused 21.5% of inoculated *Cedrus deodara* to wilt and some died, while none of the seedlings inoculated with PWNs from Nanjing, China showed any symptoms (Jiao et al. 1996); however, another field inoculation of Masson pine with the Nanjing PWN and Japanese PWN, respectively, showed that the Nanjing PWN was more virulent than the Japanese PWN (Jiao et al. 1996). In fact, many *C. deodara* trees in Nanjing City, China are street trees, but none of them have died of the disease, while Japanese black pine trees in Nanjing have been severely damaged. The difference in the dominant bacterial species associated with the PWN in different regions may explain why *C. deodara* in Japan can be affected by PWN (Dropkin et al. 1981), while not in China. This might be because when the PWN invades a new region they adopt new bacteria from the local flora. The fact that *C. deodara* is not be affected in China can be explained by the complex disease theory (Zhao et al. 2003), but it is difficult to explain it by the one-pathogen theory. To confirm the complex disease theory, however, further experimental evidence is required.

Table V.1 Identification of 24 bacteria strains carried by the pine wood nematode (*Bursaphelenchus xylophilus*) in China and the phytotoxicity (means of 6 replicates) of their cell-free culture filtrates to black pine seedlings in vitro

Isolate[a]	Sample origin	Species	Similarity[b] (%)	Phytotoxicity (days to wilting)
GcM5-1A	Guangdong, Conghua	*Pseudomonas fluorescens*	88.0	4.2
ZpB2-1A	Zhejiang, Pinghu	*P. fluorescens*	70.8	4.5
GcM6-1A	Guangdong, Conghua	*P. putida*	83.1	4.2
ZpB1-2A	Zhejiang, Pinghu	*P. putida*	97.4	4.5
HeM-139A	Hubei, Enshi	*Pseudomonas* sp.	99.9	9.0
ZpB1-2B	Zhejiang, Pinghu	*Pseudomonas* sp.	93.7	4.5
HeM-127B	Hubei, Enshi	*Pseudomonas* sp.	99.9	7.3
HeM-2A	Hubei, Enshi	*Pseudomonas* sp.	99.8	9.0
HeM142B	Hubei, Enshi	*Pseudomonas* sp.	77.8	9.3
AhM2D	Anhui, Hanshan	*P. cepacia*	60.0	9.2
JnB619	Jiangsu, Nanjing	*P. cepacia*	98.7	4.2
HeM3	Hubei, Enshi	*Pantoea* sp.	99.7	12.0
HeMA	Hubei, Enshi	*Pantoea* sp.	99.9	10.3
JnB1B	Jiangsu, Nanjing	*Pantoea* sp.	99.7	12.1
AcM1A	Anhui, Chaohu	*Pantoea* sp.	99.9	13.2
ZpB1-1A	Zhejiang, Pinghu	*Pantoea* sp.	91.0	9.8
GcM2-3B	Guangdong, Conghua	*Pantoea* sp.	99.9	3.5
JnM1B	Jiangsu, Nanjing	*Peptostreptococcus asaccharolyticus*	86.5	12.3
AcB1C	Anhui, Chaohu	*P. asaccharolyticus*	86.5	12.8
ZpB2-3A	Zhejiang, Pinghu	*Enterobacter amnigenus*	88.5	13.2
AcB2C	Anhui, Chaohu	*Buttiauxella agrestis*	99.4	12.5
AhM2A	Anhui, Hanshan	*B. agrestis*	99.4	13.0
GcM3-2A	Guangdong, Conghua	*Serratia marcescens*	99.9	3.8
GcM6-2B	Guangdong, Conghua	*S. marcescens*	99.8	4.2

Phytotoxicity indicates the days needed to show wilt symptoms after inoculation of axenic *Pinus thunbergii* seedlings with the bacterial strain + axenic *B. xylophilus* (modified from Zhao et al. 2003)

[a]The first two letters indicates the sample origin and the third one indicates the host species: B, Japanese black pine, *P. thunbergii*; M, Masson pine, *P. massoniana*

[b]The value is given by ABT Expression, with a fully automated identifier (bioMerieux Inc., Marcy-Etoile, France), and expressed by seven different categories of results: excellent identification (≥99.9), very good identification (99.0–99.8), good identification (90.0–98.9), acceptable identification (80.0–89.9), and low discrimination (<79.9)

27.3 The Species of Bacteria Associated with the PWN and Differences in their Ability to Produce Toxin Production

The PWN has been reported to lose its pathogenicity (Cao 1997; Kawazu and Kaneko 1997). It was also reported that a bacterium in the genus *Pseudomonas* was associated with the PWN (Oku et al. 1979; Higgins et al. 1999). Kawazu and his colleagues isolated bacteria from PWNs and identified three bacteria, *Bacillus*

Table V.2 Bacterial species carried by wild pine wood nematode (*Bursaphelenchus xylophilus*) and their phytotoxicity. Phytotoxicity indicated the days needed to show wilt symptom after the inoculation of axenic seedlings, *Pinus thunbergii*, with the bacterial strain + axenic *B. xylophilus* (modified from Zhao et al. 2003)

Isolate[a]	Species	Similarity (%)	Phytotoxicity (days to wilting)
NJG4	*Pseudomonas fluorescens*	82.8	4.2
NJH1	*Pseudomonas fluorescens*	84.8	4.5
NJE2	*Pseudomonas putida*	71.9	4.0
NJK2	*Pseudomonas putida*	80.1	4.5
NJG2	*Pseudomonas* sp.	68.4	4.5
NJG3	*Pseudomonas* sp.	68.4	7.3
NJI1	*Escherichia coli*	98.5	13.2
JnB619	*Pseudomonas cepacia*	98.7	4.2
NJPn7	*Shingomonus paucimobililis*	99.4	4.5
NJP4	*Enterobacter amnigenus*	85.0	12.8
JnB1B	*Pantoea* sp.	99.7	12.1

[a] Sample origin: *NJ* Nanjing, *Jn* Jiangsu, Nanjing

cereus, *B. subtilis* and *B. megaterium*, which were toxic to both the callus and seedlings of Japanese black pine (Kawazu and Kaneko 1997; Kawazu 1998). Three bacteria isolated from experimental plants have been identified as two strains of *Pseudomonas fluorescens* biotype I, *P. fluorescens* biotype II and a species of the genus *Pantoea* (Han et al. 2003). Zhao et al. (2003) isolated and identified 24 bacterial strains from living PWNs, which were isolated from diseased pine trees in a China wide survey (Table V.1). They also isolated and identified 11 bacterial strains from living PWNs collected near Nanjing (Table V.2). The results indicated that most were phytotoxin producers and the predominant species belonged to the genus *Pseudomonas*.

27.4 Mutualistic Symbiosis Between the PWN and Certain *Pseudomonas* Bacteria

Kawazu and his colleagues reported that the toxic products from bacteria in the genus *Bacillus* associated with the PWN were shown, in bioassays with pine seedlings and callus, to be phenylacetic acid. Bacterial cells grown on a Nutrient Broth medium, which contained nutrients from animal products, were more toxic than those on vegetatble-based medium. The result led him to presume that dead PWNs might provide nutrients to the bacteria and promote production of the toxin (Kawazu et al. 1996a, b; Kawazu and Kaneko 1997; Kawazu 1998, 1999).

Guo et al. (2006) reported that inoculation experiments with aseptic seedlings and callus of *Pinus thunbergii* revealed that the PWN affected the reproduction and pathogenicity of *Burkholderia cepacia* (previously, *Pseudomonas cepacia*) strain B619 carried by the PWN. Both the secretions from living nematodes and substances released from dead nematodes promoted the reproduction and pathogenicity of the strain. This was thought to result from providing the nematodes essential metabolites or nutritive substances to the bacteria. The stimulatory effect of living nematodes was stronger than that of dead ones. This effect decreased after treatment of the secretions at high temperature. This indicated that the bioactive substance(s) was heat-labile.

According to the results from greenhouse and field inoculations (Tables V.3, V.4), Zhao et al. (2003) proposed a hypothesis that pine wilt disease is a complex,

Table V.3 Greenhouse test of pathogenicity of axenic pine wood nematode (PWN) and a phytotoxin-producing bacteria strain carried by the nematode using 3-year-old black pine trees (modified from Zhao et al. 2003)

Inoculum			N	Final condition	Recovery of pathogen		
Nematode	Bacterium	Strain			Bacteria	Nematodes/g	Fungus
Axenic PWN	None	None	16/20	Healthy	–	1.2	–
			4/20	Dead	+	264.1	–
Axenic PWN	*Pseudomonas fluorescens*	GcM5-1A	20	Dead	+	201.6	–
None	*Pseudomonas fluorescens*	GcM5-1A	20	Healthy	–	0	–
Wild PWN	–	–	20	Dead	+	220.4	–
Control			20	Healthy	–	0	–

"+" and "–" were used to indicate positive and negative, respectively, in bacterium and fungus recovery

Table V.4 Field test of pathogenicity of axenic pine wood nematode (PWN) and a phytotoxin-producing bacteria strain carried by the nematode (modified from Zhao et al. 2003)

Inoculum		N	Symptom	Recovery		
Nematode	Bacterium			Bacteria	PWN[a]	Fungus
Axenic PWN	None	3	–	–	+	–
		1	–	–	–	–
		2	+++	+	++	–
		1	+++	+	+++	–
		1	++++	+	+++	+
Axenic PWN	*Pseudomonas fluorescens*	1	–	–	–	–
		1	+	–	–	–
		6	++++	+	+++	–
None	*Pseudomonas fluorescens*	8	–	–	–	–
Wild PWN	—	1	+	–	–	–
		5	++++	+	+++	–
		2	++++	+	+++	+
Control		8	–	–	–	–

"+" and "–" were used to indicate positive and negative, respectively, in bacterium and fungus recovery

2: "–" no PWN found; "+": 0 < number of PWN per gram dry wood <20; "++": 20 < number of PWN per gram dry wood <200; "+++": number of PWN per gram dry wood >200

which is induced by both the PWN and the phytotoxin-producing bacteria that it carries. In field tests they also found that the PWN was remained axenic in trees, which had been inoculated with axenic PWNs about 10 months earlier and which showed no disease symptoms just before sampling. It was also found that the number of nematodes per gram of dry wood of healthy-looking tree was much less than those from those showing wilt symptoms. Preliminary experiments of culturing PWN (from sterilized PWN eggs) on the callus of Japanese black pine with viable or dead cells of the bacterium strain GcM5-1A originally isolated from PWN, or with its culture filtrate, showed that PWN growth and reproduction were significantly enhanced (Wang 2004). These results imply that certain bacteria associated with the nematode and their metabolites could support the nematode's growth and reproduction.

Zhao and Lin (2005) examined the interactions between PWN isolates from Nanjing, China and various strains of bacteria isolated from PWNs that had been collected from the main epidemic provinces in China. To obtain axenic PWNs, they developed a new method, sterilizing the eggs of the nematode as follows: nematodes together with the medium were washed with a physiological salt solution (0.9% w/v NaCl) on a piece of filter paper which was placed on a funnel and left there for 24 h. Dead nematodes and the medium were removed and 10 ml of liquid containing living nematodes at the bottom were collected and centrifuged at 800*g* for 6 min. The supernatant was discarded and the nematode-containing precipitate was washed 6 times with a 0.9% sterilized NaCl solution. A volume of 2 ml of the nematode suspension were transferred onto a plate with PDA medium and the nematodes were left there for 12 h for oviposition. Then the plate was washed with 10 ml of sterilized NaCl solution and the water was centrifuged at 800*g* for 6 min. The supernatant was discarded and the nematodes and eggs were collected in 1.5 ml of the precipitate. Then 1.5 ml of 30% hydrogen peroxide was added to the precipitate. The mixture was kept at 20°C for 10 min and centrifuged at 800*g* for 6 min. The precipitate was washed with sterilized 0.9% NaCl solution five times after discarding the supernatants. The eggs were collected with a micro-syringe under a microscope and were ready for use. The sterility of eggs was checked on NB medium (3 g of nutrient broth, 10 g of peptone, 5 g of NaCl, 12 g of agar powder, pH 7.2–7.4). If bacterial colonies appeared in the medium, the set of eggs was discarded.

The authors tried several methods reported in the literature to obtain axenic PWN, but they were not bacteria free when checked on NB medium. As said above, the bacteria are carried by the PWNs on the surface of their body wall, so it is difficult to obtain axenic PWNs; however, it is important to reach solid conclusions with real axenic PWN in experiments made to determine the nematode's pathogenicity and other issues related to pine wilt.

To accurately measure the reciprocal effects between the PWN and its associated bacteria, five loops (0.5 cm in diameter) of the bacterial colony are put into 10 ml sterilized water in an aseptic glass tube to make a suspension of a strain. Two axenic PWN eggs, approximately 0.01 g callus of Japanese black pine, 0.05 ml of suspension of the bacterial strain and 2 ml of aseptic water were put into a small Petri dish (2 cm in diameter). The same preparation without any bacterium was used as a control. Four small Petri dishes were put into a bigger dish (8.5 cm in diameter). Eight replicates (small dishes) were made for each bacterial strain and the control. The dishes were then cultured at 25°C. When one male and one female PWN hatched from the two eggs in a small Petri dish, the offspring could be propagated. Thus, the probability of producing a second generation from the eight replicates was 50%, if the conditions in the treatment were suitable for reproduction. The dishes were observed every day under sterile conditions from the fifth to the ninth day, the number of newly hatched juveniles was recorded, and they were removed with a sterilized hooked needle and discarded. The two adults were left in the small dish to continue the culture. The mean number of juveniles produced per pair of nematodes in the replicates was calculated between the dishes with a bacterial strain and those without (control). The results are shown in Table V.5 (experiment 1).

A method similar to that described above was used to determine the effects of PWNs on the reproduction of the bacteria. The concentration of a specific strain of bacterium was measured by counting bacterial colony numbers on a plate with Nutrient Broth medium. The results are also shown in Table V.5 (experiment 2). The results indicated that a mutualistic relationship existed between PWNs and its associated bacteria in the genus *Pseudomonas*. Presumably there was a reciprocal exchange of nutrients between the PWNs and some strains of the bacteria. Other findings supporting this hypothesis were that adult nematodes in the control treatment in experiment 1 (rearing axenic PWN without any bacterium) were smaller than those from dual cultures with mutualistic bacteria. Culturing PWN with the callus and dead cells or the culture filtrate of the bacterial strain GcM5-1A significantly promoted PWN growth and reproduction (Wang 2004). Mutualistic symbiosis between nematodes and bacteria is known from entomopathogenic nematodes of the families Steinernematidae and Heterorhabditidae, which are associated with bacteria of the genera *Xenorhabdus* and *Photorhabdus*, respectively (Forst and Nealson 1996). In entomopathogenic cases, bacterium symbionts can serve as a food source or provide essential nutrients that are required for efficient nematode proliferation (Akhurst and Dunphy 1993).

However, there may be also other benefits to the symbionts. For example, the mutualistic bacteria could be carried by PWNs to the pine host and be protected from the PWN within the host tree, because inoculation experiments have shown that mutualistic bacteria could not invade the host through wounds without being carried by PWNs (Zhao et al. 2003). Perhaps the mutualistic symbiosis of PWNs and the bacteria might be the result of coevolution acting over a long time. It is difficult to imagine that PWNs and an accidentally contaminating bacterium could establish such a mutualistic association. According to the theory of coevolution, long-term close contact often results in co-speciation and co-adaptation among interacting symbionts (Paracer and Ahmadjian 2000). In Table V.5, the strains GcM5-1A, ZpB2-1A and HM 2, ZM4A all belong to *Pseudomonas fluorescens*, but the former two positively affect the PWN and are positively affected by PWNs, and the other two strains show no such traits. The PWN has evolved to be one of the two (the other one is *B. cocophilus*) currently known plant pathogenic species in the genus *Bursaphelenchus*. Its pathogenic characteristics are presumably the

Table V.5 Reciprocal interactions on reproduction between pine wood nematode (PWN) and its accompanying bacterial strains (modified from Zhao and Lin 2005)

Strain	Species	Experiment 1		Experiment 2	
		Mean of newly hatched juveniles per pair of nematodes (* significantly different from the control in Dunnett's test at α = 0.05)	Replicate(s) with new juveniles	Effects of PWN on bacterium reproduction, +: promote (sig. value); –: inhibit (sig. value); ±: no significant effects (sig. value, α = 0.01)	Relationship between nematode and bacterium
GcM5-1A	*Pseudomonas fluorescens*	22.00*	4	+ (<0.001)	Mutualistic symbiosis
ZpB2-1A	*P. fluorescens*	16.50*	4	+ (<0.001)	Mutualistic symbiosis
GcM6-2A	*P. putida*	83.75*	4	+ (<0.001)	Mutualistic symbiosis
GcM6-1A	*P. putida*	29.00*	2	+ (<0.001)	Mutualistic symbiosis
ZpB1-2A	*P. putida*	38.25*	4	+ (<0.001)	Mutualistic symbiosis
GcM2-3A	*P. putida*	32.00*	5	+ (0.019)	Mutualistic symbiosis
HeM-139A	*Pseudomonas* sp.	52.50*	2	+ (0.017)	Mutualistic symbiosis
HeM2A	*Pseudomonas* sp.	80.00*	4	+ (<0.001)	Mutualistic symbiosis
HeM1A	*Pseudomonas* sp.	76.25*	4	+ (<0.001)	Mutualistic symbiosis
GcM1-3A	*P. cepacia*	39.00*	4	+ (0.003)	Mutualistic symbiosis
HeM-127B	*Pseudomonas* sp.	0	0	+ (<0.001)	Unclear
HeM142B	*Pseudomonas* sp.	0	0	± (<0.550)	Unclear
HM 2	*P. fluorescens*	0	0	– (0.010)	Unclear
ZM4A	*P. fluorescens*	0	0	– (0.012)	Unclear
SH1C	*Pantoea* sp.	0	0	+ (0.030)	Unclear
HM 3	*Pantoea* sp.	0	0	+ (0.007)	Unclear
NM1A	*Pantoea* sp.	0	0	+ (<0.001)	Unclear
NH1B	*Pantoea* sp.	0	0	+ (<0.001)	Unclear
ZpB1-1A	*Pantoea* sp.	0	0	+ (0.002)	Unclear
AM1A	*Pantoea* sp.	0	0	– (<0.001)	Unclear
HeM-127A	*Pantoea* sp.	0	0	± (0.698)	Unclear
ZM2C	*Pantoea* sp.	0	0	– (0.002)	Unclear
HeM B	*Peptostreptococcus asaccharolyticus*	0	0	± (0.200?)	Unclear
NM1B	*P. asaccharolyticus*	0	0	± (0.561)	Unclear
ZM3A	*Enterobacter coli*	0	0	± (0.861)	Unclear
ZM1C	*Serratia arcescence*	0	0	+ (0.017)	Unclear
AH1A	*Enterobacter cloacae*	0	0	+ (<0.001)	Unclear
ZpB4-2B	*Staphylococcus sciuri*	0	0	– (<0.001)	Unclear
ZpB2-3A	*Enterobacter amnigenus*	0	0	– (0.004)	Unclear
Control	None	12.50	4	–	–

result of mutualistic symbiosis with bacteria. It is believed that the more highly evolved the symbiosis the longer time the symbionts have had to adapt to each other, and the more specific is the association (Paracer and Ahmadjian 2000).

The association between the PWN and bacteria is a good model to study co-evolution, because among the bacteria carried by the PWN there are different species in different stages of co-evolution. PWNs and bacteria can be cultivated separately on artificial media. Further research is needed to reveal the relationship between PWN and accompanying bacteria at the molecular level. This will be of significance not only for understanding the etiology of the disease, but also for its practical control.

27.5 Concluding Remarks

Insects, fungi, bacteria, environmental factors and human activities have been studied in relation to pine wilt disease. The bacteria associated with PWNs are still not very well known, as there are reports of possible bacteria-PWN associations from Japan, China and Korea, but not from North America and Europe. Detailed research on the interactions between the PWN and its associated bacteria is just starting, whereas research on many other aspects of symbiosis has produced much information, for example, on the molecular mechanism of symbiosis, the gene(s) responsible for symbiosis. Further research on the bacterial species associated with the PWN in different regions of the world and will provide knowledge to better understand pine wilt disease.

28
Nematode Fauna and Fungal Flora in Infected Pine Trees

Rina Sriwati

28.1 Introduction

In Japan the pine wood nematode (PWN) is transmitted mainly by the Japanese pine sawyer, *Monochamus alternatus*, from wilt-killed to healthy pine trees (Mamiya and Enda 1972; Morimoto and Iwasaki 1972). The adult beetles of *M. alternatus* carry a great number of PWNs in their tracheae when they emerge from PWN-killed pines in early summer. Newly emerging adults fly to healthy trees and feed on the bark of young twigs for maturation of their reproductive organs (maturation feeding). At that time, the PWNs on the vector beetles are transmitted to healthy trees and invade them through the feeding wounds made by the beetles. A small number of PWNs disperse widely in the infected trees and causes cessation of oleoresin flow. Thereafter, PWNs propagate dramatically and the trees show wilting symptoms, releasing volatiles such as ethanol, and terpenes (reviewed by Kishi 1995). Mature beetles are attracted to these wilting trees and oviposit in them. The eggs hatch within a week and the larvae feed on the inner bark and outermost sapwood, and then in autumn bore into the sapwood to form pupal chambers (PCs). The number of PWNs reaches its maximum from autumn to winter, and then decreases gradually (Mamiya et al. 1973; Fukushige and Futai 1987). The PC of *M. alternatus* beetles is one of the most important places for PWNs, because as Mamiya (1972) reported numerous PWNs aggregated around the PCs of *M. alternatus* in wilt-killed pine trees and that the beetles emerging in the subsequent year harbored many nematodes on their bodies.

PWNs, which are transmitted to healthy trees by vector beetles, feed on the parenchyma cells of the trees and on fungi such as species of *Pestalotia* and *Rhizosphaera* which are sparsely distributed in living trees. When the host tree is diseased,

Laboratory of Environmental Mycoscience, Graduate School of Agriculture, Kyoto University. Kitashirakawa Oiwake-cho, Sakyo-ku, Kyoto 606-8502, Japan

Present address: Plant Pests and Diseases Department, Agricultural Faculty, Syiah Kuala University, Banda Aceh 23111, Indonesia

Tel.: +62-651-7551977, e-mail: rin_aceh@yahoo.com

food sources of PWNs must be replaced with various wood-inhabiting fungi such as blue-stain fungi (Kobayashi 1975b; Kobayashi et al. 1975), though such fungi as *Trichoderma* spp., which also inhabit dead pines, are unsuitable for PWN reproduction (Kobayashi et al. 1975; Fukushige 1991; Maehara and Futai 2000).

Under field conditions, dying trees in general are rapidly invaded by various wood-decaying microorganisms (Shigo 1967) and intense competition among such microorganisms brings about a succession of microbial flora and fauna. Abiotic environmental conditions, especially temperature, moisture and substrate and biotic factors greatly affect the succession of organisms.

This chapter aims to discuss the changes in nematode fauna and in fungal flora in pine trees after infection by the PWN, and to clarify their relationships in the population dynamics of the PWN.

28.2 Seasonal Changes in the Nematode Fauna in Pine Wilt Affected Pine Trees

When trees become diseased, the physical conditions within the trees, such as water content and temperature, change dramatically (Shigo 1967). This invokes corresponding changes in microbial flora in the trees, indicating that the nematode fauna should also change promptly. An earlier study clarified the seasonal changes in the numbers of PWNs and other free-living nematodes, and their interrelationships (Fukushige and Futai 1987); however, this study did not show the species composition of "free-living nematodes" and so the interaction between the PWN and each species of free-living nematodes remained unclear.

A wide variety of nematode species have been found to be associated with dead *Pinus koraiensis* trees in Primorye, Russia (Kruglik 2003), most were mycophagous or saprophagous nematodes. Among them, nematodes belonging to the order Aphelenchida were quite abundant in the dead wood of the Russian trees. Sriwati et al. (2006) investigated the nematode fauna in the stems of pine trees bimonthly after infection with the PWN, clarified their seasonal changes and the interrelations among the PWN and the other species comprising the nematode fauna, and then examined the effect of PCs of *M. alternatus* on the population density of the nematode species inhabiting dead pine trees. Fifteen species of nematodes were identified, including mycophagous species, such as two species of Tylenchida and three species of *Bursaphelenchus* (*B. xylophilus*, *B. sinensis*, and *B. eproctatus*), and nine saprophagous species. Two species of Tylenchida were considered to be mycophagous because they reproduced on fungus mycelium, which grew from the wood piece on PDA medium. Interestingly, *B. eproctatus* might be an entomoparasite, because the species has an indistinct anus and rectum or lacked ones, which is a characteristic of nematodes in the entomoparasitic genera *Ektaphelenchus* and *Cryptaphelenchus* (Sriwati et al. 2008). During all seasons, *B. xylophilus* was the most abundant species, followed by Diplogastridae sp. 1. The seasonal change of the population densities of *B. xylophilus* and Diplogastridae sp. 1 are shown in Fig. V.1.

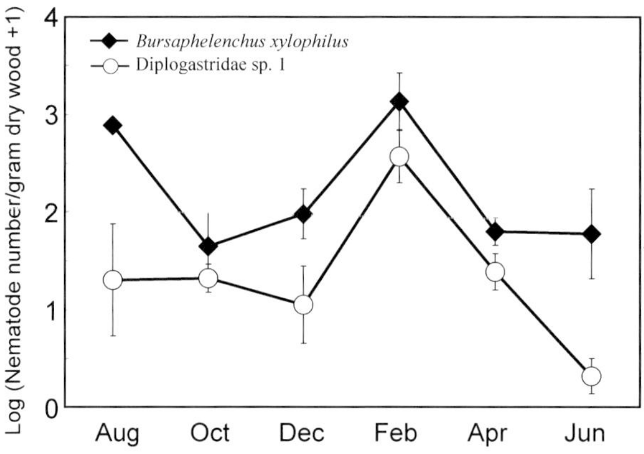

Fig. V.1 Seasonal changes in the populations of pine wood nematode (*Bursaphelenchus xylophilus*) and Diplogastridae sp. 1 in dead *Pinus thunbergii* trees. Each *bar* indicates the standard error ($n = 3$) (from Sriwati et al. 2006, with permission)

28.2.1 Relationship of the Nematode Fauna to Population Dynamics of the PWN in Infected Pine Trees

The number of PWNs decreased in December, increased in February, and then decreased again until June when the experiment terminated (Fukushige and Futai 1987; Sriwati et al. 2006). Futai et al. (1986) studied population changes of both *B. xylophilus* and other free-living nematodes in insecticide-treated and non-treated pine logs and showed that the population density of *B. xylophilus* was rather positively correlated with that of free-living nematodes. Sriwati et al. (2006) also showed that the population size of Diplogastridae sp. 1 changed synchronously with that of *B. xylophilus* in dead pine trees throughout the experimental period (Fig. V.1); however, the correlation between those two populations in the wood samples varied from tree to tree (Fig. V.2). Even in the same sampling season, except for February, no significant correlation was evident between populations of the two nematode species. In some trees during December and June, neither *B. xylophilus* nor Diplogastridae sp. 1 was found; consequently, no correlation analyses could be made. These results suggest that there might be a positive relationship between *B. xylophilus* and Diplogastridae sp. 1 at tree level, but the distribution of either nematode among wood pieces was more or less random.

Kanzaki et al. (2000, 2002) isolated both *Diplogasteroides psacotheae* (Rhabditida: Diplogastridae) and *B. conicaudatus* from adults of the yellow-spotted longicorn beetle, *Psacothea hilaris* (Coleoptera: Cerambycidae) from fig trees, *Ficus carica*. Another study by Kanzaki and Futai (2002d) also found these two nematodes to be sympatric in their vector's body and in their host fig trees. Differences in their food preferences in host trees and in the part of the vector body must enable *D. psacotheae* and *B. conicaudatus* to be sympatric. A recent study also found that a large population of both Diplogastridae sp. 1 and PWN congregated around both PC and tunnels of *M. alternatus*, and perhaps differences in the feeding habits of these two nematodes enabled them to be sympatric (Sriwati et al. 2006).

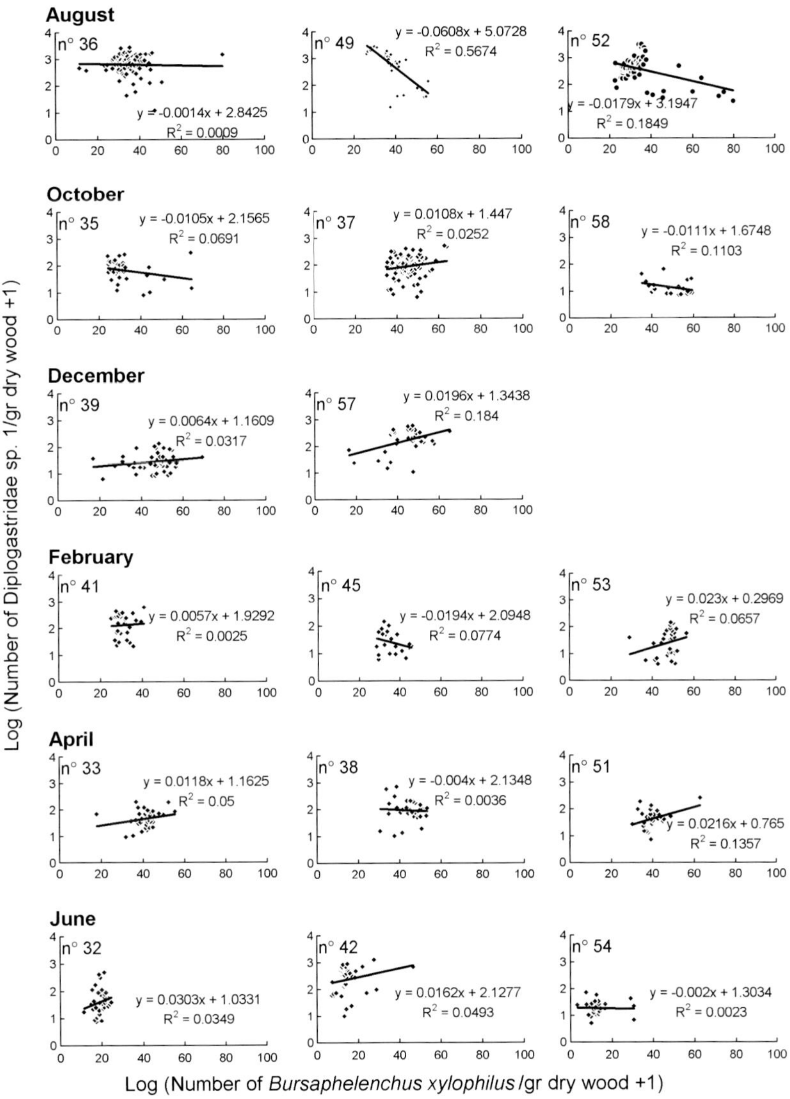

Fig. V.2 Relationship between the numbers of *Bursaphelenchus xylophilus* and Diplogastridae sp. 1 in dead *Pinus thunbergii* trees (from Sriwati et al. 2006, with permission)

28.2.2 Effect of Pupal Chambers on the Number of Nematodes

The number of nematodes is affected also by the PCs of *M. alternatus*. Figure V.3 shows the ratio of the numbers of nematodes in wood samples with PCs and tunnel fractions, or both, to those in wood samples without them. Several mycophagous

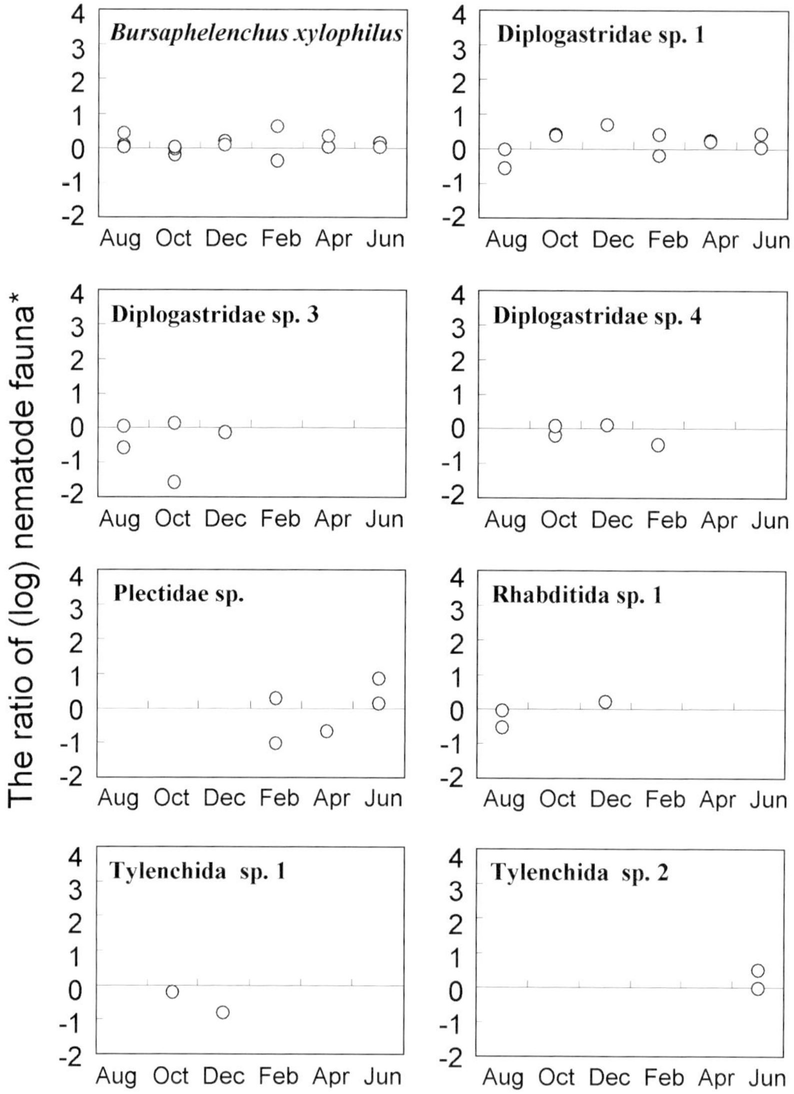

Fig. V.3 Effect of the pupal chambers (PC) of *Monochamus alternatus* in *Pinus thunbergii* wood on the population density of the dominant nematode species therein.* the ratio of logarithmic nematode numbers in wood samples containing PC or tunnel fractions to those in wood samples without them (from Sriwati et al. 2006, with permission)

nematodes such as species of Tylenchida, two species of *Bursaphelenchus* and PWN were detected, but only PWN preferred to aggregate around PC. In most cases, however, not only mycophagous PWN but also saprophagous Diplogastridae sp. 1 was more abundant around PC and tunnels than elsewhere in the wood (Sriwati et al. 2006). The PCs of *M. alternatus* are suitable places for PWNs to reproduce in infested trees due to blue-stain fungi and some fungal flora prevailed.

When the number of PWNs that aggregate around the PCs is high, the number of PWNs carried by beetles emerging from the PCs increases (Maehara et al. 2005). Kobayashi et al. (1974, 1975), Fukushige (1991) and Maehara and Futai (2000) reported that the blue-stain fungi, which prevailed around PCs would serve as suitable food for *B. xylophilus*. Further, intense proliferation of blue-stain fungi on the walls of PCs increased the number of PWN aggregating around these PCs (Maehara et al. 2005). PCs and tunnels of *M. alternatus* might provide suitable humidity and sufficient nutrients for the growth and reproduction of fungi and other microbes, and thus serve as food for nematodes. Many factors could affect the microclimate of PCs and tunnels. Environmental conditions and other fungal species besides blue-stain fungi might also play an important role in determining the suitability for propagation of the PWN, and *Monochamus* beetles must introduce many organisms, including various fungi, into pine trees.

28.3 Seasonal Changes in the Fungal Flora in Infected Pine Trees

Some researchers have studied changes in the fungal flora in dead pine trees (Kobayashi et al. 1974, 1975; Fukushige and Futai 1987; Wang et al. 2005c; Sriwati et al. 2007). Among the fungal species isolated from wilt-killed pine trees, some are known to be suitable food sources for the PWN, for example, *Ophiostoma minus*, *Ceratocystis* sp., *Diplodia* sp., and *Pestalotia* sp., *Phialophora repens*, *Sphaeropsis sapinea*, while others are unsuitable, for example, *Cephalosporium* sp., *Penicillium* sp., *Trichoderma* spp., *Verticillium* sp. (Kobayashi et al. 1974, 1975; Fukushige 1991; Maehara and Futai 2000; Sriwati et al. 2007). Under laboratory conditions, Maehara and Futai (1996, 1997) demonstrated that each fungal species that proliferated around the PC of *M. alternatus*, affected not only PWN multiplication but also the number of PWN carried by a vector beetle. These findings clearly indicate that fungal flora in a dead pine tree might be one of the most determinative biotic factors for the multiplication and distribution of PWN inside the tree. Sriwati et al. (2007) found 18 species of fungi in dead pine trees after inoculation with the PWN, and among them *P. repens*, *S. sapinea*, two *Pestalotiopsis* spp., and *Rhizoctonia* sp. were frequently isolated and were considered as dominant fungi (Table V.6). Although the dominant fungi were constantly detected over the experimental period, the composition of fungal species varied slightly among seasons as reported by Kobayashi et al. (1974, 1975), Fukushige and Futai (1987), and Kuroda and Ito (1992). For example, Kuroda and Ito (1992) reported that the fungal species detected from PWN-inoculated pines were the same with those in healthy trees during the 4 weeks after inoculation. The species detected both from healthy and inoculated pine trees were *Pestalotiopsis* spp., *Nigrospora* spp., *Cladosporium* spp., and *Phomopsis* spp. They found a blue-stain fungus, *Ceratocystis* sp., and bacteria 5 weeks after nematode inoculation. Other researchers have shown that the minor fungi disappeared when pine trees were completely killed in December, and fungal flora of the pine trees gradually changed

Table V.6 Fungi isolated from dead pine trees inoculated with pine wood nematode and healthy trees (from Sriwati et al. 2007, Nematology, with permission)

Isolated from dead pine trees	Isolated from healthy pine trees
Aspergillus sp.	*Aspergillus* sp.
Aureobasidium sp.	*Gliocladium* sp.
Fusarium sp. 1	*Mucor* sp.
Fusarium sp. 2	*Penicillium* sp. 1
Gliocladium sp.	*Penicillium* sp. 2
Mucor sp.	*Penicillium* sp. 3
Mortierella sp.	*Pestalotiopsis* sp. 1
Penicillium sp. 1	*Pestalotiopsis* sp. 2
Penicillium sp. 2	*Phialophora* sp.
Penicillium sp. 3	*Sphaeropsis sapinea*
Pestalotiopsis sp. 1	*Trichoderma* sp. 1
Pestalotiopsis sp. 2	*Trichoderma* sp. 2
Phialophora repens	*Trichoderma* sp. 3
Rhizoctonia sp.	
Sphaeropsis sapinea	
Trichoderma sp. 1	
Trichoderma sp. 2	
Trichoderma sp. 3	

until June, the end of the experiment (Sriwati et al. 2007). Kobayashi et al. (1974) recorded one species of *Diplodia* as one of the common fungi in the wood of dead pine trees affected by *B. lignicolus* (=*B. xylophilus*). They also found that the *B. lignicolus* multiplied well on its mycelia grown on PDA plate medium. A similar result was detected by Sriwati et al. (2007), who frequently isolated *S. sapinea*, which is regarded as a synonym of *Diplodia* (Denman et al. 2000). De Wet et al. (2003) also suggested that *S. sapinea* should revert to its former name of *Diplodia pinea*.

A close affinity between the PWN and blue-stain fungi has been reported in several studies (Kobayashi et al. 1974, 1975; Fukushige 1991; Maehara and Futai 2000; Maehara et al. 2005). Kobayashi et al. (1974, 1975) considered that a kind of blue-stain fungus might be transmitted by *Monochamus* beetles when they feed on young shoots of healthy pine trees in early summer.

28.3.1 Influence of Each Fungus on the Presence and Reproduction of the PWN

The suitability of fungi for the reproduction of the PWN under laboratory conditions has been well investigated by many researchers. For instance, Kobayashi et al. (1974, 1975) and Fukushige (1991), and Maehara and Futai (2000) compared the propagation of PWN on the mycelial colonies of fungi isolated from both healthy and dead pine trees. They revealed that PWNs fed and propagated well on some of the fungi isolated from such trees, while they neither fed upon nor reproduced on other fungi isolated from dead pine trees.

To evaluate the influence of each fungus on the presence and reproduction of PWNs under field conditions, Sriwati et al. (2007) calculated the "Nematode population ratio (NPR)". To calculate the average PWN number over fungal positive pieces harvested from each tree, sum of the number of PWN extracted from them was divided by the sum of their dry weight. Likewise, a harmonic average of PWN number over fungal negative wood pieces was calculated. These two averages were denoted as Nf and N0, respectively. They estimated the influence of each fungus on the population growth and settlement of PWN using NPR, which is defined as the ratio of Nf to N0. This value shows the relative abundance of PWN yielded on samples with a given fungal species, and becomes >1 when a given fungus facilitates the growth or settlement of PWN population; on the other hand, if growth or settlement is adversely affected, it becomes <1 (Fig. V.4). NPR of four dominant fungi, *P. repens*, *S. sapinea*, *Pestalotiopsis* spp., and *Rhizoctonia* sp., were slightly >1 every season, suggesting that these fungi facilitate the population growth of PWN. In contrast the NPR of *Trichoderma* sp. and *Penicillium* sp. were <1 and =1, respectively. *Trichoderma* sp. suppressed the population growth of PWN, and *Penicillium* sp. influenced the population growth neither positively nor negatively. Thus, these nematodes preferably aggregated to dominant fungi over the experimental period, but the promotive effects of each fungus on the presence or reproduction of PWNs changed from season to season and from tree to tree.

28.3.2 Cohabitation of PWN and Fungal Species in PWN-Inoculated Tree

The relationship between the presence of fungi and that of the PWN has been studied by Sriwati et al. (2007). They calculated the "Index of cohabitation ability (ICA)" by using the formula:

$$ICA = \mathrm{Log}\frac{(\mathrm{An}+0.01)\times(\mathrm{B0}+0.01)}{(\mathrm{A0}+0.01)\times(\mathrm{Bn}+0.01)}$$

where

A0 = total number of wood pieces which contained only the fungus per tree,
An = total number of wood pieces which contained both fungus and nematode per tree,
B0 = total number of wood pieces, which contained neither fungus nor nematode per tree,
Bn = total number of wood pieces which contained only nematode per tree.

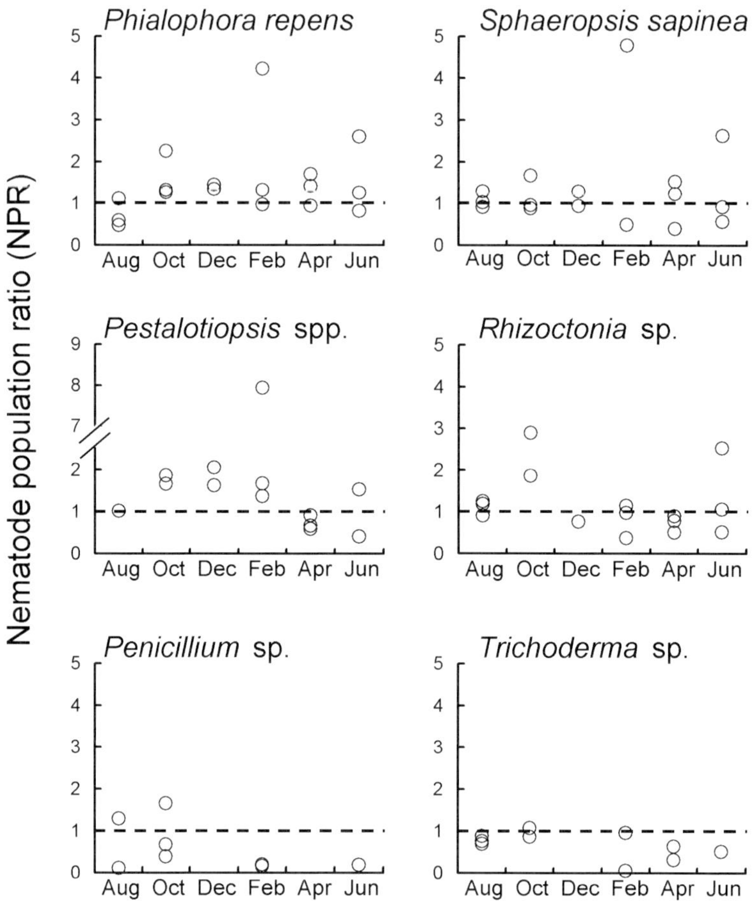

Fig. V.4 The effect of individual fungi on the distribution of pine wood nematodes in dead *Pinus thunbergii* trees. *NPR* is the ratio of *Nf* to *N0*, where *Nf* and *N0* are average nematode numbers over fungus positive and wood pieces lacking fungi, which were harvested from each pine tree (from Sriwati et al. 2007, Nematology, with permission)

This index shows the degree of co-occurrence of PWN with a given fungus, a positive ICA indicating a tendency of co-occurrence, while a negative ICA indicates repulsion. ICA values were slightly higher than zero for three dominant fungi, *P. repens*, *S. sapinea*, and *Pestalotiopsis* spp., while those for *Penicillium* and *Rhizoctonia* species were around zero, and that for *Trichoderma* sp. was slightly lower than zero (Fig. V.5). Thus, three dominant fungi tended to cohabit with the PWN, while *Trichoderma* sp. showed repelling effects toward the PWN. Neither *Penicillium* sp. nor *Rhizoctonia* sp. had a special relationship with the nematode regarding their distribution. Thus, this suggests that the dominant fungi in PWN-killed pine trees by promoted the reproduction and/or the settlement of the nematode.

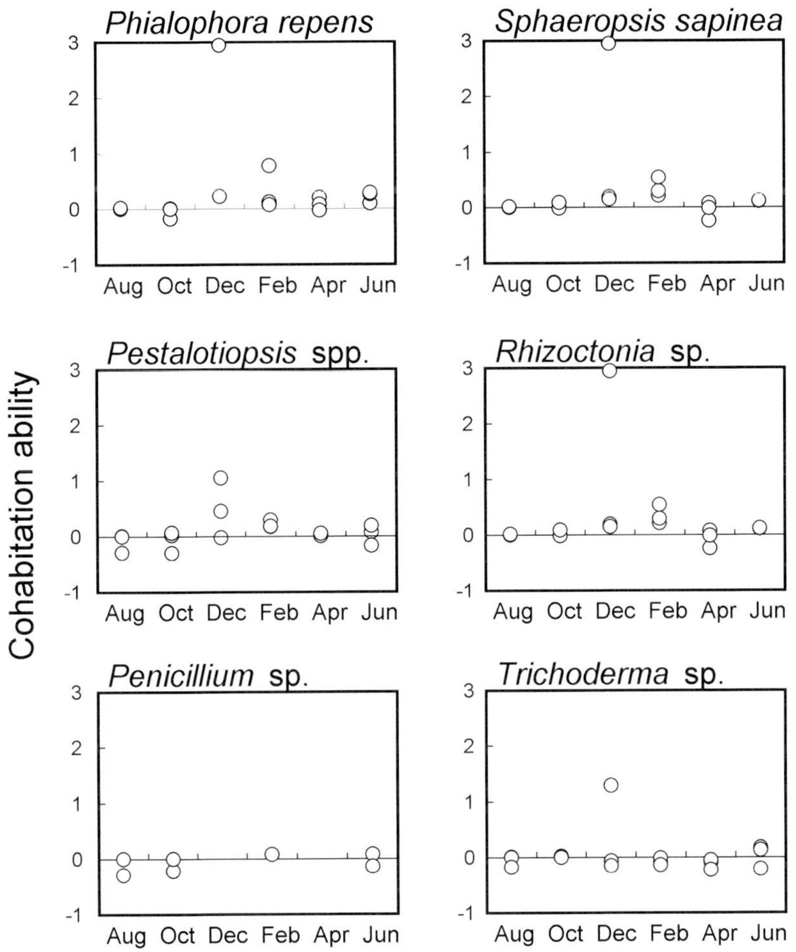

Fig. V.5 Cohabitation ability between the pine wood nematode and six fungi. ICA = Log [(An + 0.01) × (B0 + 0.01)]/[(A0 + 0.01) × (Bn + 0.01)], where *A0* is the number of wood pieces per *Pinus thunbergii* tree containing only the fungus; *An* is the total number of wood pieces per tree containing both the fungus and nematodes; *B0* is the number of wood pieces per tree with neither fungi nor nematodes; *Bn* is the number of wood pieces per tree containing only nematodes (from Sriwati et al. 2007, Nematology, with permission)

As suitable food, some fungi seem to promote PWN propagation, which might be one of the reasons why the PWN has a high cohabiting tendency with these dominant fungi. In the case of other fungi, their cohabiting tendency was lower. This might result from their unsuitability as food for the PWN or perhaps result from a repellent they might produce.

These speculations based on the results from the field experiment should be confirmed under laboratory conditions. Sriwati et al. (2007), therefore, also examined PWN propagation on 18 fungal species isolated from dead or dying pine

Table V.7 Population numbers of pine wood nematode on 19 fungal species cultured on PDA (from Sriwati et al. 2007, Nematology, with permission)

Fungi tested	Population density (log no. of nematode + 1)		
	5 days	10 days	15 days
	After inoculation		
Aspergillus sp.	0.000 ± 0.000 a	0.000 ± 0.000 a	0.000 ± 0.000 a
Gliocladium sp.	0.202 ± 0.096 a	0.000 ± 0.000 a	0.000 ± 0.000 a
Mortierella sp.	0.000 ± 0.000 a	0.000 ± 0.000 a	0.000 ± 0.000 a
Mucor sp.	0.000 ± 0.000 a	0.000 ± 0.000 a	0.000 ± 0.000 a
Penicillium sp. 1	1.604 ± 0.071 bcdefg	1.270 ± 0.111 b	0.000 ± 0.000 a
Trichoderma sp. 1	0.000 ± 0.000 a	0.000 ± 0.000 a	0.000 ± 0.000 a
Trichoderma sp. 2	0.000 ± 0.000 a	0.000 ± 0.000 a	0.000 ± 0.000 a
Trichoderma sp. 3	0.000 ± 0.000 a	0.000 ± 0.000 a	0.000 ± 0.000 a
Fusarium sp. 2	0.804 ± 0.091	1.544 ± 0.086 b	0.228 ± 0.160 a
Penicillium sp. 2	1.484 ± 0.047 bcdef	1.952 ± 0.069 cde	1.587 ± 0.073 bc
Penicillium sp. 3	1.744 ± 0.072 cdefghi	2.173 ± 0.021 de	1.861 ± 0.021 bc
Aureobasidium sp.	1.658 ± 0.062 bcdefgh	1.833 ± 0.070 cd	1.889 ± 0.018 cde
Rhizoctonia sp.	1.371 ± 0.090 bcde	2.392 ± 0.095	2.172 ± 0.048 de
Fusarium sp. 1	1.349 ± 0.077 bcde	2.981 ± 0.072 f	2.485 ± 0.091
Phialophora repens	1.838 ± 0.083 defghi	3.081 ± 0.027 f	5.135 ± 0.073 fg
Pestalotiopsis sp. 1	2.055 ± 0.050 fghij	3.480 ± 0.040	5.421 ± 0.072 fgh
Pestalotiopsis sp. 2	2.226 ± 0.065 hij	3.828 ± 0.036 g	5.650 ± 0.070 gh
Sphaeropsis sapinea	1.832 ± 0.116 defghi	4.022 ± 0.106 g	6.156 ± 0.036 i
Botrytis cinerea	1.945 ± 0.016 efghij	3.785 ± 0.018 g	6.230 ± 0.018 i

Values are means ± SD. Means followed by the same letters in a column are not significantly different at 5% level (Tukey–Kramer's test)

trees under laboratory conditions (Table V.7). PWN propagated well on the colonies of the dominant fungi detected from PWN-killed pine trees except for *Rhizoctonia* spp., that is, *S. sapinea, P. repens*, and two *Pestalotiopsis* spp., grown on PDA medium. Especially, PWNs multiplied as well on *S. sapinea* as on *Botrytis cinerea*, which is commonly used to rear PWNs in the laboratory. Conversely, little or no reproduction was observed on the colonies of three species of *Trichoderma*, two species of *Fusarium*, three species of *Penicillium*, and each species of *Mucor*, *Mortierella*, *Gliocladium*, *Aspergillus*, *Rhizoctonia*, and *Aureobasidium*. Some researchers also reported that some species of *Trichoderma* and *Penicillium* are unsuitable for PWN propagation (Kobayashi et al. 1974, 1975; Fukushige 1991; Maehara et al. 2000). Therefore, it is thought that PWN can survive and propagate well in a dead pine tree by feeding on the dominant fungi.

28.4 Concluding Remarks

From a practical point of view, the population dynamics of PWNs in dead pine trees have been a major research concern, because the density of PWNs in such trees seems to influence the number of PWNs carried by a vector beetle,

M. alternatus. In addition, this number may determine the ability of each beetle to spread pine wilt. From a biological viewpoint, the population changes of PWNs could be regarded as part of sequential microbial relationships occurring in the process of wood degradation.

The number of the PWNs decreased from October to December, increased in February, and then decreased again until June when the experiment terminated (Fukushige and Futai 1987; Sriwati et al. 2006). This fluctuation in the population density of *B. xylophilus* seemed to synchronize with that of Diplogastridae sp. 1; However, correlations between the numbers of the two nematodes varied considerably both among trees and seasons. Although only little is known about the relationship of these two populations of nematodes, there is a tendency for the nematode community to have little interrelation with PWN population size (Sriwati et al. 2006).

PCs and tunnels of *Monochamus* beetle might provide suitable humidity and sufficient nutrients for the growth and reproduction of fungi and other microbes, and therefore for those of nematodes feeding on such microbes. Kobayashi et al. (1974, 1975), Fukushige (1991), Maehara and Futai (2000), and Maehara et al. (2005) reported that the blue-stain fungi, which prevailed around PCs would serve as food for *B. xylophilus*. However, the fact that Sriwati et al. (2007) did not isolate this group of fungal species in their experiment may be due to artificial inoculation of PWN. Instead they found that some of the dominant fungi such as *S. sapinea*, *Pestalotiopsis* spp., and *P. repens* play important roles in determining the propagation and distribution of PWNs in dead pine trees. Thus, not only blue-stain fungi but also other fungal flora should determine the distribution and population change of PWNs in the natural field, as various fungal species besides blue-stain fungi have been found as suitable food for PWN. This result strongly suggests that dominant fungi in PWN-killed pine trees served as adequate food for PWN propagation. The dominant fungi were detected also in healthy pine trees (Table V.6) and might be neither pathogenic nor parasitic to pine trees.

The other factors that affect the population dynamics of nematode fauna are still unclear. As Shigo (1967) stated, when trees become diseased, the physical conditions within the trees, such as water content, temperature, and so on must change dramatically. The transition of season should also affect environmental conditions. Thus, many other biotic and abiotic factors could affect the microhabitat conditions of pine wood, and further studies are needed to explore other factors that affect PWN population and distribution in host pine trees.

29
Interactions of Pine Wood Nematodes, Wood-Inhabiting Fungi, and Vector Beetles

Noritoshi Maehara

29.1 Introduction

The pine wood nematode (PWN), *Bursaphelenchus xylophilus*, the causal agent of pine wilt disease (Kiyohara and Tokushige 1971), is vectored from wilt-killed to healthy pine trees by cerambycid beetles of the genus *Monochamus* (Mamiya and Enda 1972; Morimoto and Iwasaki 1972; Linit 1988) and kills healthy trees, but other organisms beside the PWN, pine sawyers and pine trees, are involved in the disease complex.

Pine wilt-killed wood is often stained blue or blue-black (Fig. V.6). This discoloration is caused by blue-stain fungi, for example, fungi in the genus *Ophiostoma*. Other fungi also occur in dead pine wood (Kobayashi et al. 1974, 1975). As time passes mushrooms of wood-decay fungi sometimes appear on dead pine trees or their wood, for example, *Cryptoporus volvatus* (Fig. V.7) and *Trichaptum abietinum*. Since the PWN is mycophagous, wood-inhabiting fungi play an important role in pine wilt.

Pines are well-known ectomycorrhizal trees and have a mutualistic relationship with ectomycorrhizal fungi. In pine forests, the mushrooms of many ectomycorrhizal fungi regularly occur, such as *Tricholoma matsutake*, *Suillus luteus*, *Rozites caperata*, *Rhizopogon rubescens*, and *Pisolithus tinctorius*. Thus, mycorrhizal fungi should play an important role in the disease.

In this chapter, the role of wood-inhabiting fungi in pine wilt disease will be discussed in relation to the PWN and vector beetles. Then follows a brief discussion on the role of mycorrhizae in pine wilt disease.

Tohoku Research Center, Forestry and Forest Products Research Institute, 92-25 Nabeyashiki, Shimo-Kuriyagawa, Morioka 020-0123, Japan

Tel.: +81-19-648-3962, Fax: +81-19-641-6747, e-mail: maehara@ffpri.affrc.go.jp

Fig. V.6 Blue-stained *Pinus densiflora* logs (triangular areas) (see Color Plates)

Fig. V.7 *Cryptoporus volvatus* mushrooms on *Pinus densiflora* logs (see Color Plates)

29.2 Pine Sawyer and PWN

The PWN is an entomophilic nematode that depends on insects during part of its life cycle and utilizes insects as vectors (phoretic association), as opposed to entomogenous nematodes, which take in nutrients and develop within the bodies of insects (Mamiya 2003b). The PWN is vectored from wilt-killed to healthy pine trees by *Monochamus* beetles, for example, *M. alternatus*, though numerous beetles of various genera in the families Cerambycidae, Curculionidae and Scolytidae also colonize wilt-killed pine trees (Kishi 1995). This raises the question as to why *Monochamus* beetles play a role as the principal vectors of PWN. Moreover, some *Monochamus* beetles carry large numbers of nematodes, but others carry none, that is, the number of nematodes carried by a *Monochamus* beetle ranges from 0 to 200,000 (Kishi 1995). Why is the range so great?

29.3 PWN and Fungi

29.3.1 Feeding Habits of the PWN

Nematodes in the order Aphelenchida, including the PWN, are mycophagous (see Chap. 9). Many species of mycophagous nematodes feed on fungi in the galleries of wood boring beetles in the families Cerambycidae, Scolytidae, and Curculionidae, and are carried (vectored) to new habitats by these borers (Rühm 1956). The PWN is also mycophagous as it can be reared on fungi such as *Botrytis cinerea*.

The PWN is a plant-parasite and as a pathogen it kills pine trees. Thus, not only can the PWN be cultured on fungi but also on pine callus (Tamura and Mamiya 1979; Iwahori and Futai 1990) and aseptic pine seedlings (Kuroda and Mamiya 1986). Moreover, unsaturated fatty acids contained in the parenchyma cells of pines enhance the reproductive rate and the survivability of the PWN (Mamiya 1990b); therefore, PWN is considered to feed on the parenchyma cells of pines. Another entomophilic nematode in the order Aphelenchida, the red ring nematode (*Bursaphelenchus cocophilus*), cause of red ring disease is also a plant-parasite and as such it kills coconut palm trees (*Cocos nucifera*) (Griffith 1987). This nematode is carried by a palm weevil (*Rhynchophorus palmarum*).

Plant-parasitic and mycophagous nematodes have stylets, which they insert into hyphae or parenchyma cells to absorb the cytoplasm. The inner diameter of the stylets varies from 0.2 to 1.0 μm. The inner diameter of the PWN stylet is 0.7 μm [measured from Fig. 1 in Mamiya and Kiyohara (1972)] and PWN feeds using the stylet (Tanaka 1974).

29.3.2 Reproduction of PWN in Pine Trees

The PWN is carried by *Monochamus* beetles and enters healthy pine trees through maturation feeding wounds, created by the beetles, and soon after entering the trees some PWNs quickly spread within the host's branches, roots and trunks. It is thought that the PWNs feed on living parenchyma cells at that time because, as mentioned above, they can reproduce on pine callus.

When pine wilt symptoms appear, for example, discoloration and wilting of the host's needles, wood-inhabiting fungi such as blue-stain fungi in the genus *Ophiostoma*, grow rapidly (Kuroda and Ito 1992). Blue-stain fungi cannot decompose cellulose and lignin, but it can use absorbable polysaccharides. They stain wood blue or blue-black (Fig. V.6). Kobayashi et al. (1974, 1975), Fukushige and Futai (1987), Wingfield (1987b), and Maehara et al. (2005) isolated blue-stain fungi more often from PWN-infested or -inoculated pine trees than from unaffected trees, that is, they are the dominant fungi in pine wilt-killed trees. The PWN can feed on these fungi and reproduce well in dead wood (Kobayashi et al. 1974, 1975).

29.3.3 Fungus Suitability for PWN Reproduction

Although the PWN is mycophagous, it does not feed on all fungi. Temporal changes in PWN populations were examined on nine species of fungi growing in pine-branch segments (Maehara and Futai 2000). On the blue-stain fungus *Ophiostoma minus*, *Macrophoma* sp., and *Trichoderma* sp. 1, the PWN populations increased quickly for 4 weeks after nematode inoculation, and then increased slightly or remained high (Fig. V.8). In contrast, the population of PWNs decreased on *Verticillium* sp. and *Trichoderma* sp. 3. Thus, some fungi were more suitable and others unsuitable for PWN reproduction (Kobayashi et al. 1974, 1975; Fukushige 1991; Wang Y et al. 2005; Sriwati et al. 2007). See Chap. 28 for the succession of fungi in pine trees after infection with the PWN and suitability of the fungi for PWN reproduction.

29.3.4 Nematophagous Fungi

As stated above, some fungi serve as food for the PWN, but other fungi that feed on nematodes are classed as nematophagous. They are divided into three groups depending on the manner in which they attack their host: ectoparasites, endoparasites, and parasites of nematode eggs or cysts (Tamura 1973; Mitsui 1983, 1992; Dix and Webster 1995).

Ectoparasites are nematode-trapping fungi and capture their prey using a variety of structures: sticky hyphae, sticky knobs, adhesive networks, and non-constricting or constricting rings. After capture, hyphae ramify throughout the nematode body

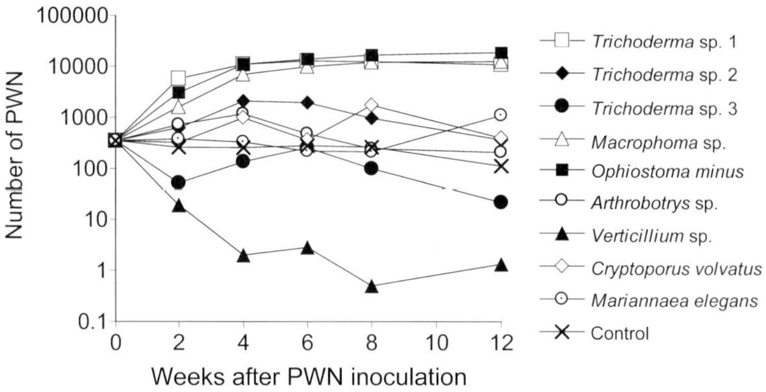

Fig. V.8 Pine wood nematode (PWN) population changes on nine species of fungi related to pine trees (modified from Maehara and Futai 2000, with permission)

and digest the contents. Ectoparasites include fungi in the genera *Arthrobotrys*, *Dactylaria*, and *Monacrosporium*. Mamiya and Tamura (1976) and Tamura (1980) found *Arthrobotrys* and *Dactylaria* on PWNs isolated from around the pupal chambers of *M. alternatus*. In contrast to ectoparasites, endoparasites have no special trapping structures. They produce conidia which may be ingested by the nematodes or which may be attached to the nematode's cuticle. Conidia germinate and germ tubes penetrate through the nematode's gut wall or the cuticle. Fungi in the genera *Verticillium*, *Harposporium*, *Drechmeria*, and *Nematoctonus* are typical endoparsites. Fukushige (1991) isolated *Verticillium* from PWNs collected from an affected pine tree. PWN populations decrease on both *Verticillium* and *Trichoderma* because the former is an endoparasite and kills nematodes, while the latter is unsuitable for nematode reproduction (see data in Sect. 29.3.3). The third group of nematophagous fungi parasitizes *Meloidogyne* eggs (species of *Verticillium*, *Paecilomyces*, *Dactylella*, *Rhopalomyces*), or cysts of *Heterodera* and *Globodera* (species of *Verticillium* and *Paecilomyces*).

The wood-inhabiting, gill fungus *Pleurotus ostreatus* immobilizes and digests nematodes (Thorn and Barron 1984; Barron and Thorn 1987), including the PWN (Mamiya et al. 2005). As such it is unsuitable for PWN reproduction (Dozono 1974). Mamiya et al. (2005) also reported that *Trichaptum abietinum* and *Cryptoporus volvatus* had limited ability to prey on the PWN.

29.4 Pine Sawyer and Fungi

Kobayashi et al. (1974, 1975) examined fungi present: (1) in the pupal chambers of *M. alternatus*, (2) on the body surface of adult *M. alternatus* beetles just after emergence from chambers, (3) on the surface of pine twigs fed upon by adult

Fig. V.9 Blue-stain fungus on the elytra of the Japanese pine sawyer, *Monochamus alternatus* (see Color Plates)

beetles, and (4) in the wood of wilt-killed trees. A blue-stain fungus [identified as a species of *Ceratocystis* by Kobayashi et al. (1974, 1975)] was isolated from all four kinds of samples, that is, the blue-stain fungus on the surface of pupal chambers which adheres to the surface of the bodies of adult *M. alternatus* (Fig. V.9) where it is transmitted to the twigs of healthy trees, and grows in wood when the trees are weakened by PWNs. As mentioned above, PWN can feed on this fungus and reproduce in wilt-killed trees.

Kobayashi et al. (1974, 1975) also found that another blue-stain fungus (which they identified as a species of *Verticicladiella*) was not transmitted by *M. alternatus*. Masuya et al. (1998, 1999) isolated blue-stain fungi from the bark beetles *Tomicus piniperda* and *Tomicus minor* (Scolytidae) and their galleries in weakened pine trees. The conclusion was that bark beetles transmit blue-stain fungi to wilt-killed pines.

29.5 Life Cycle of the PWN

The life cycle of the PWN is divided into dispersal and propagative forms (Mamiya 1975a). The dispersal form is composed of third-stage dispersal juveniles (J_{III}) and fourth-stage dispersal juveniles (J_{IV}). A high population density is a prerequisite for the occurrence of J_{III} (Mamiya et al. 1973; Kiyohara and Suzuki 1977; Ishibashi and Kondo 1977; Tamura 1986; Forge and Sutherland 1996). J_{III} moults to J_{IV} in the presence of insect vectors in the genus *Monochamus* (Morimoto and Iwasaki 1973; Warren and Linit 1993; Maehara and Futai 1996; Necibi and Linit 1998), and the J_{IV} transfers from infested wood to the *Monochamus* vector beetles. Consequently the J_{IV} stage is an important in the PWN life cycle. Also see Chap.13.

29.6 Interactions Among PWN, Wood-Inhabiting Fungi, and the Pine Sawyer

29.6.1 Importance of the Number of PWNs Carried by a Pine Sawyer

As stated above, the PWN is vectored from wilt-killed to healthy pines by *Monochamus* beetles; however, some beetles carry many nematodes and others no nematodes, even if these beetles emerge from the same pine tree. The number of nematodes that enter a healthy tree is directly proportional to the number of nematodes carried by *Monochamus* beetles (Togashi 1985). The rate of disease development is directly related to the number of nematodes in the inoculum (Kiyohara et al. 1973). Thus, beetles carrying more nematodes can kill pine trees easier. To understand the dynamics of pine wilt disease development, it is important to identify the factors affecting the number of nematodes carried by an individual beetle, which ranges from 0 to over 200,000.

29.6.2 Factors Affecting the Number of PWNs Carried by a Pine Sawyer

Many studies have been carried out to determine why the number of nematodes carried by beetles varies so much (reviewed by Kishi 1995). *Monochamus alternatus* emerging from extremely dry or wet pupal chambers (Morimoto and Iwasaki 1973; Maehara et al. 2005), or logs (Terashita 1975; Kobayashi et al. 1976; Togashi 1989g; Fukushige 1990) carried relatively few nematodes. There were differences in the numbers of nematodes carried by *Monochamus* beetles among individual pine trees from which the beetles emerged, although the factors causing these differences are unknown (Maehara et al. 2005). *Monochamus alternatus* that emerged from pupal chambers constructed just beneath the bark carried fewer nematodes than beetles emerging from chambers constructed in the sapwood, because the conditions beneath the bark might be unsuitable for the nematode (Maehara et al. 2005). Large *Monochamus* beetles carried more nematodes (Linit et al. 1983; Humphry and Linit 1989; Aikawa and Togashi 1998). The virulence of the PWNs affected the number of nematodes carried by a *Monochamus* beetle (Aikawa et al.

2003b). Also see Chap. 13. The number of nematodes carried by *Monochamus* beetles is affected by: (1) reproduction of nematodes in dead pine trees, (2) aggregation of nematodes around pupal chambers of vector beetles, and (3) transfer of nematodes to beetles. Therefore, the reproduction of nematodes is required first of all, and kinds and abundance of fungi in dead wood are important for the reproduction because the PWN is mycophagous.

29.6.3 Effect of Wood-Inhabiting Fungi on the Number of PWNs Carried by a Pine Sawyer

In a field survey, the relationship between the intensity of blue-stain in wood around pupal chambers of *Monochamus* beetles and the number of nematodes carried by the beetles that emerged from the chambers was examined (Maehara et al. 2005). The presence of blue-stain fungi can be monitored by examining the blue-stain intensity of wood without isolating the fungi from the wood, because the fungi could be isolated from over 90% of blue-stained wood. The number of nematodes aggregating around such chambers increased with an increase in the intensity of blue-stain on the pupal chamber walls of *M. alternatus* because the PWN feeds on blue-stain fungi and multiplies, and thus the number carried by beetles which emerged from the chambers also increased (Table V.8).

Maehara and Futai (1996) devised artificial pupal chambers in which PWNs transfer to *Monochamus* beetles. Small (2.5 × 2.5 × 5 cm) wood blocks are obtained from the bole of pine trees and a hole is bored into the top of each block to simulate a *Monochamus* pupal chamber (Fig. V.10). After autoclaving the blocks, one or two fungi are inoculated into each hole. When the mycelium of these fungi covers each block, PWNs are inoculated into the hole and reproduce; whereupon, a mature *Monochamus* larva is inserted into the hole. This larva becomes a pupa and then an adult, and a nematode laden adult emerges from each block.

Table V.8 Effect of blue-stain intensity of *Monochamus alternatus* pupal chamber walls in pine wilt-killed trees on the numbers of pine wood nematodes (PWN) aggregating around the chambers and carried by the beetles (modified from Maehara et al. 2005, Nematology, with permission)

Intensity of blue-stain	Number of observations	Number of PWN aggregating around pupal chambers	Number of PWN carried by a beetle
0	0	—	—
1	4	153.5 ± 168.4	139.0 ± 168.6
2	12	1661.6 ± 4675.7	1522.8 ± 4662.5
3	33	4421.6 ± 7395.4	4247.3 ± 7329.5

Values are means ± SD

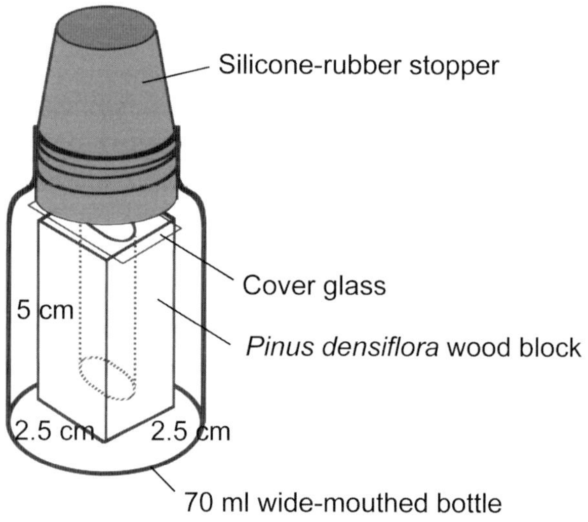

Fig. V.10 Simulated pupal chamber of the Japanese pine sawyer, *Monochamus alternatus* (modified from Maehara and Futai 1997)

Initially using artificial chambers, the effect of individual fungi in wood on the number of nematodes carried by *Monochamus* beetles was examined (Maehara and Futai 1996). The fungi used were the blue-stain fungus *O. minus* and *Trichoderma* sp. 3, which were both isolated from wilt-killed *Pinus densiflora*. The former fungus is suitable and the latter fungus is unsuitable for PWN reproduction. After the emergence of *M. alternatus*, the number of nematodes carried by the beetle was examined. *Monochamus* beetles that emerged from wood blocks inoculated with *O. minus* carried much more nematodes than those emerging from blocks inoculated with *Trichoderma*. This was because: (1) nematodes reproduced much better in *O. minus*-inoculated blocks than in *Trichoderma*-inoculated blocks; (2) the percentages of J_{III} and J_{IV} to the total nematode population were much higher in the former than in the latter.

The above authors (Maehara and Futai 1997) then examined the effect of two fungi cohabiting wood on the number of nematodes carried by *Monochamus* beetles. *Verticillium* sp., an endoparasite of nematodes (see Sect. 29.3.4), was used in addition to *O. minus* and *Trichoderma* sp. 3. They inoculated the blocks with *Ophiostoma* and *Trichoderma*, or with *Ophiostoma* and *Verticillium*, either simultaneously or with a delay between the two inoculations. Control blocks were inoculated with one fungus only. Nine fungal treatments were divided into three groups (Table V.9). In blocks inoculated with *Verticillium*, regardless of the inoculation order, the number of nematodes carried by a beetle was much smaller because this fungus kills nematodes and nematode populations did not build up. In blocks inoculated with *Trichoderma*, the number was also small, except for blocks inoculated first with *Ophiostoma* then with *Trichoderma*. *Trichoderma* suppresses nematode propagation not by killing nematodes but by outcompeting *O. minus*, the food source

Table V.9 Effect of fungi and the inoculation order on the propagation of pine wood nematodes (PWN) and on the numbers of nematodes carried by *Monochamus alternatus*, which emerged from artificial pupal chambers in wood blocks (modified from Maehara and Futai 1997)

Fungi and inoculation order	Number of beetles	Total number of PWN	Number of PWN carried by a beetle
Verticillium sp. alone	8	0.3 ± 0.5	0 ± 0
Ophiostoma minus then *Verticillium* sp.	5	0.6 ± 0.9	0 ± 0
Verticillium sp. then *O. minus*	6	0.7 ± 1.0	0.3 ± 0.8
O. minus and *Verticillium* sp. simultaneously	5	3.6 ± 3.0	1.2 ± 1.6
Trichoderma sp. 3 alone	9	11914.9 ± 3634.6	34.9 ± 62.9
O. minus and *Trichoderma* sp. 3 simultaneously	12	16482.5 ± 8663.6	85.8 ± 103.1
Trichoderma sp. 3 then *O. minus*	9	12172.3 ± 4489.5	87.9 ± 93.7
O. minus alone	10	65579.0 ± 10155.5	2759.0 ± 1885.7
O. minus then *Trichoderma* sp. 3	12	36187.5 ± 17290.3	2979.2 ± 2395.7

Values are means ± SD

of nematodes. Beetles emerged from blocks inoculated with only *O. minus* or inoculated first with *Ophiostoma* then with *Trichoderma* carried many nematodes because nematode populations developed very well and the percentages of J_{III} and J_{IV} were high.

PWNs reproduce well in wilt-killed trees colonized by blue-stain fungi such as *Ophiostoma*. When *Monochamus* beetles are present, many J_{IV} develop and transfer to the vector beetles. Other fungi, including species of *Trichoderma* and *Verticillium* that are unsuitable for nematode propagation, often occur in killed trees, in which case nematodes reproduce poorly and only a few J_{IV} occur and transfer to the vector. Thus, the predominant fungi in wilt-killed trees are paramount in determining the number of nematodes carried by beetles emerging from such trees.

29.6.4 Microbial Control of PWN by Fungi

In an attempt to reduce the number of nematodes carried by *Monochamus* beetles, Maehara et al. (2006) tried to change the mycoflora and also to prevent blue-stain fungi from spreading within the wood of wilt-killed pines. They did this by inoculating other fungi, which are unsuitable for PWN reproduction, into the dead pine logs in autumn. After the emergence of *Monochamus* beetles the following summer, the number of nematodes carried by beetles was examined. In the first trial, *Trichoderma* sp. 2, *Pycnoporus coccineus* (a wood-decaying fungus), and *Trichoderma* sp. 3 inoculations tended to decrease the number of nematodes carried by a beetle. In the second trial, *Trichoderma* sp. 3, *Verticillium* sp., and *Trichoderma* sp. 2 inoculations seemed to decrease the number, although *P. coccineus* had no effect. To find the best fungus for suppressing the number of nematodes carried by a beetle, Maehara (2008) made a study to determine which *Trichoderma* spp. would be the most detrimental for PWN reproduction. Selected *Trichoderma* spp. were inoculated into wilt-killed pine logs. Although several *Trichoderma* isolates decreased the number of nematodes carried by the beetles emerging from logs, *Trichoderma* sp. 3 was the most effective.

Biological and microbial controls of *Monochamus* beetles (vector) have been attempted in several ways (see Chap. 35). In addition, microbial control of the PWN (the pathogen) is needed to prevent the transmission of nematodes from vectors to healthy pines (Maehara et al. 2007).

29.6.5 Effect of Blue-Stain Fungi on the Number of PWNs Carried by Several Species of Beetles

As stated above, many beetles in various genera in the families Cerambycidae, Curculionidae and Scolytidae colonize wilt-killed pine trees, but the PWN is vectored from wilt-killed to healthy pines only by *Monochamus* beetles. Maehara and Futai (2002) examined four species of beetles, which emerged from wilt-killed pines: *M. alternatus*, *Acanthocinus orientalis* (Cerambycidae), *Pissodes obscurus* (Curculionidae), and *Trogossita japonica* (Trogossitidae). Numerous PWN aggregated around pupal chambers of *M. alternatus* in wilt-killed trees and the beetles carried many nematodes; however, few nematodes aggregated around chambers of the other three beetles and these beetles carried few or no nematodes. All the *M. alternatus* pupal chambers were confined to the sapwood. However, all *P. obscurus* pupal chambers, two-thirds of the *A. orientalis* chambers, and about a quarter of the *T. japonica* chambers were just beneath the bark; therefore, this might affect the number of nematodes around *M. alternatus* chambers. The percentage of chambers with intense blue-stain was higher for *M. alternatus* than for the other beetles (Fig. V.11). This would cause more nematodes to aggregate around *M. alternatus* chambers, because blue-stain fungi are suitable for PWN reproduction. The propagative juveniles and J_{III} of the nematode aggregated around *M. alternatus* chambers in response to unsaturated fatty acids deposited by fourth-instar larvae of the beetle, that is, linoleic, palmitoleic and oleic acids (Miyazaki et al. 1977a, b). One-week-old cultures of nematodes were also attracted to homogenates of, and lipid extracts from, fourth-instar larvae of *Monochamus carolinensis* (Bolla et al. 1989). As such, few nematodes aggregated around chambers of beetles other than *Monochamus*, because the beetles might not deposit any or deposit less nematode attractive compounds, but more investigations are needed.

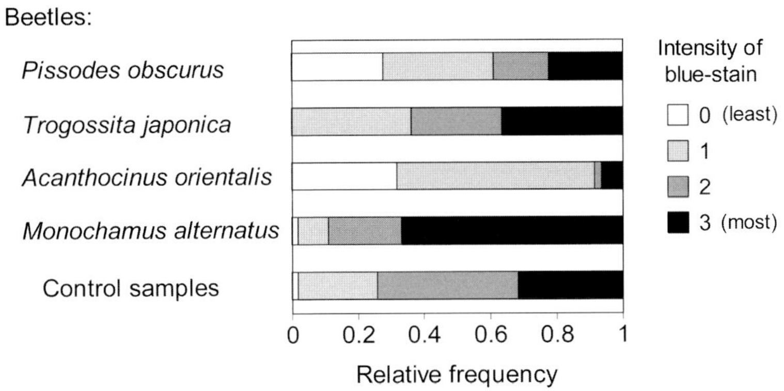

Fig. V.11 Intensity of blue-stain in wood around pupal chambers of four species of beetles in logs from wilt-killed *Pinus densiflora* trees. Control samples were collected from the center of the logs. Blue-stain intensity from 0 = no stain to 3 = most heavily stain (modified from Maehara and Futai 2002, Nematology, with permission)

29.7 Effect of Mycorrhizae on Pine Wilt Disease

Fungi related to pine trees are not only wood-inhabiting fungi but also ectomycorrhizal fungi. As mentioned above, these fungi produce mushrooms in pine forests. Ectomycorrhizae perform many functions, such as the improvement of host resistance to root disease (Marx 1969; Buscot et al. 1992) and the enhancement of nutrient and water uptake (Finlay and Read 1986; Bledsoe 1992). In turn, ectomycorrhizae are given photosynthetic products by host trees.

Kikuchi et al. (1991) examined the effect of ectomycorrhizae as a resistance factor of pine trees to pine wilt disease. The inoculation of pine seedlings with the ectomycorrhizal fungi, *S. luteus* and *R. rubescens*, resulted in a slight increase of resistance to subsequent infection by the PWN. Mycorrhizal infection improved the growth of seedlings and decreased mortality caused by the PWN. Akema and Futai (2005) investigated ectomycorrhizal development in a pine stand in relation to its location on a slope and the effect of mycorrhizae on the mortality of pine trees from pine wilt disease. The development of pine roots and mycorrhizae was greater on the upper part of the slope. Tree mortality was clearly biased and more trees survived on the upper part than on the middle and lower parts. The abundant mycorrhizae on the upper part would enhance the water uptake of pines, mitigate drought stress, and thereby decrease mortality. On the other hand, Ichihara et al. (2001) reported that ectomycorrhizal development in young pine trees was suppressed by inoculation of PWN. Ectomycorrhizae also compensated for the harmful effect of acid mist on pine seedlings, but in turn, acid mist retarded mycorrhizal development (Maehara et al. 1993).

29.8 Concluding Remarks

In this chapter, the interactions among the PWN, wood-inhabiting fungi, and vector beetles of the PWN have been reviewed. So, do wood-inhabiting fungi affect healthy pine trees? Masuya et al. (2003) inoculated healthy *P. densiflora* trees with several species of blue-stain fungi isolated from pine trees and associated bark beetles, and examined the ability of these fungi to induce desiccation in the xylem. They reported that the most virulent fungus was *Leptographium wingfieldii*, but did not determine the ability of this fungus to kill mass-inoculated trees, more studies are called for.

Thus, fungi, such as wood-inhabiting fungi and mycorrhizal fungi, interact with the PWN, vector beetles of the PWN and pine trees, and play important roles in pine wilt disease. For example, wood-inhabiting fungi affect: (1) the reproduction of PWNs in pine wilt-killed trees, (2) the occurrence of the dispersal form (J_{III} and J_{IV}) of PWN, and (3) the number of PWNs carried by vector beetles; that is, they determine the dynamics of pine wilt disease. Based on this information, microbial control of the PWN by fungi should be attempted as mentioned in Sect. 29.6.4. This could prove to be fruitful when the interactions among the PWN, wood-inhabiting fungi, and vector beetles in pine wilt disease are better understood.

30
Concluding Remarks

In this part, we have reviewed nematodes and microorganisms, such as bacteria and fungi, related to pine wilt disease, but we have not reported on entomopathogenic fungi, bacteria, viruses, or protozoa associated with *Monochamus* beetles as these are reviewed in Chap. 35. In Chap. 27, Zhao reviews the possibility that mutualistic symbiosis exists between the PWN and its associated bacteria, showing that they enhance one another's reproduction. In contrast, some PWN-associated bacteria have no beneficial effects but rather they are detrimental to the PWN. Zhao suggests that the mutualistic symbiosis between the PWN and some bacteria might result from coevolution over a long time, because it is difficult to imagine that PWN and accidentally contaminating bacteria could establish such a mutualistic association. As Zhao points out, intensive research on the interactions between the PWN and bacteria has just started and further research is needed.

The population dynamics of the PWN in wilt-killed trees is important because the abundance of the PWN affects the number of PWNs carried by *Monochamus* beetles. As well, the number determines the ability of vector beetles to spread pine wilt disease. In Chap. 28, Sriwati explains seasonal changes in nematode fauna and populations and the interrelations between the PWN and other species of nematodes. The PWN and a species of Diplogastridae were the most prevalent nematodes isolated. The population changes of the PWN appeared to be related to those of the Diplogastridae nematode; however, correlations between the numbers of the two nematodes varied considerably both among trees and seasons. In general, the non-PWN community had little interrelation to PWN population abundance.

In Chaps. 28 and 29, Sriwati and Maehara review the important roles of wood-inhabiting fungi in pine wilt disease. These fungi determine the distribution and population changes of the PWN, the occurrence of the dispersal form of the PWN, and then the number of PWNs carried by vector beetles; that is, the spread of pine wilt disease. It is very interesting that the relationship between nematodes and fungi determines the dynamics of an epidemic in forests. Based on this knowledge microbial control of the PWN using antagonistic fungi is being attempted. In Chap. 29, Maehara also reported the important role of ectomycorrhizal fungi in pine wilt disease.

Thus, we see not only the relationships among host, pathogen, and vector, but also the interactions among various other microbes in pine wilt disease. In the future when more information will be available on such interactions we could develop an approach leading to the successful biological control of the PWN and vectors. In addition, we will also gain important knowledge about invasive pests, which often cause great damage to domestic ecosystems. We will further clarify the various interactions in forests, and will be surprised by their complexity.

Part VI
Management of Pine Wilt Disease

31
Introduction

In Japan pine wilt disease was especially serious after World War II, and Dr. R. L. Furniss, an entomologist working for the General Headquarters of the Occupation Forces proposed that the problem should be controlled by felling, stripping and burning damaged pine trees. Although the cause of this epidemic disease was not known at that time his proposal was based on the speculation that the disease is associated with insects under the pine bark. His speculation was right, and with the use of his recommendations pine wilt decreased in the following few decades.

These methods of Furniss are still effective for controlling the vector and are similar to the methods used today. After discovery of the pine wood nematode (PWN) as the cause of this disease, it became clear that three organisms are involved; pine trees as the host, the PWN as the pathogen, and pine sawyer beetles as the vector. Control measures against this disease have been systematically developed, and studies on resistance breeding of pines and integration of disease management methods have been carried out as a long-term approach to control. Conventional control methods are largely directed against the vector insect, and the pathogen.

Trunk injection is the only practical control method for the PWN. The possible increase of the PWN in the injected trees can be prevented even when *Monochamus alternatus* adults feed on twigs and the PWNs depart from the adult beetles and enter the tree through the maturation feeding scars. Chemicals similar to anthelmintics, such as morantel tartrate (e.g., Greenguard®), are used for this purpose. Using this method results in excellent disease prevention, little adverse environmental effects, and is effective for the entire feeding period of adults. However, its high cost and risk of causing decay following the drilling of tree trunks for injecting the material, and a rather short efficacy period (2–4 years), are its main drawbacks. Furthermore, this method can be used only for important or irreplaceable trees growing at famous temples, shrines, golf courses and the like, and its use in forests is not practical.

For controlling the vector, *M. alternatus*, two approaches can be used. In the first, a preventive spray can be applied to avoid maturation feeding of *M. alternatus* adults and prevent transmission of the PWN to host trees. Insecticides are sprayed onto the foliage of healthy pine trees by using airplanes, helicopters, and unmanned helicopters or from ground-based power sprayers. Spraying is usually done just before adult vector emergence and 14 days after the first emergence of adults (depending on the insecticide formulation), and this is the only practical way to protect pine forests from pine wilt today. The disease is well controlled with this method, but it is difficult using this method alone to eliminate the disease. The harmful effects on non-target organisms are a concern. Preventive spraying is often performed inadequately due to limited budgets. Recently, the number of sprays has often been reduced from two to one, which gives inefficient control of late-emerging adults. Some new insecticide formulations are sprayed only once at the beginning of adult emergence; however, resulting from variations in the emergence date of adults it is better to spray again.

A second approach for controlling *M. alternatus* is the extermination of insect-infested trees. Pine trees damaged by pine wilt are beyond recovery, and they must be completely disposed of to eliminate the source of the vector and pathogen in the following year. There are several ways to dispose of infested trees. Conventional chemicals (insecticide sprays, fumigation) and sometimes physical (burning, burying, or chipping) measures are used to dispose of dead pine trees infested with the PWN. Pine wilt-affected trees should be cut down and removed from the forest, or if this is too difficult, the trees should be completely treated not to allow emergence of the adults in the following year. In fact, small branches are often left in the forest or in some worse case scenarios, dead trees are felled and left in the forest without any treatment. From these pieces of woods, surviving beetles can emerge and spread the disease. As such, disposal of dead trees is necessary, but difficult.

Thirty-seven years have passed since the PWN was first shown to be the pathogen of pine wilt disease (Kiyohara and Tokushige 1971), and several research projects have been carried out on the problem. Some novel control measures and strategies for a long-term approach have been studied, as have conventional control measures. The authors in this part have reviewed the integration of individual control tactics, the present status of new control techniques in China, breeding of resistant pines, biological control in general and control using entomopathogenic nematodes.

32
Integrated Pest Management of Pine Wilt Disease in Japan: Tactics and Strategies

Naoto Kamata

In 1971, the pine wood nematode (PWN), *Bursaphelenchus xylophilus*, was found to be the causal agent of pine wilt disease in Japan and it was shown that the vector is the Japanese pine sawyer, *Monochamus alternatus*. Pine wilt is the most serious forest disease in Japan and elsewhere in Southeast Asia. Here I review the tactics and strategies that have been developed for managing this extremely serious tree disease. Several strategies must be combined to manage epidemics of infectious plant diseases as is done for human diseases. These strategies are: (1) reduce the reservoirs of infection, (2) isolate the hosts from the pathogen and (3) increase the resistance of host populations. There are several ways to carry out these strategies. The most effective is to decrease reservoirs of infection and eradication of pine trees killed by pine wilt, referred to here as "diseased trees". Control of vector insects is an effective way to isolate susceptible host trees from the pathogen. Concerning an increase in host resistance, susceptible pine trees can be replaced with resistant pines or other tree species. Another technique is to inject a nematicide into the trunks of susceptible pines. Each of these practices is called a control tactic. In contrast, a control strategy is a combination of several tactics.

32.1 Control Tactics

32.1.1 Eradication: Reducing Reservoirs of Infection

The most effective measure to control PWN damage is to kill vector beetles, that is, the Japanese pine sawyer, in the larval galleries or pupal chambers before PWN transmission occurs. The practices that are important for such eradication, but are

University Forest in Chichibu, University Forests, Graduate School of Agricultural and Life Sciences, The University of Tokyo. 1-1-49 Hinodamachi, Chichibu 368-0034, Japan

Tel.: +81-494-23-9620, Fax: +81-494-22-0272, e-mail: kamatan@uf.a.u-tokyo.ac.jp

apt to be neglected are: (1) small branches larger than 2 cm in diameter should be completely treated as well as trunks because vector beetles can emerge from such materials, and (2) eradication should be completed before the adult pine sawyers emerge.

32.1.1.1 Physical Controls

Physical controls are major tactics among the various eradication procedures. Here diseased trees are felled and then treated physically such as “cut and crush”, “cut and burn”, and “cut and use for making charcoal”.

32.1.1.1.1 Cut and Crush

A regulation of the Ministry of Agriculture, Forestry, and Fisheries of Japan (MAFF) specifies two sizes of chips that should results from the crushing of diseased trees, that is, <6 mm using an ordinary chipper at a factory, and <15 mm using a mobile chipper in the field. The <6 mm size is based on the size of the Japanese pine sawyer pupal chambers (6–9 mm wide). The latter is founded on the results of field tests, in which 100% sawyer mortality was obtained for overwintering larvae when the maximum size of chips was 80 × 60 × 16 mm.

32.1.1.1.2 Cut and Burn and Use for Making Charcoal

Overwintering pine sawyer larvae can be killed by carbonizing the wood into charcoal even after the pupal chambers have formed. Complete mortality of vector beetles can be achieved by charring the outer 1 cm of the wood because most pupal chambers are formed approximately 1 cm beneath the bark of pine trees. When burning infested wood debris caution must be used to prevent forest fires and high soil temperatures, which stimulate spore germination of the root rot pathogen *Rhizina undulata*, the cause of *Rhizina* root rot, a very serious disease of pines. One possibility is to use portable incinerators for burning and making the charcoal.

32.1.1.1.3 Cut and Bury

Other post felling options include burying cut logs in the ground or soaking them in water. A soil covering of >15 cm can completely prevent emergence of sawyer beetle adults. The cut-and-bury-in-the-ground option is possible for coastal pine stands with a sandy soil. In contrast, the cut-and-soak-in-water option is not practical because it takes over 100 days to kill vector beetles in infested logs soaked in seawater.

32.1.1.2 Chemical Controls

Chemical controls are also major tactics for eradication. When diseased trees are detected in accessible locations, they are felled and the branches and logs are treated with chemical insecticide. This, the so-called cut and chemical application treatment; can be used for inaccessible, dead trees such as those on cliffs or steep slopes. In such situations spot spraying from a helicopter is the only practical application method.

32.1.1.2.1 Cut and Chemical Application: Spray and Fumigation

Felled trees are cut into sections after the branches are removed, then the infested materials are treated with a chemical insecticide. The numbers of pine sawyer larvae tend to be higher in those parts of the tree with thinner bark such as upper portions of the main trunk and branches. For the same reason, higher numbers of larvae tend to infest lower portions of the main trunk in pine species with thinner bark, for example, *Pinus densiflora*, compared to trees species with thicker bark such as the Japanese black pine, *P. thunbergii*.

One or several of Chlorpyriphos-methyl, pyridaphenthion, prothiophos, or BPMC (2-sec-butylphenyl methylcarbamate) chemicals are the major components of the insecticides used in the cut-and-spray method. First, the active constituents are dissolved in an organic solvent. Emulsion formulations are prepared by mixing this solution with an emulsifier and the insecticide is applied by dilution in water. For instance, oil solutions are applied by diluting with kerosene. Approximately 100% mortality of the pine sawyer is possible by spraying chemicals at the end of October when larvae live beneath the host bark or are constructing pupal chambers (fall treatment). When chemicals are applied from November to March (winter treatment) or after April (spring treatment) the chemicals do not reach the pupal chambers even when they are sprayed onto the bark surface. This is because the entrance of the pupal chamber is plugged with excelsior produced by the pine sawyer larvae when excavating the tunnel in which they overwinter. Consequently, an oil solution of chemicals with a high permeability should be used for both "winter" and "spring" treatments.

In contrast, there is no restriction regarding the application season for fumigation. Metam-ammonium carbam NCS or Carbam sodium are major component of the formulations used for fumigation. Almost 100% pine sawyer mortality is possible by fumigation with the fumigation killing both the nematode and its insect vector. Fumigants are easy to apply because they are classified as normal (Based on pesticide toxicity, Japanese MAFF pesticide legislation ranks each pesticide as: either poisonous, deleterious, or normal substances. Normal pesticides are recognized as the safest). One disadvantage of fumigation is its cost as chemical formulations for fumigation are more expensive than those used as sprays though the difference in the cost between the two is not so great compared to total costs. The procedure is to first put pruned branches on the forest floor after removing the humus layer. Pine wilt affected logs are piled on the branches and treated with the chemicals and the pile is covered with a PCV sheet. Soil is placed around the edge of the PCV sheet to seal it. Fumigation should last at least 7 days.

32.1.1.2.2 Spot Sprays

Chemical insecticide is sprayed over infested pine trees from a helicopter. This tactic was first used in remote areas or on steep slopes, where normal cut-and-treat methods could not be carried out; however, since 1987, spot spraying has been employed even in accessible areas to develop an advanced method of preventive spraying against pine wilt. When just a few affected pine trees occur over a wide area, it is possible, but inefficient, to spray from the ground because of inaccessibility. Aerial spraying is also not a good choice because of the relatively greater adverse impact upon ecosystems compared to the benefits of preventing pine wilt. In such cases, spot-spraying using a helicopter is efficient from the viewpoint of cost and is less damaging to the environment. Chemical insecticide is sprayed onto a pine-wilt killed tree and adjacent healthy pines within <15 m from the dead tree. This kills pine sawyers adults emerging from the affected tree.

32.1.1.3 Biological Control

Bird and insect predators, entomopathogenic fungi and entomophilic nematodes have been studied for biological control of pine wilt. Only the great spotted woodpecker, *Dendrocopos major*, and an entomopathogenic fungus, *Beauveria bassiana*, are used in Japan as biological control agents against the disease. However, the great spotted woodpecker is effective only where the disease is endemic, but not where it is epidemic. A more complete discussion on biological control of pine wilt disease is given in Chap. 35.

32.1.2 Preventive Sprays

32.1.2.1 Chemical Insecticides

One or several MEP (Fenitrothion), MPP (Fenthion), NAC (NAC sevin denapon carbaryl), or Thiacloprid compounds are the major components of insecticides used for preventive sprays against pine sawyers. Fenitrothion and Fenthion are organic phosphorus insecticides while NAC is a carbamate insecticide. Those inhibit cholinesterase activity, which relates to impulse transmission in insect nerve systems; however, as in other noenicotinoid insecticides, Thiacloprid is a neonicotinoid insecticide, which causes nervous hyper-excitability by combining with the acetylcholine receptor of insects. This kills the insect.

Mammals and birds have enzymes that can break down these chemicals into non-toxic substances and excrete them in their feces; consequently, these materials do not accumulate in the bodies of mammals or birds. In fact, BHC bioaccumulates, but Fenitrothion does not. No Fenitrothion was detected in the internal organs of birds in sprayed areas; however, there are some reports of negative effects of these chemicals on birds: The population levels of bulbul, *Hypsipetes amaurotis*, birds decreased by 39% after spraying NAC (Tanaka 1989). Oral intake of Fenitrothion caused, liver fat modification, small bowel lesions, and decreased clutch size for the quail, *Coturnix japonica*. Thus Fenthion is not used for aerial spraying because of its high toxicity to birds.

Fenitrothion reportedly caused chemical injury or mortality to the Japanese cypress, *Chamaecyparis obtusa* (Kobayashi 1981). Fenitrothion changes to sumioxon through photochemical reactions especially by UV rays. Sumioxon is a highly toxic but unstable compound. It is believed that an aerial spray of Fenitrothion is not so harmful to birds or mammals, including humans, because sumioxon is detoxified soon after being ingested. However, recently in Japan Fenitrothion usage for preventative spraying has decreased sharply because of the general increased awareness by humans about environmental issues. Farmers cannot market agricultural products from which unsafe levels of pesticide residue are found. The tolerance for pesticide residue for each agricultural product is defined in the food product health law.

NAC formulations have efficient levels of insecticide efficiency and residual efficacy if they are applied appropriately; however, their effects vary greatly because NAC tends to be carried away by rain runoff so that residual efficacy is lost by several tens of millimeters of precipitation.

Thiacloprid is a newly registered pesticide, effective against insects belonging to the Lepidoptera, Hemiptera, and Coleoptera. One advantage of Thiacloprid is its reduced effect on ecosystems because of its minimal impact on fish and pollinating insects. Fish toxicity of Thacloprid is ranked A. Hymenopteran pollinators can be released the day following spraying. Use of Thiacloprid as a prevention spray is increasing because of its low impact on ecosystems.

32.1.2.2 Aerial Spraying

Aerial spraying is done by using an airplane or helicopter with the insecticide being applied to tree canopies from above to kill adult pine sawyer beetles, which are visiting pine trees for maturation feeding. Though there have been a few examples, which showed significant differences in pine tree mortality between aerial spraying and no spraying, some reports showed that the mortality of pine trees decreased greatly after starting aerial spraying or conversely increased greatly after stopping aerial spraying, indicating that aerial spraying is effective.

Aerial spraying is effective especially when vast areas of pine forests require preventive spraying, mainly where pine wilt is evenly distributed across the area; however, great care must be taken to protect the health of humans, animals, agriculture, and fisheries. As the effective period of registered insecticide formulations is 3 weeks after spraying, aerial application should be done twice each season, at the beginning and at the peak period of the adult vector emergence. The duration is 1.5–2 months for adult pine sawyer emergence and 2.5–3 months for maturation feeding; but adult emergence decreases sharply after peaking. As such sufficient preventive effects can be obtained if the insecticide is effective for approximately 2 months.

However, recently aerial spraying has decreased because of its negative impact on ecosystems. The executive branch of government that conducts aerial spraying should establish a public administration that values the safety of humans and ecosystems to avoid needless conflict. The concentration of insecticides in the air following aerial application and impact on organisms living in and around sprayed areas should be monitored. As well, such data should be published.

32.1.2.3 Ground Spraying

The basic concepts are the same for both ground and other methods of spraying. Usually, by using a high pressure spray gun with a power sprayer the insecticide can reach 15–20 m above ground level and it can reach 40 m above ground level using a special type of tailed mist blower called a "Spouter", which was developed for preventive spraying against the pine sawyer in the tops of tree crowns (Model VSM-1000/1010, Maruyama, Japan; Model-IJ30, Marunaka, Japan). Although the "Spouter" was originally a brand name, it is now commonly used for tail mist blowers used in preventive spraying against pine wilt in tree tops in general.

Ground spraying is a more time-consuming than aerial spraying as drift of the insecticide outside the pine stand is reduced more than for aerial spraying; ground spraying is usually conducted once a year, but to obtain the same effect a greater amount of insecticide is used. However, increasing the frequency of ground spraying increases its efficiency, even if the amount of insecticide used for each spray decreases.

Sawyer beetle adults prefer young branches and shoots for maturation feeding, which occurs mainly in the upper forest canopy so it is important to apply insecticide to this area. The numbers of PWNs transmitted from pine sawyer adults to pine trees is relatively small during the first 7 days following adult emergence. As such the insecticide should be applied at the beginning of adult emergence; however, a major problem is to estimate the timing of adult emergence, which differs among years and locations. It is not easy to change the date of spraying based on annual or seasonal variations in climate. The reported emergence of PWNs from pine sawyer adults extends over more than 2 months, suggesting that it is impossible to prevent PWN infection completely even if an insecticide is sprayed at precisely the correct time. However, preventive sprays can decrease the probability of infection.

32.1.3 Host Tree Resistance to PWN

32.1.3.1 Breeding for Resistance to the PWN

In 1978 resistance breeding to the PWN started in Japan at the initiative of the National Tree Breeding Center, Forestry Agency, Japanese MAFF and by 1984, 92 Japanese red pine, *Pinus densiflora*, and 16 Japanese black pine, *P. thunbergii*, trees had been selected as being pine wilt resistant. Seed orchards have been established in several prefectures since 1985, and their progeny have been used widely since 1992 (for details, see Chap. 34).

32.1.3.2 Vaccination

Preventive "vaccination" is a method wherein a nematicide or a chemical formulation is injected into a tree to prevent the reproduction or to kill PWNs within the tree. These formulations are injected into holes made on the trunk of pine trees. This method is highly effective if the formulation is injected properly. There are several advantages of this method since the "vaccine" is effective for several years and is not influenced by precipitation or other natural elements. As well, this method has no negative impact on the environment or humans and it can be used for all trees, which are accessible to forest workers.

Preventive vaccination has been used in some pine wilt disease management programs in "pine stands worthy of protection" since 1994. In these programs, preventive vaccination is applied to pine stands in which a preventive spray is not applicable, and stands in which pine tree mortality is not acceptable. For example, recreation areas, and pine stands containing endangered species. However, preventive vaccination is not a good choice for use in large pine forests because it is more expensive and labor-intensive than aerial and ground spraying. Other disadvantages of this method include the problem of chemical injury (phytotoxicity) since, for example, the inoculation wound sometimes injuries the cambium, and causes discoloration beneath the bark, cracks the bark, or interrupts water movement or results in infection by wood rotting or blue stain fungi. Preventive vaccination should be conducted once every 3–4 years depending on the effective period of each formulation. Inoculation injuries accumulate with each inoculation although vaccine formulations have been improved to decrease the incidence of chemical injury and also the effective period of the materials has been prolonged. The number of vaccine inoculations should be minimized under a long-term management strategy.

Since 2006, the Agency for Cultural Affairs, Ministry of Education, Culture, Sports, Science and Technology (MEXT) have issued instructions to local governments to actively use preventive vaccination to protect important pine trees from pine wilt.

Several types of vaccine formulations are registered for use. The main component of these is one of the following: (1) *Morantel tartrate*, an ordinary (see definition above) substance with rank A for fish toxicity, which inhibits PWN reproduction; (2) *Levamisol hydrochloride*, an ordinary substance with rank A for fish toxicity, which also inhibits PWN reproduction; (3) *Mesulfenfos*, a deleterious substance with rank A for fish toxicity, which acts as a nematicide for against the PWN; and (4) *Nemadectin*, a deleterious substance with rank C for fish toxicity, which inhibits PWN reproduction.

32.1.4 Silvicultural Management

"Pine stands worthy of protecting" are usually valuable from a public viewpoint, but are uncared-for because of their low economic valuc. In these stands there are many suppressed trees and dead branches in which pine sawyer adults lay eggs. Such material which is available for pine sawyer activities should be removed from "pine stands worthy of protection" by sanitation, including improvement cutting, forest thinning, pruning dead branches, and logging.

32.2 Control Strategy

32.2.1 History of Pine Wilt Disease Management and Related Laws in Japan

According to a "5-year Special Law for PWD (pine wilt disease) Management" enacted in 1973, and a "5-year Special Measures Law for PWD Prevention" enacted in 1977 and revised in 1982, 1987, and 1992, billions of yens of government funding has been spent on aerial and ground spraying and eradication. As a result, the annual incidences of the disease in the early 1990s decreased to almost half that of 1979, a peak year. However, approximately a million cubic meters of pine trees were still killed by pine wilt every year so the special law was significantly amended by a 1992 revision, which decreased coverage of the national government's subsidy for preventive spraying and eradication. Pine trees around "pine stands worthy of protection" were eliminated and converted to other tree species to decrease the populations of pine sawyers invading the protected pine stands.

"Pine stands worthy of protection" are those that have important public benefit and those located at the periphery of a PWN epidemic. The "5-year Special Measures Law for PWD Prevention" expired on 31 March 1997; thereafter, the "Forest Pests and Diseases Control Law" was partially revised to involve PWD management that had been done during the past "5-year Special Measures Law for PWD Prevention".

Many control tactics have been studied and developed since 1971 when PWN was discovered as the causal agent of pine wilt disease; however, pine wilt disease management strategies have not been well studied, although researchers have been involved in government decision-making on pine wilt disease management strategies since the late 1990s, which has generated several success stories (Yoshida 2006).

32.2.2 Possible Pine Wilt Disease Management Strategy in Japan

The following strategies should help maintain the existence of pine stands in Japan: (1) complete eradication of the PWN and preventing its reinvasion; (2) eradication of the sawyer beetle vector or to depress its reproduction; (3) replace susceptible pine trees with highly resistant pines; and (4) reduce the incidence of the PWN to a low level, allowing pine stands to survive.

The first strategy is the most straightforward against invasive pests, trying to restore pine stands to their original condition before PWN invasion; however, it is almost impossible to eliminate PWNs that became established almost 100 years ago and spread across Japan. It is possible to eliminate PWNs at a local level but impossible to prevent PWN reinvasion completely. As shown in an example on Okinoerabu Island, it is possible to eliminate the pine wilt disease from an isolated island, but the possibility of PWN reinvasion still exists (Muramoto 1998).

The second strategy is also does not ensure complete success because of the high potential of pine sawyer reproduction and dispersal, the absence of potent natural enemies, and limitations on the use of chemical insecticides.

The third strategy tries to redirect pine stand management so that the result is the same as in North America where native pine trees are favored that are resistant to the PWN. Resistant strains of pine seedlings are available; however, it takes time to replace all susceptible pines with resistant ones. There are some reports that some of resistant pines planted in the past have been killed by the PWN. "A red queen hypothesis" predicted an evolutionary arms race between organisms. Such an evolutionary arms race between pines and PWN is presently progressing in Japan. It is difficult to predict how it takes to establish an evolutionary stable relationship between pines and the PWN in Japan. These should be taken into consideration when employing resistant strains of pine trees in pine wilt management.

The fourth strategy is the most realistic and adoptable to most locations in Japan, in which elimination of the PWN or pine sawyers is difficult. Under these circumstances, pine stands will collapse if control efforts are reduced. "Pine stands worthy of protection" should be selected, in which intensive control efforts can be achieved. Control efforts in these stands should be intensive enough almost completely eradicate the pine sawyers. Invasion of pine wilt disease from outside of such stands should be prevented as much as possible. Unless the probability of invasion from outside of the stands becomes impossible incidences of pine wilt disease will continue and control efforts will also be necessary.

32.2.3 Dispersal Ability of Japanese Pine Sawyer Beetles and the Probability of Invasion

The probability of new invasions by pine sawyer adults is needed to determine pine wilt disease management strategy(ies). Only one report is available for determining the distance-dispersal probability relationship. Fujioka (1993) determined the relationship between distance and trap captures using a mark-recapture method. Traps were deployed along a line at definite intervals. Marked sawyer adults were released at one location. The number of recaptured adults by each of the traps were determined. Actual trap captures underestimated the probability of dispersal at each distance, therefore, the relationship between distance and dispersal probability was determined by compensating for the underestimation (Yoshimura et al. 1998). The probability tends to decrease with distance from the release point and was small enough at 2 km to quarantine and eradicate new invasions so that, based on these results, widths of the barrier zone and the scrutiny zone are usually 2 km. In the case of islands, however, the probability of invasion from outside is smaller than the probability estimated from the relationship because sawyer adults cannot pass over the ocean.

32.2.4 Success Stories of Pine Wilt Disease Management Strategies

In this section, tactics that have resulted in pine wilt control are summarized by presenting several success stories. Examples include an island, coastal pine plantations, and a park in an urban area. On an island, once pine wilt is completely eliminated the incidence of wilt can be lowered without any control efforts because the island is isolated from other population sources of pine sawyers and the nematode. In contrast, eradication and prevention spraying are always needed in coastal pine plantations because it is almost impossible to decrease the probability of PWN reinvasion to zero. In the urban park, the tolerance level of pine wilt disease incidence is much lower than plantations so more intensive control efforts are required.

To keep pine wilt incidence low enough to maintain pine stands, according to the fourth strategy mentioned in Sect. 32.2.2, the following three measures are necessary: (1) elimination of sawyer beetle source populations from the area around the pine stands that are being protected; (2) complete eradication of the disease in pine stands worthy of protection; and (3) preventive spraying of pine trees in such pine stands. The details of each disease management strategy should be flexible depending on the situation.

32.2.4.1 Eradication of Pine Wilt Disease in Isolated Islands: Okinoerabu Island in Amami, Southern Japan

Okinoerabu Island, which is located between Tokunoshima Island and Yoronto Island, has 978 ha of forests (Fig. VI.1). Forty percent of the forest consists of Ryukyu pine, *P. luchuensis*. The forest had been carefully maintained because on the island forests are important as water reservoirs and as windbreakers.

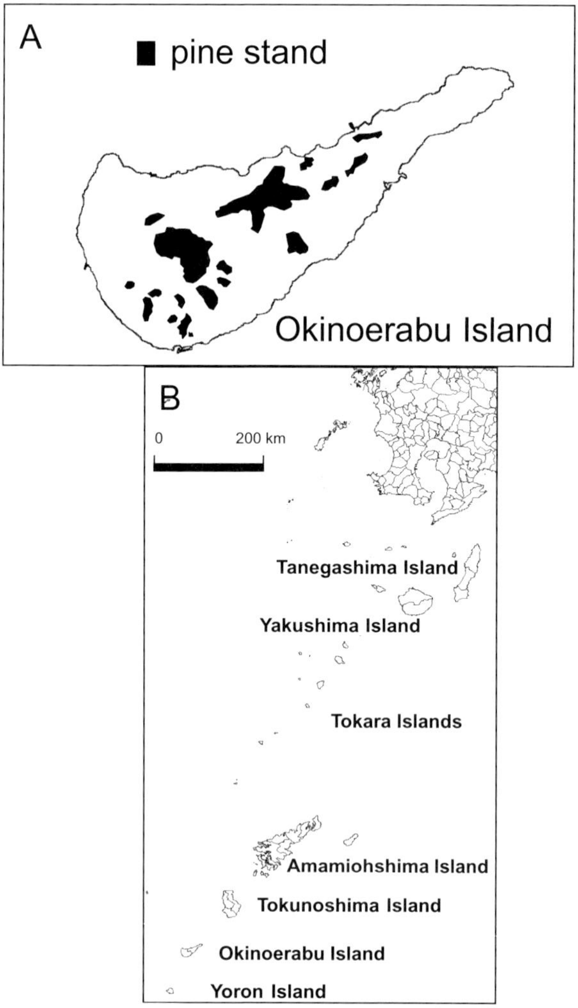

Fig. VI.1 Distribution of pine stands on Okinoerabu Island (**A**) and location of the island (**B**) (after Muramoto 1998)

P. luchuensis is susceptible to PWN, but it is not so tolerant to chloride damage as is Japanese black pine. Pine wilt disease was first found on the island in 1977. It is believed that pine wilt infested wood was brought onto the island for rebuilding after a typhoon. More than 2,800 m^3 of pine wood were killed by pine wilt in 1982. Aerial spraying of healthy pine trees with Fenitrothion started in 1982. The cut-and-burn method also started to that same year. Incidences of pine wilt decreased sharply after these two tactics were adapted. In 1989, only ca. 10 m^3 of pine trees were killed by wilt. It is believed that the PWN was completely eradicated from the island by 1995.

The success story for Okinoerabu Island is discussed from the three viewpoints mentioned in Sect. 32.2.4. The first goal was easily realized because the island is isolated enough to keep pine sawyers off the island. The second goal was realized by burning diseased trees, including their branches. Sawyer larvae living under bark on pine stumps were checked by removing the bark. Prevention spraying from a helicopter was carried out from 1982 to 1994, which achieved the third point.

32.2.4.2 Management in Pine Stands

32.2.4.2.1 Enjugahara in Wakayama Prefecture

There is a 84-ha, coastal pine stand in Enjugahama, midwestern Wakayama Prefecture, consisting of a mixture of 64,000 Japanese red pine and Japanese black pine (Fig. VI.2). In Enjugahama, pine wilt had been epidemic since 1977, but became endemic in 1984. Less than 100 pine trees were killed by the disease from 1984 to 1993; however, mortality started to increase again from 1994 affecting ca. 500 trees in 1994 and ca. 2000 in 1999. From 1977 to 1984, the pine sawyers invading from population sources in the Mt Nishiyama area were the likely cause of pine wilt disease incidences in Enjugahama. In the Mt Nishiyama area, the number of pine trees decreased greatly as the results of pine wilt and forest fires, so aerial spraying had stopped by 1986. In Enjugahama in 1989, a cut-and-chemical-application method was replaced with a cut-and-burn method and preventive, aerial spraying stopped in 1996. Only ground spraying has been used since 1996.

In 1999, the control strategy was greatly changed based on recommendations from the Forestry and Forest Products Research Institute (FFPRI) because pine wilt was expected to again increase. The strategy used the three concepts mentioned in Sect. 32.2.4, that is, (1) elimination of pine sawyer source populations around pine stands worthy of protection, (2) complete eradication in pine stands worthy of protection, and (3) preventive spraying of pine trees in pine stands worthy of protection. Regarding the first concept, pine trees that had become established at Mt Nishiyama after the collapse of pine stands in the mid-1980s were thought to be a pine sawyer population source jumping to Enjugahama. Pine trees growing in the Mt Nishiyama area were cut and removed to encourage broadleaf tree regeneration. Regarding to the second concept, the local government employed persons to find dead trees and ask local people living around Enjugahama to notify the government

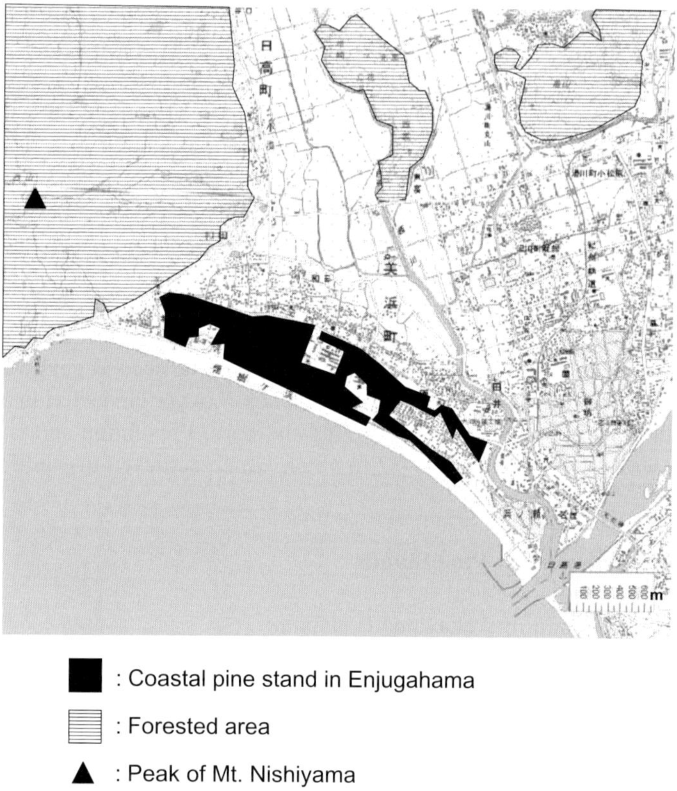

Fig. VI.2 Location of a coastal pine stand in Enjugahama, neighboring forested areas, and the peak of Mt Nishiyama (Wakayama Prefecture, Japan) (This illustration was modified by the 1 : 25,000 Scale Topographic Maps (Mio and Gobou) published by Geographical Survey Institute, Japan)

if they found any dead trees. An aerial survey was also done for diseases trees. The government then changed the eradication procedure. They educated forest workers to eradicate not only big trees but also small ones, including small branches; because Japanese pine sawyer beetle larvae prefer areas on trees with thin bark. Big logs were carried out to a road on the day of felling and then chipped in a factory. Small trunks and branches were also carried to the seashore on the day of felling and then burned at intervals. Forest roads were constructed inside the pine plantation to facilitate these operations. Regarding the third concept, the frequency of ground sprays increased from two to three a year. Two types of power sprayers, a high-pressure spray gun with a power sprayer and the "Spouter (see Sect. 32.1.2.3)" were both used depending on tree heights. Timing of the ground sprays was determined by checking adult vector emergence from infested logs kept in a chamber in the pine stands. As a result, pine wilt incidence decreased to ca. 500 trees in

2001 and ca. 100 after 2003. These numbers likely include trees that were killed by other causes besides pine wilt.

32.2.4.2.2 Nijinomatsubara in Saga Prefecture

At Nijinomatsubara there is a Japanese black pine plantation of approximately 230 ha, on Karatsu Bay, Saga Prefecture, Kyushu, Japan. The incidence of pine wilt increased greatly in 1971. Until then ground spraying and eradication were the major tactics used against pine wilt; however, the incidence decreased greatly after 1973, probably because aerial spraying has been used since then. Vast areas of pine stands existed on the slope of Mt Kagamiyama, less than 1 km from Nijinomatsubara (Fig. VI.3). Pine wilt incidence was kept low from 1975 onward, but increased again in 1991. The numbers of pine trees killed by pine wilt were 1,200–2,300 every year after 1991 so a new management strategy was applied after 1996. This included, firstly eliminating the pine sawyer populations around pine stands worthy of protection. Until 1996, most pine stands within 2 km from the edge of Nijinomatsubara collapsed because of the disease, with only one exception at the top of Mt Kagamiyama. However, only ca. 100 pine trees remained there, and the distance between the edges of the two pine stands was ca. 1.5 km so that the pine stand at Mt Kagamiyama did not likely influence pine wilt incidence at Enjugahama.

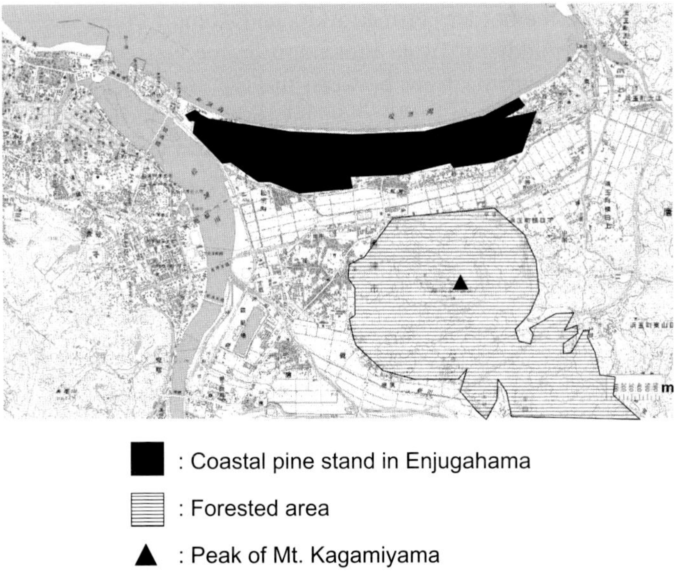

Fig. VI.3 Location of a coastal pine stand in Nijinomatsubara, a pine forest on the slope of Mt Kagamiyama, and its peak (Saga Prefecture, Japan). (This illustration was modified by the 1:25,000 Scale Topographic Map (Shisa) published by Geographical Survey Institute, Japan)

A second step included complete eradication of the disease in pine stands worthy of protection. A cut-and-burn method was employed instead of a cut-and-chemical spray method because there was little possibility of *Rhizina* root rot. Not only big trees but also small trees and branches were cut and burned immediately. Pine trees that had died from causes other than pine wilt were also eradicated in the same way. In 1996 and 1997, aerial surveying was carried out after eradicating the dead trees detected by a ground survey. In 1996, 86 dead trees were found by aerial survey and these were immediately removed. At present, dead trees are piled together after felling and then burned because the number of diseased trees is small.

A third approach was preventive spraying of pine trees in pine stands worthy of protection. The timing of aerial sprays was determined by checking adult sawyer emergence from infested logs kept in a chamber placed in the pine stands. The timing of spraying has improved every year by comparing each year of wilt incidence with the timing of sprays and the seasonal occurrence of pine sawyer adults in the chamber.

Since 1997, the number of dead pine trees has been ca. 100 every year. The PWN was not detected in most of these dead trees so that the number of pine-wilt affected trees is now less than ten per year (Nakamura 2004).

32.2.4.3 Management Efforts in Advancing Fronts of Pine Wilt

At present, pine wilt has reached the northern part of Honshu, the main island of Japan and invaded all the prefectures of Japan, except Aomori and Hokkaido. The disease has reached the border between Akita and Aomori Prefectures on the west coast, but only the middle of Iwate Prefecture on the east coast (Fig. VI.4). The differences in the advancing front between the east and west are approximately 50 min in latitude (ca. 100 km). Pine wilt invasion occurred for a longer time in Iwate Prefecture than Akita, indicating that the spread rate after invasion in each prefecture was greater in Akita than in Iwate. The great difference in summer temperatures related to tidal effects is likely responsible for the difference in the spread rate.

Another cause of the slow spread in Iwate is the intensive control efforts being used there. Japanese red pine is one of the most important timber species in Iwate so the prefecture government has made great effort to slow pine wilt spread. They set up a scrutiny zone, 2-km wide and 60-km long, at the advancing front of the disease and about 30 people are employed to look for pine-wilt affected trees. The local government has also enacted a regulation that allows such trees to be eradicated by the local government without the permission of forest owners. More than 30,000 trees have been eradicated every year in this scrutiny zone; however, a few pine wilt incidences have been found to the north of this zone. Eventually, however, the scrutiny zone cannot stop the disease spread, only slow it down.

On the west coast, pine wilt incidence was found 250 m from the border between Akita and Aomori Prefectures. In the summer of 2006, the Tohoku Regional Forestry Office and the local government of Aomori Prefecture set up two barrier zones

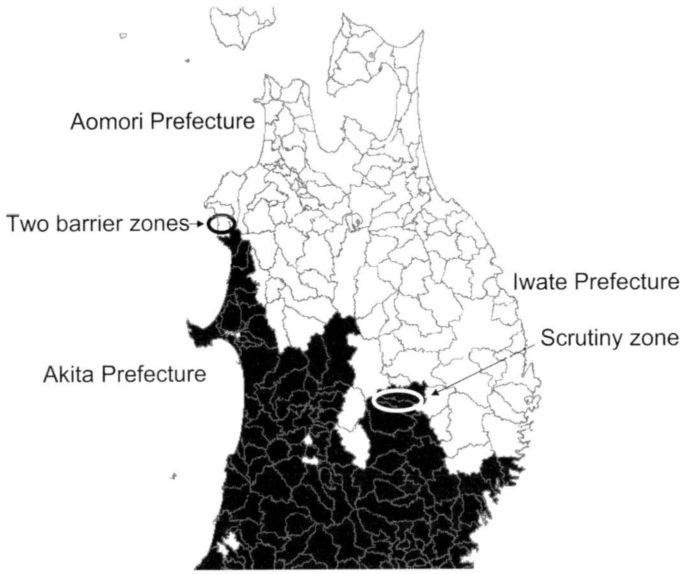

Fig. VI.4 Incidence of pine wilt in the municipality (*dark*), and approximate locations of a scrutiny zone and two barrier zones at a spreading pine wilt epidemic on the northern part of Honshu, Japan's main island

and a scrutiny zone in Aomori Prefecture near the border. One barrier zone is 1 km wide and the other 2 km. The distance between the two is about. 3 km (=width of a scrutiny zone). All pine trees inside the barrier zones were removed to decrease the probability of pine sawyers invading from Akita to Aomori Prefecture. Incidences of pine wilt disease inside the scrutiny zone are examined carefully.

32.2.4.4 Management of Pine Trees in Urban Areas

Different strategies are necessary for pine wilt management in urban areas. An example is Kenrokuen Garden in Ishikawa Prefecture. The Kenrokuen Garden is one of the three most beautiful gardens in Japan. It was founded during the Edo Era (1603–1868), and subsequently extended over generations by the Maeda family, feudal lords of Kaga. Approximately 600 pines grow inside the garden. Although pine wilt was first found in Ishikawa Prefecture in 1972, the disease did not occur at the Kenrokuen Garden until 2002. The Kenrokuen Garden was designated a cultural property specification garden by the Agency of Cultural Affairs. Among ca. 600 pine trees, 17 trees of special interest were especially important elements of its scenery. Nine prevention sprays were conducted every year up until 2000 because no pine tree mortality was tolerated in the garden. The number of sprays decreased to six per year after 2001 according to a guideline for pesticide use; however, in 2003, three pine trees in the Kenrokuen Garden were killed by pine

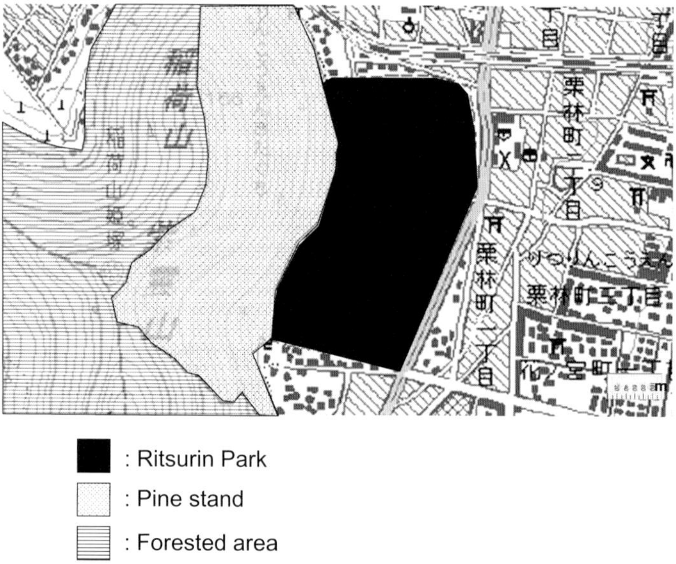

Fig. VI.5 Location of Ritsurin Park, an adjacent pine stand, and neighboring forested area (Kagawa Prefecture, Japan). (This illustration was modified by the 1 : 25,000 Scale Topographic Maps (Takamatsu-Hokubu and Takamatsu-Nanbu) published by Geographical Survey Institute, Japan)

wilt, an Otohamatsu pine and one of the Meotomatsu pines, both belonging to the 17 pine trees of special interest, and another pine tree other than these 17. The local government organized a committee to improve the pine wilt management strategy in the Kenrokuen Garden. The key points of the new strategy are as follows:

The first is to decrease the probability of a new invasion from outside. Even if eradication and prevention spraying are thoroughly carried out inside the garden, it is difficult to prevent pine wilt if pine sawyer invasion occurs frequently from outside. This is supported by the fact that pine wilt is found every year in Ritsurin Park, Kagawa Prefecture, where many wilt incidences occur every year on the adjacent mountain slopes (Fig. VI.5) although prevention sprays are conducted in the same way as in the Kenrokuen Garden. A PWN-pine sawyer management strategy is needed for both the forested areas outside the garden area as well as for gardens in urban areas, the same as for pine stands. Suspected sources of pine wilt in the Kerokuen area include several urban parks, where pine wilt is found every year; around the Kenrokuen Garden (Fig. VI.6), however, these are not a likely cause of pine wilt in Kenrokuen Garden because all of the dead pine trees are properly eradicated. It is likely that these pine trees could act as stepping stones for pine sawyer adults but it is impossible to remove these pine trees. Vaccination is carried out for pine trees in the urban parks. As for pine trees on private land, the government has asked local people around the Kenrokuen Garden to notify the government if they find any weakened or dead pine trees. The Kenrokuen staff

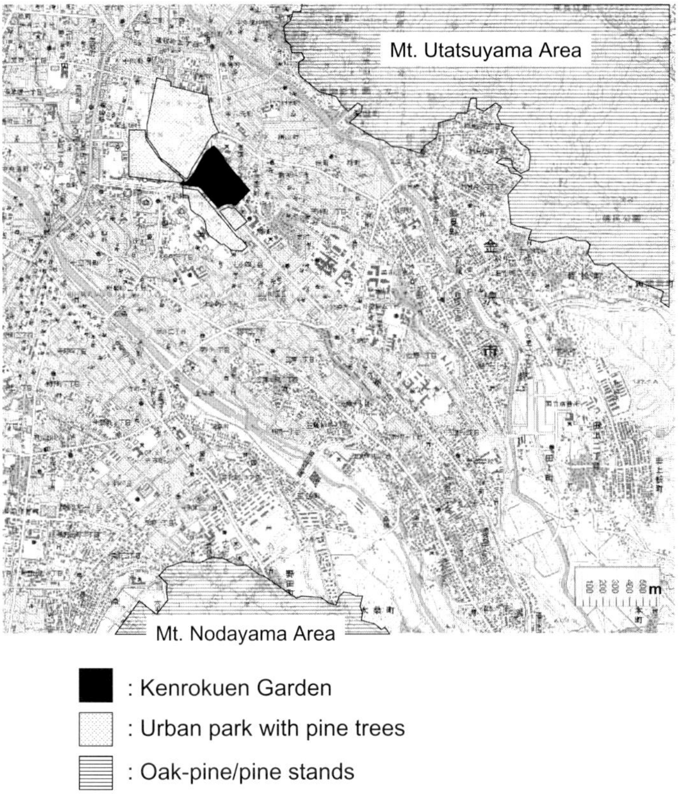

Fig. VI.6 Location of Kenrokuen Garden, adjacent urban parks, and a neighboring oak-pine stand at Mt Utasuyama and a pine stand at Mt Nodayama. (Ishikawa Prefecture, Japan). (This illustration was modified by the 1 : 25,000 Scale Topographic Map (Kanazawa) published by Geographical Survey Institute, Japan)

checks such trees to determine the cause of death. Local governments ask owners of "suspicious" pine trees to remove them properly. The most likely sources of pine wilt in Kenrokuen Garden and other urban parks are sawyer beetle populations on Mt Utatsuyama (<2 km from the Kenrokuen) and Mt Nodayama (<4 km away). Forests on Mt Utatsuyamna and Mt Nodayama are owned by one of the local governments (Ishikawa Prefecture or Kanazawa City) and private owners so it is impossible to treat them uniformly. Regarding private forests, the decision concerning pine wilt management depends upon forest owners; however, concerning forests owned by the prefecture or city government, the government has decided to protect only a few pine trees and to convert pine stands to broadleaf, natural forests. Accordingly, vaccination is used to protect pine trees worth protecting, but prevention spraying has been stopped to promote conversion from pine stands to broadleaf natural forests. As a result, the numbers of trees killed by pine wilt decreased from 381 to 168 in Mt Utatuyama and 495 to 67 in Mt Nodayama from 2003 to 2006.

The second point is complete eradication. The number of pine trees in the Kenrokuen Garden is small enough to detect and eradicate all pine-wilt affected trees. Because many people visit the garden every day, it is unlikely that diseased pine trees go unnoticed. Tree health surveys started in the Kenrokuen Garden in 2003. All pines are checked annually to detect diseased trees, making it possible to treat unhealthy pine trees properly and to prevent these unhealthy trees from attracting pine sawyers.

The third point is prevention. Vaccination is employed, but not for the 15 remaining pine trees of special interest because the probability of chemical injury is possible. Forty trees were injected in 2004 and 232 in 2005. Injections are again scheduled for these trees in 2008 and 2009 because chemical formulations are effective for 4 years. Pine trees in forests near the periphery of the park were vaccinated in the first year because of the problem of insecticide drift made it difficult to adequately spray these trees. Pine trees greater than 16 m in height, or those that grow in clumps, were vaccinated in the second year because it was difficult to adequately spray them. In Kenrokuen Garden, preventive sprays also have been conducted for vaccinated trees because such vaccinated trees could act as stepping stones for pine sawyers if the vaccinated trees were not sprayed. This strategy also depends on the policy that no pine trees mortality was tolerated in the garden. Although the frequency of preventive sprays decreased from nine a year to six, it was not a good idea to increase spray frequency considering the cost–benefit ratio. Researchers meet each spring to discuss and decide upon the spray schedule. Basically, the first spray is conducted soon after 20 May, which is a week or 10 days prior to the estimated date of the first emergence of pine sawyers. Thereafter, spraying is conducted once every 2 weeks. The fifth spray is carried out at the end of the rainy season. The last one is conducted in early August. No pine wilt has occurred in the Kenrokuen since 2004.

33 Recent Advances in the Integrated Management of the Pine Wood Nematode in China

Fuyuan Xu

33.1 Introduction

The main control strategies for pine wilt in China are use of intensive quarantines, clear cutting and methyl bromide fumigation of pines trees killed by the pine wood nematode (PWN, *Bursaphelenchus xylophilus*). As well, both aerial and ground sprays of fenitrothion insecticides can significantly reduce tree death rates by killing the beetle vectors. Other effective treatments to control the pine sawyer beetle, *Monochamus alternatus*, a cerambycid beetle, include hot water treatment of the infected wood, and for affected logs and branches high temperature treatment, submerging in water and burying in soil, and direct trapping of adult, vector beetles using terpene-baited traps. However, because of the extensive areas of affected forests these methods do not give effective control (Xu et al. 1994a; Dwinell 1997).

Recently, the main pine species damaged by the PWN in southeast China has changed from *Pinus thunbergii* to *P. massoniana*. Of the total dead pine trees, the proportion of dead *Pinus massoniana* in the forest has increased from 10 to 50% since 1985. Now the disease is reported in 11 provinces in China. Some provinces, such as Zhejiang, Anhui and Jiangsu, have suffered severe pine wilt disease. More than 5.2 million pine trees were killed by PWN in Jiangsu alone from 1982 to 2003. Present research is focused on finding a PWN resistant Masson pine and on the control of its principal vector, *M. alternatus*. In China the disease cycle is very similar to that in Japan and South Korea. It is especially important to use ecological methods to halt this cycle and control the PWN. Intensive breeding of a natural enemy, *Scleroderma guani*, to control *M. alternatus* larvae and techniques for

Forestry Academy of Jiangsu Province, Nanjing 211153, People's Republic of China

Tel.: +86-25-52744730, Fax: +86-25-52741620, e-mail: xufuyua@yahoo.com.cn

integrated management of the disease have been of major focus. The main research results of this work are summarized below.

33.2 Treatment of Dead Pine Trees to Halt the PWN Disease Cycle

The most important management tool for control is forest sanitation, especially the removal of killed pines to prevent their becoming breeding grounds for both vector beetles and the nematodes. Dead or dying trees should be removed as soon as possible and should not be stored for later use as firewood. Entire trees should be removed. There is some debate about the possibility of the nematode spreading through root grafting, so trenching around disease centers may have to be considered in areas heavily planted with black or Masson pines. Even a single infected pine tree left standing can become an infection center that will devastate other pines nearby. In China five treatments involving dead pines are used to halt the PWN disease cycle. The first is clear-cutting of affected areas to remove and treat dead pine trees. From 2000 to 2005, 1.99 million pine trees killed by pine wilt were clear-cut in Jiangsu Province alone. All of the dead pine trees were sent to nearby wood-processing factories to be chipped for making chipboard. In Jiangsu Province alone, 57,961 tons of small wood pieces were chipped, and 31,878.55 m^3 of high density chipboard were made in wood-processing factories in 5 years (Xu et al. 2008). Since 1998 Guangdong Province has carried out clear-cut 21,346.67 ha of pine wilt diseased forests, then bag-fumigation the wood with aluminium phosphide to kill pests in the wood. After 3 years of application, the damaged area was reduced from 21,346.67 to 17,100 ha, that is, a 19.9% decrease in the area damaged by the PWN (Xie and Chen 2003). Third, heating and hydraulic pressing treatment of nematode-infected pine timbers was conducted. Heating at 65–75°C for 15 h, hydraulic pressing at 9 MPa, 157–168°C for 10 min or both of *B. xylophilus*-infested pine timbers of different sizes gave good control of both *B. xylophilus* and *M. alternatus*. In particular, treatment of diseased-infested boards less than 2.8-cm thick or 10 cm × 10 cm can kill 100% of the pests (Chen et al. 2000). Fourthly, submerging wilt-affected wood in hot water, exposing it to high temperatures, natural drying and hot evaporation of vapor with the temperature and exposure time required for 100% mortality are reported for different temperature treatments. The results showed time-mortality relationships for treatment of the PWN and the vector pine sawyer beetles in dead logs (average diameter of 15.99 ± 2.36 cm) were checked and analyzed. The thermal death point of nematodes is 45°C when killed logs are submerged in hot water, and their thermal death point is 50°C when such wood is exposed to high temperature treatment. Submerging dead logs in hot water is more effective than high temperature treatment and other treatments. Exposure times in hot water were significantly reduced as the water temperature was increased above 45°C (thermal death point). When wilt-affected logs were submerged in hot

Table VI.1 Comparison of treatments to rid wood and wood products of the pine wood nematode and its vector beetle

Treatments	Methods	Temperature (°C)	Control effectiveness (%)
Chipping and processing chipboard	In processing factories	100	100
Bag-fumigation with AlP	In a plastic film bags	Field temperature	98
Heat treatment	Treated for 30 min	65	100
Heating and hydraulic pressing treatment	Heat treatment	65–75°C and hydraulic pressing at 9 MPa	100
Hot-water treatment	Treated for 2.30 ± 0.50 h	60	100

water at 60°C in a large tank for 2.30 ± 0.50 h, the mortality of *B. xylophilus* and its vector *M. alternatus*, and that of other beetles such as *Shirahoshizo insidiosus* and *Blastophagus piniperda* all reached 100%. Hot-water treatment of lumber to a core temperature of 60°C for 2.30 ± 0.50 h was sufficient to confirm that dead pine trees for wood products were free of living nematodes or their beetle vectors in Nanjing; 100% of the living nematodes or their beetle vectors were killed by this treatment. Those methods could be used for the large-scale treatment of the trees killed by the PWN (Xu et al. 1991).

Canadian and US exporters have been working with researchers to develop wood treatments that kill the nematodes present in wood chips or other unseasoned lumber. Another approach is to use, heat treatment and hot-water treatment of unseasoned lumber to a core temperature of 65°C for 30 min. This proved sufficient in ensuring that dead pine trees intended for use as wood products were free of living nematodes or their beetle vectors in Anhui and Jiangsu Provinces. Currently, bag-fumigation and microwave treatment kill significantly more weevils than other treatments when the treatment procedure is limited in a certain area. All test results were good, but the treatment costs were relatively high (Xie and Chen 2003, Jiang et al. 2006). All of the five above treatments eliminated almost 100% of the pine sawyer beetle larvae (Table VI.1). The above treatments were both very effective and safe for controlling PWNs in large, wilt-affected areas (Xu et al. 1991).

33.3 Direct Trapping of Vector Adults Using Terpene-Baited Traps

In China volatiles from a stressed host has been tested for several attractants to *M. alternatus*. Since ovipositing *M. alternatus* females prefer stressed Masson pine over other pines; host discrimination by *M. alternatus* suggests that changes in the chemical composition of pines may mediate the host preference of beetles. Volatile compounds from stressed and healthy pine stems were collected using the absorbent trap collection method (Li et al. 2004). Significant differences in absolute terpene quantities were found between stressed and healthy pines. Healthy pines released seven more terpenes than stressed pines (Yang et al. 2003). Field trials demonstrated that four terpenes identified from host pines were attractive to *M. alternatus* with (+)-α-pinene being the most attractive compound. Ethanol appeared to be an important synergistic compound causing a significant increase in the activity of attraction (Yang et al. 2003). Pine tree's odor plays a vital role in several activities of *M. alternatus* such as in the behavior of habitat localization, searching for host plants, maturation feeding, and oviposition. The principal constituent of a pine tree's volatile material is terpenes and an anaerobic fermentation product mixture. Ovipositing female *M. alternatus* prefer stressed rather than healthy *P. massoniana* trees (Yang et al. 2003). According to Ikeda et al. (1980), healthy pines also volatilize microterpenes with α-pinene as the principal constituent, and the quantity of a pine tree's volatile material increases rapidly from initial deterioration to wilting, while non-oxygen fermentation ingredients, with ethanol as the principal constituent, are produced, which attract *M. alternatus*. Host discrimination by *M. alternatus* suggests that changes in the chemical composition of pines may mediate the host preference of beetles. Volatile compounds from stressed and healthy pine stems were collected and used in a trapping experiment (Ikeda et al. 1980). Siegfried (1987) showed that traps baited with α-pinene, a β-phellandrene-limonene mixture, limonene, and β-pinene captured significantly more weevils than unbaited traps.

By analyzing the volatile material of pine trees, the principal constituents of the attractant were identified as mixtures of monoterpene hydrocarbon (α-pinene), ethanol and so on (Yang et al. 2003). In Japan, attractants have been studied since 1980, with some attractants having become commercialized products (see Chap. 35). In China, researchers have also actively carried out research and developed several attractants for *M. alternatus*. Researchers in Guangdong Province (Chen et al. 2002) used monoterpenes and 55% ethanol compounds to trap adult *M. alternatus*, and the effects of attraction were remarkable. By comparing nine formulations, researchers in Zhejiang Province (Zhao et al. 2000) determined the attractant of M99-1 with the highest activity. Its principal constituents are monoterpenes, ethanol, isopropanol and other organic solvents.

Presently, the following more efficient attractants with a broader spectrum for trapping *M. alternatus* are being used in China: (1) M99-1, developed by the Chinese Academy of Forestry (Zhao et al. 2000); (2) PEII, developed by the Guangdong Forest Research Institute (Chen et al. 2002); (3) FJ-MA-02, researched by the Fujian Forest Research Institute (Huang JS et al. 2005); and (4) MA2K05, studied by Anhui Agriculture University. It is easy to make an attractant and apply it (Wang SB et al. 2005). According to published reports, M99-1 caught 151.5 adult *M. alternatus* per trap and reduced the next generation of beetles by 1204.4 eggs on average (Zhao JN et al. 2000). M99-1, PEII and FJ-MA-02 attractants were

Table VI.2 Comparison of the effects of different attractants on adult pine sawyer beetles

Attractants	Attracted number of pine sawyer beetles	Reference
M99-1	299	Zhao et al. (2000)
PEII	298	Chen et al. (2002)
FJ-MA-02	257	Huang JS et al. (2005)
MA2K05	105	Wang SB et al. (2005)

Note: Number of pine sawyer beetles captured by per trap in one trapping season. According to the authors, trapping periods and attractant prescriptions differed among the four provinces where the studies were done

applied over a large area of China and effectively attracted adult *M. alternatus* in pine forests (Table VI.2).

The research showed that these attractants have strong trapping ability for *M. alternatus* and some trapping role for 30 other kinds of important pine beetles such as *Shirahoshizo patruelis*, *Spondylis buprestoides*, *Arthopalus rusticus*, *Massicus raddei*, *Megopis sinica*, and *Cryphalus massonianus*. Among pine-boring beetles, *Shirahoshizo patruelis* and *M. alternatus* are the dominant pests with the highest population density (Wang SB et al. 2005). Attractants play a very important role in controlling harmful forest insects. The conclusion is that attractants are an effective method for monitoring beetle population dynamics and for reducing populations of boring beetles in pine forests.

33.4 Biocontrol Methods and their Control Effectiveness

Several natural enemies for possible control of *M. alternatus* were discovered in Dongshanqiao district near Nanjing and other areas in China (Xu et al. 2004). These were mainly:

33.4.1 Natural Enemies

The parasitic insects, *Scleroderma guani* and *Dastarcus helophoroides* have been found while microbes parasitic to *M. alternatus* were mainly parasitic fungi, parasitic bacteria and parasitic nematodes. Parasitic fungi include *Beauveria bassiana*, *B. brongniartii*, *Metarhizium anisopliae*, *Isaria farinosa*, *Aspergillus flavus*, *Verticillium* spp. and *Acremorium* sp., whereas parasitic bacteria include *Serratia marcescens* and the parasitic nematode, *Steinernema feltiae*.

33.4.2 *Biocontrol Methods and their Control Efficacy*

33.4.2.1 *Dastarcus helophoroides*

Dastarcus helophoroides is also one of the main natural enemies of *M. alternatus* in China. Its parasitic rate surpassed 20% on larvae of *M. alternatus* in the Tiexin forest, Nanjing in 1988 (Xu, unpublished). Parasitism on larvae of *M. alternatus* surpassed 23.7% and on adults, pupae and mature larvae exceeded 85% in Yantai, Shandong Province. Laboratory experiments on the oviposition and development of *D. helophoroides* (Ishii 2004) showed that the number of eggs laid per female was over 2,000 per year. Larvae from eggs oviposited in April required about 27 days after oviposition to parasitize *M. alternatus*, versus only about 8 days for larvae from eggs oviposited in early July. About 73 days were required for adults to emerge. The Chinese Academy of Forestry (CAF) intensively rears, in the laboratory, more than *D. helophoroides* 20,000 annually (Huang et al. 2003).

33.4.2.2 *Beauveria bassiana* and its Effectiveness

The techniques of releasing *B. bassiana* using wood-boring beetles *C. massonianus* were studied for controlling *M. alternatus* (Xu et al. 2000). Studies using the natural population of the wood-boring beetle showed that it could effectively carry *Beauveria bassiana* to the larva canal of *M. alternatus. Beauveria bassiana* could be carried on the tarsal claws and the body of bark beetle adults. Adult bark beetles live in pine trees killed by the PWN. In August larvae of *M. alternatus* were parasitized by *B. bassiana* under the bark of pine trees while in September they parasitized the larvae in the wood of trees. Its control effectiveness can reach 45% if it is released at the optimum time during the last 10 days of June. This was higher than the 26.7% control achieved by spraying with *B. bassiana*. Using the optimum release timing and dosage of *B. bassiana* (30 g ha^{-1}), 1266.67 ha were treated in a test to control *M. alternatus*. Its control efficacy reached 41.5–69.2% in the field with a cost of only 3.33 yuan ha^{-1}. As well, the field tests showed that this product of *B. bassiana* could reduce PWN-caused mortality of pine trees by 95.0%. These results showed that release of *B. bassiana*-laden wood-boring beetles *C. massonianus* could lead to effective control of *M. alternatus* and PWN in the field (Xu et al. 2000).

33.4.2.3 Adults of *Scleroderma guani* Transfer *B. bassiana* for Controlling Pine Sawyer Beetles

In large areas of pine forest 80% of the *M. alternatus* larvae can be parasitized by *S. guani* and *B. bassiana*. This rate of control is much higher than that achieved by using either *S. guani* or *B. bassiana* alone; however, the treatments it also produced some adverse effects. One is that *B. bassiana* that can also infect *S. guani*, and the other is that the cost doubled, so its possible usage needs more study (Xu et al. 2008).

33.5 Intensive Breeding of *Scleroderma guani*, and the Effect of Release Techniques on *Monochamus alternatus* and PWN Control Over a Large Area

33.5.1 Intensive Breeding Methods of Scleroderma guani

Scleroderma guani is one of the most successful natural parasites of *M. alternatus* larvae. By studying the biology and reproduction of *S. guani*, propagation and release techniques have been devised for controlling *M. alternatus* larvae. The research determined that *S. guani* could be readily propagated on the larvae of *M. alternatus*. Following release of *S. guani* in pine forests, parasitism of the larvae of *M. alternatus* could surpass 66.8%. Using dead branches from pine wilt affected forests 0.3 million *M. alternatus* larvae were collected annually for each of 5 years (=1.5 million larvae in total) and used for intensive laboratory rearing of *S. guani*. At the beginning in 2000 it was very easy to collect 0.3 million larvae from two forests to breed *S. guani*. Since 2002 collections have had to be made in other forests in order to collect as many as 0.3 million larvae for intensive breeding of more than 10.0 million *S. guani* annually. Most *M. alternatus* larvae in those two forests have been devastated by the collection of larvae. However, morality of pine trees killed by the PWN has decreased from 18.51 and 26.26% in 2000 to 0.29 and 0.38% in 2003 (Fig. VI.7), respectively, in the two forests where the collections have been made. Subsequently it has been shown that collecting *M. alternatus* larvae in the clear-cut area was also a good way to control PWNs and it is an

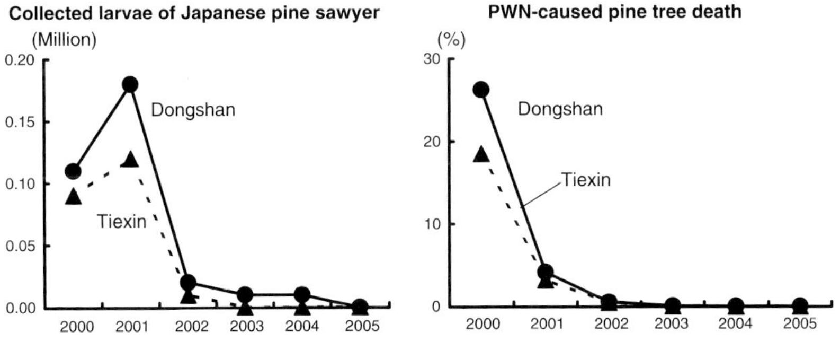

Fig. VI.7 Effect of collecting Japanese pine sawyer larvae over a 5-year period in an area after clear-cutting a stand of *Pinus massoniana* and *P. thunbergii* at the Tiexin forest; Dongshan forest farm in Jiangsu Province

Table VI.3 The intensive breeding of *Scleroderma guani* using *Monochamus alternatus* larvae as bait

Year	*M. alternatus* larvae collected (million)	Number of *S. guani* obtained (million)	Reproduce *S. guani* (times)	Increase times
2000	0.2	1.0	5.0	1.0
2001	0.3	5.0	16.7	3.34
2002	0.3	6.0	20.0	4.0
2003	0.3	10.0	33.3	6.66
2004	0.3	13.0	43.3	8.67
2005	0.3	13.0	43.3	8.67
2006	0.3	13.0	43.3	8.67

environmentally friendly way to use *M. alternatus* larva to intensively breed their natural enemy, *S. guani*. Recently, *S. guani* is being released to control *M. alternatus* larvae in forests affected by the PWN (Xu et al. 2008).

At the beginning of 2000, the laboratory could not produce the 10.0 million *S. guani* that were needed annually to control PWN in Jiangsu province. After many sterilization tests to prevent infection during breeding, intensive breeding was established and the resulting numbers of *S. guani* increased significantly. Between 2004 and 2006, 13 million female *S. guani* were produced annually using the same number (0.3 million) of *M. alternatus* larvae baites.

S. guani is produced from January to May (Table VI.3), and released during mid-July, so 13.0 million female adults were stored (in the year 2002–2007) in a cold room (at a temperature of 6°C ± 2°C) until release time.

33.5.2 Techniques for Releasing **Scleroderma guani** *and its Biocontrol Effect on* **Monochamus alternatus** *and PWN in a Large Area*

Based on numerous field tests on release density, timing and method, the "release from a single tree method" was selected. When 5,000 wasps were released using the "a single tree method" for each hectare in a test forest in mid-July the rate of parasitism of *M. alternatus* larvae was 66.82–84.21%. This demonstrated that *S. guani* provided good control of *M. alternatus*. Two generations of *S. guani* parasitize *M. alternatus* larvae in each year. After releasing the wasp, the first generation of *S. guani* parasitizes young *M. alternatus* larvae under the bark of dead pine trees in July, and the second generation of *S. guani* parasitizes third-instar larvae, which have bored into the wood of dead pine trees in late August (Xu and Xu 2002). On a 540 ha test area of pines, *S. guani* controled *M. alternatus* larvae effectively, and the percent of pine trees killed by PWN was reduced by more than 97.0% in the following year (Table VI.4). Based on these results we concluded that pine wilt disease could be effectively controlled (Xu and Xu 2002; Xu et al. 2008).

Table VI.4 Bio-control effect of *Scleroderma guani* on *Monochamus alternatus* on a large forest area

Release locations	Release area (ha)	Parasite effect on *M. alternatus* larvae		
		Number of larvae collected	Number of parasitized larvae	Parasitizum (%)
Dongjin FF-1	40.0	140	109	66.8
Dongjin FF-2	100.0	38	33	84.2
Dongjin FF-3	150.0	23	17	68.7
Dongjin FF-4	150.0	36	30	80.0
Control treatment	0	6	1	—

FF forest farm

33.6 Use of Resistant Pine Species and Resistant Masson Pine Provenances in Forest Clear-Cuts

33.6.1 The Selection of Resistant Pine Species and Resistant Masson Pine Provenances to the PWN

In a clear-cut area there are many, various-sized openings caused by the PWN. These areas need to be reforested so that they can resume their ecological function of protecting water and soil in mountainous areas. In Nanjing in 1995, following inoculation tests with PWNs, three resistant provenances were selected from 40 provenances of Massons pine, *P. massoniana*. This work focused only on the selection of resistant pine species and provenances to the PWN. For the field trial two resistant pine species, including *P. taeda*, and three resistant Masson pine provenances, GX2, GX3 and GD5 were outplanted into plantations in south Nanjing. Since 1984, the Jiangsu Provincial government has replanted 5,340 ha with *P. taeda* and 1,340 ha with resistant GX2 provenance of Masson pine. After 24 years, the replanted *P. taeda* and the GX2 provenances are still growing very well, and their selected resistance is stable and reliable (Table VI.5). All these areas, which were

Table VI.5 Growth and occurrence of resistant *Pinus taeda* and resistant Masson pine provenances planted in 1984 in a clear-cutting area

Pine species	Seedlings (million)	Plantation area (ha)	Tree growth high/ diameter (m cm^{-1})	Index of pathogenesis (IP[a])
P. taeda	5.0	5340	6.68 ± 0.68/14.58 ± 1.58	0
GX2 Masson pine provenance	2.0	1340	5.28 ± 0.39/10.68 ± 1.38	0

[a] $IP = \frac{\sum (Ni \times I)}{Nt \times I_{max}} \times 100$

Ni: Number of plants in disease grade I; I: Disease grade (one of range from 0–4); Nt: Total number of the treated plants; I_{max}: The highest disease grade (4 in the case)

previously damaged by the PWN are again fulfilling their role in soil and water conservation (Xu et al. 1999, 2004).

33.6.2 Host Preference for Maturation Feeding by **Monochamus alternatus** *on Various Pine Species and Masson Pine Provenances*

Eight conifers and 40 Masson pine provenances and three types of laboratory bioassays were used to determine the feeding consistency for, feeding areas and visitation frequency of the pine sawyer beetle (Li et al. 2004). Their work showed that *M. alternatus* adults have the highest frequency of feeding and prefer to feed on the branches of *P. massoniana* and *P. densiflora*, and demonstrated significant host selectivity on eight conifer species in an area near Nanjing. Adult feeding visitation frequencies were one to three adults per branch. The sizes of the maturation feeding areas on the branches were much larger on *P. massoniana* and *P. densiflora* than on *P. taeda*, *P. elliottii*, *Juniperus virginiana* and *Cunninghamia sinensis*. After initial selection tests, 12 Masson pine provenances were found to have smaller maturation feeding areas. The results of the second and the third selection tests showed that 12 provenances and 6 provenances also had consistent reductions in feeding by *M. alternatus*. The three Masson pine provenances (FJ 2, GD 5, GD 6) had the smallest maturation feeding areas. The reasons for this are now being studied. There was no correlation between the size of maturation feeding areas and resistance to the PWN in the 40 Masson pine provenances examined by Xu et al. (1999).

After clear-cutting, various sized openings were left in some areas of the forest. These areas need to be reforested with PWN resistant seedlings as the area is important in the conservation of water and soil.

33.7 Conclusions

After clear-cutting pine wilt disease killed pines, some dead pine branches containing *M. alternatus* larvae cannot be removed easily from the forest. We have collected more than 2.1 million larvae of *M. alternatus* from forests over 7 years (0.3 million larvae annually) and used them for intensive in vitro breeding of *S. guani*. The results showed that collecting *M. alternatus* larvae from the dead pine trees at the clear-cut area, and using them as the baits of their natural enemy, *S. guani* has proven to be an environmentally friendly way to control the pine wilt disease.

Several natural enemies of *M. alternatus* have been found near Nanjing and other areas in China. *Scleroderma guani* is one of the most successful natural enemies for controlling *M. alternatus* larvae and is easy to produce using *M. alternatus* larvae as bait. Both collecting *M. alternatus* larvae from dead trees and releasing *S. guani* as biological control agent are good ways to control *M. alternatus* and pine wilt disease in a large area because of the low risk to humans and the environment. The number of dead pine trees was also reduced significantly in the test area.

The clear-cut areas were replanted with selected, resistant pine species, such as *P. taeda,* and three resistant Masson pine provenances, GX2, GX3 and GD5. After 24 years the resistant pine species and a Masson pine provenance (GX2) have retained their resistance to the PWN. Consequently, these various control methods are being used against *M. alternatus* and pine wilt disease in Jiangsu Province.

34
Breeding for Resistance to Pine Wilt Disease

Mine Nose and Susumu Shiraishi*

34.1 Introduction

To cope with pine wilt disease the first resistance breeding program started in western Japan in 1978. In this program, resistant Japanese black pines (*Pinus thunbergii*) and Japanese red pines (*P. densiflora*) were selected. Subsequently, their progenies have come into wide use as resistant seedlings. Breeding programs are also carried out in the other parts of Japan and Anhui Province in China, and resistant pines, including Ryukyu pine (*P. luchuensis*) and Masson pine (*P. massoniana*), are also being selected. The demand for more resistant pines is increasing. In Kyushu region, second-generation breeding that creates more stable resistant cuttings has been carried out since 2004. Since resistance to the pine wood nematode (PWN) seems to be a polygenic trait, gathering genes in a single cultivar by crossing resistant clones would make resistance higher.

The studies and strategies of the breeding plans for pines' resistance to pine wilt disease will be explained in this chapter. This knowledge can be used to plan more efficient programs for other regions or other tree species in the future.

34.2 Strategies of Resistance Breeding

34.2.1 Breeding for Resistance in Forest Trees

In forest trees, breeding for resistance is one of the important methods to prevent damage from organisms such as pathogens and insect pests, and environmental stresses such as cold temperatures and drought. The resistant individuals are

Laboratory of Silviculture, Graduate School of Bioresource and Bioenvironmental Science, Kyushu University, 6-10-1 Hakozaki, Higashi-ku, Fukuoka 812-8581, Japan

*Tel.: +81-92-642-2872, Fax: +81-92-642-2872, e-mail: sushi@agr.kyushu-u.ac.jp

selected (selection breeding) or new cultivars are created by making crossings with resistant species or individuals (cross breeding). Those individuals or populations are examined for their resistance and then put to practical use. A key feature of tree breeding is that it has no direct effect on existing forests but contributes to future forests (Fujimoto 1991). Because of their long life cycle and large stature, time and effort are necessary to provide resistant trees for practical application. For these reasons, though referring to breeding strategies of short life agricultural crops is necessary, it is more important to work out specific strategies for tree breeding. Also, we need to keep in mind that while agricultural breeders have been working essentially at the single genotype level for both hosts and pathogens, tree breeders have been working at the population level (Carson and Carson 1989).

34.2.2 Breeding for Disease Resistance

Intraspecific crossing with resistant individuals and interspecific crossing with resistant allied species are effective strategies of breeding for resistance. Selecting resistant candidates from natural forests or plantations where the disease has been severe are also basic components of resistance breeding. As there may be resistant populations or individuals within a species, the surviving populations or individuals can be free of disease symptoms or less impacted by the disease. Clonal propagules or seedling progenies of selections are often tested by artificial inoculation to ascertain the level of resistance and the type of resistance responses. The inoculum may involve a mixture of pathogen genotypes or specific isolates, and the length of testing depends on the nature of the disease (Sniezko 2006). Field-testing is also a key component of resistance breeding since resistance to a given disease mainly depends upon both genotype and environmental interactions (Carson and Carson 1989). The selections are then evaluated as to whether or not resistance is expressed under a variety of field conditions and the stability of the resistance.

Accumulating not only a single resistance gene, but also different kinds of genes in a population or an individual may provide long-term utility (Namkoong 1991; Burdon 2001; Winter et al. 2005). In the case of plants in which QTL (quantitative trait loci) analysis has already made progress, MAS (marker-assisted selection) can be an effective tool (Lande and Thompson 1990; Xie and Xu 1998). We need to know that inbreeding and loss of genetic diversity within a population may reduce not only productive fitness and evolutionary potential, but also resistance to the disease. In pines, the number of seed parents in an open-pollinated orchard cannot be reduced to less than about 15 without risking an unacceptable amount of inbreeding and decline of genetic diversity (Carson and Carson 1989).

Some examples of successful resistance breeding programs include: larch (*Larix kaempferi*) to needle cast (*Mycosphaerella larici-leptolepis*), chestnut (*Castanea crenata*) to chestnut gall wasp (*Dryocosmus kuriphilus*), Japanese cedar (*Cryptomeria japonica*) to criptomeria bark midge (*Reeseliella odai*), Japanese black pine to pine needle gall midge (*Thecodiplosis japonensis*), Western white pine (*P. monticola*) to white pine blister rust (*Cronartium ribicola*), Port-Orford-ceder (*Chamaecyparis lawsoniana*) to root rot (*Phytophthora lateralis*), American elm (*Ulmus americana*) to Dutch elm disease (*Ophiostoma ulmi* & *O. novo-ulmi*), American chestnut (*Castanea dentata*) and to chestnut blight (*Cryphonectria parasitica*).

34.3 Pine Wood Nematode Resistance Breeding

In Asia, pines provide wood and pulp and as well pine species also play important roles in spiritual and cultural aspects. Since the PWN invaded Japan from North America in the early twentieth century (de Guiran and Bruguier 1989), natural forests and plantations of Japanese black and red pines have been severely damaged by the disease. The damage has expanded during the last five decades, and the volume of damaged trees surpassed 2 million m^3 in 1978. According to the record of the Japanese Forestry Agency, its peak was in 1979 when wood stock of 2.43 million m^3 was damaged. To address this situation, a PWN-resistant breeding plan has been carried out. Two methods of resistance breeding have been used: selecting highly resistant individuals within the species, and creating resistant trees by making crossings within and between species.

34.3.1 Selection Breeding

Three mechanisms, avoidance of the vector, antibiosis and tolerance to the pathogen, are important components in resistance to plant pathogens. For pine wilt disease, since there were no differences in the preference of vector beetles (*Monochamus alternatus*) to pine species, avoidance of the vector was not of necessity considered in the breeding strategy (Toda 2004). Surviving trees were selected from stands heavily damaged by the disease, and grafts were used to evaluate their resistance by artificial inoculation (Fig. VI.8). Before starting the plans, three fundamental technical studies on resistance breeding for the pine wilt disease were made in the western part of Japan.

34.3.1.1 Fundamental Research

1. Survey of PWN Isolates for Use in the Artificial Inoculation Testing

For the resistance breeding program, it was necessary to survey PWN isolates to determine their pathogenicity for use in the artificial inoculation tests. The PWN used for inoculum must have stable virulence, high reproducibility, and low host

Fig. VI.8 Inoculation of Japanese black pine to determine resistance to the pine wood nematode (see Color Plates)

preference. Twenty-two PWN isolates and an isolate of a closely related nematode (*B. mucronatus*) were collected from damaged pine trees in various areas of Japan. These isolates had different reproduction rates. They were injected into pine seedlings and to evaluate their virulence. The results showed that the virulence of the isolates varied, causing 0–97.3% pine seedling mortality (Ibaraki et al. 1978). It was also found that there was no host preference in PWNs. As a result, a highly virulent "Shimabara" isolate with a high reproduction rate was chosen for use in the inoculations. Although Kiyohara (1976) reported a decrease in virulence due to in vitro subculture, this did not occur for the "Shimabara" isolate (Toda 2004).

2. Pine Seedlings for the Artificial Inoculation Tests

In some pine species, because cutting propagation is difficult, plantlets have been raised from seeds or grafts; however, there are some problems with using seedlings for inoculation tests. Many seedlings are needed for accurate examination since their genotypes vary due to high heterozygosis in pine species, which is an allogamous (reproducing by cross-fertilization) plant. The selected resistant candidate is required to have high flowering ability. For these reasons, grafts were used in the selecting breeding program. By using graftings, inoculation tests can be done with fewer plantlets since they have the same genotype as their mother trees. The level of resistance to pine wilt disease varies greatly among pine species (Kiyohara and

Tokushige 1971). Japanese black pine is often used as the root stock because of its good rooting ability and growth. Though the scion and stock of a graft consist of different species, Toda (2004) reported that the resistance level of the scion was not influenced by that of the stock. Inoculation of 2- to 60-year-old trees with PWNs showed that the trees were damaged by pine wilt disease regardless of their age (Toda 1997). However, other reports (Mamiya 1975d, 1980) have suggested nematode invasion might change with the growth of pine trees. The PWN migrates mainly via cortical resin canals in the branches. Since the canals are generally not present in stem tissue of trees over 4 years old, nematode invasion in mature pine trees might differ from that in young seedlings (Ichihara et al. 2000b).

3. A PWN Artificial Inoculation Method (see also the column in Chap. 21)

Development of an inoculation test to evaluate the resistance level of pine trees made it possible to carry out resistance breeding for pine wilt disease. A very simple and accurate inoculation method with low damage to pine tree vigor was essential for treating mass numbers of samples (Nishimura et al. 1977; Tsuda et al. 1978). Many different methods were tried, and "a modified peeling bark method" was chosen (Fujimoto et al. 1989). In this method, the bark on a 1–1.5 cm (width) by 3–5 cm (length) section on the main, current year shoot is cut and peeled back and the slit is scratched with a saw to make it rough, and a PWN suspension is pipetted into the slit using a micropipette (Fig. VI.9). The number of nematodes in the inoculum affects the survival rate of the pine seedlings. The more PWNs, the higher the pine wilt rate (Kiyohara et al. 1973). At least 6,000 nematodes are required for uniform dispersal of PWNs inside a tree (Hashimoto et al. 1976). However, pine

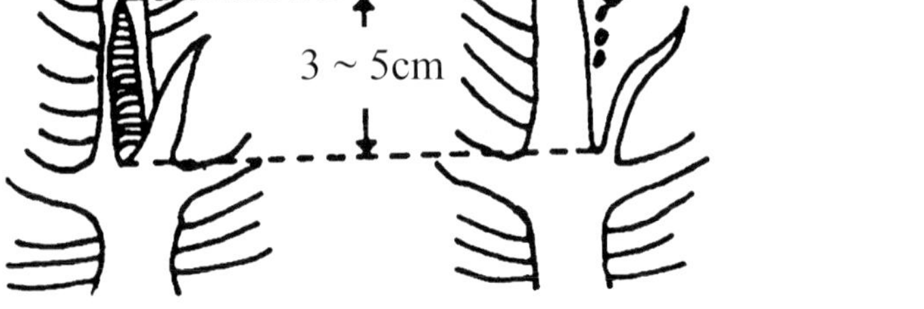

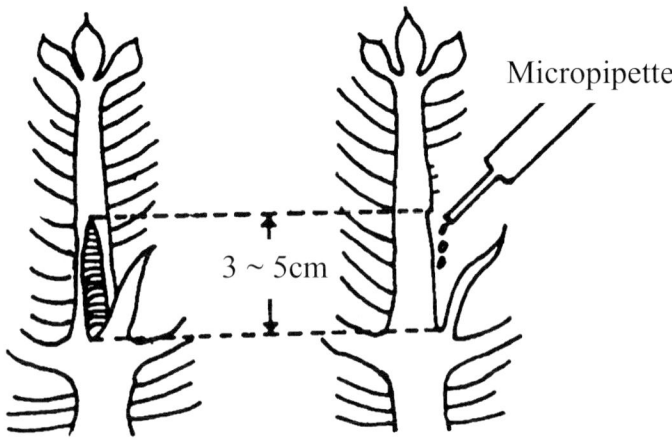

Fig. VI.9 Bark peeling inoculation method for pine wood nematode inoculations (from Fujimoto et al. 1989, with permission)

mortality is not influenced by the number of inoculation sites (Yamate and Okubo 1977). According to these results and considering the loss of nematodes prior to host invasion because of resin exudation and drying, 10,000 nematodes are injected into an inoculation site on a tree for most of resistance breeding projects.

34.3.1.2 Three Selection Breeding Plans in Japan

The first resistance breeding program against the PWN was carried out from 1978 onward in western Japan. Later, a similar breeding plan started in 1992, since major pine wilt damage increased also in the Tohoku region, the northern part of Japan. Many pine forests have been damaged in China since the 1980s (Sun 1982; Enda and Taketani 1992), and a resistance breeding program against the PWN has been carried out in Anhui Province in cooperation with Japan since 2001. The strategy based on selection breeding was adopted in these three plans (Fig. VI.10).

1. Breeding Plan for Resistance to PWN in Western Japan

This work was started in 1978 in cooperation with the national forest tree breeding institute and 14 prefectural research organizations. Firstly, resistant candidate trees were selected. Generally, disease-resistant individuals are estimated to exist in 1×10^{-3} to 1×10^{-4}% in natural pine forests (Ohba 1976). Many resistant candidates were selected from severely damaged stands since resistant individuals may exist more frequently there than in non- or only slightly damaged stands. After further consideration of tree growth and stem straightness, survivors older than 30 years in the stands, where more than 90% of trees were damaged, were selected as candidates. Japanese black pines (14,620) and Japanese red pines (11,446) were selected and their ramets were propagated by grafting (Toda 2004).

The first inoculation test to Japanese black and red pine was done from 1980 to 1982. Ten grafts of each candidate were inoculated with 10,000 "Shimabara" isolate PWNs. The candidates which showed survival (the total of non and partial damaged trees) and sound (no damage) rates equal to or higher than those for Loblolly pine (*P. taeda*) seedlings, which were used as resistant controls, were examined to reconfirm their resistivity in the second round of the tests done from 1982 to 1984. For Japanese red pine, the acceptable survival rate for the second inoculation test was the same as that of the first. For Japanese black pine, however, 60% of the Loblolly pine survival rate was adapted as the lower acceptable limit since the resistant level of Japanese black pine was lower than that for Japanese red pine. Base on the results of these two inoculation tests, 16 Japanese black pines and 92 Japanese red pines were finally selected for further use (Research and Extension Division, Forestry Agency 1994b).

After the first breeding plan, an additional selection of resistant Japanese black pine was carried out from 1995 onward in the Kyushu region. Surviving mother trees were surveyed in heavily damaged forests and open pollinated seeds were

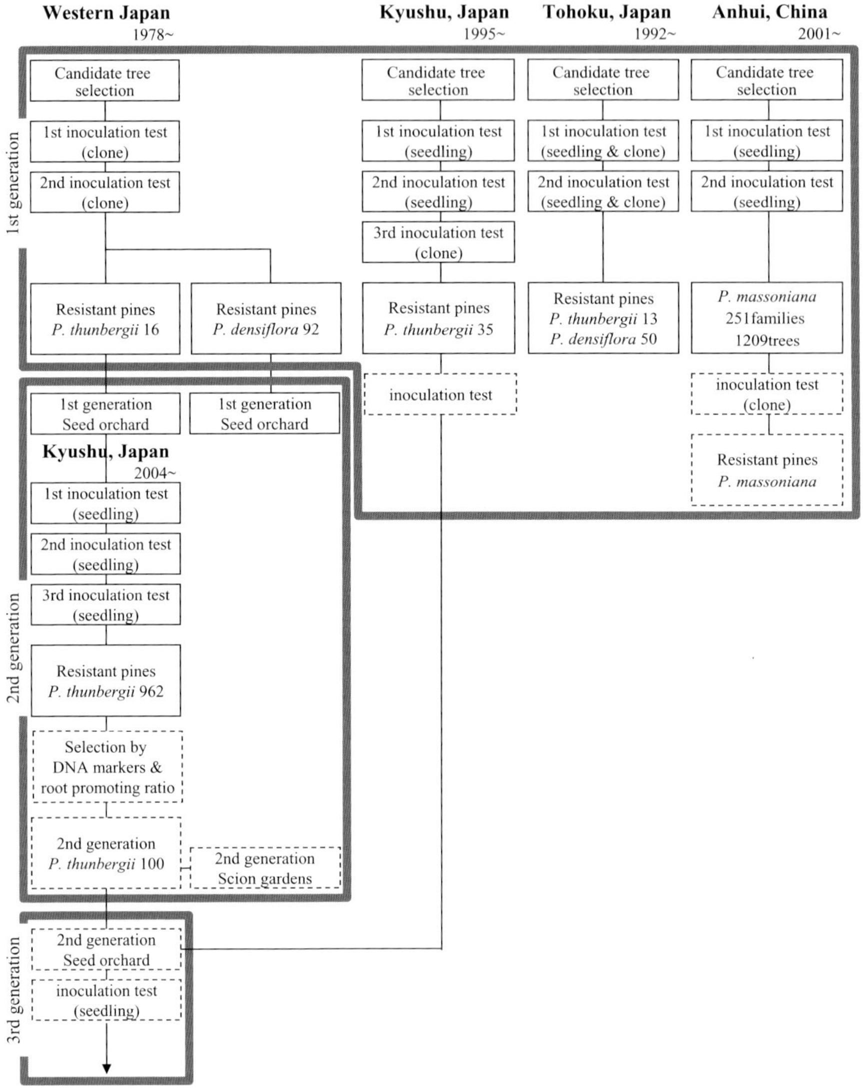

Fig. VI.10 Breeding plans to obtain pine wood nematode resistant pines in Japan and China. The actions with the solid line quadrilateral have been completed, the *dotted line* indicates that the plan put in practice or has not yet started. The figure in the quadrilateral shows the number of plus trees or candidates that have been selected for these plans

collected from the stands. The seedlings originating from the seeds were screened twice by artificial inoculation with the "Shimabara" PWN isolate (Okamura 2004). Thirty-five clones were registered as being resistant to the PWN.

2. Another Resistance Breeding Plan in Japan

As PWN damage expanded to northeastern Japan, a resistant breeding plan was started in 1992 in the Tohoku and other regions, cooperating with the national tree breeding institution and nine prefectures. Grafts or seedlings from candidate trees were used for the inoculation tests. The advantages of selections originating from seedlings are the short growing period, high seed production rate of selected seedlings (Toda 2006), and the possibility of a higher resistance level than that of the mother tree (Higashihara 2004).

In this program "the cutting-main-stem method (cut the main stem, crush it and inoculate the nematodes on the crushed area)" was used as an inoculation method in addition to "the peeling-bark method". Inoculation tests were done from mid-June to the beginning of July, since trees inoculated after the mid-July do not show pine wilt symptoms during the current year in cold districts such as the Tohoku region. The "Shimabara" and more virulent PWN isolates were used as the inoculum, and 10,000 nematodes per tree were used as the inoculum dose. Symptom development was observed for 10 weeks following inoculation (Research and Extension Division, Forestry Agency 1994a).

Instead of Loblolly pine, which is difficult to grow in cold areas, plus trees of Japanese red pine with resistance to PWN, selected in the Tohoku breeding region, were used as a reference of resistance. Survivors of two inoculation tests were assessed for resistance. As pine trees are needed which possess both resistance to pine wilt disease and are cold hearty in these districts, additional resistance breeding will be carried out during the 2006–2010 period. As of April 2007, there were 13 and 50 resistant clones of Japanese black pine and Japanese red pine, respectively, in the northern Kanto and districts along the Sea of Japan. However, the number of resistant plus trees is still not enough.

3. The Resistance Breeding Plan for PWN in Anhui Province, China

In China, the first damage from pine wilt disease was recognized at Nanjing in 1982 (Sun 1982), and the disease had spread to 14 provinces by 2004 (Toda 2006). Initially, only Japanese black pine was damaged, but by now Masson pine (*P. massoniana*) has been damaged. The forest area covers 3.32 million hectares in Anhui Province, and 42% is mostly Masson pine. The "Resistance Breeding plan for Pine Wood Nematode in Anhui Province" was carried out by Japan-China technical cooperation from 2001 to 2006. A native PWN isolate with stable virulence and high pathogenicity was used for the inoculation tests. Nineteen PWN isolates were collected from 11 forest stands in Anhui Province and their virulence was evaluated by inoculating pine seedlings. The "KS3B" PWN isolate was used for the inoculation tests, since it met the above conditions the best (Cai et al. 2003; Gao et al. 2003).

The goal of the plan is to select 200 resistant trees. Two hundred and eighty-eight mother trees were chosen from survivors in 31 heavily damaged stands. Seedlings from those trees were evaluated for their resistance by the first inoculation test and 20% of them remained healthy. The second inoculation test was carried out on the surviving seedlings the next year, and 1,209 seedlings of 251 families were finally selected (Toda 2006). To confirm that they show enough resistance for practical use, they will be propagated by grafting and used for in further inoculation tests (Cai et al. 2006).

34.3.2 Cross Breeding

Cross breeding is a method to create new varieties with desirable characteristics by repeated crossing and backcrossing. In crop breeding, inbred lines are usually created by repeatedly selfing and inbreeding, and heterosis (hybrid vigour) is created by crossing between those inbred lines. In forest tree breeding, however, since it is difficult to create inbred lines, F_1 generations are created by interspecific or intraspecific crossing and put to practical use directly, so the selection of parent species/genotypes must be paid the closest attention (Ohba 1991).

There are some reports on cross breeding for resistance to pine wilt disease. The resistance level varies greatly among and within pine species. Some exotic pine species, such as Loblolly pine, Jack pine (*P. banksiana*), Eastern White pine (*P. strobus*), and Table Mountain pine (*P. pungens*) show high resistance to pine wilt disease (Kiyohara and Tokushige 1971). Interspecific crossing with those resistant species, or intraspecific crossing among resistant individuals may create higher resistance. The high cross-affinity between species, the ability to improve, and adaptability to the growing environment are required for successful cross breeding.

34.3.2.1 Interspecific Hybrid

In Japan, two major pine species, Japanese black pine and Japanese red pine, because of their economic and cultural significance, were selected for use in the national program. There are 19 species within subsection *Sylvestres.* in the world including the two species mentioned above. Crossing the two Japanese pines with other 14 *Sylvestres* species was tried, and all 28 combinations resulted in hybrids (Furukoshi and Sasaki 1979). These F_1 hybrids were used for inoculation tests, and it became clear that the resistance level varied within *Sylvestres* species. Japanese red pine × Dwarf mountain pine (*P. mugo*), Japanese red pine × Corsican pine (*P. nigra* var. *corsicana*), Japanese black pine × Masson pine, Japanese black pine ×

Chinese pine (*P. tabulaeformis*), and Japanese black pine × Corsican pine showed higher resistance than the hybrid between Loblolly pine and Pitch pine (*P. rigida*), which is highly resistant. Moreover, feeding preference tests indicated no significant difference in the preference of the vector among the hybrids (Furukoshi and Sasaki 1979, 1983, 1985; Sasaki et al. 1982a,b).

1. Japanese Black Pine × Masson Pine

The hybrid between Japanese black pine and Masson pine showed higher resistance, higher fertility and vigorous growth among the above mentioned 28 hybrids. For these reasons, this hybrid was practically supplied to western Japan in the "Breeding plan on resistance to the pine-wood nematode" from 1983 to 1988 (Sasaki et al. 1982b). According to reports on interspecific hybrids, it is estimated that Japanese black pine has a single or a few recessive susceptible genes and Masson pine has a dominant resistance gene almost in homozygosis (Sasaki et al. 1983a,b).

2. Loblolly Pine × Pitch Pine

The hybrid between Loblolly pine and Pitch pine showed higher resistance than Japanese black pine and Japanese red pine (Ohba et al. 1977), and it is well known for its greater cold tolerance and higher ability to sprout than that of Loblolly pine (Fukuda and Iwakawa 1979); however, this hybrid was not used for plantations, because its adaptability to Japanese environmental conditions are still unknown.

3. Japanese Black Pine × Japanese Red Pine

In crossing Japanese black pines as the mother tree with Japanese red pine as the father (pollen) tree, the following observations were noted. When both parents are resistant, the resistance level of their hybrid lines was improved. Also, when either the mother tree or the pollen tree is resistant, the resistance level of the progeny was improved except for combinations of some mother trees. Moreover, selfing reduces the resistance level, and there seems to be selfing depression in resistance to the PWN (Toda 1996, 1997; Toda et al. 1997).

34.3.2.2 Intraspecific Crossing

1. Japanese Red Pine

According to the crossing test using resistant and non-resistant Japanese red pines, the survival rate of the hybrid between resistant × resistant was high in every pair (survival rate: 82.5–100%), while resistant × non-resistant varied from 25.0 to 95.0%. The specific combining ability was reported by Handa et al. (1995).

2. Japanese Black Pine

In resistance breeding in the Tohoku region, the pollen of resistant Japanese black pine selected in Western Japan was used for crossings made with pines native to Tohoku, and resistance candidates were selected from these seedlings (Toda and Terada 2001). In Kyushu, the resistance of seedlings of seven clonal seed orchards is influenced by the clone layout, and the progenies of mother trees surrounded with highly resistant clones show stable resistance (Miyahara et al. 2005). As a result of the statistical genetic analysis of three Japanese black pine families created by controlled pollination, the heritability of broad sense (the proportion of total genetic variance to phenotypic variance) and narrow sense (the proportion of additive genetic variance to phenotypic variance) were 0.985 and 0.623, respectively (Kuramoto et al. 2007). This high heritability suggests that accumulating resistant genes by repetitive crossing is an effective breeding strategy to improve the resistance to the PWN.

34.3.3 Characteristics of Resistant Pine Clones

Sixteen, Japanese black pine clones and 92 Japanese red pine clones were selected in the previous resistance breeding plan in western Japan. Thirty-four resistant seed orchards have been established using these resistant clones in 21 prefectures, mainly in Western Japan. For effective management of the clonal orchards, their various characteristics such as flowering and seed production were surveyed. Japanese black pines with genetically improved PWN resistance grow better than non-improved pines. This may be because growth and stem straightness were also taken into adequate consideration when the selections were taken from heavily damaged pine populations (Toda et al. 1993).

Resistant pines begin to flower when they are 2 years old, and all clones flower until 4 years old in Japanese red pine, while 80% of the Japanese black pine clones flower by 6 years old, but no flowering was observed in two clones until 1992. The number of seeds per cone, 14.7 in Japanese red pine and 5.8 in Japanese black pine, is lower than in general pine seed orchards (Toda et al. 1993). A shortage of pollen in the resistance seed orchards is a problem in Japanese black pine. In seed orchards consisting of resistant clones, pollen contamination from outside of orchards was investigated by using DNA molecular markers (Goto et al. 2002a,b), and its influence on resistance depression determined (Handa et al. 1995).

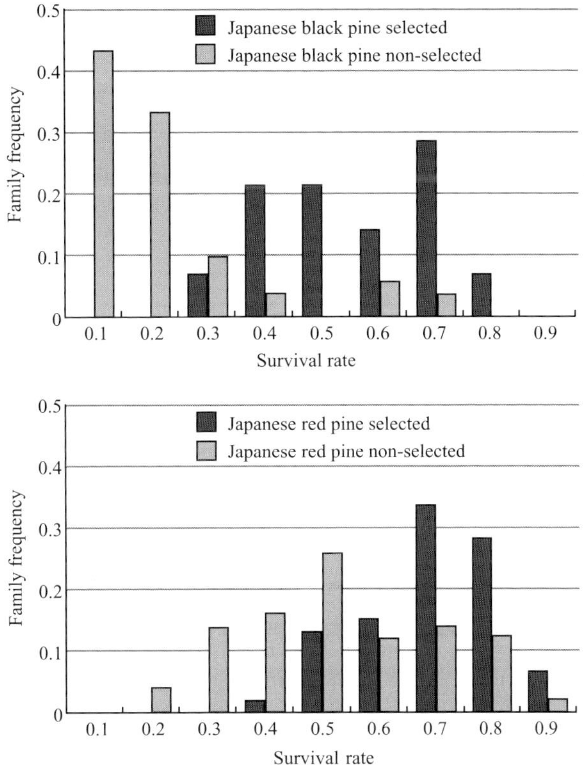

Fig. VI.11 Nursery survival rate of Japanese black pine and Japanese red pine families three months after inoculation with the pine wood nematode (from Toda and Kurinobu 2002, with permission)

Genetic gains in the resistance breeding program in Western Japan were estimated by analyzing the 10-year data of the artificial inoculation, PWN tests. The survival rates of the improved population were higher than the non-improved population (Fig. VI.11). In the improved population the rates were 0.514 in Japanese black pine and 0.650 in Japanese red pine, while in the non-improved population they were 0.166 and 0.473, respectively. The realized genetic gain was 0.348 and 0.177, respectively. The larger value in Japanese black pine may be because the susceptibility of the non-improved population is higher in this species (Toda and Kurinobu 2002). To confirm the improvement in resistance in the field, progeny test plantations were established in two places. Feeding marks made by pine sawyers were detected after 3–4 years, and about 50% of trees were attacked by sawyers by the fourth year. Although these pines were inoculated with the PWN by the pine sawyers, their survival rates were still high. Ninety percent of pines survived in both plantations for 7 years. This indicates that resistant pines selected by artificial inoculation test would also show high resistance in the field (Toda 2004).

34.4 Progress Toward Greater Resistance

34.4.1 Present Resistant Seedling Production System

Japanese black pine has played an important role in protection against wind and to control soil erosion, since it is well adapted to conditions along the coast and in the low mountains running along the coastline (Toda and Terada 2001). For these reasons, many resistant Japanese black pines have been planted in the Kyushu region, which has a long coastline. Since the resistance level differs within seedlings (second-generation) obtained from first-generation seed orchards, artificial inoculation tests are carried out to eliminate susceptible individuals before planting them as resistant seedlings. As pine wilt disease still damages pine forests, about 100,000 seedlings are produced every year in the Kyushu region.

There are some problems in the present resistant seedling supply system. Firstly in this decade, only 34% of the seedlings on average passed the inoculation test. This means that more than 60% of the cost and labor to raise the seedlings is useless. Human errors are likely to occur since the inoculation tests are done in mid-summer under the scorching sun. Also, since the inoculation test using the peeling bark method makes the main stem of seedlings thin, it is easily fractured by strong winds such as typhoons. The results of the test are thought to be influenced by weather, and the pass rate in inoculation tests goes down in high temperature and low precipitation summers (Tobase 2003; Toda 1997). Furthermore, the production of seeds changes annually. For these reasons, the costly production system that requires so much time and labor makes resistant seedlings expensive. While a normal 2-year-old pine seedling costs 50–60 yen, a resistant seedling costs about 670 yen, or about 10 times more.

34.4.2 A New Production System for Second-Generation Pine Cuttings Resistant to Pine Wilt

To solve the problems mentioned above, the development of a new production system of second-generation pines resistant to PWN started in Kyushu region in 2004. The goals of this program are to improve the resistance of Japanese black pine by gathering resistance genes, and to stabilize the supply of clonal plantlets, which contain the same genes using cutting techniques.

The number of resistance genes, resistance factors and mechanisms have not been clarified yet. Since there is variation in the progenics' resistance level, it is assumed that several genes take part in resistance. Since the multiplication of the PWN in pine branches was much different among the 16 resistant clones, it seems there are some resistance factors among them (Nose et al., data not shown).

34.4.2.1 Fundamental Research

In the new breeding plan, the most virulent PWN isolate was inoculated to select highly resistant trees from seedlings obtained from first-generation seed orchards. Of these 962 individuals survived, and about 100 clones will be selected, considering their genetic diversity and rooting ability. Plantlets of these clones, which possess highly resistance genes, will be propagated by cutting and supplied to foresters for planting. To carry out this work, three fundamental research areas were identified.

1. Isolation of the PWN from Kyushu and Preparation of the PWN Population for Inoculation Tests

In advanced breeding, the creation of new PWN populations is needed for the inoculation tests. This must be a highly virulent population with high genetic diversity and low host preference. Considering those requirements, a "Highly virulent PWN population" was made by mixing six virulent isolates chosen from different strains.

2. A New Technique to Propagate Cuttings

It is said that Japanese pine species are difficult to propagate by cutting because of their poor rooting ability, but using juvenile stools and optimizing growth conditions such as the composition of cutting beds and auxin treatment, more than 60% of the cuttings successfully rooted. By pruning the scion trees, the number of branches increased. Consequently, about 40% of the cost can be reduced compared to the present resistant seedling production system. By using this new technique, there is no decline in the resistance level by contaminant pollen coming from outside the seed orchards and no annual changes in the seed yield; therefore, it will be possible to stabilize the supply of high resistant plantlets (Ohira et al. 2007).

3. Pedigree Management of Resistant Trees Using DNA Markers

By using genetic markers to determine the genetic relationship among resistant trees, the genetic bias in the breeding population might be reduced. Especially in tree species of large stature or with a long generation time, reducing the breeding period and labor by introducing a family management system using DNA markers is extremely needed (Paterson et al. 1991). Actually, DNA markers are used in hybrid Eucalyptus (*Eucalyptus grandis* × *E. urophylla*) seed orchard management. Parents with a low reproduction rate were culled from orchards, and management procedures were adopted to minimize external pollen contamination (Grattapaglia et al. 2004).

There are two roles in genetic management with DNA markers: One is to select the next generation without reducing the genetic diversity and creating the genetic bias in the breeding population. Every clone of a seed orchard has to contribute equally to the selected next generation. For this purpose, parentage diagnosis is done using DNA markers. Another role is the clonal management of planted trees to estimate the actual in situ adaptability of each clone. The most suitable clone for each situation could be revealed by measuring the growth rate and pattern, and so on. Usually, in progeny test plantations all the planted trees are labeled to keep track of their clone or pedigree, but this is not necessary owing to DNA markers (Shiraishi 1996, 1997).

Two analysis systems, a multiplex SSR (simple sequence repeat, microsatellite) for parentage diagnosis and MuPS markers [multiplex-PCR of SCAR (sequence characterized amplified region) markers; Hisaeda et al. 2003] for clone identification, were developed in this program. SSR has been detected universally in the genomes of every organism analyzed to date (Tautz 1989; Tautz and Renz 1984), and is suitable for parentage diagnosis, which is a powerful family management tool, while MuPS is an effective tool for retrospective analysis of clones in progeny tests and clonal management of scion gardens and seed orchards.

34.5 Future of Resistance Breeding Against Pine Wilt Disease

34.5.1 Durability of Resistance

It is always unclear whether the resistance gained by breeding provides long-term usefulness. The effect of resistance breeding might often break down because of genetic shifts in the pathogen population (Burdon 2001). In rice, for example, the virulence of a brown plant hopper (*Nilaparvata lugens*) strain has continued to become stronger between 1988 and 1990, and the insect has attacked *indica* rice varieties carrying the resistant *Bph1* gene (Tanaka and Matsumura 2000). In Sugar pine (*P. lambertiana*), resistance to white blister rust that is attributed to a single dominant gene has also been overcome (Kinloch and Comstock 1981). This resulted from breeding practices based on ignorance of the resistance mechanism and the genetic variation of hosts (Burdon 2001). For long-lived organisms such as pines, it is desirable to maintain genetically diverse populations and to combine different types of resistance (resistance mechanisms) for long-lasting or durable resistance.

It is important for resistance breeding to investigate the strain of regional PWNs and their virulence. It is reported that the virulence level of the PWN varies (Kiyohara and Bolla 1990), but presently it is said that there are no geographical differences in their pathogenicity in Japan (Akiba 2006). Based on studies done in Kyushu, it has become clear that PWN strains differ among regions. There are some reports that pathogenicity of the PWN varies in North America and shows different host preference (Wingfield et al. 1983; Bolla et al. 1986, 1988).

Damage from pine wilt disease is not serious in North America, where the PWN originated, because the host and pathogen have coevolved for a long time. Since resistance breeding is an action that promotes co-evolution, it might allow the PWN and pines to coexist in countries where the PWN have invaded. Resistant trees will be expected to serve as the sources of future generations of trees and to co-evolve with the pathogen (Sniezko 2006).

34.5.2 Resistance Breeding in Future

Selection of the first generation of PWN-resistant pines is underway in the Tohoku region, Japan and Anhui province, China, and selection of the second generation and development of a resistant cutting supply system is ongoing in the Kyushu region, Japan. Greater resistance is expected by accumulating several types of resistance genes. In future generations, that is, the third and fourth generations and beyond, the accumulation of resistance genes will develop, and the level and endurance of resistance will improve greatly. To assure great improvement in future generations, the genetic diversity of the original population is extremely important. In Japan, the Seed and Seedling Law limits free movement of seeds and seedlings from the point of view of environmental adaptation. There are, by law, two appropriate seed and seedling zones for Japanese black pine and three for Japanese red pine (Fig. VI.12). For this reason, resistant pine trees can only be planted in limited areas based on their place of origin. Therefore, the number of resistant trees is not enough to maintain genetic diversity in some areas. Further efforts are required for the additional selection of resistant pines and their occasional inclusion into the breeding population to minimize inbreeding and loss of genetic diversity. The concentration and diversity of PWN-resistant genes in the pine populations are key ingredients for stabilizing pine forests against the disease.

Selection-assisted DNA markers, which are valuable for the reduction of scale and time of breeding, are essential tool in the breeding of most tree species (Peterson et al. 1991). Recently, DNA markers for Japanese black pine and Japanese red pine have been developed (Lian et al. 2000; Watanabe et al. 2006b) and used for estimating the outcross rate and paternity analysis (Lian et al. 2001). Moreover, QTL analysis of the resistance to the PWN will be made by using a linkage map constructed with DNA markers (Isoda et al. 2007). Attempts are also being made to identify resistance genes by gene-expression profiling such as EST (expressed sequence tag) (Watanabe et al. 2006a, 2007). PWN-resistant breeding will make great progress by using highly efficient selection using MAS. There are some reports that efficiency of artificial selection can be increased substantially using MAS or multiple-trait MAS (Lande and Thompson 1990; Xie and Xu 1998).

Although resistance of a selected population is certainly higher than that of a non-selected population, we must recognize that it is not absolute. The resistant trees might still be damaged by pine wilt disease when the environmental conditions are less than ideal for the pines. The resistance of trees is influenced by many

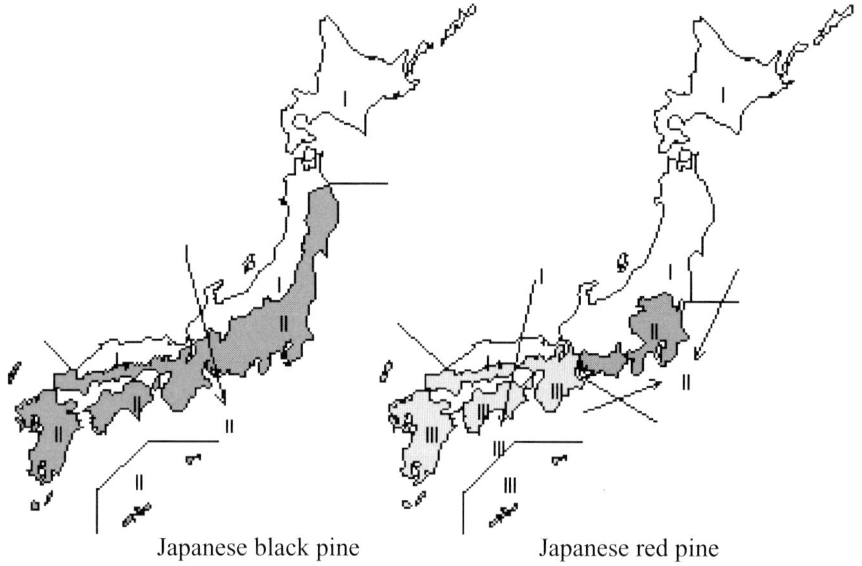

Fig. VI.12 Distributable area of seeds and plantlets of for Japanese black pine and Japanese red pine. *I–III* indicate distributable area; arrows indicate direction permitted for genetic material transfer

factors, such as environmental conditions, tree age, virulence and pathogenicity of nematodes, and the population density of the vector beetles. To maintain pine forests, it is essential to keep the forests healthy by integrated PWN control such as removal of damaged trees and aerial insecticide spraying plus the use of resistant trees (Taoda 1996; Yoshida 2005, 2006).

35 Biological Control of the Japanese Pine Sawyer Beetle, *Monochamus alternatus*

Mitsuaki Shimazu

35.1 Biological Control and Control Agents

Chemicals are traditionally used to prevent the spread of pine wilt disease, but these chemicals may pose environmental and human health hazards and as such it is necessary to study other possible controls. "Biological control" is a method of controlling or managing pests by using natural organisms such as predators, parasites or other natural mechanisms, for example, the use of natural enemies to lower the pest insect population, mating disruption with pheromones, and inoculation of avirulent viruses to protect against virulent viruses. This is the broad definition of biological control; however when the control agent is not a living organism itself, but a derivative of a living organism the term biorational control is used.

A "natural enemy" is defined as any organism located higher in the food chain than another organism and functioning as a lethal factor. Natural enemies of insects are divided into three categories: predators, parasites, and pathogens based on their pattern of aggression. Parasites as natural enemies eventually kill their host as if they are predators, and are called "parasitoids", while those not acting as lethal factors, as is the case of ascaris against humans, are not regarded as natural enemies. For that reason, most pathogens of insects can be classed as natural enemies because they are lethal to insects, while pathogens of mammals are not always natural enemies because mammals often recover from microbial infections. Various microorganisms such as viruses, bacteria, fungi, protozoan, nematodes, are known as natural enemies of insects: as are snails, insects, mites, spiders, amphibians, reptiles, birds, and mammals, but their use is limited for the biological control of *Monochamus alternatus*.

Insect Management Laboratory, Forestry and Forest Products Research Institute, 1 Matsunosato, Tsukuba 305-8687, Japan

Tel.: +81-29-829-8254, +81-29-829-8337, Fax: +81-29-874-3720, e-mail: shimazu@ffpri.affrc.go.jp

35.2 Vertebrate as Natural Enemies of *Monochamus alternatus*

Woodpeckers are important predators of wood-boring insects such as cerambycids. Yui et al. (1993) reported the predation of *M. alternatus* larvae by woodpeckers. All species of woodpeckers whose distribution coincides with that of *M. alternatus*, such as *Dendrocopos major*, *D. kizuki*, *Picus awokera*, and *D. leucotos*, prey on *M. alternatus* larvae. Of these species, *D. major* preys on *M. alternatus* more so than do the others, since it is morphologically well adapted for preying on *M. alternatus* larvae in pine wood. The predation rate of *M. alternatus* larvae in wood was higher when the numbers of *D. major* were higher than when it was less abundant (Yui et al. 1993). *Dendrocopos major* is an important natural enemy of *M. alternatus* because it consumes larvae beneath the bark on dead pine branches and trunks.

The predation effect of *D. major* on *M. alternatus* can be increased by attracting it to the target forest. A broadleaf forest or mixed forest with large broadleaved trees is ideal for attracting this bird, but as an emergency measure installation of nest boxes is effective. Yui et al. (1985) set up hardwood logs, with an entrance hole, in a pine forest and induced *D. major* to utilize the logs for breeding. They also developed a roosting nest box, which attracted *D. major* while preventing other birds, besides woodpeckers, from using them. This box had no bottom and had notches inside as steps, which allowed the woodpeckers to escape attack by snakes. When the roosting nest boxes were installed in pine forests, they were well used by *D. major* (Fig. VI.13).

Yui et al. (1993) reported that *D. major* consumes up to 64 *M. alternatus* larvae per day, and when the woodpecker numbers reach 1/5 ha^{-1}, and the incidence rate of pine wilt disease is less than 1%, the bird can kill 90% of the *M. alternatus* larvae. Predation by *D. major* is less effective in heavily infested pine forests such as in West Japan. The conclusion from such studies is that attraction of *D. major* with nest boxes will be effective only in areas where pine wilt incidence is low.

Other birds also prey on cerambycid beetles, but there are few definite records of them killing *M. alternatus*. Such birds only caught adult cerambycid when the insects were not hidden. As such, their importance as natural enemies of *M. alternatus* is low (Yui 1980).

Reptiles also prey on insects and pine wilt damage in the Ogasawara (Bonin) Islands of Japan has decreased rapidly, which is attributed to reptiles. Okochi and Kagaya-Shoda (2005) reported that the green anole, *Anolis carolinensis* (Reptilia: Squamata: Iguanidae) played an important role as a predator of *M. alternatus*. This is a very rare case of pine wilt control without the intervention of artificial control measures; however, the green anole is an invasive species and causes a serious undesirable impact on other rare insects in Ogasawara. Although it is a good example of natural control of pine wilt, the green anole should not be introduced into in other areas for biological control.

Fig. VI.13 The roosting nestbox for attraction of the woodpecker *Dendrocopos major* (Photo: courtesy of Mitsuhiro Nakamura)

35.3 Insect Parasitoids and Predators, and Other Small Animals

Nobuchi (1980) reviewed the Japanese literature on the research done on insects as natural enemies of *M. alternatus* based on the research by the Forestry and Forest Products Research Institute and the prefectural forestry research institutions of Japan. For predators, the following insects were confirmed to prey on *M. alternatus*. Adults of *Anisolabis maritima* (Dermaptera: Anisolabididae) prey on *M. alternatus* larvae in pupal chambers, while larvae of *Inocellia japonica* (Neuroptera: Inocelliidae) are predators of *M. alternatus* larvae under bark. Similarly, larvae of *Thanasinus lewisi* (Coleoptera: Cleridae) mainly feed on larvae of *M. alternatus*. Furthermore, larvae of *Temnochila japonica* (Coleoptera: Trogossitidae) also prey on *M. alternatus* larvae. Larvae of *Alaus berus* (Coleoptera: Elateridae) were found in pupal chambers of *M. alternatus* and predation of *M. alternatus* larvae in Petri dishes was confirmed. Adults of the ants (Hymenoptera: Formicidae), *Formica fusca japonica*, *Monomorium nipponense* and *Iridomyrmex itoi* prey on *M. alternatus* larvae. Arihara (1984) studied natural enemies of *M. alternatus*, and discovered some other insects besides those reported in Nobuchi (1980)'s review. He reported predation by *Alaus putridus* on all stages of *M. alternatus* and *Ampedus* sp. on larvae of *M. alternatus*. Because of their low specificity against *M. alternatus* and difficulty of rearing them in large numbers, most of these predators have not been studied as biocontrol agents for *M. alternatus*. Kishi (1995) reported that their predatory potential is not strong enough to suppress *M. alternatus*, therefore, no natural native insect enemies are available for control of *M. alternatus*, but the introduction of *Enoclerus* spp. (Coleoptera: Cleridae), which effectively kills bark beetles in Europe and America is promising.

As for parasitoids of *M. alternatus*, Nobuchi (1980) reported that two ichneumonids (Hymenoptera), *Dolycomitus* sp. and *Megarhyssa* sp. parasitize on larvae or pupae of *M. alternatus* in pupal chambers. Larvae of two braconids (Hymenoptera), *Atanycolus initiator* and *Dryctus* sp. are also parasitic on *M. alternatus* larvae. In addition, Arihara (1984) found *Spathius* sp. from *M. alternatus* eggs, *Spathius radzayanus* (Hymenoptera: Braconidae) and *Lonchaea scutellaris* (Diptera: Lonchaeidae) parasitic on larvae of *M. alternatus*. Larva of *Sclerodermus nipponicus* (Hymenoptera: Bethylidae) is an ectoparasite of various insects, and attacks larvae or pupae of *M. alternatus*, sucking its hemolynph. Enda (1994) released *S. nipponicus* on *M. alternatus* in logs in a large, screen cage during August to September and obtained 1.6–4.5% parasitism of the *M. alternatus* population. Miura et al. (2000) also released *S. nipponicus* in summer, and only four *M. alternatus* larvae appeared out of 27 entrance holes in the treatment plot, that is, 23 larvae were killed before emerging, whereas 11 larvae emerged out of 13 entrance holes of the non-treated plot; however, no effect on the *M. alternatus* population was seen with a spring release. Low temperatures in the spring may reduce the activity of *S. nipponicus* and may explain why good control was not achieved. Utilization of *Sclerodermus* is thus difficult to implement. Moreover, *Sclerodermus* species including *S. nipponicus* sting humans (Aruga 1959), and an outbreak of *S. nipponicus* at a school resulted in over 340 people being stung (Asahina 1952). Therefore, release of this insect could result in its becoming a hygienic insect pest, as such no study of this insect as a biocontrol agent is going on in Japan. However, in China, *Scleroderma guani* (=*Sclerodermus guani*), a closely related species to *S. nipponicus*, has been released in a *Cunninghamia lanceolata* (Pinales: Cupressaceae) forest for the control of the juniper bark borer, *Semanotus bifasciatus* (Coleoptera: Cerambycidae). Field experiments targeting *M. alternatus* have also been conducted in China (Enda 1992), the details are described in Chap. 33.

Dastarcus helophoroides is a Coleopteran ectoparasite of Cerambycid larvae (Fig. VI.14), and Taketsune (1982) discovered it as a parasite on *M. alternatus* larvae, however, the rate of parasitism was only 2.7–31.4% (Taketsune 1982; Inoue

Fig. VI.14 *Dastarcus helophoroides*, a Coleopteran ectoparasite of Cerambycid larvae (*Left* larvae reared on a *Galleria mellonella* larva; *Right* adult)

1993). An artificial rearing method was developed by Ogura et al. (1999), and artificially reared *D. helophoroides* were used to control *M. alternatus*. Urano (2003) inoculated *D. helophoroides* adults and hatched larvae and obtained a 63.2% parasitism rate. However, releasing the eggs in a large cage in the field resulted in 49.7% parasitism and a 45.9% survival rate of *M. alternatus* larvae in treated logs, while 96.4% of *M. alternatus* larvae survived in the control logs. Obviously, the release was effective in reducing the larval population. The rate of parasitism differed widely among logs, and tended to be lower in logs from which *M. alternatus* adults emerged soon after the release of *D. helophoroides* eggs. As such, to obtain a high percentage of parasitism placement of eggs should be synchronized with the occurrence of the susceptible host stage (Urano 2003). In a field experiment the parasitism of *M. alternatus* larvae in wood with *D. helophoroides* was 6–69% (average 34.8%) (Urano 2003). Parasitism of *M. alternatus* adults with *D. helophoroides* was low. Migration of *D. helophoroides* was supposedly low, and there was no migration from the released to non-released trees. In the released tree, parasitism was highest in the middle of the trunk, with the rate tending to be less further up the tree (Urano 2003).

A new species of nematode parasitic on *M. alternatus* adults was discovered by Enda and Nobuchi (1970); it was described as *Contortylenchus genitalicola* by

Kosaka and Ogura (1993). This nematode has two life cycles; an entomophagous stage in which it parasitizes the genital organs of *M. alternatus* adults and a mycophagous stage where it feeds on fungi under pine bark. *Monochamus alternatus* adults become sterile when parasitized by a large numbers of this nematode. Inoculation of this nematode into the trunks of dead pines infested with *M. alternatus* could lead to parasitism of this nematode on *M. alternatus* adults; but it is difficult to reduce the fertility of *M. alternatus* (Forestry Agency 1992). Therefore, to utilize this nematode for *M. alternatus* control, it would be necessary to develop a method to castrate the entire male population by mass-parasitism or to use nematode strains or closely related species, that are more potent on *M. alternatus* (Kosaka, personal communication).

The use of entomopathogenic nematodes is thought to be promising although in nature such parasitism of *M. alternatus* has not been found. Details of entomopathogenic nematodes are described in Chap. 36. Mamiya and Tamura (1983) obtained a strain of *Steinernema carpocapsae* from France and investigated its activity on *M. alternatus*. When 300–3,000 nematodes were suspended in water and dropped onto a filter paper in a plastic cup with an *M. alternatus* adult the nematodes killed more than 94% of *M. alternatus* adults within 48 h. Application of a nematode suspension directly into the mouthparts or under the elytra of *M. alternatus* adults resulted in the death of 50–70% of the beetles. When the suspension was dropped onto the entrance holes of pine logs infested with *M. alternatus* in June, larvae, pupae and adults of *M. alternatus* in pupal chambers were killed and the mortality was 87%, but when the suspension was dripped into artificially opened holes, mortality was only 43%. Mamiya and Shoji (1984) investigated the control effects of using All strain and a Mexican strain of *S. carpocapsae* against *M. alternatus*. Inoculation with the Mexican strain at a rate of 1,000 nematodes per beetle in the laboratory killed 95% of *M. alternatus* beetles, while All strain at the same concentration killed 70% of the beetles within 48 h. However, the Mexican strain and the All strain inoculated at a 1/10 concentration killed 35 and 30% of the *M. alternatus*, respectively. The Mexican strain also killed 90% of larvae or 60% of pupae of *M. alternatus* within 48 h. In a field experiment using infested pine logs, 56–61% of the *M. alternatus* larvae in wood were killed by spraying *S. carpocapsae* in April. Applications in June reduced the mortality of *M. alternatus* by 9–42% by dropping the nematode suspension into artificially opened holes, and spraying the nematodes onto the surface of the logs. Katagiri et al. (1984) inoculated *M. alternatus* larvae with *S. carpocapsae* by releasing the beetle larvae in a Petri dish with a filter paper containing 1,000 nematodes for 24 h. The larvae mortality rate was high (93–100%) within 48 h. About 45% of *M. alternatus* were killed in 9 days by spraying the Mexican suspensions onto pine logs in October, and 59% were killed within 21 days of spraying in January, resulting in 75% mortality rate with the All strain but none was killed by the Mexican strain.

Some *M. alternatus* adults are attacked by species of spiders and mites (Kishi 1995). Their importance as natural enemies is not great, and use of such arthropods seems questionable and so no studies have been done on their potential use to control *M. alternatus*.

35.4 Pathogens

35.4.1 Species of Pathogens

Microorganisms such as viruses, bacteria, fungi, and protozoa are known pathogens of insects. Viral diseases have often been used to control Lepidopteran insect pests, and many commercial microbial insecticides have been registered which contain nuclear polyhedrosis virus (NPV) or granulosis virus (GV). However, viruses do not occur as commonly on Coleopteran insects as on Lepidoptera. Some viruses are known pathogens of Coleoptera, for example, *Oryctes* nonoccluded virus on *Oryctes rhinoceros* (Coleoptera: Scarabaeidae) (Huger 1966), or entomopoxvirus on *Melolontha melolontha* (Coleoptera: Scarabaeidae) (Vago 1963), but very few viruses have been reported from members of the cerambycid. Koyama (1963) listed NPV as pathogen of the white-striped longicorn, *Batocera lineolata* (Coleoptera: Cerambycidae); but, this is the only report of NPV infecting a Coleoptera species and no actual specimen were preserved.

Bacteria are also one of a major biocontrol agent and *Bacillus thuringiensis* (BT) is the bacterium that is most widely used as a microbial insecticide in many countries. BT is a complex of various strains of the bacterium that have different host specificity. The most important characteristic of BT is its production of crystal toxins (δ-endotoxin) that damage insect midgut cells. Insecticidal activities of each BT strain are different and specific. Most commercial BT formulations target Lepidopteran insects and some target Dipteran and Coleopteran insects; but no BT strain pathogenic to Cerambycidae insects has been found.

Some other bacteria cause diseases of insects. Although they are not usually epizootic to insects, *Pseudomonas aeruginosa* (Bucher and Stephens 1957) and *Serratia marcescens* (Bucher 1960) are highly pathogenic once they enter into the insect's hemocoel. These bacteria have been found in association with various species of insects. A survey of *M. alternatus* pathogens, Shimazu and Katagiri (1981) listed *Serratia* as the most common bacterial pathogen from North to South Japan; on average, 19.1% of the *M. alternatus* specimens were infected with *Serratia*. Shimazu et al. (1983) sprayed a mixture of *S. marcescens* and the entomopathogenic fungus *Beauveria bassiana* onto the crowns of healthy pines, and then released *M. alternatus* adults. Mortality of the adults in 30 days of which was 30% was attributable to *B. bassiana* alone (1×10^7 conidia ml^{-1}), however, when mixed with *S. marcescens*, mortality increased to 51.7%.

Protozoan diseases of insects are not common, but sometimes they heavily impact insect populations. The most important characteristic of protozoan diseases is vertical transmission. They can sometimes be utilized for microbial control. In particular, microsporidia are formulated and sold commercially as microbial insecticides, mainly against locusts. Microsporidia are associated with Coleoptera including Cerambycidae (Lipa 1968), but no protozoa have ever been obtained from *M. alternatus*. Due to difficulties with artificial cultivation, their rarity on cerambycid, and the fact that they take a long time to kill their host suggests that, protozoa may not be useful for biocontrol of *M. alternatus*.

Fungi, especially Deuteromycetes (Anamorphic fungi) may be a major factor in epizootics in Coleopteran populations while for Lepidoptera, epizootics often occur as the result of viruses and fungi. Therefore, for microbial control of Coleoptera, fungi offer great promise.

Isaria farinosa was the first pathogen of *M. alternatus* recorded by Hasegawa and Koyama (1937). *Isaria farinosa* is a synonym of *Paecilomyces farinosus*, but at that time in Japan, the name "*Isaria farinosa*" meant *B. bassiana* of today (Aoki 1971), and the fungus that Hasegawa and Koyama collected was probably *B. bassiana*. Koyama (1959) published a list of causal agents responsible for epizootics in the forest pest insects of Japan, but *M. alternatus* was not included in that list. Apparently *I. farinosa* did not occur frequently on *M. alternatus*. Shimazu and Katagiri (1981) collected cadavers of *M. alternatus* from all over Japan, and isolated fungal pathogens. They found fungi in the genera *Beauveria*, *Verticillium*, and *Paecilomyces* as important fungal pathogens. Above all, *B. bassiana* was isolated from almost all localities (Fig. VI.15). The incidence of diseased *M. alternatus* was generally low in the field and no distinct epizootic attributable to any particular pathogen occurred on this insect. Even the natural infection rate of the most common pathogen, *B. bassiana*, was very low on *M. alternatus* cadavers. Togashi (1989) also recorded *B. bassiana*, *S. marcescens*, and *Verticillium* species, but their incidences on cadavers were also very low. Wang et al. (2003) reported that 1–5% of *M. alternatus* larvae in China were infected by pathogens in nature; among the isolated pathogens, about 12–19% was *B. bassiana*. *Metarhizium anisopliae* is also one of the most common entomopathogenic anamorphic fungi, however it rarely occurs on *M. alternatus*. Shimazu and Kushida (1983) found *M. anisopliae* as highly pathogenic to *M. alternatus* as well as *B. bassiana*. Soper and Olson (1963) surveyed the biota of *Monochamus* species in Maine, USA, and recorded *B. bassiana*, *I. farinosa*, *Verticillium* sp., *Aspergillus flavus*, and *Fusarium* sp. from *Monochamus* beetles, the same genus as the Japanese pine sawyer. In this case, *I. farinosa* would be the *P. farinosus* today, because *I. farinosa* was recorded separately from *B. bassiana*. Recently, Francardi et al. (2003) also studied the possibility of using *B. bassiana* for the control of *Monochamus galloprovincialis*, the PWN vector in Portugal. Practical use of *B. bassiana* to control *M. alternatus* is described later in this section and its present status in China is given in Chap. 33.

Beauveria brongniartii, another entomopathogenic fungus, is very pathogenic on adult yellow-spotted longicorn beetle, *Psacothea hilaris* and the white-spotted longicorn beetle, *Anoplophora malasiaca*. The formulation of this fungus (BiolisaKamikiri, produced by the Nitto Denko Co.) was registered as the first fungus-based insecticide in Japan targeting these Cerambycid beetles. Details of this fungus are described later.

Fig. VI.15 *Beauveria bassiana* on a *Monochamus alternatus* larva in pine wood

35.4.2 Spraying Conidial Suspensions of Beauveria bassiana

Shimazu and Kushida (1983) selected an isolate of *B. bassiana*, F-263, by bioassaying many isolates of pathogens from *M. alternatus*. Among the isolates tested this isolate caused the greatest of mortality of *M. alternatus* larvae and adults, and its median lethal concentration (LC_{50}) to *M. alternatus* larvae was 1.1×10^3 conidia ml^{-1} (gross mortality) or 2.0×10^3 conidia ml^{-1} (net mortality) (Shimazu 1994). This isolate also produced conidia abundantly on artificial media. The researchers mass-produced this fungus in plastic boxes to obtain conidia. The fungus was applied at first by spraying conidial suspension and by modifying the application technique used for applying conventional chemical insecticides. Dead pine trees were treated targeting larvae under the bark and adults emerging from the trees, and live trees were treated targeting the adults feeding on them (Shimazu and Kushida 1980; Shimazu et al. 1982, 1983). The treatments aimed at adults resulted in low mortality and required a long period to death. This delay allowed the adults to transmit PWNs and to lay eggs. For these reasons it was concluded that the treatment showed little promise against pine wilt disease. However, spraying the fungus on the surface of dead pine trees to target larvae under the bark resulted in greater mortality; that is, about 75% (Shimazu and Kushida 1980), but mortality obtained with similar experiments was lower.

In these field experiments, concentrations of fungal suspensions were about 10,000-fold higher than the LC_{50} in the laboratory; even then the mortality was too low. This may result from the fact that there was little chance for the fungus, which was sprayed onto the bark, to come into contact with the larvae under the bark. To increase mortality, the fungus was introduced directly under the bark, or when applied to the outer the bark, the fungus concentration was increased.

35.4.3 Introduction of **Beauveria bassiana** *Under the Bark*

Nobuchi (1989) used a novel method to introduce *B. bassiana* under the bark. That is, he used the pine bark beetle, *Cryphalus fulvus* to carry the *B. bassiana* conidia. This insect bores and breeds only under the bark of dead pine trees in a manner similar to that of *M. alternatus*. As well, this insect does not have a diapause and it is easy to rear artificially. Its use as a carrier for *B. bassiana* was based on the idea that when adults of this insect were released in pine wilt-infested forests after contamination with *B. bassiana* conidia, they would bore into dead pine trees, be killed by the fungus in several days and *M. alternatus* larvae inhabiting the same area would come into contact with the conidia and become infected (Nobuchi 1989). Moreover, since larvae of this insect grow in larval canals connected with the egg gallery an increase of the fungus under the bark as the result of larvae infection was expected. Enda et al. (1989) confirmed this in a large cage experiment. Field experiments releasing pine bark beetles contaminated with *B. bassiana* were also conducted at various sites in Japan, and the mortality of *M. alternatus* by *B. bassiana* was confirmed (Enda et al. 1989, 1991). Infection of *M. alternatus* with *B. bassiana* was found for several years in *Pinus thunbergii* forests in which *B. bassiana* was introduced by this method (Forestry Agency 1993). This method has the advantage in that it is not required to select and fell dead trees before applying the fungus, because *C. fulvus* feeds on almost the same part of the dead pine trees as *M. alternatus*. Some disadvantages of this method include the fact that it requires numerous workers for mass-producing the carrier, *C. fulvus*, and the fungal conidia which are applied do not always enter under the bark of dead pines, because they are carried by the bark beetles without control. The problem of mass production and mass releasing contaminated *C. fulvus* was subsequently solved by the development of an "automatic contaminating device" by Nobuchi (1993) (Fig. VI.16).

Fig. VI.16 An "automatic contaminating device" for releasing *Beauveria*-contaminated *Cryphalus fulvus* adults

To reduce diffusion of the fungus and to obtain better introduction of the fungus under the bark, Shimazu et al. (1992) developed *Beauveria* spawn chips to introduce *B. bassiana* under the bark of dead pines. These spawn chips were made of 7 × 20 mm cylindrical, commercially available wheat bran pellets added to water, autoclaved, inoculated with *B. bassiana*, and to induce conidia production on the surface. The trunk of pine trees infested with *M. alternatus* is drilled and the pellets inserted to apply the spawn chips. Tests show that about 80% of *M. alternatus* larvae are killed by *B. bassiana* when these pellets were applied in August to September at a rate of 12 pellets m^{-1}. On average, 45–100% mortality was obtained from experiments done by the Forestry and Forest Products Research Institute or prefectural forestry research institutes, and higher mortality was obtained with earlier application (Shimazu 1993). When this method was applied to standing dead trees, infected larvae were mostly found in regions where the pellets were applied, although some infected larvae were found on lower areas of the trees, but seldom in the upper areas. Mortality increased considerably using this method, but some problems exist such as high labor input for drilling the holes, the fact that the treatment was most effective near where it was applied, and the difficulty in treating small branches.

35.4.4 Control of **Monochamus alternatus** *Larvae with* **Beauveria bassiana** *on Nonwoven Fabric Bands*

Another way to increase sawyer mortality under the bark of dead trees is to increase the amount of conidia applied to the outside of the bark, but there is a limit as to the conidial density that can be applied in a suspension, that is, for spraying onto the outside of the bark. Therefore, to apply mass numbers of conidia without spraying the suspension onto the surface of the bark, the "solid state" application of conidia was tested by the Forestry and Forest Products Research Institute.

The culture medium itself or wheat bran pellets could be used as the carrier of high numbers of conidia, but these materials are difficult to apply to the surface of dead pine trees. Katagiri et al. (1983) tried to band fungus-cultured materials such as straw matting or corrugated fiberboard onto the trunks of pines, and obtained infection of overwintering larvae of *Dendrolimus spectabilis* (Lepidoptera: Lasiocampidae). Similar fungus substrates made of polyurethane foam or nonwoven fabric strips have been used to control cerambycid beetles using *B. brongniartii* (Kashio and Ujiye 1988; Hashimoto et al. 1991). Nitto Denko Co. has commercialized this type of formulation of *B. brongiartii* under the name "BiolisaKamikiri" as a control agent of *A. malasiaca* and *P. hilaris.* These substrates are suitable for applying *B. bassiana* to dead pine trees. Shimazu et al. (1995) employed nonwoven fabric as the carrier for *B. bassiana* and used it against *M. alternatus* larvae. The nonwoven fabric material is the same as "BiolisaKamikiri". It is manufactured in 45 × 5 cm, shaped bands by Nitto Denko Co. (This Department of Nitto Denko was moved to Idemitsu Kosan Co. in 2007). *Beauveria bassiana* was shake-cultured to obtain the seed culture, then it was mixed with the fresh medium and the nonwoven fabric material was soaked in the medium. The materials were incubated in plastic baskets for 3 weeks at 25°C to obtain maximum sporulation. After culture, ca. 1.8×10^8 conidia cm^{-2} were produced on the material. Such nonwoven fabric strips were applied to the bark of dead pine trunks and branches in various ways and stapled in place (Fig. VI.17). The results are shown in Table IV.6. Many *M. alternatus* larvae were infected, and their mortality was equivalent to or more than that obtained by implanting wheat bran pellets. The rate of mortality was high and good enough for a field experiment with an entomopathogenic fungus. Nonwoven fabric strips are superior to wheat bran pellets because they are easier to apply and are highly effective. Moreover, nonwoven fabric strips can be used on standing dead trees, piled dead branches and so on. At present, this method is thought to be most practical way to apply *B. bassiana* as a microbial insecticide against *M. alternatus*.

Fig. VI.17 Application of *Beauveria bassiana*-cultured nonwoven fabric strips onto a dead *Pinus thunbergii* trunk for treatment of dead trees (*Left* banding on a standing dead tree; *Right* application on logs)

Table VI.6 Number and mortality* of *Monochamus alternatus* larvae in artificially oviposited logs in 1993 (Shimazu et al. 1995)

Treatment and method	Month	Number of logs	Number of larvae				Infected (%)	
			Alive	*B. bassiana*	Other Death	Total	Measured	Corrected
Control	July	4	6.8 (1.4)	1.0 (1.0)	0.5 (0.3)	8.3 (2.0)	7.69 (7.69)	0.00
Control	August	8	5.4 (1.1)	0.9 (0.5)	0.1 (0.1)	6.4 (1.4)	12.02 (4.96)	0.00
Pellet × 1	July	4	5.5 (3.2)	6.3 (1.8)	0.3 (0.3)	12.0 (3.0)	58.74 (16.46)	55.31
Pellet × 2	July	4	0.8 (0.5)	7.8 (2.1)	0.0 (0.0)	8.5 (1.7)	87.50 (7.98)	86.46
Pellet × 1	August	8	1.5 (0.5)	4.9 (1.0)	0.0 (0.0)	6.4 (1.2)	78.23 (6.99)	75.25
Pellet × 2	August	8	1.1 (0.5)	7.0 (1.8)	0.1 (0.1)	8.3 (2.1)	87.52 (4.80)	85.81
Pellet × 2, sunlight	August	5	2.6 (1.6)	5.0 (1.0)	0.0 (0.0)	7.6 (1.7)	76.33 (14.50)	73.10
Strip, wrap	July	7	0.1 (0.1)	2.7 (0.9)	0.0 (0.0)	2.9 (1.0)	96.00 (3.38)	95.67
Strip, piled	July	8	2.8 (1.0)	3.5 (0.8)	0.1 (0.1)	6.4 (1.6)	66.14 (8.77)	63.32
Strip, wrap	August	8	0.1 (0.1)	7.1 (2.5)	0.0 (0.0)	7.3 (2.5)	96.43 (3.34)	95.94
Strip, piled	August	8	0.4 (0.3)	8.5 (2.0)	0.1 (0.1)	9.0 (2.0)	94.60 (3.59)	93.86
Strip, piled, sunlight	August	5	1.2 (0.7)	4.8 (2.5)	0.0 (0.0)	6.0 (3.2)	78.84 (9.61)	75.95

*Mean with SE in parentheses

35.4.5 *Control of* **Monochamus alternatus** *Adults with* **Beauveria bassiana** *on Nonwoven Fabric Bands*

Direct spraying *B. bassiana* does not result in high mortality of *M. alternatus* adults in a short time, but recently, it was found that *B. bassiana* may cause high mortality of *M. alternatus* adults in less time when the adults were allowed to walk on the nonwoven fabric strips containing the *B. bassiana* conidia (Okabe et al. 2001, 2002; Okitsu et al. 2000). The use of nonwoven fabric strips to apply *B. bassiana* has been registered and commercialized based on the joint work of the Forestry and Forest Products Research Institute, universities, prefectural government and a production company (Fig. VI.18). In this method of application, the adults become infected by walking on the conidial mass; thus, the virulence of the fungus cannot be measured by the conventional method of dipping the experimental insect into the conidial suspensions. Consequently, a novel bioassay technique was developed by mixing dead conidia with live conidia at different ratios to regulate inoculum density, and the virulence of dry conidia could be measured (Shimazu 2004b). Using this bioassay technique, young *M. alternatus* adults, within 4 days after

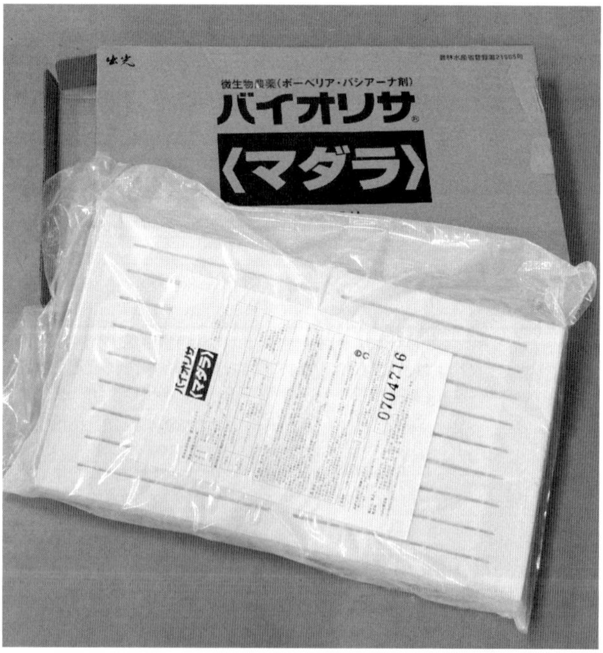

Fig. VI.18 "BiolisaMadara", a commercial formulation of the fungus *Beauveria bassiana* for control of the Japanese pine sawyer beetle *Monochamus alternatus*

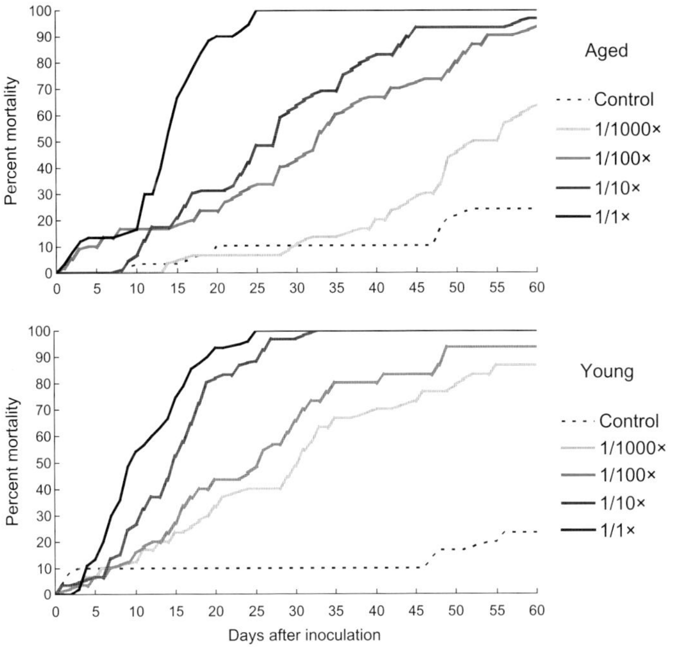

Fig. VI.19 Cumulative mortality of aged (*top*) and young (*bottom*) adult Japanese pine sawyers, *Monochamus alternatus*, inoculated with *Beauveria bassiana* conidia at different concentrations, that is, 1/1000 to 1/1×

emergence, were found to be more susceptible to *B. bassiana* than were older adults (>10 days after emergence; Fig. VI.19).

The mortality of *M. alternatus* adults was more than 90%, 15 days after emergence in field experiments with nonwoven fabric strips applied to pine sawyer-infested pine logs (Table VI.7). The peak of maturation feeding and the post-oviposition period for *M. alternatus* is 2–3 weeks after emergence; therefore, this mortality and the days to death were effective in preventing the transmission of PWNs and *M. alternatus* oviposition. Maehara et al. (2007) investigated the amount of maturation feeding and transmission of PWNs by *M. alternatus* adults exposed to the *B. bassiana* formulation. Inoculated beetles feed less than the control beetles and ceased feeding several days before death (Fig. VI.20). Among inoculated beetles carrying >1,000 nematodes, some died before nematode departure. Although the remaining heavily nematode-infested beetles lived until the beginning of nematode departure, they had stopped feeding, preventing the nematodes from entering pine twigs, that is, in inoculating host trees. It was concluded that this method is effective in preventing the transmission of PWN to healthy pine trees.

Table VI.7 Percent mortality of *Monochamus alternatus* in field experiments 15 days after being exposed to *Beauvaria bassiana* impregnated, non-woven fabric strips

Experimental site	Year				
	2003	2004		2005	
	Open[a]	Open[a]	Close[b]	Open[a]	Close[b]
Kumamoto	95	88	—	—	—
Kagoshima Univ.	52	61	100	—	95
Shiga (1)	94	95	—	100	—
Shiga (2)	97	83	96	71	91
Nihon Univ.	92	71	—	—	—
Tokyo Univ. Agr. Tech.	96	100	—	98	—
Akita	99	89	100	91	96
Shimane	93	98	94	—	98
Fukuoka	—	92	—	—	—
Saga	—	100	—	—	—
Yamaguchi	—	—	—	90	93

[a] Open = logs covered by plastic sheet with both ends open
[b] Close = logs covered by plastic sheet without opening

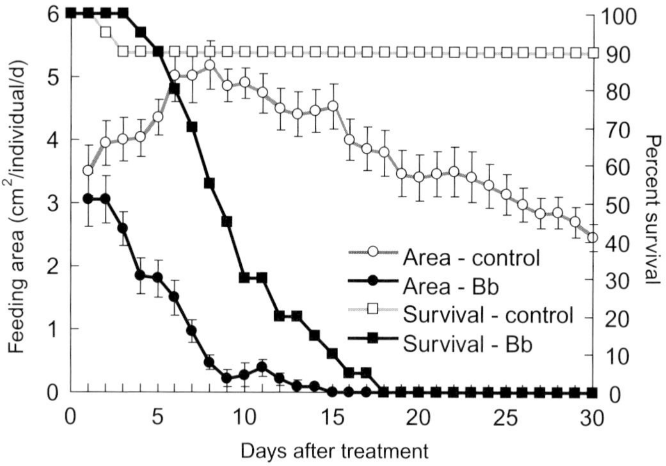

Fig. VI.20 Survival rate of the Japanese pine sawyer beetle, *Monochamus alternatus*, inoculated with the fungus *Beauveria bassiana* and the amount of maturation feeding done by the beetles on *Pinus densiflora*

35.4.6 Safety of Beauveria bassiana *Isolate F-263*

Beauveria bassiana is often used as a microbial control agent throughout the world because it is not pathogenic to mammals. Shimazu (2004a) investigated the effects of temperature on the growth of isolate F-263, and found that it does not grow at the body temperature of mammals. Inoculation tests of F-263 on several species of beneficial insects were conducted by the Forestry Agency to determine the pathogenicity of this isolate (1999, 2000, 2001). This fungus caused only slight mortality of the adult *Harmonia axyridis* (Coleoptera: Coccinellidae) and *Dastarcus helophoroides* (Coleoptera: Colydiidae). It was weakly pathogenic to *Orius sauteri* (Heteroptera: Anthocoridae) adults, and the LC_{50} was 2.5×10^7 ml^{-1} to 2.5×10^9 ml^{-1}. It is moderately pathogenic to the honeybee, *Apis mellifera* (Hymenoptera: Apidae), with a LC_{50} of 6.1×10^6 ml^{-1}, therefore, when applied commercially careful application is required to avoid direct contact with honeybees.

Experiments (Forestry Agency 1999, 2000) with direct inoculation of non-woven fabric strips with F-263 showed no effects on Japanese fir, *Abies firma* (Pinaceae), hinoki cypress, *Chamaecyparis obtusa* (Cupressaceae), sugi, *Cryptomeria japonica* (Taxodiaceae), eurya, *Eurya japonica* (Theaceae), Japanese beautyberry, *Callicarpa mollis* (Verbenaceae), *Hydrangea hirta* (Saxifragaceae), the soybean, *Glycine max* (Leguminosae), eggplant, *Solanum melongena* (Solanaceae), Japanese ginger, *Zingiber mioga* (Zingiberaceae), gold-banded lily, *Lilium auratum* (Liliaceae), a bamboo, *Pleioblastus chino* (Gramineae), and lily turf, *Liriope graminifolia* (Liliaceae).

Conidia of F-263 were mixed into forest soil to determine the effect of *B. bassiana* on soil microorganisms, and the density dynamics of soil bacteria, actinomycetes, and general fungi (Shimazu et al. 2002a). F-263 had no effect on bacteria, actinomycetes, and soil fungi. *Beauveria bassiana* conidia seemed not to germinate in forest soil, and in 12 months its population levels gradually decreased to 1/10.

Beauveria bassiana is also known to cause white muscardine on the silkworm, *Bombyx mori* (Lepidoptera: Bombycidae). The virulence of *B. bassiana* F-263 to silkworm larvae was determined by the Forestry Development Technological Institute (1990) and the LC_{50} was 9.8×10^4 conidia ml^{-1} for first instar larvae and 8.9×10^5 conidia ml^{-1} for fifth instar larvae. These LC_{50} values were almost the same as those of other wild *B. bassiana* strains to the silkworm (Kawakami 1973).

Suzuki et al. (1991) looked at the risk of accidental contamination of mulberry leaves by releasing *Beauveria*-contaminated pine bark beetle, *C. fulvus* by placing potted mulberry plants around the release point of *C. fulvus*. Mulberry leaves from the release point caused a high rate of *B. bassiana* infection of the silkworm larvae, but plants further from the release point caused little infection, and it was concluded that the risk of releasing *C. fulvus* is limited to only a very small area near the release point.

Beauveria-cultured nonwoven fabric strips hold more conidia, and the risk of dispersal of the fungus is thought to be far greater than releasing *Beauveria*-contaminated bark beetle. Shimazu et al. (2002b) investigated conidial dispersal from experimentally set nonwoven fabric strips with *B. bassiana* using selective medium plates to reconfirm presence of the *B. bassiana*. Conidial density more than 60 m from the strips was 5.2×10^2 m^{-2} day^{-1} at maximum, and they concluded that if conidia continued to disperse at same rate, it would take more than 40,000 days to achieve the lethal conidial density of silkworm larvae of 2.1×10^7 conidia m^{-2} on mulberry leaves (Wada and Miyamoto 1997, 1999). It was concluded that, contamination of mulberry leaves more than 60 m away from nonwoven fabric strips is extremely rare.

35.5 Use of Infochemicals

35.5.1 Kairomones

When a chemical substance produced by an organism gives some benefit to other organisms, which received that substance, it is called a kairomone. For example, a substance released from a plant that attracts an insect pest for feeding is a kairomone. Among chemical substances that attract *M. alternatus* adults, attractants related to oviposition have been most studied. After emergence from dead pine wood, *M. alternatus* adults feed on twigs of live pine trees and become sexually mature in 16–30 days (Enda and Nobuchi 1970). These sexually mature males and females are attracted to the trunks of dying or physiologically weakened pine trees where they mate and the females lay eggs. Chemically treated live trees or felled pine trees also have the same attraction effect for *M. alternatus* adults (Yamasaki et al. 1980; Ikeda et al. 1981). It was thought that some volatile substances from those dying or weakened trees might attract *M. alternatus* adults as kairomones. Analysis of volatiles from freshly felled logs of *Pinus densiflora* revealed that monoterpene hydrocarbons, such as α-pinene, β-pinene, β-phellandrene, β-myrcene, camphene were the major components of the oil substances and ethanol was the major component of the water-soluble extract (Ikeda et al. 1980). Monoterpenes and ethanol increased in volatility from nematode-infested *P. densiflora*, and their compositions resembled that of freshly felled logs (Ikeda and Oda 1980). Ethanol and α- and β-pinene were detected in the volatiles of *P. densiflora* injured with acetone (Ikeda et al. 1981).

Utilizing the oviposition attraction of *M. alternatus*, two attractants for this insect have been commercialized in Japan. One of them, "Hodoron" (Hodogaya Chemical Co. Ltd) mainly consists of eugenol and benzoic acid and was developed before the determination of volatiles from dying pine trees. Its production was stopped and registration expired in 2001 as commercial demand decreased. Another product is available as "Madara-Call" (Sankei Chemical Co. Ltd). It is comprised of dispensers of α-pinene and ethanol, and an exclusive trap (Fig. VI.21). These commercial attractants can collect adults of *M. alternatus* and have been used for monitoring its populations. The attraction of adults is not directly related to beetle control, and would be difficult to use as a control tool. It attracts sexually mature adult sawyers, but not newly emerged adults. Those older adults are mostly more than 16 days old and have been feeding on pine twigs and have already transmitted PWNs to healthy trees. As such these attractants do not prevent pine wilt infection in the present year; and also they do not reduce the number of eggs laid in naturally infested pines because the attractiveness of dying pine trees is far greater and most adults are attracted to naturally infested pine trees (Ikeda 1993). However, *M. alternatus* adults are less migratory during the oviposition period, and application of insecticides around a long-term volatilizing attractant in slightly damaged regions may be useful for beetle control (Ikeda 1993).

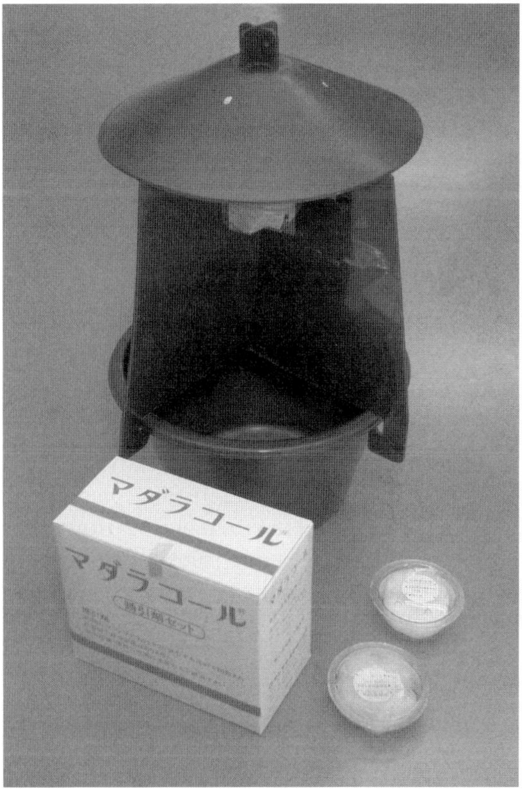

Fig. VI.21 "Madara-Call" trap utilizing the oviposition attractant for the Japanese pine sawyer *Monochamus alternatus*

35.5.2 Sex Pheromones

A pheromone is defined as a chemical that triggers an innate behavioral response in another member of the same species. It is not rare for Lepidopteran adults to fly a long distance toward the opposite sex attracted by sex pheromones, but this is not the case for *M. alternatus*. However, the adults are caught by traps containing adult beetles or a liquid in which they have been washed, which indicates the existence of a sex pheromone (Kishi and Otsu 1987; Kishi 1988a, b). Fauziah et al. (1987) suggested the presence of a male sex pheromone that attracts females of *M. alternatus*. Kim et al. (1992) found two different pheromones; a volatile pheromone to attract females produced by males, and a contact pheromone on the body surface of the female eliciting copulatory behavior in the male. To date, the chemical structures of these substances has not been determined. Males and females of *M. alternatus* meet each other when they are attracted to dying pines by kairomones, and therefore, sex pheromones have no role in long distance attraction.

Thus, it is not possible to utilize sex pheromones to attract this insect, but they may be more useful for mating inhibition than for attraction and capturing.

35.6 Concluding Remarks

Individual techniques for biological or biorational control are explained in this section. When these techniques are used singly, only *B. bassiana* may cause enough mortality to be used against *M. alternatus*. Actually, there is a natural decrease in pine wilt in areas where the green anole is prevalent, but this animal also is harmful to other rare insects. *Beauveria bassiana* can cause high mortality of *M. alternatus*, but it takes a long time to kill the adults, and cannot be used to prevent maturation feeding. As such it can only be used against larvae or adults during the disposal of infested pine wood.

As often emphasized, both preventive treatment of healthy pines and the disposal of dead trees must be done for the control of pine wilt. For treatment of healthy pines, an immediate effect is required. However, an immediate effect is not possible for ordinary pathogens and parasites and is only possible by use of predators; but because they lack specificity predators in general are difficult to use as biocontrol agents of adults.

The discovery of a new BT strain pathogenic to the adults of *M. alternatus* is expected as a novel biocontrol agent having an immediate insecticidal effect. Most BT strains are pathogenic to Lepidopteran larvae, but recently, several strains pathogenic to other orders of insects have been discovered. Among these, some strains are pathogenic to Coleoptera, although pathogenicity to Cerambycidae has not yet been found. Since many new BT strains have been found in soil and the phyllosphere discovery of BT strains pathogenic to *M. alternatus* adults will be possible. BT differs from other insect pathogens in that its insecticidal effects results from crystal toxins, which it produces and which kill insects rapidly. If a BT strain pathogenic to *M. alternatus* adults is discovered, it would rapidly kill the adults feeding on healthy pine trees by spraying the BT strain onto the foliage as preventive spraying.

Since no rapid-killing biological control agent is available, it will be impossible to manage *M. alternatus* by biological means alone. To achieve the desired objective, it will be necessary to use a combination of biological and chemical controls.

36
Potential of Entomopathogenic Nematodes for Controlling the Japanese Pine Sawyer, *Monochamus alternatus*

Long Ke Phan

36.1 Introduction

The main vector of the pine wood nematode (PWN), *Bursaphelenchus xylophilus* is the Japanese pine sawyer beetle *Monochamus alternatus* (Mamiya and Enda 1972; Morimoto and Iwasaki 1972). Control of *M. alternatus* is challenging for several reasons. Chemical control of *M. alternatus* is difficult due to the insect's behavior and because of environmental and health hazards. The use of many insecticides has led to a high level of resistance in many insect species. Alternative control measures may be practical in such situations.

Nematode parasites of insects have been known since the seventeenth century and belong to more than 30 families (Poinar 1979, 1990; Nickle 1984; Kaya and Stock 1997). The major focus of research and development in this area has been on nematode species in seven families, Mermithidae, Tetradonematidae, Allantonematodae, Phaenopsitylenchidae, Sphaerulariidae, Steinernematidae and Heterorhabditidae, because of their potential as biological control agents of insects (Kaya and Stock 1997). The entomopathogenic steinernematid and heterorhabditid nematodes possess many attributes of parasitoids and pathogens. They occupy diverse ecological nitches including agricultural fields, forests, grasslands, deserts, and ocean beaches (Hominick et al. 1996).

Entomopathogenic nematodes can be excellent biological control agents for soil-dwelling stages of many insect pests (Kaya and Koppenhöfer 1999). They are safe for most non-target organisms and for the environment. They can be mass produced, formulated and easily applied as bio-pesticides (Georgis and Kaya 1998; Georgis and Manweiler 1994), and are compatible with various insecticides and

Institute of Ecology and Biological Resources, Vietnamese Academy of Science and Technology, 18 Hoang Quoc Viet, Nghiado, Caugiay, Hanoi, Vietnam

Tel.: +84-4-2109211, Fax: +84-4-8361196, e-mail: pklong@vast.ac.vn, pkelong@yahoo.com

biological control agents (Ishibashi 1993; Thurston et al. 1994; Nishimatsu and Jackson 1998; Koppenhöfer et al. 1999, 2000; Chen et al. 2003). They also have a broad host range of insect pests in a variety of habitats (Gaugler and Kaya 1990; Kaya and Gaugler 1993).

The family Steinernematidae has 2 genera: *Steinernema* (more than 50 species) and *Neosteinerma* (1 species), and the family Heterorhabditidae has 1 genus: *Heterorhabditis* (12 species).

36.2 Entomopathogenic Nematodes

36.2.1 Life Cycle

The active stage of the nematode that invades an insect is the infective juvenile (IJ) stage. The infective juveniles (IJs) of these nematodes are non-feeding, free-living stages and occur in the soil. They can survive for long periods in the soil and may wait for the opportunity to infect a host. These IJs may be used for insect control. The IJs locate the potential host, move towards it and penetrate the host's body. The IJ enters the insect host through the mouth, anus, spiracles, or by direct penetration through the cuticle. Once inside the nematode releases symbiotic bacteria (in the genera *Xenorhadus* and *Photorhabdus* for steinernematids and heterorhabditids, respectively), when the bacteria reaches the hemocoel of a host it multiplys rapidly in the hemolymph; and usually the insect dies within 24–48 h (Poinar 1986; Kaya and Gaugler 1993). The IJs feed upon the bacteria, liquefy host tissues, and mature into adults. The life cycle is completed in a few days, and hundreds of thousands of new IJs emerge in search of new hosts (Poinar 1979). Although the bacterium is primarily responsible for the mortality of most insect hosts, the nematode also produces a toxin that is lethal to the insect (Burman 1982).

When the IJ of *Steinernema* enter the haemocoel of a host, they reach a feeding stage, feed on the bacteria, moult to the fourth stage and then to males and females of the first generation. After mating, the females lay eggs that hatch as first-stage juveniles that moult successively to second-, third-, and fourth-stage juveniles and finally to second generation males and females. The adults mate and the eggs produced by these second generation females hatch as first-stage juveniles that moult to the second stage. The late second-stage juveniles cease feeding, incorporate a pellet of bacteria in the bacterial chamber, and moult to the third stage (IJ), retaining the cuticle of the second stage as a sheath, and leave the cadaver in search of new hosts. In some hosts, the second generation is bypassed and the eggs that are laid by the first-generation adult females develop into IJ (Adams and Nguyen 2002).

The life cycle of *Neosteinernema* is similar to that of *Steinernema*, except that *Neosteinernema* has only one generation: females move out of the host into the environment, assume a spiral shape and become immobile. The females retain all eggs that hatch, and the juveniles undergo two molts to become IJ. When most of the juveniles are infective, they move vigorously, break through, and emerge from the female cadaver in the region of the tail or stoma (Nguyen and Smart 1994).

In contrast in the life cycle of *Heterorhabditis*, the IJ mature gives rise to a first generation of hermaphroditic females. These hermaphrodites lay eggs, which develop into second generation males and females, and also retain eggs which develop into IJ via *endotokia matricida* (intra-uterine birth causing maternal death). Second-generation females also lay eggs, which develop into another generation of adults, but they also retain eggs within the nematode body which again develop into IJ via *endotokia matricida*. Third-generation females do not oviposit and all their eggs develop via *endotokia matricida* into IJ (Burnell and Stock 2000).

36.2.2 Pathogenicity

IJs seek insect hosts to infect using species-specific foraging strategies that consist of suites of behavioral, morphological, physiological and ecological traits (Campbell and Kaya 2002). Entomopathogenic nematode foraging strategies are divided into two broad categories: cruiser (widely foraging) and ambusher (sit and wait) (Pianka 1966; Schoener 1971; Echkhardt 1979; Huey and Pianka 1981; McLaughlin 1989). Cruise foragers have a higher probability of finding sedentary and cryptic resources than ambush foragers, but ambush foragers are more effective than cruise foragers in finding resources (host insects) with high mobility (Lewis et al. 2006). *Steinernema carpocapsae*, *S. siamkayai* are described as classic ambush foragers, remaining near the soil surface and attaching to a mobile host. *Heterorhabditis* spp., *S. cubanum*, *S. glaseri*, *S. karii*, *S. longicaudum*, *S. oregonense*, *S. puertoricenses* and *S. sangi* TX1 are classified as cruise foragers, ranging widely and responding to volatiles from a sedentary host, and *S. ceratophorum*, *S. feltiae*, *S. monticolum*, *S. riobrave* are described as having an intermediate foraging strategy (Grewal et al. 1994; Campbell and Gaugler 1997; Lewis 2002; Phan et al. 2006). The IJs of some entomopathogenic nematodes exhibit two additional behaviors that facilitate ambush foraging: standing and jumping. Campbell and Kaya (2002) detected *S. carpocapsae*, *S. ceratophorum*, *S. scapterisci* and *S. siamkayai* with high rates of jumping; *S. abbasi*, *S. bicornutum*, *S. rarum* and *S. riobrave* with intermediate rates of jumping; *S. kushidai* and *S. monticolum* with low rates of jumping; and *S. affine*, *S. arenarium*, *S. cubanum*, *S. feltiae*, *S. glaseri*, *S. karii*, *S. kraussei*, *S. longicaudum*, *S. oregonense* and *S. puertoricense* with no jumping.

Migration is advantageous for both survival and for finding hosts. Quick migration of nematodes into the soil is essential to escape solar radiation and desiccation (Gaugler 1988). Soil texture can greatly affect nematode survival (Kung and Gaugler 1991), movement (Moyle and Kaya 1981; Georgis and Poinar

1983a,b; Schoeder and Beavers 1987) and infectivity (Georgis and Poinar 1983a,b; Moyle and Kaya 1981) by soil particle size, aeration and water content. Survival of *S. carpocapsae* and *S. glaseri* is best in sandy loam and sandy soils. A generally higher clay content results in lower nematode survival. The dispersal rate of IJs is negligible in very heavy clay soil, but the rate increases as the proportion of clay and silt decreases. Pathogenic studies results paralleled those obtained for survival (Chen 2003). Hanula (1993) demonstrated that for a biological control agent to be effective against larvae of *Otiorhynchus sulcatus*, it must penetrate to a depth of 15 cm. Nguyen and Smart (1990) argued vertical dispersal must be considered before deciding how to apply entomopathogenic nematodes in the field. Entomopathogenic nematodes migrate towards the host by responding to various types of host stimuli: temperature (Burman and Pye 1980), contact cues (Lewis et al. 1992), volatile cues (Lewis et al. 1993), and CO_2 or other attractants from insects (Schmidt and All 1978, 1979; Gaugler et al. 1980; Bird and Bird 1986; Ishibashi and Kondo 1990; Robinson 1995; van Tol and Schepman 1999; Fairbairn et al. 2000; Hui and Webster 2000). The positive effect of host presence on the migration of steinernematids was earlier demonstrated by Moyle and Kaya (1981). Miduturi (1997) and Phan et al. (2006) showed that *S. feltiae*, *S. affine* and *S. sangi* TX1 are able to migrate over 10 cm within a period of 3–7 days; the presence of a host positively affecting their migration. Wang and Ishibashi (1999) showed that 24 h after inoculation 5% of *S. carpocapsae* and 18% of *S. glaseri* juveniles penetrated into the wax worm, *Galleria mellonella* situated at 7 cm depth.

Temperature has a direct influence on the entomopathogenic nematode host-searching behavior (Byers and Poinar 1982), pathogenicity (Molyneux 1984, 1986), development (Zervo et al. 1991), and survival (Molyneux 1985; Kung and Gaugler 1991). Each species has an optimum temperature for survival. IJs are most active between 12 and 32°C; outside of this range, penetration and development are very slow (Woult 1991). Responses of the nematodes to temperature can vary with the species and isolates, possibly due to the climatic conditions at the geographic origin of each species (Molyneux 1984, 1986). Thus, Bedding (1990) suggested that collection of indigenous nematodes provides isolates more suited for inundative release against local pest insects because of their adaptation to the local climate and population regulators. Obviously, temperature is one of the factors that may hinder the use of entomopathogenic nematodes. To solve this problem, Glazer et al. (1992) added the antidesiccant Biosys 727 (20% wt/wt), natural wax (18% wt/wt) or Folicote (6% wt/wt). This reduced the concentration of *S. carpocapsae* to control noctuids *Earias insulana* and *Spodoptera littoralis* required for control from 500 and 1,000 IJ ml^{-1} to 125–250 IJ ml^{-1}, respectively. In a microplot experiment, the application of entomopathogenic nematodes at 250 IJ ml^{-1} mixed with 6% Folicote reduced *S. littoralis* damage by 46%. Similar treatments at 125 IJs ml^{-1} reduced damage caused by the Egyptian stemborer, *Earias insulana* by 76%. Treatment with nematodes alone reduced host numbers by 66% as compared with the control. In Saudi Arabia, Hanounik et al. (2000) evaluated the potential for the control of the red palm weevil, *Rhynchophorus ferrugineus* on individually, caged date palm trees in the field by *Heterorhabditis* sp. (isolate HAS-17), isolated from the Eastern Region of Saudi Arabia with or without anti-desiccants. Application of this nematode isolate along with the anti-desiccant Leaf Shield caused 87.5% mortality in adult red palm weevils compared to 65% mortality when the nematode was applied with the antidesiccant Liqua-Gel or with water only. A level of mortality of 20% occurred in control plots, which received water only. Abbas et al. (2000) used soil applications of entomopathogenic nematodes to control the red palm weevil, *R. ferrugineus*. Spraying the nematode suspension on palm trees, with or without anti-desiccants, at a rate of 8×10^6 IJ tree^{-1}, caused only 8.3–13.3% mortality of the adults, but application of nematodes in soil around the trunk, at a rate of 8×10^6 IJ tree^{-1}, caused 33.3–86.7% mortality.

36.2.3 Integrated Pest Management (IPM)

When formulating IPM strategies combining entomopathogenic nematodes with pesticides, the compatibility of these nematodes with pesticides should be considered. Pramila and Siddiqui (1999) studied the compatibility of *S. carpocapsae* with 11 chemical pesticides commonly used in India. The results showed that *S. carpocapsae* was compatible with endosulfan, phosphamidon, cypermethrin, malathion, monocrotophos, dithane, copper oxychloride, agallol and 2.4-D sodium salt, but not with phorate and bavistin. Since entomopathogenic nematodes are highly tolerant to conventional pesticides at the recommended dosage, mixed application with chemicals is recommended for the integrated control of soil pests (Ishibashi and Choi 1992). Lee et al. (2002) evaluated six Korean, entomopathogenic nematode isolates, *S. carpocapsae* Pocheon, *S. glaseri* (Dongrae strain), *S. glaseri* (Mungyeong strain), *S. longicaudum* (Gongju strain), *S. longicaudum* (Nonsan strain), and *Heterorhabditis* sp. (Gyeongsan strain) for their potential as bioinsecticides for controlling *Exomala orientalis* (Coleoptera), an important pest of turfgrass on Korean golf courses. They evaluated a reduced chemical insecticide approach that combined chlorpyrifos-methyl with these nematodes. The combination of a one-half rate of *Heterorhabditis* sp. (Gyeongsan strain) and a one-half rate of chlorpyrifos-methyl was synergistic, causing 91% mortality of *E. orientalis* compared with 69% for the full rate of *Heterorhabditis* sp. (Gyeongsan strain) or 22% for the full rate of chlorpyrifos-methyl. The combination of a one-half rate of *S. longicaudum* (Nonsan strain) and a one-half rate of chlorpyrifos-methyl caused 96.8% mortality, much more than a full rate of *S. longicaudum* (Nonsan strain) (45.9% mortality) or a full rate of chlorpyrifos-methyl (28.7% mortality). The interaction of *Heterorhabditis* sp. (Gyeongsan strain) and *S. longicaudum* (Nonsan strain) with chlorpyrifos-methyl in field trials appeared to be synergistic. Entomopathogenic nematodes were also proven to be compatible with other biological control agents. Ishibashi and Choi (1992) applied a mixture of fungivorous nematodes, *Aphelenchus avenae* and *S. feltiae*. The combination suppressed the virulence of *Rhizoctonia solani* and *Meloidogyne incognita*, and removed leaf miner-like symptoms caused by

A. avenae in cucumbers in Japan. Koppenhöfer and Kaya (1997) observed that a synergistic effect occurred against the white grub *Cyclocepha hirta* with a combined application of entomopathogenic nematode and *Bacillus thuringiensis* subsp. *japonensis*. Similarly, a combination of *S. feltiae* and *B. thuringiensis* subsp. *israelensis* caused 78 and 77% mortality of L1 and L4, respectively of the crane fly, *Tipula paludosa* (Oestergaad et al. 2006). Ansari et al. (2006) examined interaction under field conditions and compared the combination of a commercial formulation of *Heterorhabditis bacteriophora* (Nema-green®) and *Metarhizium anisopliae*. Controls were a single application of *M. anisopliae*, chlorpyrifos (Dursban 5 Granules) or *H. bacteriophora*. Field applications (surface or subsurface) were made against a mixed population of second/third-instar white grub, *Hoplia philanthus* in an infested sport field and lawn in the province of West-Flanders. In both trials, the combination of *M. anisopliae* with *H. bacteriophora* at 5×10^{12} conidia $ha^{-1} + 2.5 \times 10^{9}$ IJ ha^{-1} resulted in additive or synergistic affects, causing more than 95% grub mortality when the nematodes were applied 4 weeks after the application of fungus, whereas, the application of nematode, chlorpyrifos or fungus alone provided 39–66, 42–60% (surface) and 33–76, 82–100 or 37–65%, (subsurface) control of *H. philanthus*. They concluded that the pathogen combinations tested are compatible elements of IPM and are likely to improve the control of *H. philanthus* larvae and perhaps of other insect pests beyond what is expected from a single application of the pathogen.

36.2.4 Potential for Control of Japanese Pine Sawyer

Entomopathogenic nematodes have been demonstrated to have a high potential for controlling Japanese pine sawyer *M. alternatus*. In the laboratory, spraying 6×10^{6} and 12×10^{6} nematodes m^{-2} of *S. carpocapsae* in April or September produced 70–80% mortality of *M. alternatus* larvae (Ogawa 1988). Yamanaka (1993) showed that the application of *S. carpocapsae* (Mexican strain), which is considered an ambush forager, at a dose of 6×10^{6} and 12×10^{6} nematodes m^{-2} could kill 69.2 and 72.2% *M. alternatus* larvae, respectively, however, the dose of the nematodes that was used in these experiments was too high compared to the standard, 0.25×10^{6} nematodes m^{-2} (Gaugler and Han 2002). The thermal niche ranges for infection of *S. carpocapsae* was 15–25°C (Vega et al. 2000) in which the optimal temperatures for infection of this species was 20°C (Vega et al. 2000). The average temperature in April in Japan is below 15°C (Kishi 1995) so it is not favorable for the infection of the entomopathogenic nematode and, at this time, *M. alternatus* larvae have already bored inside their tunnels within infested trees. Phan et al. (unpublished result) observed that the virulence of a cruiser species, *S. glaseri* did not differ from that of an ambusher species, *S. carpocapsae*, but the infection rate of *S. glaseri* was slightly higher. Moreover, *S. glaseri* had greater mobility than *S. carpocapsae* and they strongly responded to the host signal (Fig. VI.22); therefore, *S. glaseri* may prove to be a good candidate for controlling *M. alternatus*.

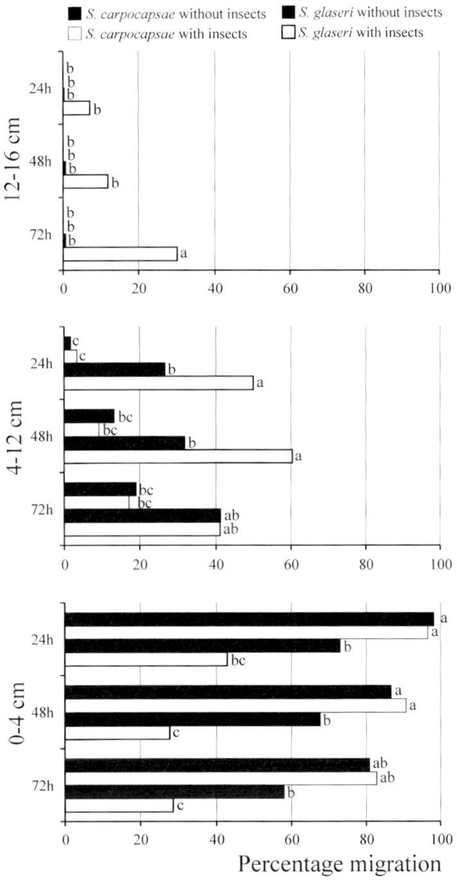

Fig. VI.22 Percentage migration third-stage larvae of *Monochamus alternatus* and no insects in 16 cm high columns of Japanese red pine sawdust containing *Steinernema glaseri* and *S. carpocapsae* at 25 ± 1°C. *Bars* followed by the *same letter* are not significantly different

36.3 Discussion and Conclusions

Monochamus alternatus are widely distributed in China, Japan, Korea, Laos, Taiwan and Vietnam (Kishi 1995; Thu 2003). *Monochamus alternatus* emerge from dead pine trees in Japan in May–July after over-wintering as larvae and pupates in the pupal chamber. Emerging adults of *M. alternatus* start feeding on the pine bark and release PWN that kill pine trees, wilt symptoms appearing two to three months later (Kishi 1995). Twenty days after emergence, the mature females of *M. alternatus* start laying eggs in the dead pine trees and these eggs hatch within 5–7 days after oviposition (Enda et al. 1970). The first- to third-stage larvae of *M. alternatus* feed in the subcortical zone, consuming the phloem (Morimoto and Iwasaki 1974a). The fourth stage larvae of *M. alternatus* start boring into the tunnel after oviposition (35–45 days) (Morimoto 1977). Thus, it is better to control *M. alternatus* when they are still under the bark (first- to third-stage larvae in late August and late September) or at emergence time (May–July).

Several methods have been used successfully for the control of *M. alternatus* larvae such as de-barking and burning, spraying of insecticides, fumigation, but these current methods are not recommended because they are time consuming, have a high application cost or increase the risk of forest fires and negatively affect the environment (Kishi 1995). Although the mortality of *M. alternatus* due to its natural enemies ranged from 0.9 to 47.3% (Forestry Agency 1984), the native habitats of these enemies are in the soil and on the ground or some of these enemies require high relative humidity for their pathogenicity (Doane 1954). Therefore, felling and de-barking diseased trees is necessary because it is easier to apply the control agents and also provides a favourable environment for natural enemies of *M. alternatus* as well as PWN.

Entomopathogenic nematodes showed high potential to control Japanese pine sawyer but very few species have been screened. Entomopathogenic nemotodes can not migrate through the wood, however, as when the beetles are feeding on the sapwood, they produce sawdust (Fig. VI.23) where the nematodes can migrate in search of the host and the nematodes are affective at the first- to third-stage larvae of Japanese pine sawyer (Fig. VI.24) (L.K. Phan, personal communication). In the

Fig. VI.23 Sawdust produced by larvae of the Japanese pine sawyer, *Monochamus alternatus*, feeding on sapwood where entomopathogenic nematodes can migrate (see Color Plates)

Fig. VI.24 Second to fifth stage (*left* to *right*) larvae of the Japanese pine sawyer beetle, *Monochamus alternatus*, killed by the entomopathogenic nematode *Steinernema carpocapsae* (see Color Plates)

future it will be necessary to screen for more effective entomopathogenic nematodes species. As no single microbial control agent provides sustainable control of insect pests or complex of pests (Lacey et al. 2001) it is better to combine entomopathogenic nematodes with other biocontrol agents that may provide a good for control of Japanese pine sawyer.

37
Concluding Remarks

The Japanese pine sawyer, *Monochamus alternatus,* the vector of PWN is native to Japan and East Asia. Originally, *M. alternatus* lived on weakened or dead pines such as dead branches or suppressed trees, and did not kill pine trees. In the past its natural population levels were far lower than today and it was a rare insect species before the PWN was introduced into Japan. Before then the population levels of the pine sawyer were limited mainly by the amount of available food, weakened trees, and thus, the importance of natural enemies was low.

In Japan and Eurasia, *Bursaphelenchus mucronatus*, a closely related species to the PWN, is naturally distributed and carried by *M. alternatus*. This nematode does not kill Asian native pines but lives on weakened pines. However, in North America *B. xylophilus* seldom kills American native pine species. In this regard it is like *B. mucronatus* that does not kill any native pine species in East Asia and lives on weakened pine trees and is carried by other species of *Monochamus* beetles. The PWN originally depended on *Monochamus* beetles. After its introduction to East Asia, the PWN could utilize *M. alternatus* as a vector, and to provide food to *M. alternatus* larvae by killing Asian native pine species. Presence of both the insect vector and pathogenic nematodes was of great benefit to both partners, that is, their relationship is symbiotic. The *M. alternatus*–PWN complex continues to kill East Asian pine species, which are highly susceptible to the PWN. The absence of important natural enemies to *M. alternatus* also benefits pine wilt.

It must be recognized that *M. alternatus*, now associated with PWN, differs in nature before its encounter with the PWN, and must be regarded as an invasive species rather than a native insect. This means that the pine wilt outbreak will never end without the use of artificial intervention.

How should we proceed against this complex? Yoshida (2006) proposed four possible courses against the pine wilt disease, the details of which are explained in Chap. 32; here only his points are summarized: (1) exterminate the PWN, (2) exterminate *M. alternatus*, (3) introduce resistant pines, and (4) continue control to keep the damage at a sub-epidemic level. However, the first two points are almost impossible to achieve except in isolated areas such as islands. Point three, that is, the introduction of resistant pines will require a long time and is not suitable for use to protect existing pine trees, and therefore only option four is suitable for practical control of pine wilt in most areas of Japan. To implement option four, pine forests that are to be saved must be designated; and then pine wilt there must be suppressed to a sub-epidemic level; and thirdly, such damage must be kept at a sub-epidemic level. To suppress the damage level in an affected pine forest, control measures such as eradication of dead pine trees infested with *M. alternatus* as sources of the subsequent infection, and preventive measures against infection for living trees, are required.

Actually pine wilt damage is still continuing. Why? Because, the control program has not been implemented properly. Some of the reasons are underestimation of damage by "today's" *M. alternatus*, budget shortfalls, a decrease in economic importance of pine trees and the indifference of forest owners. Sudden increases in pine wilt damage often occur because local governments have tended to cut the budget for pine wilt control and because control efforts have been relaxed after many years of continuous control. We must be aware that spending half the amount on the control of pine wilt disease is the same as wasting money, and we should try to control the disease with the aim of complete success.

Sometimes the mass media report the prevalence of pine wilt disease even after so much money has been spent. The general public becomes confused about the cause and the measures being used against the disease. Therefore, education of the public and administrative officials is necessary, to explain that "today's" *M. alternatus* populations did not attain equilibrium by the ecological forces of nature and to appeal for the necessity of complete control. Scientists must help in providing the knowledge about this important forest tree disease.

Addendum

It was in November 2003 that Professor Kazuyoshi Futai's book on pine wilt disease appeared before Japan's general reading public under the title *Matsugare wa mori no kansenshô*, which literally translates as "The pine wilt is an infectious disease of forests." This record of scientific endeavor reported that two culprits had been identified after decades of careful investigation on the part of capable scientists in and outside Japan during the latter half of the last century to discover the reason for pine wilt disease: the first, a kind of nematode, *Bursaphelenchus xylophilus*, which acts as a pathogen; and the second, a beetle known as *Monochamus alternatus*, which acts as a vector. The volume also offered insights into which species of pine were vulnerable to attack.

The special attraction of Professor Futai's book, however, is that it goes beyond the scientific discussion of cause and effect. The pine wilt story is part of a more sweeping narrative: that of the on-going consequences of escalation in the mobility of goods and humans that began during the latter half of the nineteenth century. Species that used to coexist within comparatively stable and small ecological communities in the past have been mobilized on an unprecedented scale, the result being numerous encounters between mutually unfamiliar specimens that perhaps should never have met – species that had undergone independent courses of evolution unique to their diverse original habitats. It is only natural that the result of these encounters should often become dramatic, as is the case with the story of the spread of pine wilt disease.

As in this story's beginning, the role of human activity will remain central to how the story will end. The search for a "happy ending" to this narrative requires not only knowledge, but deep thought and wisdom; the magnitude of the problem must be fully understood and international measures should be taken in the pursuit of attaining a new equilibrium. For this, interdisciplinary dialogue across national boundaries is indispensable. I believe that the symposium held at Kyoto University in February 2007 will be remembered in the future as the first substantial step in that direction.

Kyoto University is known for its historical commitment to academic freedom and originality. This commitment has encouraged a thriving tradition of Kyoto scholars' field sciences, where a long line of scholars have looked to the world beyond the written page in reconsidering their preconceptions and creating new ideas. In 2001, Kyoto University set out to integrate the increasingly over-diversified academic disciplines by promulgating a new ideal in the form of a new Mission Statement for the university: "To pursue harmonious coexistence within the human and ecological community on this planet". The drafting committee emphasized that this community should include non-humans ranging from animals and plants, to rocks and streams. Today, when humanity finds itself at a critical stage of its history and conventional notions of nature and humanity are under serious review, it is time, perhaps, for Kyoto University, with its rich tradition of field sciences, to aspire to the ideal espoused in its Mission Statement.

Evergreen pine trees have been treasured as a symbol of long life in East Asia for many centuries. It is natural that many people wish to see pines continue to flourish wherever they grow, be it in gardens or graveyards, in the hills sheltering cities like Kyoto or along windswept coastlines. Still, I personally feel inclined to include even the "guilty" nematode among the non-human members of our planetary community. Might *Bursaphelenchus xylophilus* find a proper niche once more – one in which it need not play such a role of powerful pathogen as to compel the ever quick-tempered humans to seek its eradication?

Toshio Yokoyama, D.Phil.
Professor and Vice-President, Kyoto University
Kyoto, Japan
May 2008

Addendum

Pine wilt disease causes serious damage to pine forests worldwide. It is now the most devastating forest disease in the People's Republic of China. Although much progress has been made in research, we still do not have effective measures to completely control the disease. Further scientific research on pine wilt disease is needed both in theoretical and in practical areas.

After more than a decade of efforts, Professor Bo Guang Zhao and his colleagues have proposed a new theory based on experimental data to explain the complex causes behind the symptoms of pine wilt disease. The new hypothesis is still developing through discussions with scientists who hold different opinions. I think it will provide new directions for clarifying the obscure aspects of the disease.

This book is an excellent example of scientists from different countries collaborating in research on a common problem in the world. I believe it can be useful for university students and scientists who are working in this field, or are going to, because it has assembled knowledge from almost every area of the disease by authors from around the world.

I am very pleased to see this book published by Springer. It represents the achievements of authors with many years' endeavors to publish this work, including the editors, Professor Bo Guang Zhao and his colleagues, Dr. K. Futai, Dr. J. R. Sutherland, Dr. Y. Takeuchi, and all of the scientists around the world who have contributed their efforts to conquering pine wilt disease. I especially thank Dr. K. Futai for his great help in editing the book when Professor Zhao needed his assistance.

Shi-Yuan Yu
Professor, Dr. of Biochemistry
President, Nanjing Forestry University
Nanjing, People's Republic of China
May 2008

References

Abad P, Tarés S, Bruguier N, De Guiran G (1991) Characterization of the relationships in thepinewood nematode species complex (PWNSC) (*Bursaphelenchus* spp.) using a heterologous *unc-22* DNA prove from *Caenorhabditis elegans*. Parasitology 102:303–308

Abbas MST, Hanounik SB, Mousa SA, Albagham SH (2000) Soil applications of entomopathogenic nematodes as a new approach for controlling *Rhynchophorus ferrugineus* on date palm. Int J Nematol 10:215–218

Adam BJ, Nguyen KB (2002) Taxonomy and systematics. In: Gaugler R (ed) Entomopathogenic nematology. CABI Publishing, Oxfordshire, pp 1–33

Adams JC, Morehart AL (1982) Decline and death of *Pinus* spp. in Delaware caused by *Bursaphelenchus xylophilus*. J Nematol 14:382–385

Agriculture, Forestry and Fisheries Research Council (1987) Studies on pathogenesis of pine wilt for exploitation of novel control techniques (in Japanese). Res Rep 180:153

Aikawa T (2006) Transmission mechanism of the pinewood nematode, *Bursaphelenchus xylophilus*: How does the nematode transfer to and depart from its insct vector? (in Japanese with English abstract). J Jpn For Soc 88:407–415

Aikawa T, Togashi K (1997) An effect of inoculum quantity of *Bursaphelenchus xylophilus* (Nematoda: Aphelenchoididae) on the nematode load of *Monochamus alternatus* (Coleoptera: Cerambycidae) in laboratory. Jpn J Nematol 27:14–21

Aikawa T, Togashi K (1998) An effect of pine volatiles on departure of *Bursaphelenchus xylophilus* (Nematoda: Aphelenchoididae) from *Monochamus alternatus* (Coleoptera: Cerambycidae). Appl Entomol Zool 33:231–237

Aikawa T, Togashi K (2000) Movement of *Bursaphelenchus xylophilus* (Nematoda: Aphelenchoididae) in tracheal system of adult *Monochamus alternatus* (Coleoptera: Cerambycidae). Nematology 2:495–500

Aikawa T, Maehara N, Futai K, Togashi K (1997) A simple method for loading adult *Monochamus alternatus* (Coleoptera: Cerambycidae) with *Bursaphelenchus xylophilus* (Nematoda: Aphelenchoididae). Appl Entomol Zool 32:341–346

Aikawa T, Kikuchi T, Kosaka H (2003a) Demonstration of interbreeding between virulent and avirulent populations of *Bursaphelenchus xylophilus* (Nematoda: Aphelenchoididae) by PCR–RFLP method. Appl Entomol Zool 38:565–569

Aikawa T, Togashi K, Kosaka H (2003b) Different developmental responses of virulent and avirulent isolates of the pinewood nematode, *Bursaphelenchus xylophilus* (Nematoda: Aphelenchoididae), to the insect vector, *Monochamus alternatus* (Coleoptera: Cerambycidae). Environ Entomol 32:96–102

Aikawa T, Kikuchi T, Kosaka H (2006) Population structure of *Bursaphelenchus xylophilus* within single *Pinus thunbergii* trees inoculated with two nematode isolates. For Path 36:1–13

Akasofu Y (1974) Infestation of *Bursaphelenchus xylophilus* and *Monochamus alternatus* in Toyama Prefecture (I) (in Japanese). Ann Rep Toyama Pref For Exp Stn 9:82–98

Akbulut S, Linit MJ (1999) Flight performance of *Monochamus carolinensis* (Coleoptera: Cerambycidae) with respect to nematode phoresis and beetle characteristics. Environ Entomol 28:1014–1020

Akbulut S, Vieira P, Ryss A, Yuksel B, Keten A, Mota MM (2006) Preliminary survey of the pinewood nematode in Turkey. EPPO Bull/Bull OEPP 36:538–542

Akbulut S, Braasch H, Baysal I, Brandstetter M, Burgermeister W (2007) Description of *Bursaphelenchus anamurius* sp. n. (Nematoda: Parasitaphelenchidae) from *Pinus brutia* in Turkey. Nematology 9:859–867

Akema T, Futai K (2005) Ectomycorrhizal development in a *Pinus thunbergii* stand in relation to the location on a slope and their effects on tree mortality from Pine Wilt Disease. J For Res 10:93–99

Akhurst RJ, Dunphy G (1993) Tripartite interactions between symbiotically associated entomopathogenic bacteria, nematodes, and their insect host. In: Beckage N, Thompson S, Federici B (eds) Parasites and pathogens of insects, vol 2. Academic Press, New York, pp 1–23

Akiba M (2006) Diversity of pathogenicity and virulence in the pinewood nematode, *Bursaphelenchus xylophilus* (in Japanese with English abstract). J Jpn For Soc 88:383–391

Akiba M, Nakamura K (2005) Susceptibility of adult trees of the endangered species *Pinus armantii* var. *amamiana* to pine wilt disease in the field. J For Res 10:3–7

Akiba M, Nakamura K, Ishihara M (2000) Adult emergence of *Monochamus alternatus* from dead *Pinus armandii* var. *amamiana* trees and its nematode load (in Japanese). Trans Ann Mtg Kyushu Br Jpn For Soc 53:103–104

Akiba M, Ishihara M, Sasaki M, Okamura M, Sahashi N (2003) Virulence and genetic structure of *Bursaphelenchus xylophilus* populations in a *Pinus densiflora* stand under the expanding injury from the pine wilt disease (in Japanese). Trans Jpn For Soc 114:756

Allison AC (1954) Protection afforded by sickle-cell trait against subtertian malarial infection. Br Med J 1:290–294

Ambrogioni L, Palmisano MA (1998) Description of *Bursaphelenchus tusciae* sp. n. from *Pinus pinea* in Italy. Nematol Mediterr 26:97–116

Anbutsu H, Togashi K (2000) Deterred oviposition response of *Monochamus alternatus* (Coleoptera: Cerambycidae) to oviposition scars occupied by eggs. Agric For Entomol 2:217–223

Anbutsu H, Togashi K (2001) Oviposition deterrent by female reproductive gland secretion in Japanese pine sawyer, *Monochamus alternatus*. J Chem Ecol 27:1151–1161

Anbutsu H, Togashi K (2002) Oviposition deterrence associated with larval frass of the Japanese pine sawyer, *Monochamus alternatus* (Coleoptera: Cerambycidae). J Insect Physiol 48:459–465

Andow DA, Kareiva PM, Levin SA, Okubo A (1990) Spread of invading organisms. Landscape Ecol 4:177–188

Anonymous (1986) European Plant Protection organization. EPPO data sheets on quarantine organism #158: *Bursaphelenchus xylophilus* (Steiner & Buhrer) Nickle et al. (Nematoda: Aphelenchoidae). EPPO Bull 16:55–60

Ansari MA, Shah FA, Tirry L, Moens M (2006) Field trials against *Hoplia philanthus* (Coleoptera: Scarabaeidae) with a combination of an entomopathogenic nematode and the fungus *Metarhizium anisopliae* CLO 53. Biol Control 39:453–549

Aoki J (1971) *Beauveria bassiana* (Bals.) Vuill. isolated from some lepidopterous species in Japan (in Japanese). Jpn J Appl Ent Zool 15:222–227

Aoyagi M, Ishibashi N (1983) Gametogenesis of the pinewood nematode, *Bursaphelenchus xylophilus*. Jpn J Nematol 13:20–25

Arakawa Y, Togashi K (2002) Newly discovered transmission pathway of *Bursaphelenchus xylophilus* from males of the beetle *Monochamus alternatus* to *Pinus densiflora* trees via oviposition wounds. J Nematol 34:396–404

Arakawa Y, Togashi K (2004) Presence of the pine wood nematode, *Bursaphelenchus xylophilus*, in the spermatheca of female *Monochamus alternatus*. Nematology 6:157–159

Arias M, Robertson L, Garcia-Alvarez A, Arcos SC, Escuer M, Sanz R, Mansilla JP (2005) *Bursaphelenchus fungivorus* (Nematoda: Aphelenchida) associated with *Orthotomicus erosus* (Coleoptera: Scolitydae) in Spain. For Path 35:375–383

Arihara T (1984) Studies on new technologies to prevent pine wilt disease—study on utilization of natural enemies of *Monochamus alternatus* (in Japanese). Bull Fukushima Pref For Exp Stn 16:1–22

Arihara T (1997a) Trials for simple assay of resistance to pine wilt disease (I) (in Japanese). Shinrin-boeki (For Pests) 46:3–8

Arihara T (1997b) Trials for simple assay of resistance to pine wilt disease (II) (in Japanese). Shinrin-boeki (For Pests) 46:224–226

Aruga Y (1959) *Sclerodermus nipponensis*, a parasitic wasp of *Ernobius mollis*, stinging the human (in Japanese). Shinrin-boeki (For Pests) 8:112–113

Asahina S (1953) On a remarkable case of the biting of a parasitic wasp, *Sclerodermus nipponensis* Yuasa in Tokyo (Hymenoptera, Bethylidae). Jpn J Med Sci Biol 6:197–199

Asai E (2002) The effects of simulated acid rain on the development of pine wilt disease. PhD thesis, Kyoto Univ, Japan, p 122

Asai E, Futai K (2005) Effects of inoculum density of pinewood nematode on the development of pine wilt disease in Japanese black pine seedlings pretreated with simulated acid rain. For Path 35:135–144

Asai E, Futai K (2006) The effect of acid rain on the defense response of pines to pinewood nematodes. In: Kamata N, Liebhold AM, Quiring DT, Clancy KM (eds) Proccedings of IUFRO Kanazawa 2003 "forest insect population dynamics and host influences", Kanazawa University, Kanazawa, pp 21–24

Babu P (1974) Biochemical genetics of *C. elegans*. Mol Gen Genet 135:39–44

Bachman ES, McClay DR (1996) Molecular cloning of the first metazoan beta-1,3 glucanase from eggs of the sea urchin *Strongylocentrotus purpuratus*. Proc Natl Acad Sci USA 93: 6808–6813

Baker AD (1962) Check lists of the nematode superfamilies Dorylaimoidea, Rhabditoidea, Tylenchoidea, and Aphelenchoidea. E.J. Brill, Leiden

Bakhetia M, Charlton W, Atkinson HJ, McPherson MJ (2005) RNA interference of dual oxidase in the plant nematode Meloidogyne incognita. Mol Plant Microbe Interact 18:1099–1106

Barras F, Gijsegem Fv, Chatterjee AK (1994) Extracellular enzymes and pathogenesis of soft-rot *Erwinia*. Annu Rev Phytopathol 32:201–234

Barron GL, Thorn RG (1987) Destruction of nematodes by species of *Pleurotus*. Can J Bot 65:774–778

Baujard P (1980) Trois nouvelle espèces de *Bursaphelenchus* (Nematoda: Tylenchida) et remarques sur le genre (in French). Rev Nématol 3:167–177

Baujard P (1989) Remarques sur les gernes des sousfamilles Bursaphelenchinae Paramonov, 1964 et Rhadinaphelenchinae Paramonov, 1964 (Nematoda: Aphelenchoididae) (in French). Rev Nématol 12:323–324

Baujard P, Boulbria A, Ham R, Laumond C, Scotto La Massèse C (1979) Premières données sur la nématofaune associée aux dépérissements du pin maritime dans l'Ouest de la France (in French). Ann For Sci 36:331–339

Beckenbach K, Smith MJ, Webster JM (1992) Taxonomic affinities and intra- and interspecific variation in *Bursaphelenchus* spp. as determined by polymerase chain reaction. J Nematol 24:140–147

Beckenbach K, Blaxter M, Webster JM (1999) Phylogeny of *Bursaphelenchus* species derived from analysis of ribosomal internal transcribed spacer DNA sequences. Nematology 1:539–548

Beckenbach K, Smith MJ, Webster JM (1992) Taxonomic affinities and intra- and interspecific variation in *Bursaphelenchus* spp. as determined by polymerase chain reaction. J Nematol 24:140–147

Bedding RA (1990) Logistics and strategies for introducing entomopathogenic nematode technology into developing countries. In: Gaugler R, Kaya HK (eds) Entomopathogenic nematodes in biological control. CRC Prees, Boca Raton, pp 233–246

Bedker PL, Blanchette RA (1988) Mortality of Scots pine following inoculation with the pinewood nematode, *Bursaphelenchus xylophilus*. Can J For Res 18:574–580

Bedker PL, Wingfield MJ, Blanchette RA (1987) Pathogenicity of *Bursaphelenchus xylophilus* on three species of pine. Can J For Res 17:51–57

Bentley MD, Mamiya Y, Yatagai M, Shimizu K (1985) Factors in *Pinus* species affecting the mobility of the pine wood nematode, *Bursaphelenchus xylophilus*. Ann Phytopathol Soc Jpn 51:556–561

Bergdahl DR (1982) Occurrence of the pinewood nematode in eastern larch. In: Appleby JE, Malek RB (eds) Proceedings of the 1982 national pine wilt disease workshop, pp 47–55

Bergdahl DR, Halik S (1999) Inoculated *Pinus sylvestris* serve as long term hosts for *Bursaphelenchus xylophilus*. In: Futai K, Togashi K, Ikeda T (eds) Sustainability of pine forests in relation to pine wilt and decline. Proceedings of international symposium, Tokyo, 27–28 October 1998. Shokado, Kyoto, Japan pp 73–78

Beschin A, Bilej M, Hanssens F, Raymakers J, Van Dyck E, Revets H, Brys L, Gomez J, De Baetselier P, Timmermans M (1998) Identification and cloning of a glucan- and liopoplysaccharide-binding protein from *Eisenia foetida* earthworm involved in the activation of prophenoloxidase cascade. J Biol Chem 273:24948–24954

Bird AF, Bird J (1986) Observations on the use of insect parasitic nematodes as a means of biological control of root-knot nematodes. Int J Parasitol 16:511–516

Blaxter M, Page AP, Rudin W, Maizels RM (1992) Nematode surface coats actively evading immunity. Parasitol Today 8:243–247

Blaxter ML, De Ley P, Garey JR, Liu LX, Schheldeman P, Vierstraete A, Vanfleteren JR, Mackey LY, Dorris M, Frisse LM, Vida JT, Thomas WK (1998) A molecular evolutionary framework for the phylum Nematoda. Nature 392:71–75

Bledsoe CS (1992) Physiological ecology of ectomycorrhizae: implications for field application. In: Allen MF (ed) Mycorrhizal functioning: an integrative plant-fungal process. Chapman & Hall, New York, pp 424–437

Bogdanowicz SM, Mastro VC, Prasher DC, Harrison RG (1997) Microsatellite DNA variation among Asian and North American gypsy moths, *Lymantria dispar* (Lepidoptera: Lymantriidae). Ann Entomol Soc Am 90:768–775

Bogdanowicz BM, Schaefer PW, Harrison RG (2000) Mitochondrial DNA variation among worldwide populations of gypsy moths, *Lymantria dispar*. Mol Phylogenet Evol 15:487–495

Bogdanowicz BM, Wallner WE, Bell J, Odell TM, Harrison RG (1993) Asian gypsy moths (Lepidoptera: Lymantriidae) in North America: Evidence from molecular data. Ann Entomol Soc Am 86:710–715

Bolla RI, Boschert M (1993) Pinewood nematode species complex: interbreeding potential and chromosome number. J Nematol 25:227–238

Bolla R, Shaheen F, Winter REK (1982a) Phytotoxin produced in *Bursaphelenchus xylophilus*-infected *Pinus sylvestris*. In: Appleby JE, Malek RB (eds) Proceedings of the 1982 national pine wilt disease workshop, pp 17–31

Bolla RI, Shaheen F, Winter REK (1982b) Phytotoxin produced in *Bursaphelenchus xylophilus* pine wilt. J Nematol 14:431

Bolla RI, Shaheen F, Winter REK (1984a) Effect of Phytotoxin from nematode-induced pine wilt on *Bursaphelenchus xylophilus* and *Ceratocystis ips*. J Nematol 16:297–303

Bolla RI, Shaheen F, Winter REK (1984b) Phytotoxins in *Bursaphelenchus xylophilus* induced pine wilt. In: Dropkin V (ed) The resistance mechanisms of pines against pine wilt disease, proceedings of the US–Japan seminar, pp 119–127

Bolla RI, Winter REK, Fitzsimmons K, Linit MJ (1986) Pathotypes of the pinewood nematode *Bursaphelenchus xylophilus*. J Nematol 18:230–238

Bolla RI, Fitzsimmons K, Winter REK (1987) Carbohydrate concentration in pine as affected by inoculation with *Bursaphelenchus xylophilus*. J Nematol 19:51–57

Bolla RI, Weaver C, Winter REK (1988) Gemonic differences among pathotypes of *Bursaphelenchus xylophilus*. J Nematol 20:309–316

Bolla RI, Nosser C, Tamura H (1989) Chemistry of response of pines to *Bursaphelenchus xylophilus*: resin acids. Jpn J Nematol 19:1–6

Boots M, Sasaki A (1999) "Small worlds" and the evolution of virulence: infection occurs locally and at a distance. Proc R Soc Lond B 266:1933–1938

Borgonie G, Jacobsen K, Cooman A (2000) Embryonic lineage evolution in nematodes. Nematology 2:65–69

Bowers WW, Hudak J, Raske AG, Magasi LP, Lachance D, Myren DT, Cerezke HF, Van Sickle GA (eds) (1992) Host and vector surveys for the pinewood nematode, *Bursaphelenchus xylophilus* (Steiner and Buhrer) Nickle (Nematoda: Aphelenchoididae) in Canada. Information Report N-X-285, Forestry Canada Newfoundland and Labrador Region

Boyd L, Guo S, Levitan D, Stinchcomb DT, Kemphues KJ (1996) PAR-2 is asymmetrically distributed and promotes association of P granules and PAR-1 with the cortex in *C. elegans* embryos. Development 122:3075–3084

Braasch H (1998) *Bursaphelenchus hofmanni* sp. n. (Nematoda: Aphelenchoididae) from spruce wood in Germany. Nematologica 44:615–621

Braasch H (2000) *Bursaphelenchus paracorneolus* sp. nov. (Nematoda: Parasitaphelenchidae) aus Koniferenhorz in Deutschland und Bemerkungen zu seiner Biologie und Verbreitung (in German). Ann Zool 50:177–182

Braasch H (2001) *Bursaphelenchus* species in conifers in Europe: distribution and morphological relationships. EPPO Bull 31:127–142

Braasch H (2004) A new *Bursaphelenchus* species (Nematoda: Parasitaphelenchidae) sharing characters with Ektaphelenchidae from the People's Republic of China. Zootaxa 624:1–10

Braasch H, Braasch-Bidasak R (2002) First record of the genus *Bursaphelenchus* Fuchs, 1937 in Thailand and description of *B. thailandae* sp. n. (Nematoda: Parasitaphelenchidae). Nematology 4:853–863

Braasch H, Burgermeister W (2002) *Bursaphelenchus rainulfi* sp. n. (Nematoda: Parasitaphelenchidae), first record of the genus *Bursaphelenchus* Fuchs, 1973 from Malaysia. Nematology 4:971–978

Braasch H, Enzian S (2004) The pinewood nematode problem in Europe—present situation and outlook. In: Mota M, Vieira P (eds) The pinewood nematode, *Bursaphelenchus xylophilus*. Nematology monographs and perspectives, vol 1. E.J. Brill, Leiden, pp 77–91

Braasch H, Schmutzenhofer H (2000) *Bursaphelenchus abietinus* sp. n. (Nematoda: Parasitaphelenchidae) associated with fir bark beetles (*Pityokteines* spp.) from declining silver fir trees in Austria. Russ J Nematol 8:1–6

Braasch H, Burgermeister W, Pastrik KH (1995) Differentiation of three *Bursaphelenchus* species by means of RAPD–PCR. Nachrichtenblatt des Deutschen Pflanzenschutzdienstes 47:310–314

Braasch H, Burgermeister W, Harmey MA, Michalopoulos-Skarmoutsos H, Tomiczek C, Caroppo S (2000) Pest risk analysis of pinewood nematode related *Bursaphelenchus* species in view of South European pine wilt and wood imports from Asia. Final Report of EU Research Project Fair CT 95-0083

Braasch H, Schönfeld U, Polomski J, Burgermeister W (2004) *Bursaphelenchus vallensianus* sp. n.—a new species of the *Bursaphelenchus sexdentati* group (Nematoda: Parasitaphelenchoididae). Nematologia Mediterranea 32:71–79

Braasch H, Gu J, Burgermeister W, Zhang J (2005) *Bursaphelenchus doui* sp. n. (Nematoda: Parasitaphelenchidae) in packaging wood from Taiwan and South Korea—a new species of the xylophilus group. Russ J Nematol 12:19–27

Braasch H, Brandstetter M, Burgermeister W (2006a) Supplementary characters of *Bursaphelenchus lini* Braasch, 2004 (Nematoda: Parasitaphelenchidae) and remarks on this nematode. Zootaxa 1141:55–61

Braasch H, Burgermeister W, Schönfeld U, Metge K, Brandstetter M (2006b) *Bursaphelenchus hildegardae* sp. n. (Nematoda: Parasitaphelenchidae)—a new species belonging to the "eggersi" group. J Nematode Morphol System 9:27–38

Braasch H, Burgermeister W, Schröder T, Apel K-H, Brandstetter M (2006c) Vorkommen von *Bursaphelenchus eremus* Rühm, 1956 (Nematoda: Parasitaphelenchidae) an Eichen in Brandenburg/Deutschland und Ergänzungen zur Artbeschreibung (in German). Nachrichtenblatt des Deutschen Pflanzenschutzdienstes 58:148–153

Braasch H, Gu J, Burgermeister W, Brandstetter M, Metge K (2006d) *Bursaphelenchus africanus* sp. n. (Nematoda: Parasitaphelenchidae)—found in packaging wood exported from South Africa to Ningbo/China. J Nematode Morphol System 9:71–81

Braasch H, Gu J, Brandstetter M (2007) *Bursaphelenchus burgermeisteri* sp. n. (Nematoda: Parasitaphelenchidae) in packaging wood from Japan—a second species of the "africanus" group. J Nematode Morphol System 10:39–48

Brenner S (1974) The genetics of *Caenorhabditis elegans*. Genetics 77:71–94

Brzeski MW, Baujard P (1997) Morphology and morphometrics of *Bursaphelenchus* (Nematoda: Aphelenchoididae) species from pine wood of Poland. Ann Zool 47:305–319

Bucher GE (1960) Potential bacterial pathogens of insects and their characteristics. J Insect Pathol 2:172–195

Bucher GE, Stephens JM (1957) A disease of grasshoppers caused by the bacterium *Pseudomonas aeruginosa* (Schroeter) Migula. Can J Microbiol 3:611–625

Burdon DR (2001) Genetic diversity and disease resistance: some considerations for research, breeding, and deployment. Can J For Res 31:596–606

Burgermeister W, Gu J, Braasch H (2005a) *Bursaphelenchus arthuri* sp. n. (Nematoda: Parasitaphelenchidae) in packaging wood from Taiwan and South Korea—a new species belonging to the fungivorus group. J Nematode Morphol System 8:7–17

Burgermeister W, Metge K, Braasch H, Buchbach E (2005b) ITS–RFLP patterns for differentiation of 26 *Bursaphelenchus* species (Nematoda: Parasitaphelenchoididae) and observations on their distribution. Russ J Nematol 13:29–42

Burman M (1982) *Neoaplectana carpocapsae*: toxin production by axenic insect parasitic nematodes. Nematologica 28:62–70

Burman M, Pye AE (1980) *Neoaplectana carpocapsae*: Movements of the nematode populations on a thermal gradient. Exp Parasitol 49:258–265

Burnell AM, Stock SP (2000) *Heterorhabditis, Steinernema* and their bacterial symbionts-lethal pathogens of insects. Nematology 2:31–42

Buscot F, Weber G, Oberwinkler F (1992) Interactions between *Cylindrocarpon destructans* and ectomycorrhizas of *Picea abies* with *Laccaria laccata* and *Paxillus involutus*. Trees 6:83–90

Byer JA, Poinar GO (1982) Location of insect hosts by nematode, *Neoaplectana carpocapsae*, in response to temperature. Behaviour 79:1–10

Cai W, Xi Q, Toda T (2003) The actual condition of damage of pine wilt disease and starting resistant breeding (in Japanese). Shinrin-boeki (For Pests) 52:4–10

Cai W, Gao J, Xu L, Xi Q, Toda T (2006) Development of selection methods for resistant candidates of pine wilt disease in Anhui province—the state of breeding for resistance to pine wilt disease in Anhui, Japan International Cooperation Agency technical cooperation project—the report of completed technology developmental projects announcement (in Japanese), pp 21–25

Campbell JF, Gaugler R (1997) Inter-specific variation in entomopathogenic nematode foraging strategy: dichotomy or variation along a continuum? Fundam Appl Nematol 20:393–398

Campbell JF, Kaya HK (2002) Variation in entomopathogenic nematodes (Steinernematidae and Heterorhabditidae) infective-stage jumping behaviour. Nematology 4:471–482

Canny MJ (2001) Contributions to the debate on water transport. Am J Bot 88:43–46

Cao Y (1997) Studies on toxin of pine wilt disease caused by pine wood nematode (in Chinese with English abstract). PhD thesis, Nanjing For Univ, China, pp 1–46

Cao Y, Shen BK (1996) Studies on toxicity of extraction of pine wood nematodes cultured in artificial media (in Chinese with English abstract). J Nanjing For Univ 20:13–16

Çao AX, Liu XZ, Zhu SF, Lu BS (2005) Detection of the pinewood nematode, *Bursaphelenchus xylophilus*, using a real-time polymerase chain reaction assay. Phytopathology 95:566–571

Carling DE (1984) Some insect associates of the pinewood nematode in eastern Virginia. Can J For Res 14:826–829

Carpita NC, Gibeaut DM (1993) Structural models of primary cell walls in flowering plants: consistency of molecular structure with the physical properties of the walls during growth. Plant J 3:1–30

Carson SD, Carson MJ (1989) Breeding for resistance in forest trees—a quantitative genetic approach. Annu Rev Phytotathol 27:373–395

Castagnone-Sereno P, Castagnone C, François C, Abad P (2008) Satellite DNA as a versatile genetic marker for *Bursaphelenchus xylophilus*. In: Mota M, Vieira P (eds) Pine wilt disease: a worldwide threat to forest ecosystems. Springer, The Netherlands, p 30

Cesari M, Marescalchi O, Francardi V, Mantovani B (2005) Taxonomy and phylogeny of European *Monochamus* species: first molecular and karyological data. J Zool Syst Evol Res 43:1–7

Cheng HR, Lin MS, Qian RJ (1986) A study on morphological diagnosis and pathogenicity of the pine wood nematode. J Nanjing Agri Univ 2:55–59

Chen MR, Yu HB, Fang TS (2002) Comparison of trapping effect between two kinds of attractants on *Monochamus alternatus* (in Chinese). For Pest Dis (6):3–4

Chen Q, Rehman S, Smant G, Jones JT (2005) Functional analysis of pathogenicity proteins of the potato cyst nematode *Globodera rostochiensis* using RNAi. Mol Plant Microbe Interact 18:621–625

Chen SL (2003) Infectivity and persistence of entomopathogenic nematodes used to control the cabbage root fly *Delia radicum*. PhD thesis, University of Ghent, Belgium

Chen SL, Han X, Moens M (2003) Effect of chlorpyrifos on infectivity and survival of *Steinernema feltiae*. Russ J Nematol 11:1–6

Chen Y, Tan JJ, Feng ZX (2003) Effects of oxytetracycline hydrochloride on pine wood nematode disease. J Sichuan For Sci Technol 25:7–10

Chen YH, Ye JR, Wei CJ (2005) Reactions of pine seedling and in vitro seedling twig to the infection of *Bursaphelenchus xylophilus*. J Nanjing For Univ 29 (Natural Science edition): 19–22

Chen ZL, Zhang XH, Chai XM, Zhao JN, He ZH, Jiang P (2000) Stoving and hydraulic pressing treatment of *Bursaphelenchus xylophilus*-infected pine timbers (in Chinese with English astract). For Pest Dis (6):15–18

Chi SY, Han ZM, He YQ (2006) Studies on the pathogenicity of 10-year-old black pine inoculated with aseptic pine wood nematode (in Chinese with English abstract). Scientia Silvae Sinicae 42(10):71–73

Chida T, Sato H (1981) Adult occurrence time of *Monochamus alternatus* in Ichinoseki, Iwate Prefecture (in Japanese). Trans Ann Mtg Jpn For Soc Tohoku Br 33:144–145

Christiansen E, Solheim H (1990) The bark beetle-associated blue-stain fungus *Ophiostoma polonicum* can kill various spruces and Douglas fir. Eur J For Path 20:436–446

Chung Y-J, Lee S-M, Kim D-S, Choi K-S, Lee S-G, Park C-G (2003) Measurement and within-tree distribution of larval entrance and adult emergence holes of Japanese pine sawyer, *Monochamus alternatus* (Coleoptera: Cerambycidae) (in Korean). Korean J Appl Entomol 42:315–321

Cobb NA (1919) A newly discovered nematode (*Aphelenchus cocophilus* n. sp.) connected with a serious disease of the coconut palm. West Ind Bull 17:203–210

Conn HJ (1977) Biological stains. In: Lillie RD (ed) A handbook on the nature and uses of the dyes in the biological laboratory. 9th edn. Williams & Wilkins, Baltimore, p 692

Coutts MP (1977) The formation of dry zones in the sapwood of conifers 11—the role of living cells in the release of water. Eur J For Pathol 7:6–12

Daub M, Schroeder T, Sikora R (2008) Distribution, migration behavior and population dynamics of *B. xylophilus* in young *Pinus sylvestris* trees during early wilt at controlled optimum temperature. In: Mota M, Vieira P (eds) Pine wilt disease: a worldwide threat to forest ecosystems. Springer, The Netherlands, p 44

Dautova M, Rosso MN, Abad P, Gommers FJ, Bakker J, Smant G (2001) Single pass cDNA sequencing—a powerful tool to analyse gene expression in preparasitic juveniles of the southern root-knot nematode *Meloidogyne incognita*. Nematology 3:129–139

de Boer JM, McDermott JP, Davis EL, Hussey RS, Popeijus H, Smant G, Baum TJ (2002) Cloning of a putative pectate lyase gene expressed in the subventral esophageal glands of *Heterodera glycines*. J Nematol 34:9–11 de Boer JM, Yan Y, Wang X, Smant G, Hussey RS, Davis EL, Baum TJ (1999) Developmental expression of secretory beta-1,4-endoglucanases in the subventral esophageal glands of *Heterodera glycines*. Mol Plant Microbe Interact 12:663–669

de Guiran G, Boulbria A (1986) Le nématode des pins. Caractéristiques de la souche française et risque d'introduction et d'extension de *Bursaphelenchus xylophilus* en Europe (in French). EPPO Bulletin 16:445–452

de Guiran G, Bruguir N (1989) Hybridization and phylogeny of the pine wood nematode (*Bursaphelenchus* spp.). Nematologica 35:321–330

De Ley IT, De Ley P, Vierstraete A, Karssen G, Moens M, Vanfleteren J (2002) Phylogenetic analyses of *Meloidogyne* small subunit rDNA. J Nematol 34:319–327

De Ley P, Blaxter ML (2002) Systematic position and phylogeny. In: Lee DL (ed) The biology of nematodes. Taylor & Francis, London & New York

Denman S, Crous PW, Tailor JW, Kang JC, Pascoe I, Wingfield MJ (2000) An overview of the taxonomic history of *Botryosphaeria*, and a re-evaluation of its anamorphs based on morphology and ITS rDNA phylogeny. Stud Mycol 45:129–140

Devdariani TG (1974) New nematode species of a small, black spruce-capricorn beetle (*Monochamus sutor* L) (in Gerorgian). Bull Acad Sci Georgian SSR 76:709–712

Devdariani TG, Kakulia GA, Khavatashili DD (1980) New species of nematode of small maple Cpricorn beetle (*Rhopalopus macropus*) (in Gerorgian). Bull Acad Sci Georgian SSR 98:457–459

De Wet J, Burgess T, Slippers B, Preisig O, Wingfield BD, Wingfield MJ (2003) Multiple gene genealogies and microsatellite marker reflect relationship between morphotypes of *Sphaeropsis sapinea* and distinguish a new species of *Diplodia*. Mycol Res 107:557–566

DGRF, Direcção-Geral de Recursos Florestais (2006) Programa Nacional de Luta Contra o Nemátode do Pinheiro (PROLUNP). http://www.dgrf.min-agricultura.pt/prolunp/html/home-final.htm

Dimopoulos G, Richman A, Muller HM, Kafatos FC (1997) Molecular immune responses of the mosquito *Anopheles gambiae* to bacteria and malaria parasites. Proc Natl Acad Sci USA 94:11508–11513

Dix NJ, Webster J (1995) Fungal ecology. Chapman & Hall, London, p 549

Doane CC (1954) *Beauveria bassiana* as a pathogen of *Scolytus multistriatus*. Ann Mtg Entom Soc Am 52:109–111

Dolinski C, Baldwin JG, Thomas WK (2001) Comparative survey of early embryogenesis of Secernentea (Nematoda), with phylogenetic implications. Can J Zool 79:82–94

Dong GY, Mo MH, Cheng JH, Zhang KQ (2000) The morbidity stability of different Lampteromyces japonicus strains. J Yunnan Univ 22:365–368

Dorris M, De Ley P, Blaxter ML (1999) Molecular analysis of nematode diversity and the evolution of parasitism. Parasitol Today 15:188–193

Doyle EA, Lambert KN (2002) Cloning and characterization of an esophageal-gland-specific pectate lyase from the root-knot nematode *Meloidogyne javanica*. Mol Plant Microbe Interact 15:549–556

Dozono Y (1974) Reproduction of *Bursaphelenchus lignicolus* on the colony of various kinds of fungi (in Japanese). Trans Ann Mtg Kyusyu Br Jpn For Soc 27:161

Dozono Y, Kiyohara T (1971) Population growth of *Bursaphelenchus xylophilus* on fungal mycelia at several temperatures (in Japanese). Bull Kyushu Br Jpn For Soc 25:160–161

Dozono Y, Yoshida N (1974) Application of the logistic curve for the population growth of pine wood nematode, *Bursaphelenchus lignicolus*, on the cultures of *Botrytis cinerea*. J Jpn For Soc 56:146–148

Dropkin VH (1982) Nematology research in Missouri. In: Appleby JE, Malek RB (eds) Proceedings of the 1982 national pine wilt disease workshop, pp 7–10

Dropkin VH, Foudin AS (1979) Report of occurrence of *Bursaphelenchus lignicolus*-induced pine wilt disease in Missouri. Plant Dis Reptr 63:904–905

Dropkin VH, Linit M (1982) Pine wilt: a disease you should know. J Arboricult 8:1–6

Dropkin VH, Foudin A, Kondo E, Linit M, Smith M, Robbins K (1981) Pinewood nematode: a threat to U. S. forests? Plant Dis 65:1022–1027

Duan Y, Kerdelhué C, Ye H, Lieutier F (2004) Genetic study of the forest pest *Tomicus piniperda* (Col., Scolytinae) in Yunnan province (China) compared to Europe: new insights for the systematics and evolution of the genus *Tomicus*. Heredity 93:416–422

Dwinell LD (1985a) Relative susceptibilities of five pine species to three populations of the pinewood nematode. Plant Dis 69:440–442

Dwinell DL (1985b) First report of pinewood nematode (*Bursaphelenchus xylophilus*) in Mexico. Plant Dis 77:846

Dwinell LD (1993) First report of pinewood nematode (*Bursaphelenchus xylophilus*) in Mexico. Plant Dis 77:846

Dwinell LD (1997) The pinewood nematode: Regulation and mitigation. Annu Rev Phytopathol 35:153–166

Dwinell LD, Nickle WR (1989) An overview of the pine wood nematode ban in North America. General Technical Report SE-55, North American Forestry Commission Publication No.2, Southeastern Forest Experiment Station, Forest Service, United States Department of Agriculture

Dwinell LD, Magnusson C, Tomminen J (1994) Evaluation of a Swedish steam-dryer for treatment of *Bursaphelenchus xylophilus* in pine chips. EPPO Bull 24:805–811

Ebine S (1980) Wilt disease in cedar tree caused by the pinewood nematode and behaviors of Japanese pine sawyer (in Japanese). Shinrin-boeki (For Pests) 29:201–205

Ebine S (1981) Wilt disease in spruce fir caused by the pinewood nematode (in Japanese). Shinrin-boeki (For Pests) 30:117–119

Eckhardt RC (1979) The adaptive syndromes of two guilds of insectivorous birds in the Colorado Rocky Mountains. Ecol Monogr 49:129–149

Edwards OR, Linit MJ (1992) Transmission of *Bursaphelenchus xylophilus* through oviposition wounds of *Monochamus carolinensis* (Coleoptera: Cerambycidae). J Nematol 24:133–139

Eisenback JD, Vieira P, Ryss A, Mota M (2008) Taxonomic databases for *Bursaphelenchus* and other aphelenchoid nematodes. In: Mota M, Vieira P (eds) Pine wilt disease: a worldwide threat to forest ecosystems. Springer, The Netherlands, p 23

Enda N (1972) Insect vectors of the pine wood nematode and the number of nematodes in the insect vectors (in Japanese). Trans Ann Mtg Kanto Br Jpn For Soc 24:31

Enda N (1977) Boarding of *Bursaphelenchus xylophilus* on *Monochamus alternatus* and its departure from vectors. In: Studies on the control of pine wilt disease. Secretariat of Agriculture, Forestry and Fisheries Research Council, Ministry of Agriculture, Forestry and Fisheries, Tokyo, pp 83–85

Enda N (1980) Duration of *Monochamus alternatus* Hope from pupation to emergence (in Japanese). Trans Ann Mtg Kanto Br Jpn For Soc 32:91–92

Enda N (1985) Flight performance of *Monochamus alternatus* adults investigated by flight mill technique (in Japanese). Trans Mtg Jpn For Soc 96:517–518

Enda N (1989) The state and measure of pine wilt disease in Korea (in Japanese). Shinrin-boeki (For Pests) 38:2–6

Enda N (1992) Control of *Monochamus alternatus* using *Sclerodermus* in China (in Japanese). Shinrin-beoki (For Pests) 41:126–131

Enda N (1994) Biological control of *Semanotus japonicus* and *Monochamus alternatus* using *Sclerodermus nipponensis* (in Japanese). Trans Jpn For Soc 105:178

Enda N (2006) Major pests and diseases on pine trees 6. Pests of weakened, dead and recently felled trees (in Japanese). Ringyo-to-Yakuzai 176:1–12

Enda N, Ikeda T (1983) Role of volatiles of a pine tree as emerging stimulants for attracting the pine wood nematode from the pine sawyer (in Japanese). Trans Mtg Jpn For Soc 94:479–480

Enda N, Kitajima H (1990) Rearing of adults and larvae of the Taiwanese pine sawyer *Monochamus alternatus* Hope (Coleoptera, Cerambycidae) in artificial diets (in Japanese). Trans Mtg Jpn For Soc 101:503–504

Enda N, Makihara H (1982) Estimation of the population of pine wood nematode held in the pine sawyer adult from that on its antenae (in Japanese). Trans Ann Mtg Kanto Br Jpn For Soc 34:145–146

Enda N, Makihara H (2006) Biology of the genus *Monochamus*, especially Japanese pine sawyer, *M. alternatus* (Coleoptera: Cerambycidae) (2) Bionomics of *M. alternatus* in East Asia (in Japanese). Shinrin-boeki (For Pests) 55:11–21

Enda N, Nobuchi A (1970) Studies on subcortical insects in pine trees: ovary development and parasitic nematode (in Japanese). Trans Mtg Jpn For Soc 81:274–276

Enda N, Taketani A (1992) The damage and measure of pine wilt disease in China (in Japanese). Shinrin-boeki (For Pests) 41:10–15

Enda N, Nobuchi A, Makihara H (1970) Studies on the wood-boring beetles of the pines—ovarian maturation and the presence of parasitic nematodes (in Japanese). Trans Mtg Jpn For Soc 81:274–276

Enda N, Igarashi M, Fukuyama K, Nobuchi A (1989) Control of *Monochamus alternatus* Hope (Coleoptera; Cerambycidae) by the entomogenous fungus, *Beauveria bassiana* Vuillemin (Deuteromycotina; Hyphomycetes), carried by *Cryphalus fulvus* Niijima (Coleoptera; Scolytidae) (a preliminary report) (in Japanese). Trans Mtg Jpn For Soc 100: 579–580

Enda N, Gotoh T, Fukuyama K, Tsuchiya D (1991) Control of *Monochamus alternatus* Hope (Coleoptera; Cerambycidae) by the entomogenous fungus, *Beauveria bassiana* Vuillemin (Deuteromycotina; Hyphomycetes) carried by *Cryphalus fulvus* Niijima (Coleoptera; Scolytidae) in Izu-Ohshima Island (in Japanese). Trans Mtg Jpn For Soc 102:281–282

Escuer M, Arias M, Bello A (2004) The genus *Bursaphelenchus* (Nematoda) in Spain. In: Mota M, Vieira P (eds) The pinewood nematode, *Bursaphelenchus xylophilus*. Nematology monographs and perspectives, vol 1. E.J. Brill, Leiden, pp 93–99

Evans HF, McNamara DG, Braasch H, Chadoeuf J, Magnusson C (1996) Pest Risk Analysis (PRA) for the territories of the European Union (asPRA area) on *Bursaphelenchus xylophilus* and its vectors in the genus *Monochamus*. EPPO Bull/Bull OEPP 26:199–249

Evans S, Evans H, Ikegami M (2008) Modelling PWN-induced wilt expression: a mechanistic approach. In: Mota M, Vieira P (eds) Pine wilt disease: a worldwide threat to forest ecosystems. Springer, The Netherlands, p 39

Evert R (2006) Esau's plant anatomy: meristems, cells, and tissues of the plant body: their structure, function, and development. 3rd edn. Wiley, New Jersey, USA

Fairbairn JP, Fenton A, Norman RA, Hudson PJ (2000) Re-assessing the infection strategies of the entomopathogenic nematode *Steinernema feltiae* (Rhabditidae: Steinernematidae). Parasitology 121:211–216

Fanelli E, Di Vito M, Jones JT, De Giorgi C (2005) Analysis of chitin synthase function in a plant parasitic nematode, *Meloidogyne artiellia*, using RNAi. Gene 349:87–95

Fang Y, Zhuo K, Zhao J (2002a) Description of *Bursaphelenchus aberrans* n. sp. (Nematoda: Parasitaphelenchida) isolated from pine wood in Guangdong Province, China. Nematology 4:791–794

Fang Y, Zhuo K, Zhao J (2002b) Description of *Bursaphelenchus dongguanensis* sp. n. in China (Nematoda: Aphelenchoididae). J Huazhong Agric Univ 21:109–111

Farjon A (1990) Pinaceae. Drawings and descriptions of the genera *Abies*, *Cedrus*, *Pseudotsuga*, *Larix* and *Picea*. Koelts Scientific Books, Koningstein, p 330

Farrell BD (2001) Evolutionary assembly of the milkweed fauna: cytochrome oxidase I and the age of *Tetraopes* Beetles. Mol Phylogenet Evol 18:467–478

Fauziah BA, Hidaka T, Tabata K (1987) The reproductive behavior of *Monochamus alternatus* Hope (Coleoptera: Cerambycidae). Appl Entomol Zool 22:272–285

Ferris VR, Ferris JM, Faghihi J (1993) Variation in spacer ribosomal DNA in some cyst-forming species of plant parasitic nematodes. Fundam Appl Nematol 16:177–184

Finlay RD, Read DJ (1986) The structure and function of the vegetative mycelium of ectomycorrhizal plants. II. The uptake and distribution of phosphorus by mycelial strands interconnecting host plants. New Phytol 103:157–165

Fire A, Xu S, Montgomery MK, Kostas SA, Driver SE, Mello CC (1998) Potent and specific genetic interference by double-stranded RNA in *Caenorhabditis elegans*. Nature 391: 806–811

Floyd R, Abebe E, Papert A, Blaxter M (2003) Molecular barcodes for soil nematode identification. Mol Ecol 11:839–850

Forge TA, Sutherland JR (1996) Population dynamics of the pine wood nematode, *Bursaphelenchus xylophilus*, in excised branch segments of western North American conifers. Fundam Appl Nematol 19:349–356

Forest Development Technological Institute (1990) Reports on Researches Commissioned for 1989. Biological control of pine wilt disease (in Japanese). The Forestry Development Technological Institute, Tokyo, p 71

Forestry Agency (ed) (1984) An integrated research for novel techniques to prevent the mortality of pine trees (in Japanese). Achievements in the large-dimension projects 2

Forestry Agency (1992) Reports on Researches Commissioned for 1991. Biological control of pine wilt disease (in Japanese). The Forestry Development Technological Institute, Tokyo, p 76

Forestry Agency (1993) Reports on Researches Commissioned for 1998. Biological control of pine wilt disease (in Japanese). The Forestry Development Technological Institute, Tokyo, p 70

Forestry Agency (1999) Reports on Researches Commissioned for 1998. Biological control of pine wilt disease (in Japanese). The Forestry Development Technological Institute, Tokyo, p 70

Forestry Agency (2000) Reports on Researches Commissioned for 1999. Biological control of pine wilt disease (in Japanese). The Forestry Development Technological Institute, Tokyo, p 59

Forestry Agency (2001) Reports on Researches Commissioned for 2000. Biological control of pine wilt disease (in Japanese). The Forestry Development Technological Institute, Tokyo, p 56

Forst S, Nealson K (1996) Molecular biology of the symbiotic-pathogenic bacteria *Xenorhabdus* spp. and *Photorhabdus* spp. Microbiol Rev 60:21–43

Francardi V, Rumine P, deSilva J (2003) On microbial control of *Monochamus galloprovincialis* (Olivier) (Coleoptera Cerambycidae) by means of *Beauveria bassiana* (Bals.) Vuillemin (Deuteromycotina Hyphomycetes). Redia 86:129–132

François C, Castagnone C, Boonham N, Tomlinson J, Lawson R, Hockland S, Quill J, Viera P, Mota M, Castagnone-Sereno P (2007) Satellite DNA as a target for TaqMan real-time PCR detection of the pinewood nematode, *Bursaphelenchus xylophilus*. Mol Plant Pathol 8:803–809

Franklin MT, Hooper DJ (1962) *Bursaphelanchus fungivorus* n. sp. (Nematoda: Aphelenchoidea) from rotting gardenia buds infected with *Botrytis cinerea* Purs. ex. Fr. Nematologica 8:136–142

Fuchs AG (1930) Neue an Borken- und Rüsselkäfer gebundene Nematoden, halbparasitische und Wohnungseimieter (in German). Zool Jahrb Abteilung Syst Öekol Geogr Tiere 59:505–646

Fuchs AG (1937) Neue parasitische und halbparasitische Nematoden bei Borkenkäfern und einige andere Nematoden. I. Teil (in German). Zool Jahrb Abteilung Syst Öekol Geogr Tiere 70:291–380

Fujimoto F (1991) Forest tree breeding projects—resistant breeding. In: Ohoba K, Katsuta M (eds) Forest tree breeding (In Japanese). Buneido Publishing co., Ltd, Tokyo, pp 187–201

Fujimoto F, Toda T, Nishimura K, Yamate H, Fuyuno S (1989) Breeding project on resistance to pine-wood nematode—an outline of the research and the achievement of the project for 10 years (in Japanese with English abstract). Bull For Tree Breed Inst 7:1–84

Fujioka H (1993) A report on the habitat of *Monochamus alternatus* Hope in Akita prefecture (in Japanese with English abstract). Bull Akita For Tech Cent 2:40–56

Fujioka H, Yanbe T (1991) Dispersal of released adults of *Monochamus alternatus* (in Japanese). Trans Mtg Jpn For Soc 102:291–292

Fujishita A (1978) Infestation of pine wilt disease in Shizuoka Prefecture (in Japanese). Trans Ann Mtg Chubu Br Jpn For Soc 26:193–198

Fukuda K (1993) Development of pine wilt disease in Japanese black pine (*Pinus thunbergii*) seedlings inoculated with *Bursaphelenchus xylophilus* and *B. mucronatus* under the restricted conditions of photosynthesis and transpiration (in Japanese). Trans Jpn For Soc 104:641–646

Fukuda K (1997) Physiological process of the symptom development and resistance mechanism in pine wilt disease. J For Res 2:171–181

Fukuda K, Hogetsu T, Suzuki K (1992a) Cavitation and cytological changes in xylem of pine seedlings inoculated with virulent and avirulent isolates of *Bursaphelenchus xylophilus* and *B. mucronatus*. J Jpn For Soc 74:289–298

Fukuda K, Hogetsu T, Suzuki K (1992b) Photosynthesis and water status in pine wood nematode-infected pine seedlings. J Jpn For Soc 74:1–8

Fukuda K, Hogetsu T, Suzuki K (1994) Ethylene production during symptom development of pine-wilt disease. Eur J For Path 24:193–202

Fukuda T, Iwakawa M (1979) The character and pine wood nematode resistance of hybrids from Loblolly pine and Masson pine (in Japanese). For Tree Breed 111:43–46

Fukushige H (1990) The number of *Bursaphelenchus xylophilus* carried by *Monochamus alternatus* and some possible factors regulating the number. Jpn J Nematol 20:18–24

Fukushige H (1991) Propagation of *Bursaphelenchus xylophilus* (Nematoda: Aphelenchoididae) on fungi growing in pine-shoot segments. Appl Entomol Zool 26:371–376

Fukushige H, Futai K (1987) Seasonal changes in *Bursaphelenchus xylophilus* populations and occurrence of fungi in *Pinus thunbergii* trees inoculated with the nematode. Jpn J Nematol 17:8–16

Furukoshi T, Sasaki M (1979) On the study of the hybridization among the species belong to *Sylvestres* subsect (II) phenotypic characteristics of hybrid seedlings at 0–1 stage (in Japanese). Trans Mtg Jpn For Soc 90:233–234

Furukoshi T, Sasaki M (1983) Interspecific hybridization in *Pinus* in the subsection *Sylvestres* LOUD, and breeding for resistance to pine wood nematode (in Japanese). For Tree Breed 129:1–6

Furukoshi T, Sasaki M (1985) Interspecific hybridization in *Pinus* in the subsection *Sylvestres* LOUD. Bull Fort Tree Breed Inst 3:21–35

Furuno T (1980) Exotic *Pinus* spp. damaged by the pinewood nematode (in Japanese). Trans Ann Mtg Kansai Br Jpn For Soc 31:241–243

Furuno T (1982) Studies on the insect damage upon the pine-species imported in Japan (No. 7) on the withering of the pines by the pine wilt (in Japanese with English abstract). Bull Kyoto Univ For 54:16–30

Furuno T, Futai K (1983) Growth of pines inoculated with pine wood nematode (*Bursaphelenchus xylophilus*), especially on the growth for three years after inoculation (in Japanese with English abstract). Bull Kyoto Univ For 55:1–19

Furuno T, Uenaka K (1979) Studies on the insect damage upon the pine-species imported in Japan (No. 6) on the feeding of Japanese pine sawyer adult, *Monochamus alternatus* Hope (in Japanese with English abstract). Bull Kyoto Univ For 51:12–22

Furuno T, Watanabe H, Uenaka K (1977) Studies on the insect damage upon the pine-species imported in Japan (No. 4) On the Japanese pine sawye, *Monochamus alternatus* Hope infecting Loblolly pine, *Pinus taeda* Linn., and Lace-bark pine, *P. bungeana* Zucc. (in Japanese with English abstract). Bull Kyoto Univ For 49:8–19

Furuno T, Futai K, Nakai Y (1984) Survival of *Pinus thunbergii* × *P. massoniana* (F_2), *P. thunbergii* × *P. tablaeformis* (F_1), *Pinus thunbergii* × *P. kasia* (F_1) and *P. yunnanensis* inoculated with *Bursaphelenchus xylophilus* (in Japanese). Trans Ann Mtg Kansai Br Jpn For Soc 35: 154–157

Futai K (1979) Responses of two species of *Bursaphelenchus* to the extracts from pine segments and to the segments immersed in different solvents. Jpn J Nematol 9:54–59

Futai K (1980a) Developmental rate and population growth of *Bursaphelenchus lignicolus* (Nematoda: Aphelenchoididae) and *Bursaphelenchus mucronatus*. Appl Entomol Zool 15:115–122

Futai K (1980b) Host preference of *Bursaphelenchus lignicolus* (Nematoda: Aphelenchoididae) and *B. mucronatus* shown by their aggregation to pine saps. Appl Entmol Zool 15:193–197

Futai K (1984) Tannin accumulation in the stems of pine seedlings infected with *Bursaphelenchus xylophilus* or *B. mucronatus* (in Japanese). Trans Mtg Jpn For Soc 95:473–474

Futai K (1985a) Factors determining the affinity between pine wood nematodes and their host pines. III. Host specific aggregation and invasion of *Bursaphelenchus xylophilus* (Nematoda: Aphelenchoididae) and *B. mucronatus*. Mem Coll Agric, Kyoto Univ 126:35–43

Futai K (1985b) Factors determining the affinity between pine wood nematodes and their host pines. IV. Host resistances shown at the time of pine wood nematode invasion. Mem Coll Agric, Kyoto Univ 126:45–53

Futai K (1999) The epidemic manner of pine wilt spread in a Japanese red pine stand (in Japanese). For Res, Kyoto 71:9–18

Futai K (2003a) Role of asymptomatic carrier trees in epidemic spread of pine wilt disease. J For Res 8:253–260

Futai K (2003b) Abnormal metabolites in pine wood nematode-inoculated Japanese black pine. Jpn J Nematol 33:45–56

Futai K, Furuno T (1979) The variety of resistances among pine species to pine wood nematode, *Bursaphelenchus lignicolus* (in Japanese with English abstract). Bull Kyoto Univ For 51:23–26

Futai K, Okamoto T (1989) Ecological studies on the infection sources of pine wilt. III. Analysis of spatial patterns of pine wilt (in Japanese). Trans Mtg Jpn For Soc 100:549–550

Futai K, Sutherland JR (1989) Pathogenicity and attraction to host extracts of Canadian pinewood nematodes: studies with Scots pine, western larch, and black spruce seedlings. Can J For Res 19:1256–1261

Futai K, Nakai I, Fukiharu T, Akai T (1986) Ecological studies on the infection sources of pine wilt (I) population dynamics of pine wood nematodes in the withered stems of Japnese red pine (in Japanese with English abstract). Bull Kyoto Univ For 57:1–13

Gao B, Allen R, Maier T, McDermott J, Davis E, Baum T, Hussey R (2002) Characterisation and developmental expression of a chitinase gene in *Heterodera glycines*. Int J Parasitol 32:1293

Gao J, Xu L, Toda T (2003) The programs of breeding for resistance to pine wood nematode in Masson pine in Anhui China (in Japanese). For Tree Breed 208:38–41

Gao JB, Cai WB, Xu LY, Xi QJ, Tada TD (2005) Selection on isolated species for *Bursaphelenchus xylophilus* of inoculation in resistant breeding. China For Sci Technol 19:24–27

Garcia-Alvarez A, Robertson L, Mansilla J, Bello A, Arias M (2008) Potential insect vectors of *Bursaphelenchus* spp. (Nematoda: Parasitaphelenchidae) in Spanish pine forests. In: Mota M, Vieira P (eds) Pine wilt disease: a worldwide threat to forest ecosystems. Springer, The Netherlands, pp 33–34

Gaugler R (1988) Ecological considerations in the biological control of soil-inhabiting insects with entomopathogenic nematodes. Agric Ecosyst Environ 24:351–360

Gaugler R, Han R (2002) Production technology. In: Gaugler R (ed) Emtomopathogenic nematology. CABI Publishing, Oxfordshire, pp 289–310

Gaugler R, Kaya HK (1990) Entomopathogenic Nematodes in Biological Control. CRC Press, Boca Raton, Florida, USA

Gaugler R, LeBeck L, Nakagaki B, Boush GM (1980) Orientation of the entomopathogenic nematodes, *Neoaplectana carpocapsae*, to carbon dioxide. Environ Entomol 8:649–652

Ge MH, Xu FY, Zhang P, Xu WL, Xu KQ (1999) Studies on the effects of *Pinus massoniana* and *Pinus thunbergii* resistance to pine wood nematode (PWN) induced by hormone, calcium, salicylic acid and ammonium. J Jiangsu For Sci Technol 26:7–12

Georgis R, Kaya HK (1998) Advances in entomopathogenic nematode formulation. In: Burges HD (ed) Formulation of microbial biopesticides, beneficial microorganisms nematodes and seed treatments. Kluwer, Dordrecht, pp 289–308

Georgis R, Manweiler SA (1994) Entomopathogenic nematodes: a developing biological control technology. Agric Zool Rev 6:63–94

Georgis R, Poinar GO (1983a) Effect of soil texture on distribution and infectivity of *Neoaplectana carpocapsae* (Nematoda: Steinernematidae). J Nematol 15:308–311

Georgis R, Poinar GO (1983b) Effect of soil texture on distribution and infectivity of *Neoaplectana glaseri* (Nematoda: Steinernematidae). J Nematol 15:319–322

Gerber K, Giblin-Davis RM (1990) Association of the red ring nematode and other nematode species with the palm weevil, *Rynchophorus palmarum*. J Nematol 22:143–149

Gernandt DS, López GG, García SO, Liston A (2005) Phylogeny and classification of *Pinus*. Taxon 54:29–42

Gheysen G, Jones JT (2006) Molecular aspects of plant-nematode interactions. In: Perry RN, Moens M (eds) Plant nematology. CABI, Wallingford, pp 234–254

Giblin RM (1985) Association of *Bursapelenchus* sp. (Nematoda: Aphelenchoididae) with nitidulid beetles (Coleoptera: Nitidulidae). Rev Nématol 8:369–375

Giblin RM, Kaya HK (1983) *Bursaphelenchus seani* n. sp. (Aphelenchoididae), a phoretic associate of bees in the genus *Halictus* (Hymenoptera: Halictidae). Rev Nématol 6:39–50

Giblin RM, Swan JL, Kaya HK (1984) *Bursaphelenchus kevini* n. sp. (Aphelenchida: Aphelenchoididae), an associate of bees in the genus *Halictus* (Hymenoptera: Halictidae). Rev Nématol 7:177–187

Giblin-Davis RM, Verkade SD (1988) Solarization for nematode disinfection of small volumes of soil. Ann Appl Nematol 2:41–45

Giblin-Davis RM, Mundo-Ocampo M, Baldwin JG, Gerber K, Griffith R (1989) Observation on the morphology of the red ring nematode, *Rhadinaphelenchus cocophilus* (Nemata: Aphelenchoididae). Rev Nématol 12:285–292

Giblin-Davis RM, Norden BB, Batra SWT, Eickwort GC (1990) Commensal nematodes in the glands, genitalia, and brood cells of bees (Apoidea). J Nematol 22:150–161

Giblin-Davis RM, Mundo-Ocampo M, Baldwin JG, Norden BB, Batra SWT (1993) Description of *Bursaphelenchus abruptus* n. sp. (Nematoda: Aphelenchoididae), an associate of a digger bee. J Nematol 25:161–172

Giblin-Davis RM, Hazir S, Center BJ, Ye W, Keskin N, Thorp R, Thomas WK (2005) *Bursaphelenchus anatolius* n. sp. (Nematoda: Aphelenchoididae), an associate of bees in the genus *Halictus*. J Nematol 37:336–342

Giblin-Davis RM, Kanzaki N, Ye W, Center BJ, Thomas WK (2006a) Morphology and systematics of *Bursaphelenchus gerberae* n. sp. (Nematoda: Parasitaphelenchidae), a rare associate of the palmweevil, *Rhynchophorus palmarum* in Trinidad. Zootaxa 1189:39–53

Giblin-Davis RM, Kanzaki N, Ye W, Mundo-Ocampo M, Baldwin JG, Thomas WK (2006b) Morphology and description of *Bursaphelenchus platzeri* n. sp. (Nematoda: Parasitaphelenchidae), an associate of nitidulid beetles. J Nematol 38:150–157

Gilbert M, Grégire J-C, Freise JF, Heitland W (2004) Long-distance dispersal and human population density allow the prediction of invasive patterns in the horse chestnut leafminer *Cameraria ohridella*. J Anim Ecol 73:459–468

Glazer I, Nakache Y, Klein M (1992) Use of entomopathogenic nematodes against foliage pests. Hassadeh 72:626–630

Goellner M, Smant G, Boer JMd, Baum TJ, Davis EL (2000) Isolation of beta-1,4-endoglucanase genes from *Globodera tabacum* and their expression during parasitism. J Nematol 32:154–165

Goldstein B, Hird SN (1996) Specification of the anteroposterior axis in *Caenorhabditis elegans*. Development 122:1467–1474

Goldstein B, Frisse LM, Thomas WK (1998) Embryonic axis specification in nematodes: evolution of the first step in development. Curr Biol 8:157–160

Goldstein P (1981) Sex determination in nematodes. In: Zuckerman BM, Rohde RA (eds) Plant parasitic nematodes, vol III. Academic press, New York, pp 37–60

Gönczy P, Rose LS (2005) Asymmetric cell division and axis formation in the embryo. In: The *C. elegans* Research Community (ed) Wormbook (15 October 2005). doi:10.1895/wormbook.1.30.1, http://www.wormbook.org

Goto S, Miyahara F, Ide Y (2002a) Identification of the male parents of half-sib progeny from Japanese black pine (*Pinus thunbergii* Parl.) clonal seed orchard using RAPD markers. Breed Sci 52:71–77

Goto S, Miyahara F, Ide Y (2002b) Monitoring male reproductive success in a Japanese black pine clonal seed orchard with RAPD markers. J For Res 32:983–988

Grattapaglia D, Ribeiro VJ, Rezende GDSP (2004) Retropective selection of elite parent trees using paternity testing with microsatellite markers: an alternative short term breeding tactic for *Eucalyptus*. Theor Appl Genet 109:192–199

Grewal PS, Lewis EE, Gaugler R, Campbell JF (1994) Host finding behaviour as a predicator of foraging strategy in entomopathogenic nematodes. Parasitology 108:207–215

Griffith R (1987) Red ring disease of coconut palm. Plant Dis 71:193–196

Grishok A (2005) RNAi mechanisms in *Caenorhabditis elegans*. FEBS Lett 579:5932–5939

Gu J, Braasch H, Burgermeister W, Brandstetter M, Zhang J (2006a) Description of *Bursaphelenchus yongensis* sp. n. (Nematoda: Parasitaphelenchidae) isolated from *Pinus massoniana* in China. Russ J Nematol 14:91–99

Gu J, Braasch H, Burgermeister W, Brandstetter M, Zhang J (2006b) Records of *Bursaphelenchus* spp. intercepted in imported packaging wood at Ningbo, China For Path 36:323–333

Gu J, Zhang J, Braasch H, Burgermeister W (2005) *Bursaphelenchus singaporensis* sp. n. (Nematoda: Parasitaphelenchidae) in packaging wood from Singapore—a new species of the *B. xylophylus* group. Zootaxa 988:1–12

Guo DS, Cong PJ, Li L, Zhao BG (2002) Determination of bacterial number carried by a pine wood nematode and culture of sterilized nematodes on calli of *Pinus thunbergii* (in Chinese with English abstract). J Qingdao Univ 4:29–31

Guo DS, Du GC, Li L, Zhao BG, Li RG (2007) Flagellin of P*seudomonas fluorescens* GcM5-1A carried by pine wood nematodes is toxic to suspension cells and seedlings of *Pinus thunbergii* (submitted to Nematology)

Guo DS, Zhao BG, LI RG (2006) Effect of pine wood nematode on the propagation and pathogenicity of its carrying bacterial strain (in Chinese with English abstract). Chin Appl Environ Biol 12:523–527

Guo QQ, Guo DS, Zhao BG, Jie X, Li RG (2007) Two cyclic dipeptides from *Pseudomonas fluorescens* GcM5-1A carried by pine wood nematode and their toxicities to Japanese black pine suspension cells and seedlings in vitro (submitted to J Nematol)

Haag ES (2005) The evolution of nematode sex determination: *C. elegans* as a reference point for comparative biology. In: The *C. elegans* Research Community (ed) *Wormbook* (29 December 2005) doi:10.1895/wormbook.1.120.1, http://www.wormbook.org

Hafren J, Daniel G, Westermark U (2000) The distribution of acidic and esterified pectin in cambium, developing xylem and mature xylem of *Pinus sylvestris*. Iawa J 21:157–168

Hagiwara Y, Ogawa S, Takeshita H (1975) Range expansion of pine wilt disease (in Japanese). Trans Ann Mtg Kyushu Br Jpn For Soc 28:153–154

Halik S, Bergdahl DR (1994) Long-term survival of *Bursaphelnchus xylophilus* in living *Pinus sylvestris*. Eur J For Path 24:357–363

Hamaguchi K, Matsumoto T, Maruyama M, Hashimoto Y, Yamane S, Itioka T (2007) Isolation and characterization of eight microsatellite loci in two morphotypes of the Southeast Asian army ant, *Aenictus laeviceps*. Mol Ecol Notes 7:984–986

Han H, Han BY, Chung YJ, Shin SC (2008) A simple PCR-RFLP for identification of Bursaphelenchus spp. Collected from Korea. Plant Pathol J 24:159–163

Han J-H, Yoon C, Shin S-C, Kim G-H (2007) Seasonal occurrence and morphological measurements of pine sawyer, *Monochamus saltuarius* (Coleoptera: Cerambycidae). J Asia Pac Entomol 10:63–67

Han ZM, Hong YD, Zhao BG (2003) A study on pathogenicity of bacteria carried by pine wood nematodes. J Phytopathol 151:683–689

Hanawa F, Yamada T, Nakashima T (2001) Phytoalexins from *Pinus strobus* bark infected with pinewood nematode, *Bursaphelenchus xylophilus*. Phytochemistry 57:223–228

Handa T, Katou K, Ueki T, Hanetani K (1995) Decline of pine wood nematode resistance of resistants's children caused by non-resistant pollen contamination (in Japanese). Trans Jpn For Soc 106:295–296

Hanounik SB, Saleh MME, Abuzuhairah RA, Alheji M, Aldhahir H, Alijarash Z (2000) Efficacy of entomopathogenic nematodes with antidesiccants in controlling the red palm weevil, *Rhynchophorus ferrugineus* on date palm trees. Int J Nematol 10:131–134

Hanula JL (1993) Vertical distribution of black vine weevil (Coleoptera: Curculionidae) immatures and infection by entomogenous nematodes in soil column and field soil. J Econ Entomol 86:340–347

Hao D, Ma F, Wang Y, Zhang Y, Dai H (2006) Electroantennogram and behavioral responses of *Monochamus alternatus* to the volatiles from *Pinus thunbergii* with different physiological status (in Chinese with English abstract). Chin J Appl Ecol 17:1070–1074

Hara N, Futai K (2001) Histological changes in xylem parenchyma cells and the effects on tracheids of Japanese black pine inoculated with pine wood nematode, *Bursaphelenchus xylophilus* (in Japanese with English abstract). J Jpn For Soc 83:285–289

Hara N, Takeuchi Y (2006) Histological analysis for mechanism of pine wilt disease (in Japanese with English abstract). J Jpn For Soc 88:364–369

Harmey JH, Harmey MA (1994) DNA profiling of *Bursaphelenchus* species. Gene 145:227–230

Harris TS, Sandall LJ, Powers TO (1990) Identification of single *Meloidogyme* juveniles by polymerase chain reaction amplification of mitochondrial DNA. J Nematol 22: 518–524

Harvey SC, Viney ME (2001) Sex determination in the parasitic nematode *Strongyloides ratti*. Genetics 158:1527–1533

Hasegawa K, Koyama R (1937): Studies on pathogens of forest insects and their application (preliminary report) (in Japanese). Bull For Exp Stn Imperial Household 3:1–26

Hasegawa K, Miwa S, Futai K, Miwa J (2004) Early embryogenesis of the pinewood nematode *Bursaphelenchus xylophilus*. Dev Growth Differ 46:153–161

Hasegawa K, Mota MM, Futai K, Miwa J (2006) Chromosome structure and behavior in *Bursaphelenchus xylophilus* (Nematoda: Parasitaphelenchidae) germ cells and early embryo. Nematology 8:425–434

Hashimoto H (1975) Recent and future studies on *Bursaphelenchus xylophilus*—with reference to the interrelation between the nematode and the host pine during the process of infection and pathogenesis —(in Japanese). Shinrin-boeki (For Pests) 10:189–192

Hashimoto H (1979) Changes of respiration in Japanese black pine inoculated with pine wood nematode (in Japanese). Trans Ann Mtg Kyusyu Br Jpn For Soc 32:261–262

Hashimoto H (1980a) Histochemical changes of parenchyma cells of pine trees resulted from inoculation with the pine wood nematode (in Japanese). Trans Ann Mtg Kyusyu Br Jpn For Soc 33:163–164

Hashimoto H (1980b) Changes of cambial activity in Japanese black pine inoculated with pine wood nematode, *Bursaphelenchus lignicolus* (in Japanese). Trans Mtg Jpn For Soc 91:367–370

Hashimoto H, Dozono Y (1973) Propagation of the pine wood nematode in pine logs (in Japanese). Trans Ann Mtg Kyusyu Br Jpn For Soc 26:185–186

Hashimoto H, Dozono Y (1975) Migration and reproduction of the pine wood nematode in resistant and susceptible pines (in Japanese). Trans Mtg Jpn For Soc 86:301–302

Hashimoto H, Kiyohara K (1973) Migration of pine wood nematode in pine trees (III) (in Japanese). Trans Ann Mtg Kyushu Br Jpn For Soc 26:330–332

Hashimoto H, Sanui T (1974) Influence of inoculum quantity of *Bursaphelenchus xylophilus* on wilting disease development in *Pinus thunbergii* trees (in Japanese). Trans Mtg Jpn For Soc 85:251–253

Hashimoto H, Dozono Y, Kiyohara T, Suzuki K (1976) Nematode density in the pine trees inoculated with variety of pine wood nematode consistency to different number of branches (in Japanese). Trans Ann Mtg Kyushu For Assoc 29:203–204

Hashimoto S, Sakaguchi N, Kashio T, Gyotoku Y, Kai I, Narahara M (1991) A carrier for cultures of the entomogenous fungus, *Beauveria brongniartii*, for the biological control of the whitespotted longicorn beetle, *Anoplophora malasiaca* (in Japanese). Proc Assoc Pl Prot Kyushu 37:170–174

Hauben L, Steenackers M, Swings J (1998) PCR-based detection of the causal agent of watermark disease in willows (*Salix* spp.). Appl Environ Microbiol 64:3966–3971

Hazir C, Giblin-Davis RM, Keskin N, Ye W, Kanzaki N, Center BJ, Hazir S, Kaya HK, Thomas WK (2007) *Bursaphelenchus debrae* n. sp. (Nematoda: Parasitaphelenchidae), an associate of the bee *Halictus brunnescens* in Turkey. Nematology 9:777–789

Hazir S, Stock SP, Kaya HK, Koppenhöfer AM, Keskin N (2001) Developmental temperature effects on five geographic isolates of the entomopathogenic nematodes *Steinernema feltiae* (Nematoda: Steinernematidae). J Invertebr Pathol 77:243–250

He K, Xu ZQ, Dai PL (2006) The parasitizing behavior of Scleroderma guani Xiao et Wu (Hymenoptera: Bethylidae) wasps on Tenebrio molitor pupae. Acta Entomol Sin 49:454–460

Hedgecock EM, Culotti JG, Hall DH, Stern BD (1987) Genetics of cell and axon migration in *Caenorhabditis elegans*. Development 100:365–382

Heil M, Bostock RM (2002) Induced systemic resistance (ISR) against pathogens in the context of induced plant defenses. Ann Bot 89:503–512

Henrissat B, Bairoch A (1993) New families in the classification of glycosyl hydrolases based on amino acid sequence similarities. Biochem J 293 (Pt 3):781–788

Higashihara T (2004) Development of pine wood resistant varieties in Tohoku Regional Breeding Office (in Japanese). For Tree Breed Technol News 20:2–3

Higgins DF, Harmy MA, Jones DL (1999) Pathogenicity related gene expression in *Bursaphelenchus xylophilus*. In: Futai K, Togashi K, Ikeda T (eds) Sustainability of pine forests in relation to pine wilt and decline. Proceedings of international symposium, Tokyo, 27–28 October 1998. Shokado, Kyoto, Japan, pp 23–28

Hillis WE (1987) Heartwood and tree exudates. Springer, Berlin, p 268

Hinode Y, Shuto Y, Watanabe H (1987) Stimulating effects of β-myrcene on molting and multiplication of the pine wood nematode, *Bursaphelenchus xylophilus*. Agric Biol Chem 51:1393–1396

Hirai S, Fukuda K, Hogetsu T, Suzuki K (1994) Migration of pinewood nematode (*Bursaphelenchus xylophilus*) in current-year stem of Japanese black pine (*Pinus thunbergii*) (in Japanese). Trans Jpn For Soc 105:481–482

Hisaeda K, Shiraishi S, Hujisawa Y, Miyahara H, Ishimatu M, Ieiri R, Sasaki Y, Mituki Y, Kawauchi H (2003) MuPS (multiplex-PCR of SCAR markers) types of the plus trees and native Japanese cedar (in Japanese). Bull Kyushu Univ For 84:59–71

Hodgkin J (2002) Exploring the envelope: systematic alteration in the sex-determination system of the nematode *Caenorhabditis elegans*. Genetics 162:767–780

Holdeman QL (1980) The pinewood nematode (*Bursaphelenchus lignicolus* Mamiya and Kiyohara, 1972) and the associated pine wilt disease of Japan. California Department of Food and Agriculture, Sacramento

Hominick WM, Reid AP, Bohan DA, Briscoe BR (1996) Entomopathogenic nematodes: biodiversity, geographical distribution and the Convention on Biological Diversity. Biocontrol Sci Technol 6:317–331

Hong TY, Cheng CW, Huang JW, Meng M (2002) Isolation and biochemical characterization of an endo-1,3-beta-glucanase from *Streptomyces sioyaensis* containing a C-terminal family 6 carbohydrate-binding module that binds to 1,3-beta-glucan. Microbiology 148:1151–1159

Hope IA (2002) Embryology, developmental biology and the genome. In: Lee DL (ed) The biology of nematodes. Taylor & Francis, London, pp 121–145

Hoshizaki K, Sano S, Sakuraba H, Tabuchi N, Yoshida M, Oikawa Y, Makita A, Kobayashi K (2005) A practical protection from pine-wilt disease through conversion of infected trees to charcoal: strategy for reduction of disease-vectors and a case for a coastal pine forest, northern Japan (in Japanese with English abstract). Tohoku J For Sci 10:82–89

Hotta T, Hashimoto H, Masuda T (1975) Influence of temperature on the population dynamics and pathogenicity of *Bursaphelenchus xylophilus* (in Japanese). Trans Mtg Jpn For Soc 86: 303–304

Houthoofd W, Jacobsen K, Mertens C, Vangestel S, Cooman A (2003) Embryonic cell lineage of the marine nematode *Pellioditis marina*. Dev Biol 258:57–69

Hoyar U, Burgermeister W, Braasch H (1998) Identification of *Bursaphelenchus* species (Nematoda, Aphelenchoididae) on the basis of amplified ribosomal DNA (ITS–RFLP). Nachrichtenbl Dtsch Pflanzenschutzdienstes 50:273–277

Hu KJ, Wang QL, Yang BJ (1995) A study on pathogenicity of different strains of both *B. xylophilus* and *B. mucronatus*. In: Yang BJ (ed) Epidemiology and treatment of the pine wood nematode in China. Forest Publishing Company of China, Beijing, pp 50–53

Huang G, Dong R, Allen R, Davis EL, Baum TJ, Hussey RS (2005) Developmental expression and molecular analysis of two *Meloidogyne incognita* pectate lyase genes. Int J Parasitol 35:685–692

Huang HH, Xu ZF, Yang ZQ (2003) *Dastarcus helophoroides*—a natural enemy of *Monochamus alternatus*. J Guandong For Sci Technol 3:11–15

Huang JS, He XY (2001) The present research situation of pine wood nematode disease in our country and the approach to precaution countermeasures in Fujian province. J Fujian For Sci Technol 28:12–17

Huang JS, He XY, Yang X, Chen JW Chen HM (2005) Effect and application of introducing *Monochamus alternatus* in forest with FJ-MA-02 introduction agent. J Fujian For Sci Technol 32:1–5

Huey RB, Pianka ER (1981) Ecological consequences of foraging mode. Ecology 62:991–999

Huger AM (1966) A virus disease of the Indian rhinoceros beetle, *Oryctes rhinoceros* (Linnaeus), caused by a new type of insect virus, *Rhabdionvirus oryctes* gen. n., sp. n. J Invertebr Pathol 8:38–51

Hui E, Webster JM (2000) Influence of insect larvae and seedling roots on the host-finding ability of *Steinernema feltiae* (Nematoda: Steinernematidae). J Invertebr Pathol 75:152–162

Humphry SJ, Linit MJ (1989) Effect of pinewood nematode density on tethered flight of *Monochamus carolinensis* (Coleoptera: Cerambycidae). Environ Entomol 18:670–673

Hunt DJ (1993) Aphelenchida, Longidoridae and Trichodridae—their systematics and bionomics. CAB International, Wallingford

Hunt DJ, Hague NGM (1974) A redescription of *Parasitaphelenchus oldhami* Rühm, 1956 (Nematoda: Aphelenchoididae) a parasite of two elm bark beetles: *Scolytus scolytus* and *S. multistriatus*, together with some notes on its biology. Nematologica 20:174–180

Hunt DJ (2008) A check list of the Aphelenchoidea (Nematode Tylenchina). J Nematode Morph Syst 10:99–135

Hunt MD, Neuenschwander UH, Delaney TP, Weymann KB, Friedrich LB, Lawton KA, Steiner HY, Ryals JA (1996) Recent advances in systemic acquired resistance research—a review. Gene 179:89–95

Hutchison DW, Templeton AR (1999) Correlation of pairwise genetic and geographical distance measures: inferring the relative influences of gene flow and drift on the distribution of genetic diversity. Evolution 53:1898–1914

Hwang CF, Williamson VM (2003) Leucine-rich repeat-mediated intramolecular interactions in nematode recognition and cell death signaling by the tomato resistance protein Mi. Plant J 34:585–593

Ibrahim SK, Perry RN, Burrows PR, Hooper DJ (1994) Differenciation of species and populations of *Aphelenchoides* and *Ditylenchus augustus* using a fragment of ribosomal DNA. J Nematol 26:412–421

Ibaraki C, Ohoba K, Toda T, Hashimoto H, Kiyohara T (1978) Difference of virulence to Japanese black pine seedlings among 23 race pine wood nematode. Trans Ann Mtg Kyushu For Assoc 31:211–212

Ichihara Y (2000) Migration path of the pine wood nematode in host tissue (in Japanese). Newsletter of Tohoku Research Center 462:1–4

Ichihara Y, Fukuda K, Suzuki K (2000a) Early symptom development and histological changes associated with migration of *Bursaphelenchus xylophilus* in seedling tissues of *Pinus thunbergii*. Plant Dis 84:675–680

Ichihara Y, Fukuda K, Suzuki K (2000b) The effect of periderm formation in the cortex of *Pinus thunbergii* on early invasion by the pinewood nematode. For Path 30:141–148

Ichihara Y, Fukuda K, Suzuki K (2001) Suppression of ectomycorrhizal development in young *Pinus thunbergii* trees inoculated with *Bursaphelenchus xylophilus*. For Path 31: 141–147

Ido N, Kobayashi K (1977) Dispersal of *Monochamus alternatus*. In: Studies on the control of pine wilt disease. Secretariat of Agriculture, Forestry and Fisheries Research Council, Ministry of Agriculture, Forestry and Fisheries, Tokyo, pp 87–88

Ido N, Takeda J (1972) Some observations on biology and morphology of the pine sawyer (in Japanese). Trans Ann Mtg Kansai Br Jpn For Soc 23:180–182

Ido N, Takeda J (1975) A few remarks on the oviposition and life span of *Monochamus alternatus* adults (in Japanese). Trans Mtg Jpn For Soc 86:337–338

Ido N, Takeda J, Kobayashi K, Taketani A, Hosoda R (1975) Research on dispersal of *Monochamus alternatus* adults (in Japanese). Trans Mtg Jpn For Soc 86:341–342

Igarashi M (1977) Biology of *Monochamus alternatus* in Tohoku district II (in Japanese). Bull Tohoku Branch Gov For Exp Stn 18:126–133

Igarashi M (1980) Woodpecker predation on hibernating larvae of *Monochamus alternatus* (Coleoptera: Cerambycidae) (in Japanese). Trans Mtg Jpn For Soc 91:363–364

Ikeda T (1993) Attractants of *Monochamus alternatus* (in Japanese). In: Kobayashi F, Taketani A (eds) Forest insects. Yokendo, Tokyo, pp 128–129

Ikeda T, Oda K (1980) The occurrence of attractiveness for *Monochamus alternatus* Hope (Coleoptera: Cerambycidae) in nematode-infected pine trees. J Jpn For Soc 62:432–434

Ikeda T, Ohtsu M (1992) Detection of xylem cavitation in field-grown pine trees using the acoustic emission technique. Ecol Res 7:391–395

Ikeda T, Suzaki T (1984) Influence of pine wood nematodes on hydraulic conductivity and water status in *Pinus thunbergii*. J Jpn For Soc 66:412–420

Ikeda T, Enda N, Yamane A, Oda K, Toyoda T (1980a) Attractants for the Japanese pine sawyer, *Monochamus alternatus* Hope (Coleoptera: Cerambycidae). Appl Entomol Zool 15:358–361

Ikeda T, Oda K, Yamane A, Enda N (1980b) Volatiles from pine logs as the attractant for Japanese pine sawyer *Monochamus alternatus* Hope (Coleoptera: Cerambycidae). J Jpn For Soc 62:150–152

Ikeda T, Yamane A, Enda N, Matsuura K, Oda K (1981) Attractiveness of chemical treated pine trees for *Monochamus alternatus* Hope (Coleoptera: Cerambycidae). J Jpn For Soc 63:201–207

Ikeda T, Toda T, Tajima M (1994) Histological responses of different resistant families of *Pinus thunbergii* and *P. densiflora* to an invasion of pine wood nematode. Ann Phytopathol Soc Jpn 60:540–542

Inoue E (1993) *Dastarcus helophoroides*, a natural enemy of *Monochamus alternatus* (in Japanese). Shinrin-boeki (For Pests) 42:171–175

Irei H, Miyagi T, Gushilken Y, Nakahira Y, Mori T, Kameyama N, Nakamura K, Akiba M, Sahashi N, Ishihara M (2004) Epidemiology of pine wilt disease in Ryukyu pines in Okinawa Island IV. Number of adult occurrence and its prevalence of *Monochamus alternatus* in Okinawa Island (in Japanese). Trans Jpn For Soc 115:719

Irle T, Schierenberg E (2002) Developmental potential of fused *Caenorhabditis elegans* oocytes: generation of giant and twin embryos. Dev Genes Evol 212:257–266

Ishibashi N (1993) Integrated control of insect pests by *Steinernema carpocapsae*. In: Bedding R, Akhurst R, Kaya HK (eds) Nematodes and the Biological Control of Insect Pests. CSRIO, East Melbourne, pp 105–113

Ishibashi N, Choi DR (1992) Possible simultaneous/integrated biological control of soil pests by mix application of entomopathogenic nematodes and fungivorous nematodes. In: Ooi FAC, Lim GS, Teng PS (eds) Proceedings of the 3th international conference on plant protection in the tropics, Genting Highlands, Malaysia, 20–23 March 1990, Kuala Lumpur, Malaysia. Malaysian Plant Protection Society 6, pp 10–16

Ishibashi N, Kondo E (1977) Occurrence and survival of the dispersal forms of pine wood nematode, *Bursaphelenchus lignicolus* Mamiya and kiyohara. Appl Entomol Zool 12:293–302

Ishibashi N, Kondo E (1990) Behaviour of Infective Juveniles. In: Gaugler R, Kaya HK (eds) Entomopathogenic nematodes in biological control. CRC Press, Boca Raton, pp 139–148

Ishibashi N, Aoyagi M, Kondo E (1978) Comparison of the gonad development between the propagative and dispersal forms of pine wood nematode, *Bursaphelenchus lignicolus* (Aphelenchoididae). Jpn J Nematol 8:28–31

Ishida K, Hogetsu T (1997) Rile of resin canals in the early stages of pine wilt disease caused by the pine wood nematode. Can J Bot 75:346–351

Ishida K, Hogetsu T, Fukuda K, Suzuki K (1993) Cortical responses in Japanese black pine to attack by the pine wood nematode. Can J Bot 71:1399–1405

Ishii K, Kurinobu S, Ohba K, Furukoshi T (1981) Resistance in hybrids within the subsection *Sylvestres* of the genus *Pinus* against the pinewood nematode (in Japanese). Trans Mtg Jpn For Soc 92:291–292

Ishii S (2004) Oviposition and development of *Dastarcus helophoroides* Fairmaire (=*D. longulus* Sharp) in the laboratoray. Appl For Sci 13:49–53

Ishikawa M, Shuto Y, Watanabe H (1986) β-Myrcene, a potent attractant component of pine wood for the pine wood nematode, *Bursaphelenchus xylophilus*. Agric Biol Chem 50:1863–1866

Ishikawa M, Kaneko A, Kashiwa T, Watanabe H (1987) Participation of β-myrcene in the susceptibility and/or resistance of pine trees to the pine wood nematode, *Bursaphelenchus xylophilus*. Agric Biol Chem 51:3187–3191

Islam SQ, Ichiryu J, Sato M, Yamasaki T (1997) D-Catechin: an oviposition stimulant for the cerambycid beetle, *Monochamus alternatus*, from *Pinus densiflora*. J Pestic Sci 22:338–341

Isoda K, Watanabe A, Kuramoto T (2007) Mapping of Japanese red pine and Japanese black pine using microsatellite markers (in Japanese). Trans Jpn For Soc 118:O18

Issa Z, Grant WN, Stasiuk S, Shoemaker CB (2005) Development of methods for RNA interference in the sheep gastrointestinal parasite, *Trichostrongylus colubriformis*. Int J parasitol 35:935–940

Ito K (1982) The tethered Flight of the Japanese pine sawyer, *Monochamus alternatus* Hope (Coleoptera: Cerambycidae). J Jpn For Soc 64:395–397

Iwahori H, Futai K (1990) Propagation and effects of the pine wood nematode on calli of various plants. Jpn J Nematol 20:25–36

Iwahori H, Futai K (1993) Lipid peroxidation and ion exudation of pine callus tissues inoculated with pinewood nematodes. Jpn J Nematol 23:79–89

Iwahori H, Futai K (1995) Comparative movement speed of pathogenic and nonpathogenic isolates of *Bursaphelenchus* nematodes. Appl Entomol Zool 30:159–167

Iwahori H, Futai K (1996) Changes in nematode population and movement in pine seedlings with the development of pine wilt disease: suppressive effect of water extracts from infected pine seedlings on the movement of nematodes. Appl Entomol Zool 31:11–20

Iwahori H, Tsuda K, Kanzaki N, Izui K, Futai K (1998) PCR–RFLP and sequencing analysis of ribosomal DNA of *Bursaphelenchus* nematodes related to pine wilt disease. Fundam Appl Nematol 21:655–666

Iwahori H, Kanzaki N, Futai K (2000) A simple polymerase chain reaction-restriction fragment length polymorphism-aided diagnosis method for pine wilt disease. For Pathol 30:157–164

Izumi S, Okamoto H (1990) Larval sound production and growth of the Japanese pine sawyer, *Monochamus alternatus* Hope (in Japanese with English abstract). J Jpn For Soc 72:181–187

Izumi S, Ichikawa T, Okamoto H (1990) The character of larval sounds of the Japanese pine sawyer, *Monochamus alternatus* Hope (in Japanese with English abstract). Jpn J Appl Entomol Zool 34:15–19

Jaubert S, Laffaire JB, Abad P, Rosso MN (2002) A polygalacturonase of animal origin isolated from the root-knot nematode *Meloidogyne incognita*. FEBS Lett 522:109–112

Jiang JH, Gao TH, Chen FM, Cao XY (2005) Pathogenicity on the non-host plants and effects on the lipoxygenase in the pine needle cells of *Pinus thunbergii* of toxins from a bacterium strain carried by pine wood nematode (in Chinese with English abstract). For Pest Dis (4):1–3

Jiang LY, Sheng CS, Ma SA, Wang L, Shi J (2006) Study on processing timber infected with pine wood nematode using microwave. J Nanjing For Univ (Nat Sci) 30:87–90

Jiao GY, Shen PG, Li HM, Xu PF (1996) Study on pathogenicity of pine wood nematode of Nanjing and Japan to *Cedrus deodara* (in Chinese with English abstract). Plant Quarantine 10:193–195

Jikumaru S, Togashi K (2000) Temperature effects on the transmission of *Bursaphelenchus xylophilus* (Nemata: Aphelenchoididae) by *Monochamus alternatus* (Coleoptera: Cerambycidae). J Nematol 32:110–116

Jikumaru S, Togashi K (2001) Transmission of *Bursaphelenchus mucronatus* (Nematoda: Aphelenchoididae) through feeding wounds by *Monochamus saltuarius* (Coleoptera: Cerambycidae). Nematology 3:325–333

Jikumaru S, Togashi K (2003) Boarding abilities of *Bursaphelenchus mucronatus* and *B. xylophilus* (Nematoda: Aphelenchoididae) on *Monochamus alternatus* (Coleoptera: Cerambycidae). Nematology 5:843–849

Jikumaru S, Togashi K (2004) Inhibitory effect of *Bursaphelenchus mucronatus* (Nematoda: Aphelenchoididae) on *B. xylophilus* boarding adult *Monochamus alternatus* (Coleoptera: Cerambycidae). J Nematol 36:95–99

Jikumaru S, Togashi K, Taketsune A, Takahashi F (1994) Oviposition biology of *Monochamus saltuarius* (Coleoptera: Cerambycidae) at a constant temperature. Appl Entomol Zool 29:555–561

Jones JP, McGawley EC, Birchfield W (1982) Current status of the pinewood nematode in Louisiana. In: Appleby JE, Malek RB (eds) Proceedings of the 1982 national pine wilt disease workshop, pp 66–73

Jones JT, Furlanetto C, Kikuchi T (2005) Horizontal gene transfer from bacteria and fungi as a driving force in the evolution of plant parasitism in nematodes. Nematology 7:641–646

Kaisa TR (2003) Redescription of *Bursaphelenchus talonus* (Thorne, 1935) Massey, 1956 (Nematoda: Parasitaphelenchidae) and designation of lectotypes. Zootaxa 269:1–7

Kaisa TR (2005) Proposal of *Parasitaphelenchus dongguanensis* (Fang, Zhao & Zhuo 2002) n. comb. (Nematoda: Parasitaphelenchidae). Zootaxa 839:1–8

Kakuliya GA (1967) New nematode genus *Devibursapehelenchus* Kakuliya gen. n. (Nematoda: Aphelenchoididae) (in Gerorgian). Bull Acad Sci Georgian SSR 47:439–443

Kakuliya GA, Devdariani TG (1965) New nematode species *Bursapehelenchus teratospicularis* Kakuliya et Debdariani, sp. nov. (Nematoda: Aphelenchoidea) (in Gerorgian). Bull Acad Sci Georgian SSR 38:187–191

Kamata N (1996) Development of a barrier zone to stop the invasion of pine wilt disease in Japan. In: Chinese Society of Forestry (ed) Proceedings, International symposium on pine wilt disease caused by pine wood nematode. 31 October–5 November 1995, Beijing, pp 81–90

Kanbe T, Sakaue D, Suzuki K, Togashi K (2000) Effect of heat-shock treatments on the pinewood nematode (in Japanese). Proc Kanto Conf Jpn For Soc 51:111–112

Kaneko N, Kawazu K, Kanzaki H (1998) Difference in mobility of the pine wood nematode, *Bursaphelenchus xylophilus*, between two isolates, OKD-a and OKD-3, with different pathogenicity (in Japanese with English summary). Jpn J Nematol 28:17–21

Kaneko S (1989) Effect of light intensity on the development of pine wilt disease. Can J Bot 67:1861–1864

Kaneko S, Zinno Y (1986) Development under different light sources of pine-wilt disease caused by *Bursaphelenchus xylophilus* on the seedlings of Japanese red pine (in Japanese). J Jpn For Soc 68:208–209

Kanzaki N (2006) Taxonomy and systematics of *Bursaphelenchus* nematodes (in Japanese). J Jpn For Soc 88:392–406

Kanzaki N, Futai K (1996) Field study on symptomless carrier of pine wilt disease. Jpn J Nematol 26:46

Kanzaki N, Futai K (2002a) A PCR primer set for determination of phylogenetic relationships of *Bursaphelenchus* species within *xylophilus* group. Nematology 4:35–41

Kanzaki N, Futai K (2002b) Phylogenetic analysis of the phoretic association between *Bursaphelenchus conicaudatus* (Nematoda: Aphelenchoididae) and *Psacothea hilaris* (Coleoptera: Cerambycidae). Nematology 4:759–771

Kanzaki N, Futai K (2002c) Observation on the arangement of caudal papillae of *Bursaphelenchus conicaudatus* and *B. fraudulentus*. Jpn J Nematol 32:21–23

Kanzaki N, Futai K (2002d) Life history of *Rhabdontolaimus psacotheae* n. sp. (Diplogasterida: Diplogasteridae), and its habitat segregation from *Bursaphelenchus conicaudatus* (Aphelenchida: Aphelenchoididae). Jpn J Nematol 32:60–67

Kanzaki N, Futai K (2003a) Description and phylogeny of *Bursaphelenchus luxuriosae* n. sp. (Nematoda: Aphelenchoididae) isolated from *Acalolepta luxuriosa* (Coleoptera: Cerambycidae). Nematology 5:565–572

Kanzaki N, Futai K (2003b) Application of molecular phylogenetic analysis to the evolution and co-speciation of entomophilic nematodes. Russ J Nematol 11:107–117

Kanzaki N, Futai K (2005) Description of *Bursaphelenchus parvispicularis* n. sp. (Nematoda: Parasitaphelenchidae) isolated from dead oak tree, *Quercus mongolica* v. *grosseserrata*. Nematology 7:751–759

Kanzaki N, Futai K (2007) Isolation of *Bursaphelenchus sinensis* (Nematoda: Parasitaphelenchidae) from dead Japanese black pine, *Pinus thunbergii* Parl. in Japan. J Nematode Morphol System 10:127–134

Kanzaki N, Tsuda K, Futai K (2000) Description of *Bursaphelenchus conicaudatus* n. sp. (Nematoda: Aphelenchoididae), isolated from the yellow-spotted longicorn beetle, *Psacothea hilaris* (Coleoptera: Cerambycidae) and fig trees, *Ficus carica*. Nematology 2:165–168

Kanzaki N, Minagawa N, Futai K (2002) Description of *Rhabdontolaimus psacotheae* n. sp. (Diplogasterida: Diplogasteridae), isolated from the yellow-spotted longicorn beetle *Psacothea hilaris* (Coleoptera: Cerambycidae) and fig trees, *Ficus carica*. Jpn J Nematol 32:7–12

Kanzaki N, Maehara N, Masuya H (2007) *Bursaphelenchus clavicauda* n. sp. (Nematoda: Parasitaphelenchidae) isolated from *Cryphalus* sp. emerged from a dead *Castanopsis cuspidata* (Thunb.) Schottky var. *sieboldii* (Makino) Nakai in Ishigaki Island, Okinawa, Japan. Nematology 9:759–769

Kashio T, Ujiye T (1988) Evaluation of the entomogenous fungus *Beauveria tenella*, isolated from the yellowspotted longicorn beetle, *Psacothe hylaris* for the biological control of the whitespotted longicorn beetle *Anoplophora malasiaca* (in Japanese). Proc Assoc Pl Prot Kyushu 34:190–193

Katagiri K, Shimazu M (1980) Micro-organisms pathogenic to *Monochamus alternatus* Hope33 (in Japanese). Shinrin-boeki (For Pests) 29:28

Katagiri K, Shimazu M, Kushida T (1983) A trial of microbial control of *Dendrolimus spectabilis* using *Beauveria bassiana* without direct spraying of conidial suspension (in Japanese). Trans Mtg Jpn For Soc 94:165

Katagiri K, Mamiya Y, Shimazu M, Tamura H, Kushida T (1984) A spray application of *Steinernema feltiae* on pine logs infested with the pine sawyer, *Monochamus alternatus*, and its mortality induced by the nematode (in Japanese). Trans Mtg Jpn For Soc 95: 479–480

Kato R, Okudaira T (1977) Range expansion of the pinewood nematode in Aichi Prefecture (in Japanese). Trans Ann Mtg Chubu Br Jpn For Soc 26:159–164

Katsuyama N, Sakurai H, Tabata K, Takeda S (1989) Effect of age of post-feeding twig on the ovarian development of Japanese pine sawyer, *Monochamus alternatus* (in Japanese with English abstract). Res Bull Faculty College Agric, Gifu Univ 54:81–89

Kawabata K (1979) Movement of *Monochamus alternatus* among islands (in Japanese). Trans Ann Mtg Kyushu Br Jpn For Soc 32:281–282

Kawaguchi E (2006) Relationship between the anatomical characteristics of cortical resin canals and migration of *Bursaphelenchus xylophilus* (in Japanese with English abstract). J Jpn For Soc 88:240–244

Kawai M, Shoda-Kagaya E, Maehara T, Zhou Z, Lian C, Iwata R, Yamane A, Hogetsu T (2006) Genetic structure of pine sawyer *Monochamus alternatus* (Coleoptera: Cerambycidae) populations in Northeast Asia: consequences of the spread of pine wilt disease. Environ Entomol 35:569–579

Kawakami K (1973) Studies on the disease of the silkworm *Bombyx mori* L., with special references to the invasion of causative fungi and pathological changes of infected larvae (in Japanese). Bull Seric Exp Stn 25:347–370

Kawakami K (1978) On an entomogenous fungus *Beauveria tenella* (Delacroix) Siemaszko isolated from the yellow-spotted longicorn beetle, *Psacothea hylaris* Pascoe (in Japanese). Bull Seric Exp Stn 27:445–467

Kawazu K (1998) Pathogenic toxins of pine wilt disease. Kagaku To Seibutsu 36:120–124

Kawazu K (1990) Change in constituents of pine wood by infection of pine wood nematode (in Japanese). Nippon Nogeikagaku Kaishi 64:1262–1264

Kawazu K, Kaneko N (1997) Asepsis of the pine wood nematode isolate OKD-3 causes it to lose its pathogenicity. Jpn J Nematol 27:76–80

Kawazu K, Zhang H, Yamashita H, Kanzaki H (1996a) Relationship between the pathogenicity of the pine wood nematode, *Bursaphelenchus xylophilus*, and pheny acetic acid production. Biosci Biotech Biochem 60:1413–1415

Kawazu K, Zhang H, Kanzaki H (1996b) Accumulation of benzoic acid in suspension cultured cells of *Pinus thunbergii* Parl. in response to phenylacetic acid administration. Biosci Biotech Biochem 60:1410–1412

Kawazu K, Kaneko N, Hiraoka K, Yamashita H, Kanzaki H (1999) Reisolation of the pathogens from wilted red pine seedlings inoculated with the bacterium carrying nematode, and the cause of difference in pathogenicity among pine wood nematode isolates (in Japanese with English abstract). Sci Rep Fac Agric Okayama Univ 88:1–5

Kaya HK, Gaugler R (1993) Entomopathogenic nematodes. Annu Rev Entomol 38:181–206

Kaya HK, Koppenhöfer AM (1999) Biology and ecology of insecticidal nematodes. In: Polavarapu S (ed) Optimal use of insecticidal nematodes in pest management. Rutgers University, New Jersey, pp 1–8

Kaya HK, Stock SP (1997) Techniques in insect nematology. In: Lacey LA (ed) Manual of Techniques in Insect Pathology. Academic Press, London, pp 281–324

Kemphues KJ, Strome S (1997) Fertilization and establishment of polarity in the embryo. In: Riddle DL, Blumenthal T, Meyer BJ, Priess JR (eds) *C. elegans* II. Cold Spring Harbor Laboratory Press, New York, pp 335–359

Kerdelhué C, Roux-Morabito G, Forichon J, Chambon J-M, Robert A, Lieutier F (2002) Population genetic structure of *Tomicus piniperda* L. (Curculionidae: Scolytinae) on different pine species and validation of *T. destruens* (Woll.). Mol Ecol 11:183–494

Kichiya T, Makihara H (1991) Observation in Tôhoku district on the daily activity of mature adults of the Japanese pine sawyer, *Monochamus alternatus* Hope (Coleoptera: Cerambycidae) (in Japanese). Trans Mtg Jpn For Soc 102:293

Kikuchi J, Tsuno N, Futai K (1991) The effect of mycorrhizae as a resistance factor of pine trees to the pinewood nematode (in Japanese with English abstract). J Jpn For Soc 73:216–218

Kikuchi T, Jones JT, Aikawa T, Kosaka H, Ogura N (2004) A family of glycosyl hydrolase family 45 cellulases from the pine wood nematode *Bursaphelenchus xylophilus*. FEBS Lett 572:201–205

Kikuchi T, Shibuya H, Jones JT (2005) Molecular and biochemical characterization of an endo-beta-1,3-glucanase from the pinewood nematode *Bursaphelenchus xylophilus* acquired by horizontal gene transfer from bacteria. Biochem J 389:117–125

Kikuchi T, Shibuya H, Aikawa T, Jones JT (2006) Cloning and characterization of pectate lyases expressed in the esophageal gland of the pine wood nematode *Bursaphelenchus xylophilus*. Mol Plant Microbe Interact 19:280–287

Kim DS, Lee S-M, Chung Y-J, Choi K-S, Moon Y-S, Park C-G (2003) Emergence ecology of Japanese pine sawyer, *Monochamus alternatus*, a vector of pinewood nematode, *Bursaphelenchus xylophilus* (in Korean). Korean J Appl Entomol 42:307–313

Kim GH, Takabayashi J, Takahashi S, Tabata K (1992) Function of pheromones in mating behavior of the Japanese pine sawyer beetle, *Monochamus alternatus* Hope. Appl Entomol Zool 27:489–497

Kim YS, Ryu JH, Han SJ, Choi KH, Nam KB, Jang IH, Lemaitre B, Brey PT, Lee WJ (2000) Gram-negative bacteria-binding protein, a pattern recognition receptor for lipopolysaccharide and beta-1,3-glucan that mediates the signaling for the induction of innate immune genes in *Drosophila melanogaster* cells. J Biol Chem 275:32721–32727

Kimble JE, Hirsh D (1979) Post-embryonic cell lineage of the hermaphrodite and male gonads in *Caenorhabditis elegans*. Dev Biol 70:396–417

Kimble J, Henderson S, Crittenden S (1998) Notch/Kin-12 signaling transduction by regulated protein silencing. Trends Biochem Sci 23:353–357

Kinloch BB, Comstock M (1981) Race of Cronartium ribicola virulent to major gene resistance in sugar pine. Plant Dis 65:604–605

Kinn DN (1986) Heat-treating wood chips: a possible solution to pine wood nematode contamination. Tappi J 69:97–98

Kinn DN (1987) Incidence of pinewood nematode dauerlarvae and phoretic mites associated with long-horned beetles in central Louisiana. Can J For Res 17:187–190

Kiritani K (1998) Exotic insects in Japan. Entomol Sci 1:291–298

Kishi Y (1976) Prediction of emergence of the pine sawyer adult through continuous observations of larval growth by the log-cutting method (in Japanese). Shinrin-boeki (For Pests) 25:96–98

Kishi Y (1978) Invasion of pine trees by *Bursaphelenchus lignicolus* M. & K. (Nematoda: Aphelenchoididae) from *Monochamus alternatus* Hope (Coleoptera: Cerambycidae) (in Japanese). J Jpn For Soc 60:179–182

Kishi Y (1980a) Studies on new control method against pine wilt disease (in Japanese). Bull Ibaraki Pref For Stn 13:34–35

Kishi Y (1980b) Studies on pine wilt caused by the pinewood nematode in Ibaraki (in Japanese). Annu Rep Ibaraki For Res Inst 13:83

Kishi Y (1988a) Study on pheromone of *Monochamus alternatus* (Coleoptera: Cerambycidae) (II) relation between its age and attraction (in Japanese). Trans Mtg Jpn For Soc 99:501–502

Kishi Y (1988b) Study on pheromone of *Monochamus alternatus* (Coleoptera: Cerambycidae) (III) notes on experiments in net rooms (in Japanese). Trans Ann Mtg Kanto Br Jpn For Soc 40:179–180

Kishi Y (1995) The pine wood nematode and the Japanese pine sawyer. Thomas, Tokyo, p 302

Kishi Y, Otsu S (1987) Preliminary study on pheromone of *Monochamus alternatus* (Coleoptera: Cerambycidae) (in Japanese). Trans Mtg Jpn For Soc 98:537–538

Kishi Y, Hayasaka Y, Yokomizo Y, Takeda J (1982) Variance of the shortest distance between pine inner bark and pupal chambers of *Monochamus alternatus* Hope (Coleoptera: Cerambycidae) (in Japanese with English abstract). J Jpn For Soc 64:239–241

Kiyohara T (1976) The decrease of pathogenicity of pine wood nematode, *Bursaphelenchus lignicolus*, induced by the extended subculturing on the fungal mat of *Botrytis cinerea* (in Japanese with English abstract). Jpn J Nematol 6:56–59

Kiyohara T (1982) Induced resistance in pine wilt disease (in Japanese). Trans Ann Mtg Kyushu Br Jpn For Soc 35:161–162

Kiyohara T (1984) Pine wilt resistance induced by prior inoculation with avirulent isolate of *Bursaphelenchus xylophilus*. In: Dropkin V (ed). The resistance mechanisms of pines against pine wilt disease, proceedings of the US-Japan seminar, pp 178–186

Kiyohara T (1989) Etiological study of pine wilt disease (in Japanese with English abstract). Bull For For Prod Res Inst 353:127–176

Kiyohara T (1997) Pathogenicity and life cycle of the pine woot nematode. In: Tamura T (ed) The pine wilt disease. The review of history and recent researches. Zenkoku Shinrin Byochujugai Bojo Kyokai (National Forest Pests Control Association), Tokyo, pp 26–43

Kiyohara T, Bolla RI (1990) Pathogenic variability among populations of the pinewood nematode, *Bursaphelenchus xylophilus*. For Sci 36:1061–1076

Kiyohara K, Dozono Y (1986) The relationship between the pathogenicity and the population growth of pine wood nematode (in Japanese). Trans Ann Mtg Kyushu Br Jpn For Soc 39:157

Kiyohara T, Kusunoki M (1987) Study on induced resistance of pine wilt disease: effect of pre-inoculation methods on resistance induction (in Japanese). Trans Ann Mtg Kyushu Br Jpn For Soc 40:191–192

Kiyohara T, Suzuki K (1975) Population changes of *Bursaphelenchus lignicolus* in *Pinus thunbergii* after inoculation (in Japanese). Trans Mtg Jpn For Soc 86:296–298

Kiyohara T, Suzuki K (1977) Population changes of *Bursaphelenchus lignicolus* in *Pinus thunbergii* after inoculation (II) (in Japanese). Trans Ann Mtg Kyushu Br Jpn For Soc 30:243–244

Kiyohara T, Tokushige Y (1971) Inoculation experiments of a nematode, *Bursaphelenchus* sp. onto pine trees (in Japanese with English abstract). J Jpn For Soc 53:210–218

Kiyohara T, Dozono Y, Hashimoto H, Ono K (1973) Correlation between number of inoculated nematodes and disease occurrence in pine wilt disease (in Japanese). Trans Ann Mtg Kyushu Br Jpn For Soc 26:191–192

Kiyohara T, Hashimoto H, Fujimoto Y (1983) Variation in the virulence of the pinewood nematode (in Japanese). Trans Ann Mtg Kyushu Branch Jpn For Soc 36:189–190

Kiyohara T, Ikeda T, Kusunoki M (1989) Induction of pine wilt resistance by prior inoculation with microorganism (in Japanese). Trans Ann Mtg Kyushu Branch Jpn For Soc 42: 173–174

Kiyohara T, Ikeda T, Kusunoki M (1990) Induction of pine wilt resistance by prior inoculation with microorganism: comparison of induced resistance between two *Pinus* species (in Japanese). Trans Ann Mtg Kyushu Br Jpn For Soc 43:131–132

Kiyohara T, Kosaka H, Aikawa T (1998) Intra- and inter-specific relationship among isolates of *Bursaphelenchus xylophilus* and *B. mucronatus* based on the result of mating experiment (in Japanese). For Forest Prod Res Inst Kenkyu Seika Sensyu 1997:6–7

Kiyohara T, Kosaka H, Aikawa T, Ogura N, Tabata K (1999) Experiments of induced resistance to pine wilt disease in pine forest. In: Futai K, Togashi K, Ikeda T (eds) Sustainability of pine forests in relation to pine wilt and decline. Proceedings of international symposium, Tokyo, 27–28 October 1998, Shokado, Kyoto, Japan, pp 103–104

Knowles K, Beaubien Y, Wingfield MJ, Baker FA, French DW (1983) The pinewood nematode new in Canada. For Chron 59:40

Kobayashi F (1981) Observations on the chemical injury of Hinoki, Japanese Cypress (*Chamaecyparis obtuse* Sieb. et Zucc.), by the spraying of fenitrothion on the Toyohashi national forest. J Jpn For Soc 63:60–63

Kobayashi F, Okuda M, Taketani A, Hosoda R (1971) Daily assessment of disorder in oleoresin exudation of pine trees and visit of pine borers (in Japanese). Ann Rep Kansai Br For Forest Prod Res Inst 12:117–120

Kobayashi F, Yamane A, Ikeda T (1984) The Japanese pine sawyer beetle as the vector of pine wilt disease. Annu Rev Entomol 29:115–135

Kobayashi H, Yamane A, Iwata R (2003) Mating behavior of the pine sawyer, *Monochamus saltuarius* (Coleoptera: Cerambycidae). Appl Entomol Zool 38:141–148

Kobayashi K (1975) Relationship between the degree of the mortality of pine trees and the number of *Monochamus alternatus* (in Japanese). Shinrin-boeki (For Pests) 24:206–208

Kobayashi K, Hosoda R (1978) Methods of estimating the number of *Bursaphelenchus lignicolus* carried by a living adult of *Monochamus alternatus* (in Japanese). Trans Mtg Jpn For Soc 89:301–302

Kobayashi K, Okuda, M, Hosoda R (1976) Influence of wood size and humidity on the emergence of Japanese pine sawyer and the number of pine wood nematodes per insect (in Japanese). Trans Mtg Jpn For Soc 87:239–240

Kobayashi T (1975) The pine wood nematode, fungi and their vector pine sawyer (in Japanese). Shinrin-boeki (For Pests) 24:199–202

Kobayashi T, Sasaki K, Mamiya Y (1974) Fungi associated with *Bursaphelenchus lignicolus*, the pine wood nematode (I) (in Japanese with English abstract). J Jpn For Soc 56:136–145

Kobayashi T, Sasaki K, Mamiya Y (1975) Fungi associated with *Bursaphelenchus lignicolus*, the pine wood nematode (II) (in Japanese with English abstract). J Jpn For Soc 57:184–193

Kodan A, Kuroda H and Sakai F (2001) Simultaneous expression of stilbene synthase genes in Japanese red pine (*Pinus densiflora*) seedlings. J Wood Sci 47:58–62

Koiwa T, Takahashi K, Yomogida H, Abe Y (2004) Pine wilt disease in *Picea abies* (in Japanese). Trans Jpn For Soc 115:731

Kojima K (1960) Ecological studies of the family Cerambycidae as found in Japan. On the ovipositing-habits of the adult and the food-habits of the larva (in Japanese with English abstract). Gensei 10:21–47

Kojima K, Nakamura S (1986) Food plants of cerambycid beetles (Cerambycidae, Coleoptera) in Japan. Hiba Society of Natural History, Shobara, p 336

Kojima K, Okabe M (1960) Food plants of Japanese cerambycidae. Kobundo, Kochi, p 330

Kojima K, Kamijyo A, Masumori M (1994) Cellulase activities of pine-wood nematode isolates with different virulences. J Jpn For Soc 76:258–262

Kolossova NV (1997) *Bursaphelenchus eroshenkii* sp. n. (Nematoda: Aphelenchoididae) from the Russian Far East, with a key to some species of *Bursaphalenchus* Fuchs, 1973. Russ J Nematol 6:161–164

Kondo E (1986) SEM observations on the intratracheal existence and cuticle surface of the pine wood nematode, *Bursaphelenchus xylophilus*, associated with the Cerambycid beetle, *Monochamus carolinensis*. Appl Entomol Zool 21:340–346

Kondo E, Ishibashi N (1978) Ultrastructural differences between the propagative and dispersal forms in pine wood nematode, *Bursaphelenchus lignicolus*, with reference to the survival. Appl Entomol Zool 13:1–11

Kondo K, Foundin A, Linit M, Smith M, Bolla R, Winter R, Dropkin V (1982) Pine wilt disese-nematological, entomological, and biochemical investigations. Univ Mo Columbia Agric Exp Stn SR282:1–56

Koppenhöfer AM, Kaya HK (1997) Additive and synergistic interactions between *Bacillus thuringiensis* Buibui strain and entomopathogenic nematodes. Biol Control 8:131–137

Koppenhöfer AM, Choo HY, Kaya HK, Lee DW, Gelernter WD (1999) Increased field and greenhouse efficacy against scarab grubs with a combination of an entomopathogenic nematode and *Bacillus thuringiensis*. Biol Control 14:37–44

Koppenhöfer AM, Brown IM, Gaugler R, Grewal PS, Kaya HK, Klein MG (2000) Synergism of entomopathogenic nematodes and imidacloprid against white grubs: greenhouse and field evaluation. Biol Control 19:245–251

Korenchenko EA (1980) New species of nematodes from the family Aphelenchoididae, parasites of stem pests of the Dahurian Larch (in Russian). Zool Zhurnal 59:1768–1780

Körner H (1954) Die Nematodenfauna das vergehenden Holzes und ihre Beziehungen ze den Insekten (in German). Zoologische Jahrbücher, Abteilung für Systematik, Öekologie Geogr Tiere 82:245–353

Kosaka H, Ogura N (1990) Rearing of the Japanese pine sawyer, *Monochamus alternatus* (Coleoptera: Cerambycidae) on artificial diets. Appl Entomol Zool 25:532–534

Kosaka H, Ogura N (1993) *Contortylenchus genitalicola* n. sp. (Tylenchida: Allantonematidae) from the Japanese pine sawyer, *Monochamus alternatus* (Coleoptera: Cerambycidae). Appl Entomol Zool 28:423–432

Kosaka H, Kiyohara T, Aikawa T, Ogura N, Tabata K (1998) Re-examination of the induced resistance of pine trees to pine wilt disease by prior inoculation with avirulent isolates of the pine wood nematode. In: Abstract of 7th international congress of plant pathology, 9–16 August 1998, Edinburgh, 3.7.68

Kosaka H, Irei H, Aikawa T, Ogura N (2001a) A high value of developmental zero in overwintering larvae of *Monochamus alternatus* from Okinawa (in Japanese). Trans Jpn For Soc 112:317

Kosaka H, Aikawa T, Ogura N, Tabata K, Kiyohara T (2001b) Pine wilt disease caused by the pine wood nematode: the induced resistance of pine trees by the avirulent isolates of nematode. Eur J Plant Pathol 107:667–675

Koyama R (1959) Studies on the epizootic diseases of forest insects in Japan (1) list of pathogen (in Japanese). Bull Gov For Exp Stn 112:23–31

Koyama R (1963) A revised list of microbes associated with forest insects in Japan. Mushi 37:159–165

Kozhemyako VB, Rebrikov DV, Lukyanov SA, Bogdanova EA, Marin A, Mazur AK, Kovalchuk SN, Agafonova EV, Sova VV, Elyakova LA, Rasskazov VA (2004) Molecular cloning and

characterization of an endo-1,3-beta-D-glucanase from the mollusk *Spisula sachalinensis*. Comp Biochem Physiol 137:169–178
Kruglik IA (2003) Distribution of nematodes inside trunk and branches of the dead 200 years old pine *Pinus koraiensis*. Russ J Nematol 1:140–141
Kruglik IA, Eroshenko AS (2004) *Bursaphelenchus fuchsi* sp. n. (Nematoda: Bursaphelenchidae)—new nematode species from wood of pine *Pinus koraiensis*, Primorsky Territory. In: Sonin MD (ed) Parazitichesk ie nematody rastenii i nasekomykh. Nauka, Moscow, pp 96–99
Kulinich O (2004) Survey for the pine wood nematode in Russia. In: Mota M, Vieira P (eds) The pinewood nematode, *Bursaphelenchus xylophilus*. Nematology monographs and perspectives, vol 1. E.J. Brill, Leiden, pp 65–75
Kung SP, Gaugler R (1991) Effect of soil temperature, moisture and relative humidity on entomopathogenic nematodes persistence. J Invertebr Pathol 57:242–249
Kuniyoshi S (1974) Occurrence of the pinewood nematode in Okinawa Prefecture (in Japanese). Shinrin-boeki (For Pests) 23:40–42
Kuramoto T, Ohira M, Hiraoka Y, Taniguchi T, Kashiwagi M, Inoue U, Hukuda T, Saitou S, Murayama T, Ueda M, Okamura M, Hoshi H, Fujisawa Y (2007) Genetic analysis of pine families resistance to pine wood nematode (in Japanese). Trans Jpn For Soc 118:O16
Kurashvili BE, Kakulia GA, Devdariani TG (1980) Parasitic nematodes of the bark-beetles in Georgia (in Russian). Metsniereba, Tbilisi
Kuroda H (1999) New control strategies for pine wilt disease (in Japanese). Wood Research and Technical Notes 35:32–46
Kuroda H, Kuroda K (1998) Expression of stilbene synthase gene and pine wilt disease. In: Abstract of 7th international congress of plant pathology, 9–16 August 1998, Edinburgh UK
Kuroda K (1987) Characteristics of enzymes produced by the pine wood nematode: lytic enzymes of polysaccharides and proteins (in Japanese). Ann Mtg Jpn For Soc 98:128
Kuroda K (1989) Terpenoids causing tracheid-cavitation in *Pinus thunbergii* infested by the pine wood nematode (*Bursaphelenchus xylophilus*). Ann Phytopathol Soc Jpn 55:170–178
Kuroda K (1991) Mechanism of cavitation development in the pine wilt disease. Eur J For Path 21:82–89
Kuroda K (1995) Acoustic emission technique for the detection of abnormal cavitation in pine trees infected with pine wilt disease. In: International symposium on pine wilt disease caused by pine wood nematode (Beijing, China), Proceedings, pp 53–58
Kuroda K (2004) Inhibiting factors of symptom development in several Japanese red pine (*Pinus densiflora*) families selected as resistant to pine wilt. J For Res 9:217–224
Kuroda K (2008) Defense systems of *Pinus densiflora* cultivars selected as resistant to pine wilt disease. In Mota M, Vieira P (eds) Pine wilt disease: a worldwide threat to forest ecosystems. Springer, The Netherlands, p 47
Kuroda K, Ito S (1992) Migration speed of pine wood nematode and activities of other microbes during the development of pine-wilt disease in *Pinus thunbergii* (in Japanese with English abstract). J Jpn For Soc 74:383–389
Kuroda K, Mamiya Y (1984) Anatomical observations on the disease development of the pine wilt after inoculation (in Japanese). Trans Mtg Jpn For Soc 95:471–472
Kuroda K, Mamiya Y (1986) Behavior of the pine wood nematode in pine seedlings growing under aseptic conditions (in Japanese). Trans Mtg Jpn For Soc 97:471–472
Kuroda K, Yamada T, Mineo K, Tamura H (1988) Effects of cavitation on the development of pine wilt disease caused by *Bursaphelenchus xylophilus*. Ann Phytopathol Soc Jpn 54:606–615
Kuroda K, Yamada T, Ito S (1991a) *Bursaphelenchus xylophilus* induced pine wilt: factors associated with resistance. Eur J For Path 21:430–438
Kuroda K, Yamada T, Ito S (1991b) Development of the pin-wilt disease in *Pinus densiflora* from the stand point of water conduction (in Japanese with English abstact). J Jpn For Soc 73:69–72

Kuroda K, Ohira M, Okamura M, Fujisawa Y (2007) Migration and population growth of the pine wood nematode (*Bursaphelenchus xylophilus*) related to the symptom development in the seedlings of Japanese black pine (*Pinus thunbergii*) families selected as resistant to pine wilt (in Japanese with English abstract). J Jpn For Soc 89:241–248

Kusunoki M (1987) Symptom developments of pine wilt disease-Histological observations with electron microscope. Ann Phytopathol Soc Jpn 53:622–629

Kwon SD (2006) Changes in Korean pine forests (in Korean). Mon Infor For Sci 181:16–17

Kwon TS, Lim JH, Sim SJ, Kwon YD, Son SK, Lee KY, Kim YT, Park JW, Shin CH, Ryu SB, Lee CK, Shin SC, Chung YJ, Park YS (2006) Distribution patterns of *Monochamus alternatus* and *M. slatuarius* (Coleoptera: Cerambycidae) in Korea. J Korean For Soc 95:543–550

Lacey LA, Frutos R, Kaya HK, Vail P (2001) Insect pathogens as biological control agents: do they have future? Biol Control 21:230–248

Lahl V, Sadler B, Schierenberg E (2006) Egg development in parthenogenetic nematodes: variations in meiosis and axis formation. Int J Dev Biol 50:393–398

Lai YX, Liu JD, Xu QY, Wang YH, Zhou CM (2003) Trials on the parasitism of *Beauveria bassiana* or *Verticillium lecanii* on larvae of *Monochamus alternatus* Hope. J Jiangsu For Sci Technol 30:7–9

Lande R, Thompson R (1990) Efficiency of the marker-assisted selection in the improvement of quantitative traits. Genetics 124:743–756

Leal I (2008) An effective PCR-based diagnostic method for the detection of *Bursaphelenchus xylophilus* (Nematoda: Aphelenchoididae) in wood samples from lodgepole pine. In: Mota M, Vieira P (eds) Pine wilt disease: a worldwide threat to forest ecosystems. Springer, The Netherlands, p 29

Lee DW, Choo HL, Kaya HK, Lee SM, Smitley DR, Shin HK, Park CG (2002) Laboratory and field evaluation of Korean entomopathogenic nematode isolates against the Oriental beetle *Exomala orientalis* (Coleoptera: Scarabaeidae). J Econ Entomol 95:918–926

Lee MJ (1986) Resistance of pine species in Taiwan to pinewood nematode. G J Chin For 19: 27–33

Lee SM, Choo HY, Park NC, Moon YS, Kim JB (1990) Nematodes and insects associated with dead trees, and pine wood nematode detection from part of *Monochamus alternatus* (in Korean). Korean J Appl Entomol 29:14–19

Lee SM, Chung YJ, Moon YS, Lee SG, Lee DW, Choo HY, Lee CK (2003a) Insecticidal activity and fumigation conditions of several insecticides against Japanese pine sawyer (*Monochamus alternatus*) larvae (in Korean). J Korean For Soc 92:191–198

Lee SM, Chung YJ, Lee SG, Lee DW, Choo HY, Park CG (2003b) Toxic effects of some insecticides on the Japanese pine sawyer, *Monochamus alternatus* (in Korean). J Korean For Soc 92:305–312

Lee SM, Chung YJ, Kim DS, Choi KS, Kim YG, Park CG (2004) Adult morphological measurements: an indicator to identify sexes of Japanese pine sawyer, *Monochamus alternatus* Hope (Coleoptera: Cerambycidae) (in Korean). Korean J Appl Entomol 43: 85–89

Lee SY, Wang RG, Soderhall K (2000) A lipopolysaccharide- and beta-1,3-glucan-binding protein from hemocytes of the freshwater crayfish *Pacifastacus leniusculus*—purification, characterization, and cDNA cloning. J Biol Chem 275:1337–1343

Lewis EE (2002) Behavioural ecology. In: Gaugler R (ed) *Entomopathogenic Nematology*. CABI publishing, Oxfordshire, pp 205–223

Lewis EE, Gaugler R, Harrison R (1992) Entomopathogenic nematodes host finding: response to host contact cues by cruiser and ambusher foragers. Parasitology 105:309–315

Lewis EE, Gaugler R, Harrison R (1993) Response of cruiser and ambusher entomopathogenic nematodes (Steinernematidae) to host volatile cues. Can J Zool 71:765–769

Lewis EE, Campbell J, Griffin C, Kaya H, Peters A (2006) Behavioral ecology of entomopathogenic nematodes. Biol Control 38:66–79

Li GH, Zhang KQ (2001) A new nematicidal basidiomycetes. J Yunnan Univ 23:149–152

Li RJ, Xu FY, Zhang P, Wang LF, Zhao ZD (2004) Host preference by *Monochamus alternatus* (Hope) during maturation feeding on pine species and masson pine provenances. Chin For Sci Technol 5:112–117

Li YC, Zhao YX (2006) Thinking of some problems for pine wood nematode disease treatment work. J Fujian For Sci Tech 33:96–101

Li XP, Wu RQ, Xia MZ, Zhao BG (2000) Relationship between molecular structure and nematicidal activity of two alkaloids, aloperine and Δ^{11}-dehydroaloperine. J Nanjing For Univ 24:78–81

Lian C, Hogetsu T (2002) Development of microsatellite markers in black locust (*Robinia pseudoacacia*) using a dual-suppression-PCR technique. Mol Ecol Notes 2:211–213

Lian C, Miwa M, Hogetsu T (2000) Isolation and characterization of microsatellite loci from the Japanese red pine, *Pinus densiflora*. Mol Ecol 9:1186–1188

Lian C, Miwa M, Hogetsu T (2001) Outcrossing and paternity analysis of *Pinus densiflora* (Japanese red pine) by microsatellite polymorphism. Heredity 87:88–98

Lian C, Gueng QF, Wadud MA, Shimatani K, Hogetsu T (2006) An improved technique for isolating codominant compound microsatellite markers. J Plant Res 119:417–419

Liebhold AM, MacDonald WL, Bergdahl D, Mastro VC (1995) Invasion by exotic forest pests: a threat to forest ecosystems. For Sci Monogr 30:1–49

Lietzke SE, Yoder MD, Keen NT, Jurnak F (1994) The three-dimensional structure of pectate lyase E, a plant virulence factor from *Erwinia chrysanthemi*. Plant Physiol 106:849–862

Lieutier F, Lamond C (1978) Nématodes parasites et associés à *Ips sexdentatus* et *Ips typographus* (Coleoptera, Scolytidae) en région parisienne (in French). Nematologica 24:184–200

Linit MJ (1988) Nematode–vector relationships in the pine wilt disease system. J Nematol 20:227–235

Linit MJ (1989) Temporal pattern of pinewood nematode exit from the insect vector *Monochamus carolinensis*. J Nematol 21:105–107

Linit MJ (1990) Transmission of pinewood nematode through feeding wounds of *Monochamus carolinensis* (Coleoptera: Cerambycidae). J Nematol 22:231–236

Linit MJ, Tamura H (1987) Relative susceptibility of four pine species to infection by pinewood nematode. J Nematol 19:44–50

Linit MJ, Kondo E, Smith MT (1983) Insects associated with the pinewood nematode, *Bursaphelenchus xylophilus* (Nematoda: Aphelenchoididae), in Missouri. Environ Entomol 12:467–470

Lipa JJ (1968) *Nosema lepturae* sp. n., a new microsporidian parasite of *Leptura rubra* L. (Coleoptera: Cerambycidae). Acta Protozool 5:269–272

Liston A, Robinson WA, Piñero D, Alvarez-Buylla E (1999) Phylogenetics of *Pinus* (Pinaceae) based on nuclear ribosomal DNA internal transcribed spacer region sequences. Mol Phyl Evol 11:95–109

Liu GL, Pang H, Zhou CQ (2003) Biocontrol of pinewood nematode and its insect vector. Chin J Biol Control 19:193–196

Loof PAA (1964) Free-living and plant parasitic nematodes from Venezuela. Nematologica 10:201–300

Lu Q, Wang WD, Liang J, Yan DH, Jia XZ, Zhang XY (2005) Potential suitability assessmeng of *Bursaphelenchus xylophilus* in China. For Res 18:460–464

Luchi N, Capretti P, Pinzani P, Orlando C, Pazzagli M (2005) Real-time PCR detection of *Biscogniauxia mediterranea* in symptomless oak tissue. Lett Appl Microbiol 41:61–68

Luzzi MA, Wilkinson RC, Tarjan AC (1984) Transmission of the pinewood nematode, *Bursaphelenchus xylophilus*, to slash pine trees and log bolts by a cerambycid beetle, *Monochamus titillator* in Florida. J Nematol 16:37–40

Ma CC, Kanost MR (2000) A beta 1,3-glucan recognition protein from an insect, Manduca sexta, agglutinates microorganisms and activates the phenoloxidase cascade. J Biol Chem 275:7505–7514

Mack RN, Simberloff D, Lonsdale WM, Evans H, Clout M, Bazzaz FA (2000) Biotic invasions: causes, epidemiology, global consequences, and control. Ecol Appl 10:689–710

Maeda I, Kohara Y, Yamamoto M, Sugimoto A (2001) Large-scale analysis of gene function in *Caenorhabditis elegans* by high-throughput RNAi. Curr Biol 11:171–176

Maehara N (2008) Reduction of *Bursaphelenchus xylophilus* (Nematoda: Parasitaphelenchidae) population by inoculating *Trichoderma* spp. into pine wilt-killed trees. Biological Control 44: 61–66

Maehara N, Futai K (1996) Factors affecting both the numbers of the pinewood nematode, *Bursaphelenchus xylophilus* (Nematoda: Aphelenchoididae), carried by the Japanese pine sawyer, *Monochamus alternatus* (Coleoptera: Cerambycidae), and the nematode's life history. Appl Enomol Zool 31:443–452

Maehara N, Futai K (1997) Effect of fungal interactions on the numbers of the pinewood nematode, *Bursaphelenchus xylophilus* (Nematoda: Aphelenchoididae), carried by the Japanese pine sawyer, *Monochamus alternatus* (Coleoptera: Cerambycidae). Fundam Appl Nematol 20:611–617

Maehara N, Futai K (2000) Population changes of the pinewood nematode, *Bursaphelenchus xylophilus* (Nematoda: Aphelenchoididae), on fungi growing in pine-branch segments. Appl Entomol Zool 35:413–417

Maehara N, Futai K (2001) Presence of the cerambycid beetles *Psacothea hilaris* and *Monochamus alternatus* affecting the life cycle strategy of *Bursaphelenchus xylophilus*. Nematology 3:455–461

Maehara N, Futai K (2002) Factors affecting the numbers of *Bursaphelenchus xylophilus* (Nematoda Aphelenchoididae), carried by several species of beetles. Nematology 4:653–658

Maehara N, Kikuchi J, Futai K (1993) Mycorrhizae of Japanese black pine (*Pinus thunbergii*): protection of seedlings from acid mist and effect of acid mist on mycorrhiza formation. Can J Bot 71:1562–1567

Maehara N, Hata K, Futai K (2005) Effect of blue-stain fungi on the number of *Bursaphelenchus xylophilus* (Nematoda: Aphelenchoididae) carried by *Monochamus alternatus* (Coleoptera: Cerambycidae). Nematology 7:161–167

Maehara N, Tsuda K, Yamasaki M, Shirakikawa S, Futai K (2006) Effect of fungus inoculation on the number of *Bursaphelenchus xylophilus* (Nematoda: Aphelenchoididae) carried by *Monochamus alternatus* (Coleoptera: Cerambycidae). Nematology 8:59–67

Maehara N, He X, Shimazu M (2007) Maturation feeding and transmission of *Bursaphelenchus xylophilus* (Nematoda: Parasitaphelenchidae) by *Monochamus alternatus* (Coleoptera: Cerambycidae) inoculated with *Beauveria bassiana* (Deuteromycotina: Hyphomycetes). J Econ Entomol 100:49–53

Magnusson C, Kulinich OA (1996) A taxonomic reappraisal of the original description, morphology and status of *Bursaphelenchus kolymensis* Kolentchenko, 1980 (Aphelenchida: Aphelenchoididae). Russ J Nematol 4:155–161

Magnusson C, Thunes KH, Haukeland S, Okland B (2004) Survey of the pinewood nematode, *Bursaphelenchus xylophilus*, in Norway in 2000. In: Mota M, Vieira P (eds) The pinewood nematode, *Bursaphelenchus xylophilus*. Nematology monographs and perspectives, vol. 1 E.J. Brill, Leiden, pp 101–112

Makihara H (1997) Vector insects and their life history (in Japanese). In: Tamura T (ed) The pine wilt disease. The review of history and recent researches. Zenkoku Shinrin Byochujugai Bojo Kyokai (National Forest Pests Control Association), Tokyo, pp 44–64

Makihara H (1998) Vectors of pine wilt disease and their life histories. In: Tamura H (ed) The pine wilt disease-history and current information. Kyobunsha, Tokyo, Japan, pp 44–64

Makihara H (2004) Two new species and a new subspecies of Japanese Cerambycidae (Coleoptera). Bull For Forest Prod Res Inst 3:15–24

Makihara H, Enda N (2005) Biology of the genus *Monochamus*, especially Japanese pine sawyer, *M. alternatus* (Coleoptera: Cerambycidae) (1) on global view of *Monochamus alternatus* (in Japanese). Shinrin-boeki (For Pests) 54:255–265

Malakhov VV (1994) Development. In: Hope WD (ed) Nematodes. Structure, development, classification, and phylogeny. Smithsonian Institution Press, Washington, DC, pp 134–174

Mamiya Y (1972) Aggregation of *Bursaphelenchus xylophilus* around pupal chambers of *Monochamus alternatus* (in Japanese). Trans Ann Mtg Kanto Br Jpn For Soc 24:30

Mamiya Y (1975a) The life history of the pine wood nematode, *Bursaphelenchus lignicolus* (in Japanese with English abstract). Jpn J Nematol 5:16–25

Mamiya Y (1975b) The extraction efficiency of *Bursaphelenchus xylophilus* by Baermann method (in Japanese). Shinrin-boeki (For Pests) 24:115–119

Mamiya Y (1975c) Population dynamics of *Bursaphelenchus xylophilus* within pine trees (in Japanese). Shinrin-boeki (For Pests) 24:192–196

Mamiya Y (1975d) Behavior of pine wood nematodes in pine wood in early stages of the disease development (in Japanese). Trans Mtg Jpn For Soc 86:285–286

Mamiya Y (1976a) Pine wilting disease caused by the pine wood nematode, *Bursaphelenchus lignicolus*, in Japan. Jpn Agric Res Q 10:206–212

Mamiya Y (1976b) The numbers of *Bursaphelenchus lignicolus* associated with disease development of pine seedlings (in Japanese). Trans Mtg Jpn For Soc 87:225–226

Mamiya Y (1980) Inoculation of the first year pine (*Pinus densiflora*) seedlings with *Bursaphelenchus lignicolus* and histopathology of diseased seedlings (in Japanese with English abstract). J Jpn For Soc 62:176–183

Mamiya Y (1982a) Population growth of *Bursaphelenchus xylophilus* inoculated to excised branches of *Pinus thunbergii* and *P. taeda* (in Japanese). Trans Ann Mtg Kanto Br Jpn For Soc 34:147–148

Mamiya Y (1982b) Pine wilt and pine wood nematode: histopathological aspects of disease development. Proceeding of the third international workshops of host-parasite interactions in forestry, pp 153–160

Mamiya Y (1983) Pathology of the pine wilt disease caused by *Bursaphelenchus xylophilus*. Annu Rev Phytopathol 21:201–220

Mamiya Y (1984a) The pine wood nematode. In: Nickle WR (ed) Plant and insect nematodes. Marcel Dekker, New York, pp 589–626

Mamiya Y (1984b) Resistance of *Pinus* spp. against *Bursaphelenchus xylophilus* and *B. mucronatus* focusing on the host age (in Japanese). Trans Mtg Jpn For Soc 95:475–476

Mamiya Y (1985) Initial pathological changes and disease development in pine trees induced by the pine wood nematode, *Bursaphelenchus xylophilus*. Ann Phytopathol Soc Jpn 51: 546–555

Mamiya Y (1988) History of pine wilt disease in Japan. J Nematol 20:219–226

Mamiya Y (1990a) Behavior of the pine wood nematode, *Bursaphelenchus xylophilus*, and disease development in pine tree s (in Japanese). Nippon Nogeikagaku Kaishi 64:1243–1246

Mamiya Y (1990b) Effects of fatty acids added to media on the population growth of *Bursaphelenchus xylophilus* (Nematoda: Aphelenchoididae). Appl Entomol Zool 25:299–309

Mamiya Y (2003a) Anatomical observations on the exiting behavior of the dispersal fourth stage juveniles of the pine wood nematode, *Bursaphelenchus xylophilus*, from tracheae of the adult pine sawyer, *Monochamus alternatus*. Jpn J Nematol 33:41–43

Mamiya Y (2003b) Entomogenous nematode and entomophilic nematode (in Japanese). In: Ishibashi N (ed) Biology of nematodes. University of Tokyo Press, Tokyo, pp 165–180

Mamiya Y (2004) Pine wilt disease in Japan. In: Mota M, Vieira P (eds) The pinewood nematode, *Bursaphelenchus xylophilus*. Nematology monographs and perspectives, vol 1. E.J. Brill. Leiden, pp 9–20

Mamiya Y, Enda N (1972) Transmission of *Bursaphelenchus lignicolus* (Nematoda: Aphelenchoididae) by *Monochamus alternatus* (Coleoptera: Cerambycidae). Nematologica 18: 159–162

Mamiya Y, Enda N (1979) *Bursaphelenchus mucronatus* n. sp. (Nematoda: Aphelenchoididae) from pine wood and its biology and pathogenicity to pine trees. Nematologica 25: 353–361

Mamiya Y, Kiyohara T (1972) Description of *Bursaphelenchus lignicolus* n. sp. (Nematoda: Aphelenchoididae) from pine wood and histopathology of nematode-infested trees. Nematologica 18:120–124

Mamiya Y, Shoji T (1984) Application of *Neoaplectana carpocapsae* on pine logs with the pine sawyer, *Monochamus alternatus* (in Japanese). Trans Ann Mtg Kanto Br Jpn For Soc 36:135–136

Mamiya Y, Shoji T (1985) Pathogenicity of the pinewood nematode to Japanese larch (in Japanese). Trans Mtg Jpn For Soc 96:84

Mamiya Y, Tamura H (1976) A kind of nematode-trapping fungi, *Dactylella leptospora*, found in wood around pupal chambers of *Monochamus alternatus* (in Japanese). Shinrin-boeki (For Pests) 25:147–149

Mamiya Y, Tamura H (1983) Mortality of the pine sawyer, *Monochamus alternatus* induced by the entomogenous nematode *Steinernema feltiae* (in Japanese). Trans Ann Mtg Kanto Br Jpn For Soc 35:163–164

Mamiya Y, Kobayashi T, Jinno Y, Enda N, Sasaki K (1973) Disease development of pine trees naturally infected with *Bursaphelenchus lignicolus* (in Japanese). Trans Mtg Jpn For Soc 84:332–334

Mamiya Y, Futai K, Kosaka H, Kanzaki N (2004) Pinewood nematode. In: The Japanese Nematological Society (ed) Experimental methods in nematology, The Japanese Nematological Society, Tsukuba, pp 134–153

Mamiya Y, Hiratsuka M, Murata M (2005) Ability of wood-decay fungi to prey on the pinewood nematode, *Bursaphelenchus xylophilus* (Steiner and Buhrer) Nickle. Jpn J Nematol 35: 21–30

Marx DH (1969) The influence of ectotrophic mycorrhizal fungi on the resistance of pine roots to pathogenic fungi and soil bacteria. Phytopathology 59:153–163

Massey CL (1964) The nematode parasites and associates of the fir engraver beetle, *Scolytus ventralis* Le Conte, in New Mexico. J Insect Pathol 6:133–155

Massey CL (1966) The nematode parasites and associates of *Dendroctonus adjunctus* (Coleoptera: Sclytidae) in New Mexico. Ann Entomol Soc Am 59:424–440

Massey CL (1971a) Nematode associate of several species of *Pissodes* (Coleoptera: Curculionidae) in the United States. Ann Entomol Soc Am 64:162–169

Massey CL (1971b) *Omemeea maxbassiensis* n. gen., n. sp. (Nematoda: Aphelenchoididae) from galleries of the bark beetle *Lepersinus californicus* Sw. (Coleoptera: Scolytidae) in North Dakota. J Nematol 3:289–291

Massey CL (1974) Biology and taxonomy of nematode parasites and associates of bark beetles in the United States. USDA Agriculture Handbook No. 446, USDA Forest Service, United States Governmental Printing Office, Washington DC

Masuya H, Kaneko S, Yamaoka Y (1998) Blue stain fungi associated with *Tomicus piniperda* (Coleoptera: Scolytidae) on Japanese red pine. J For Res 3:213–219

Masuya H, Kaneko S, Yamaoka Y, Osawa M (1999) Comparison of Ophiostomatoid fungi associated with *Tomicus piniperda* and *T. minor* in Japanese red pine. J For Res 4:131–135

Masuya H, Kaneko S, Yamaoka Y (2003) Comparative virulence of blue-stain fungi isolated from Japanese red pine. J For Res 8:83–88

Matsubara I (1976) Observations of the epidemic mortality of pine trees in Chiba prefecture (in Japanese). Trans Mtg Jpn For Soc 87:307–308

Matsueda A (1975) Infestation of *Bursaphelenchus xylophilus* and *Monochamus alternatus* in Ishikawa Prefecture (in Japanese). Bull Ishikawa For Exp Stn 6:43–62

Matsunaga K, Togashi K (2004) A simple method for discriminating *Bursaphelenchus xylophilus* and *B. mucronatus* by species-specific polymerase chain reaction primer pairs. Nematology 6:273–277

Matsunaga K, Togashi K, Jikumaru S (2004) A PCR-aided method for discrimination of the two species, *Bursaphelenchus xylophilus* and *B. mucronatus* (in Japanese). Trans Jpn For Soc 115: F44

McLaughlin RL (1989) Search modes of birds and lizards: evidence for alternative movement patterns. Am Nat 133:654–670

Melakeberhan HaW, J M (1990) Effect of *Bursaphelenchus xylophilus* on the assimilation and translocation of 14C in *Pinus sylvestris* J Nematol 22:506–512

Meldal BH, Debenham NJ, De Ley P, De Ley IT, Vanfleteren JR, Vierstraete AR, Bert W, Borgonie G, Moens T, Tyler PA, Austen MC, Blaxter ML, Rogers AD, Lambshead PJ (2007) An improved molecular phylogeny of the Nematoda with special emphasis on marine taxa. Mol Phylogenet Evol 42:622–636

Metge K, Bugermeister W (2006) Intraspecific variation in isolates of *Bursaphelenchus xylophilus* (Nematoda: Aphelenchoididae) revealed by ISSR and RAPD fingerprints. J Plant Dis Prot 113:275–282

Metge K, Burgermeister W (2008) Analysis of *Bursaphelenchus xylophilus* (Nematoda: Aphelenchoididae) provenances using ISSR and RAPD fingerprints. In: Mota M, Vieira P (eds) Pine wilt disease: a worldwide threat to forest ecosystems. Springer, The Netherlands, p 29

Michalopoulos-Skarmoutsos H, Skarmoutsos G, Kalapanida M, Karageorgus A (2004) Surveying and recording of nematodes of the genus *Bursaphelenchus* in conifer forests in Greece and pathogenicity of the most important species. In Mota M, Vieira P (eds) The pinewood nematode, *Bursaphelenchus xylophilus*. Nematology Monographs and Perspectives, vol 1. E.J. Brill, Leiden, pp 113–126

Miduturi JS (1997) Bionomics of naturally occurring entomopathogenic nematodes in Belgium. PhD thesis, University of Ghent, Ghent, Belgium

Mineo K (1974) Inoculation of the pinewood nematode on exotic pine species (in Japanese). Trans Ann Mtg Kansai Br Jpn For Soc 25:312–313

Mineo K (1983) The departure of *Bursaphelenchus xylophilus* nematodes from *Monochamus alternatus* and their invasion into the tree body (in Japanese). Proc Kanto Conf Jpn For Soc 34:259–261

Mineo K, Kontani S (1973) Damage of *Pinus pinaster* by the pinewood nematode (in Japanese). Shinrin-boeki (For Pests) 22:227–229

Mitsui Y (1983) The distribution, and physiological and ecological nature of the nematode trapping fungi in Japan (in Japanese with English abstract). Bull Nat Inst Agr Sci Ser C 37:127–211

Mitsui Y (1992) Fungal parasites of nematodes (in Japanese). In: Nakasono K (ed) Progress in nematology. The Japanese Nematological Socoety, Ibaraki, pp 262–266

Miura K, Okamoto Y, Abe T, Nakashima Y (2000) Parasitic characteristics of *Sclerodermus nipponicus* and *Dastarcus helophoroides* to *Monochamus alternatus* (in Japanese). Shinrin-boeki (For Pests) 49:225–230

Miwa J (1986) Making of the roundworm II (in Japanese). Kagaku 56:162–170

Miwa J, Schierenberg E, Miwa S, von Ehrenstein G (1980) Genetics and mode of expression of temperature-sensitive mutations arresting embryonic development in *Caenorhabditis elegans*. Dev Biol 76:160–174

Miyahara F, Kuramoto T, Ohira M, Okamura M, Hiraoka Y, Mori Y, Miyasaki J, Yoshimoto K, Tobase M, Kusano Y, Ochiai T, Mitsuki Y, Koyama T, Toda T (2005) Resistance of each resistant family produced from resistant Japanese black pine orchards (in Japanese). Trans Jpn For Soc 116:1D14

Miyazaki M, Oda K, Yamaguchi A, Yamane A, Enda N (1974) Pine bark constituents related to feeding response by Japanese sawyer adult, *Monochamus alternatus* Hope (I) bioassays in small vessels and some properties of the biting factors. J Jpn For Soc 56:239–246

Miyazaki M, Oda K, Yamaguchi A (1977a) Behaviour of *Bursaphelenchus lignicolus* to unsaturated fatty acids (in Japanese with English abstract). J Jpn Wood Res Soc 23:255–261

Miyazaki M, Oda K, Yamaguchi A (1977b) Deposit of fatty acids in the wall of pupal chamber made by *Monochamus alternatus* (in Japanese with English abstract). J Jpn Wood Res Soc 23:307–311

Miyazaki M, Yamaguchi A, Oda K (1978a) Behaviour of *Bursaphelenchus lignicolus* in response to carbon dioxide (in Japanese with English abstract). J Jpn For Soc 60:203–208

Miyazaki M, Yamaguchi A, Oda K (1978b) Behaviour of *Bursaphelenchus lignicolus* in response to carbon dioxide released by respiration of *Monochamus alternatus* pupa (in Japanese with English abstract). J Jpn For Soc 60:249–254

Mo MH, Zhou W, Zhao ML, Zhang KQ (2002) Capacity of Arthrobotrys spp. to capture nematodes *Bursaphelenchus xylophlus* and *B. mucronatus* in vitro. Acta Microbiol Sin 29:13–16

Mock KE, Bentz BJ, O'neill EM, Chong JP, Orwin J, Pfrender ME (2007) Landscape-scale genetic variation in a forest outbreak species, the mountain pine beetle (*Dendroctonus ponderosae*). Mol Ecol 16:553–568

Molyneux AS (1984) The influence of temperature on the infectivity of heterorhabditid and steinernematid nematode for larvae of the sheep blowfly, *Lucilia cuprina*. In: Baily P, Swincer D (eds) Proceeding of the fourth Australian applied entomological research conference, Adelaide, pp 344–351

Molyneux AS (1985) Survival of infective juveniles of *Heterorhabditis* spp. and *Steinernema* spp. (Nematoda: Rhabditida) at various temperatures and the subsequent infectivity for insects. Rev Nématol 8:165–170

Molyneux AS (1986) *Heterorhabditidae* spp. and *Steinernema* spp. temperature and aspects of behavior and infectivity. Exp Parasitol 62:169–180

Mori T, Inoue T (1983) Changes of stem respiration during the development of pine wilt disease caused by *Bursaphelenchus xylophilus* (in Japanese). Trans Mtg Jpn For Soc 94: 307–308

Mori T, Inoue T (1986) Pine-wood nematode-induced ethylene production in pine stems and cellulase as an inducer (in Japanese with English abstract). J Jpn For Soc 68:43–50

Morimoto K (1977) Pine wilt disease caused by the pinewood nematode and its control method (in Japanese). Comprehensive commentaries on forestry researches 58

Morimoto K, Iwasaki A (1972) Role of *Monochamus alternatus* (Coleoptera: Cerambycidae) as a vector of *Bursaphelenchus lignicolus* (Nematoda: Aphelenchoididae) (in Japanese with English abstract). J Jpn For Soc 54:177–183

Morimoto K, Iwasaki A (1973) Studies on the pine sawyer (IV). Biology of the pine sawyer and the pine wood nematode in the pupal cell (in Japanese). Trans Ann Mtg Kyushu Br Jpn For Soc 26:199–200

Morimoto K, Iwasaki A (1974a) Studies on the Japanese pine sawyer (X). Larval molting (in Japanese). Trans Mtg Jpn For Soc 85:227–228

Morimoto K, Iwasaki A (1974b) Studies on the Japanese pine sawyer (IX). Density effects on the emergence rate (in Japanese). Trans Mtg Jpn For Soc 85:299–300

Morimoto K, Mamiya Y (1977) Pine wilt disease and control method (in Japanese). Japan Forestry Technology Association, Tokyo

Morimoto K, Iwasaki A, Taniguchi A (1972) Studies on *Monochamus alternatus* (Coleoptera: Cerambycidae) XIV—relationship between proportion of beetles staying on a tree in a day and air temperature (in Japanese). Trans Ann Mtg Kyushu Br Jpn For Soc 28:199–200

Morooka N, Tsunoda H, Tatuno T (1983) Studies on the fungi participating in the pine wilt (in Japanese). J Antibact Antifung Agents 11:423–431

Morris DW, Diffendorfer JE, Lundberg P (2004) Dispersal among habitats varying in fitness: reciprocating migration through ideal habitat selection. Oikos 107:559–575

Mota M (2004) Occurrence of the pine wood nematode, *Bursaphelenchus xylophilus* in Portugal and perspectives of the disease spread in Europe. In: Cook RC, Hunt DJ (eds) Nematology monographs and perspectives, vol 2. E.J. Brill, Leiden, pp 837–841

Mota M, Vieira P (eds) (2004) The pinewood nematode, *Bursaphelenchus xylophilus*. Nematology monographs & perspectives, vol 1. E.J. Brill, Leiden, p 291

Mota, M, Vieira, P (eds) (2008) Pine wilt disease: a worldwide threat to forest ecosystems. Springer, The Netherlands. xviii + p 405

Mota MM, Braasch H, Bravo MA, Penas AC, Burgermeister W, Metge K, Sousa E (1999) First report of *Bursaphelenchus xylophilus* in Portugal and in Europe. Nematology 1: 727–734

Mota MM, Takemoto S, Takeuchi Y, Hara N, Futai K (2007) Comparative studies between Portuguese and Japanese isolates of the pinewood nematode, *Bursaphelenchus xylophilus*. J Nematol 38:429–433

Moyle PL, Kaya HK (1981) Dispersal and infectivity of entomopathogenic nematode, *Neoaplectana carpocapsae* Weiser (Rhabdtida: Steinernematidae) in sand. J Nematol 13:295–300

Munro E, Nance J, Priess JR (2004) Cortical flow powered by asymmetrical contraction tansport PAR protein to establish and maintain anterior-posterior polarity in the early *C. elegans* embryo. Dev Cell 7:413–424

Muramoto M (1998) Ending of pine wilt disease in Okinoerabu island, Kagoshima Prefecture. In: Futai K, Togashi K, Ikeda T (eds) Sustainability of pine forests in relation to pine wilt and decline: proceedings of international symposium Tokyo, 27–28 October 1998. Shokado, Kyoto, Japan, pp 193–195

Myers RF (1982) Susceptibility of pines to pinewood nematode in New Jersey. In: Appleby JE, Malek RB (eds) Proceedings of the 1982 national pine wilt disease workshop, pp 38–46

Myers RF (1984) Comparative histology and pathology in conifers infected with pine wood nematode, *Bursaphelenchus xylophilus*. In: Dropkin V (ed) The resistance mechanisms of pines against pine wilt disease, proceedings of the US–Japan seminar, pp 91–95

Myers RF (1988) Pathogenesis in pine wilt caused by pine wood nematode, *Bursaphelenchus xylophilus*. J Nematol 20:236–244

Nakai I, Kitagawa S, Akita Y, Nakane I, Shibata S (1995) The distribution and its chronical change of a pine wilt in pine forest stand at Tokuyama Experimental Station of Kyoto University Forest at Yamaguchi Prefecture. Rep Kyoto Univ For 28:1–9

Nakamura H, Tsutsui N, Okamoto H (1995a) Oviposition habit of the Japanese pine sawyer, *Monochamus alternatus* Hope (Coleoptera: Cerambycidae) I. Factors affecting the vertical distribution of oviposition scars in a pine tree. Jpn J Entomol 63:633–640

Nakamura H, Tsutsui N, Okamoto H (1995b) Oviposition habit of the Japanese pine sawyer, *Monochamus alternatus* Hope (Coleoptera: Cerambycidae) II. Effect of bark thickness on making of oviposition scars. Jpn J Entomol 63:739–745

Nakamura K (2004) Traits in the occurrence of dead trees of *Pinus thunbergii* in a seacoast pine forest where pine wilt disease has been successfully controlled. Trans Jpn For Soc 115:729

Nakamura K, Okochi I (2002) Longevity and ovarian status of the adult *Monochamus alternatus* Hope fed on non-pine tree species (in Japanese with English abstract). J Jpn For Soc 84:21–25

Nakamura K, Yoshida N (2004) Successful control of pine wilt disease in Fukiage-hama seacoast pine forest in southwestern Japan. In: Mota M, Vieira P (eds) The pinewood nematode, *Bursaphelenchus xylophilus*: proceedings of an international workshop, University of Évora, Portugal, 20–22 August 2001. E.J. Brill, Leiden, pp 269–281

Nakamura K, Akiba M, Sahashi N, Kameyama N, Mikami A, Motoshige T, Ito S, Mizuma Y, Mori T, Maeda D, Irei H, Kiyuuna C, Nakahira Y, Miyagi T, Gushiken T (2005) Epidemiology of pine wilt disease in Ryukyu pines in Okinawa Island VI. Results of the four-year consecutive study (in Japanese). Trans Jpn For Soc 116:PA171

Nakane I (1976) The departure pattern of *Bursaphelenchus xylophilus* dauer larvae from insect body (in Japanese). Proc Kanto Conf Jpn For Soc 27:252–254

Namkoong G (1991) Maintaining genetic diversity in breeding for resistance in forest trees. Annu Rev Phytopathol 29:325–342

Naves P, Kenis M, Sousa E (2005) Parasitoids associated with *Monochamus galloprovincialis* (Oliv.) (Coleoptera: Cerambycidae) within the pine wilt nematode-affected zone in Portugal. J Pest Sci 78:57–62

Naves PM, Camacho S, De Sousa EM, Quartau JA (2006a) Entrance and distribution of the pinewood nematode *Bursaphelenchus xylophilus* on the body of its vector *Monochamus galloprovincialis* (Coleoptera: Cerambycidae). Entomol Gener 29:71–80

Naves P, Sousa E, Quartau JA (2006b) Feeding and oviposition preferences of *Monochamus galloprovincialis* for certain conifers under laboratory conditions. Entomol Exp Appl 120:99–104

Naves P, Sousa E, Quartau JA (2006c) Reproductive traits of *Monochamus galloprovincialis* (Coleoptera: Cerambycidae) under laboratory conditions. Bull Entomol Res 96:289–294

Naves PM, Camacho S, De Sousa EM, Quartau JA (2007) Transmission of the pine wood nematode *Bursaphelenchus xylophilus* through feeding activity of *Monochamus galloprovincialis* (Col., Cerambycidae). J Appl Entomol 131:21–25

Nawa U (1937) On cerambycid beetles (in Japanese). Konchu-Sekai 41:418–419

Necibi S, Linit MJ (1998) Effect of *Monochamus carolinensis* on *Bursaphelenchus xylophilus* dispersal stage formation. J Nematol 30:246–254

Nguyen KB, Smart GC (1990) Vertical dispersal of *Steinernema scapterisci*. J Nematol 22:574–578

Nguyen KB, Smart GC (1994) *Neosteinernema longicurvicauda* n. gen., n. sp. (Rhabditida: Steinernematidae), a parasite of the termite *Reticulitermes flavipes* (Koller). J Nematol 26:162–174

Nickle WR (1970) A taxonomic review of the genera of Aphelenchoidea (Fuchs) Thorne, 1949 (Nematoda: Tylenchida). J Nematol 2:375–392

Nickle WR (1984) History, development, and importance of insect nematology. In: Nickle WR (ed) Plant and insect nematodes. Marcel Dekker, New York, pp 627–653

Nickle TL, Coburn MJ (1988) Pinewood nematode control, in wood chips, using hot water treatment. J Nematol 20:651

Nickle WR, Golden AM, Mamiya Y, Wergin WP (1981) On the taxonomy and morphology of the pine wood nematode, *Bursaphelenchus xylophilus* (Steiner & Buhrer, 1934) Nickle 1970. J Nematol 13:385–392

Nikolaou S, Hartman D, Presidente PJA, Newton SE, Gasser RB (2002) HcSTK, a *Caenorhabditis elegans* PAR-1 homologue form the parasitic nematode, *Haemonchus contortus*. Int J Parasitol 32:749–758

Nikolaou S, Hartman D, Nisbet AJ, Presidente PJA, Gasser RB (2004) Genomic organization and expression analysis for *hcstk*, a serine/threonine protein kinase gene of *Haemonchus contortus*, a comparison with *Caenorhabditis elegans par-1*. Gene 343:313–322

Ning T, Fang YL, Tang J, Sun JH (2004) Advances in research on *Bursaphelenchus xylophilus* and its key vector *Monochamus* spp. Entomol Knowl 41:97–104

Ning T, Fang YL, Tang J, Sun JH (2005) Current status of monitoring strategies and control techniques for *Bursaphelenchus xylophilus* and its vector *Monochamus alternatus*. Chin Bull Entomol 42:264–269

Nishimatsu T, Jackson JJ (1998) Interaction of insecticides, entomopathogenic nematodes, and larvae of the western corn root-worm (Coleoptera: Chrysomelidae). J Econ Entomol 91:410–418

Nishimura K, Ohoba K, Risen Y, Matsunaga K, Imamura M (1977) Development of artificial inoculation method for breeding for resistance to pine wilt disease. Trans Ann Mtg Kyushu For Assoc 30:61–62

Nishimura M (1973) Daily observation on behaviors of Japanese pine sawyer adult, *Monochamus alternatus* Hope (in Japanese with English abstract). J Jpn For Soc 73:100–104

Nobuchi A (1976) Fertilization and oviposition of *Monochamus alternatus* Hope (in Japanese). Trans Mtg Jpn For Soc 87:274–248

Nobuchi A (1980) Natural enemies of *Monochamus alternatus* (in Japanese). Shinrin-boeki (For Pests) 29:23–28

Nobuchi A (1989) A trial of microbial control of *Monochamus alternatus* utilizing *Cryphalus fulvus* as a carrier of the pathogen (in Japanese). Shinrin-boeki (For Pests) 38:133–137

Nobuchi A (1993) An automatic releasing equipment of *Beauveria*-contaminated bark beetle for microbial control of *Monochamus alternatus* (in Japanese). Shinrin-boeki (For Pests) 42: 213–217

Nobuchi T, Tominaga T, Futai K, Harada H (1984) Cytological study of pathological changes in Japanese black pine (*Pinus thunbergii*) seedlings after inoculation with pinewood nematode (*Bursaphelenchus xylophilus*). Bull Kyoto Univ For 56:224–233

Nowell W (1919) The red ring or "root" disease of coconut palms. W I Bull 17:189–202

Ochi K (1969) Ecological studies on cerambycid injurious to pine trees (II) Biology of two *Monochamus* (Coleoptera: Cerambycidae) (in Japanese with English abstract). J Jpn For Soc 51:188–192

Ochi K, Katagiri K (1979) Mortality factors in survival curves of *Monochamus alternatus* Hope (Coleoptera: Cerambycidae) population on the dead pine trees (in Japanese with English abstract). Bull Gov For Exp Stn 303:125–152

Ochiai M, Ashida M (2000) A pattern-recognition protein for beta-1,3-glucan—the binding domain and the cDNA cloning of beta-1,3-glucan recognition protein from the silkworm, *Bombyx mori*. J Biol Chem 275:4995–5002

Odani K, Yamamoto N, Nishiyama Y, Sasaki S (1984) Action of nematode on the development of pine wilt disease. In: Dropkin V (ed) The resisitance mechanism of pine against pine wilt disease, proceedings of the US-Japan seminar, pp 128–140

Odani K, Sasaki S, Nishiyama Y, Yamamoto N (1985) Early symptom development of the pine wilt disease by hydrolytic enzymes produced by the pine wood nematodes—cellulase as a possible candidate of the pathogen. J Jpn For Soc 67:366–372

Oestergaard J, Belau C, Strauch O, Ester A, van Rozen K, Ehlers RU (2006) Biological control of *Tipula paludosa* (Diptera: Nematocera) using entomopathogenic nematodes (*Steinernema* spp.) and *Bacillus thuringiensis* subsp. israelensis. Biol Control 39: 525–531

Ogawa S (1988) Biological and physical control of the pine wilt disease. Bull Fukuoka ken For Exp Stn 35:56–73

Ogawa S, Hagiwara Y (1975) Wilt disease of *Pinus elliotti* caused by the pinewood nematode (in Japanese). Shinrin-boeki (For Pests) 24:161–163

Ogura N, Nakashima T (2002) In vitro occurrence of dispersal fourth stage juveniles in *Bursaphelenchus xylophilus* co-incubated with *Monochamus alternatus*. Jpn J Nematol 32:53–59

Ogura N, Tabata K, Wang W (1999) Rearing of the colydiid beetle predator, *Dastarcus helophoroides*, on artificial diet. Biocontrol 44:291–299

Ogura T, Kishi Y, Kondo H, Ebine S (1983) Pathogenicity of the pinewood nematode on several Pinaceae species (in Japanese). Trans Mtg Jpn For Soc 94:467–468

Ohba K (1976) Breeding for resistance to pine wilt disease (in Japanese). For Tree Breed 99:1–6

Ohba K (1982) Resistance to pinewood nematode (in Japanese). In: Forest pest controlling techniques. Zenkoku Shinrin Byochujugai Kyokai (National Forest Pests Contol Association), Tokyo, Japan, pp 320–327

Ohba K (1991) Strategies of tree breeding (in Japanese). In: Ohba K, Katsuta M (eds) Tree breeding, Buneido Publishing Co., Ltd, Tokyo, pp 9–61

Ohba K, Nishimura K, Toda T, Risen Y (1977) Difference of survival ratios among half-sib families inoculated with pine wood nematodes (in Japanese). Trans Kyushu For Assoc 30:67–68

Ohba K, Furukoshi T, Kurinobu S, Ishii K (1984) Susceptibility of subtropical pine species and provenances to the pine wood nematode. J Jpn For Soc 66:465–468

Ohira M, Miyahara F, Mori Y, Miyazaki J, Masaki S, Yamada Y, Shiraishi S (2007) Development of new nursing techniques of Japanese black pine seedlings by propagating pine wilt disease resistants by cuttings (in Japanese). For Tree Breed Special number:29–32

Ohyama N, Kaminaka H (1975a) Abnormal assimilation and respiration in Japanese black pine seedlings inoculated with pinewood nematode (I) (in Japanese). Trans Kyusyu Br Jpn For Soc 28:105–106

Ohyama N, Kaminaka H (1975b) Abnormal assimilation and respiration in Japanese black pine seedlings inoculated with pinewood nematode (II) (in Japanese). Trans Mtg Jpn For Soc 86:200–201

Ohyama N, Shiraishi S, Takagi T (1986) Characteristics in the graftings of the resistant pine against wood nematode (in Japanese with English abstract). For Tree Breed 140: 17–21

Okabe T, Nakashima K, Takai K, Suzuki T, Higuchi T (2001) Biological control of the Japanese pine sawyer, *Monochamus alternatus* by *Beauveria bassiana* (in Japanese). Trans Ann Mtg Kyushu Br Jpn For Soc 54:115–116

Okabe T, Takai K, Suzuki T, Higuchi T (2002) Biological control of the Japanese pine sawyer, *Monochamus alternatus* by *Beauveria bassiana* (II)—fecundity of the female adults contaminated with *Beauveria bassiana* (in Japanese). Kyushu J For Res 55:73–74

Okamoto H, Otani H and Takai S (1986) Effect of ceratoulmin (CU) on electrogenic pumps of elm cell membrane and ion loss from cells. Phytopathology 76:116

Okamura M (2004) Additional selection of pine wood nematode resistants in Japanese black pine in Kyushu (in Japanese). Tree Breed Technol News 20:6–7

Okitsu M, Kishi Y, Takagi Y (2000) Control of adults of *Monochamus alternatus* Hope (Coleoptera: Cerambycidae) by application of non-woven fabric strips containing *Beauveria bassiana* (Deutyeromycotina: Hyphomycetes) on infested tree trunks (in Japanese with English abstract). J Jpn For Soc 82:276–280

Okochi I, Kagaya-Shoda E (2005) Diminished damages of pine wilt disease in Ogasawara Islands (in Japanese). Shinrin-boeki (For Pests) 54:46–50

Oku H (1984) Biological activity of toxic metabolites isolated from pine trees naturally infected by pine wood nematodes. In: Dropkin V (ed) The resistance mechanisms of pines against pine wilt disease, proceedings of the US–Japan seminar, pp 110–118

Oku H (1988) Role of phytotoxins in pine wilt disease. J Nematol 20:245–251

Oku H (1990) Phytotoxins in pine wilt disease. Nippon Nogeikagaku Kaishi 64:1254–1257

Oku H, Shiraishi T, Kurozumi S (1979) Participation of toxin in wilting of Japanese pines caused by a nematode. Naturwissenschaften 66:210–211

Oku H, Shiraishi T, Ouchi S, Kurozumi S, Ohta H (1980) Pine wilt toxin, the metabolite of a bacterium associated with a nematode. Naturwissenschaften 67:198–199

Oku H, Yamamoto H, Ohta H, Shiraishi T (1985) Effect of abnormal metabolites isolated from nematode-infected pine on pine seedlings and pine wood nematodes. Ann Phytopathol Soc Jpn 51:303–311

Oku H, Shiraishi T, Chikamatsu K (1989) Active defense as a mechanism of resistance in pine against pine wilt disease. Ann Phytopathol Soc Jpn 55:603–608

Orui Y (1996) Discrimination of the main *Pratylenchus* species (Nematoda: Pratylenchidae) in Japan by PCR–RFLP analysis. Appl Entomol Zool 31:505–514

Page AP, Hamilton AJ, Maizels RM (1992) Toxocara canis: monoclonal antibodies to carbohydrate epitopes of secreted (TES) antigens localize to different secretion-related structures in infective larvae. Exp Parasitol 75:56–71

Paine TD, Raffa KF, Harrington TC (1997) Interactions among scolytid bark beetles, their associated fungi, and live host conifers. Annu Rev Entomol 42:179–206

Palmisano AM, Ambrogioni L, Tomiczek C, Brandstetter M (2004) *Bursaphelenchus sinensis* sp. n. and *B. thailandae* Braasch et Braasch-Bidasak in packaging wood from China. Nematol Mediterr 32:57–65

Pan HY (2000) Strategies of managing pine wilt disease now in our country (in Chinese). For Pest Dis (6):44–47

Panesar TS, Peet FG, Sutherland JR, Sahota TS (1994) Effects of temperature, relative humidity and time on survival of pinewood nematodes in wood chips. Eur J For Path 24:289–299

Paracer S, Ahmadjian V (2000) Symbiosis, an introduction to biological associations, 2nd edn. Oxford University Press, Oxford, pp 3–13

Park IK, Park JY, Kim KH, Choi KS, Choi IH, Kim CS, Shin SC (2005) Nematicidal activity of plant essential oils and components from garlic (*Allium sativum*) and cinnamon (*Cinnamomum verum*) oils against the pine wood nematode (*Bursaphelenchus xylophilus*). Nematology 7:767–774

Parkinson J, Mitreva M, Hall N, Blaxter M, McCarter JP (2003) 400000 nematode ESTs on the Net. Trends Parasitol 19:283–286

Parkinson J, Mitreva M, Whitton C, Thomson M, Daub J, Martin J, Schmid R, Hall N, Barrell B, Waterston RH, McCarter JP, Blaxter ML (2004) A transcriptomic analysis of the phylum Nematoda. Nat Genet 36:1259–1267

Paterson AH, Tanksley SD, Sorrells ME (1991) DNA markers in plant improvement. Adv Agron 46:39–90

Penas AC, Dias LS, Mota M (2002) Precision and selection of extraction methods of aphelenchid nematodes from maritime pine wood, *Pinus pinaster* L. J Nematol 34:62–65

Penas AC, Correia P, Bravo MA, Mota M, Tenreiro R (2004) Species of *Bursaphelenchus* Fuchs, 1937 (Nematoda: Parasitaphelenchidae) found in maritime pine in Portugal. Nematology 6:437–453

Penas AC, Metge G, Mota MM, Valadas V (2006a) *Bursaphelenchus antoniae* sp. n. (Nematoda: Parasitaphelenchidae) associated with *Hylobius* sp. from *Pinus pinaster* in Portugal. Nematology 8:659–669

Penas AC, Bravo MA, Naves P, Bonifácio L, Sousa E, Mota MM (2006b) Species of *Bursaphelenchus* Fuchs, 1937 (Nematoda: Parasitaphelenchidae) and other nematode genera associated with insects from *Pinus pinaster* in Portugal. Ann Appl Biol 148:121–131

Pereira P, Roque P (2008) Spatial modelling of *Bursaphelenchus xylophilus* in Portugal. In: Mota M, Vieira P (eds) Pine wilt disease: a worldwide threat to forest ecosystems. Spinger, The Netherlands, p 38

Phan KL, Tirry L, Moens M (2006) Pathogenic potential of six isolates of entomopathogenic nematodes (Rhabditida: Steinernematidae) from Vietnam. Biocontrol 50:477–491

Pianka ER (1966) Convexity, desert lizards and spatial heterogeneity. Ecology 47:1055–1059

Poinar GO (1979) Nematodes for biological control of insects. CRC Press, Boca Raton

Poinar GO (1986) Entomogenous nematodes. In: Franz BD (ed) Biological plant and health protection. Gustav Fisher Verlag, Stuttgart, pp 95–121

Poinar GO (1990) Taxonomy and biology of Steinernematidae and Heterorhabditidae. In: Gaugler R, Kaya HK (eds) Entomopathogenic nematodes in biological control. CRC Press, Boca Raton, pp 23–61

Popeijus H, Overmars H, Jones J, Blok V, Goverse A, Helder J, Schots A, Bakker J, Smant G (2000) Degradation of plant cell walls by a nematode. Nature 406:36–37

Pramila G, Siddiqui MR (1999) Compatibility studies on *Steinernema carpocapsae* with some pesticidal chemicals. Ind J Entomol 61:220–225

Price RA, Liston A, Strauss SH (1998) Phylogeny and systematics of *Pinus*. In: Richardson DM (ed) Ecology and biogeography of *Pinus*. Cambridge University Press, Cambridge, pp 49–68

Qin L, Kudla U, Roze EH, Goverse A, Popeijus H, Nieuwland J, Overmars H, Jones JT, Schots A, Smant G, Bakker J, Helder J (2004) Plant degradation: a nematode expansin acting on plants. Nature 427:30

Quinn AE, Georges A, Sarre SD, Guarino F, Ezaz T, Graves JAM (2007) Temperature sex reversal implies sex gene dosage in a reptile. Science 316:411

Rautapaa J (1986) Experiences with *Bursaphelenchus* in Finland. Conference on pest and disease problems in European Forests. EPPO Bull 16:453–456

Research and Extension Division, Forestry Agency (1994a) Notifications related to forest tree breeding projects (in Japanese). Japan Forest Tree Breeding Association, Tokyo, pp 133–139

Research and Extension Division, Forestry Agency (1994b) Notifications related to forest tree breeding projects (in Japanese). Japan Forest Tree Breeding Association, Tokyo, pp 149–154

Riga E, Beckenbach K, Webster JM (1992) Taxonomic relationships of *Bursaphelenchus xylophilus* and *Bursaphelenchus mucronatus* based on interspecific and intraspecific cross-hybridization and DNA analysis. Fundam Appl Nematol 15:391–395

Robbins K (1982) Distribution of the pinewood nematode in the United States. In: Appleby JE, Malek RB (eds) Proceedings of the 1982 national pine wilt disease workshop, pp 3–6

Robinson AF (1995) Optimal release rates for attracting *Meloidogyne incognita*, *Rotylenchulus reniformis*, and other nematodes to carbon dioxide in sand. J Nematol 27:42–50

Rodrigues JM (2008) Eradication program for the pinewood nematode in Portugal. In: Mota M, Vieira P (eds) Pine wilt disease: a worldwide threat to forest ecosystems. Springer, The Netherlands, p 3

Roehrdanz RL (1993) An improved primer for PCR amplification of mitochondrial DNA in a variety of insect species. Insect Mol Biol 2:89–91

Rosso MN, Favery B, Piotte C, Arthaud L, Boer JMd, Hussey RS, Bakker J, Baum TJ, Abad P (1999) Isolation of a cDNA encoding a beta-1,4-endoglucanase in the root-knot nematode *Meloidogyne incognita* and expression analysis during plant parasitism. Mol Plant Microbe Interact 12:585–591

Rosso MN, Dubrana MP, Cimbolini N, Jaubert S, Abad P (2005) Application of RNA interference to root-knot nematode genes encoding esophageal gland proteins. Mol Plant Microbe Interact 18:615–620

Rozen S, Skaletsky H (2000) Primer3 on the WWW for general users and for biologist programmers. In: Krawetz S, Misener S (eds) Bioinformatics methods and protocols: methods in molecular biology. Humana Press, Totowa, pp 365–386

Rühm W (1956) Die Nematoden der Ipiden: Parasitologische Schriftenreihe, 6 (in German). Veb Gustav Fischer Verlag, Jena

Rühm W (1960) Ein beitrag zur Nomenklatur und Systematik einige mit Scolytiden vergesellschafteter Nematodenarten (in German). Zool Anzeiger 164:201–213

Rühm W (1964) Ein Beitrag zur Vergesellschaftung zwischen Nematoden und Insekten (*Pelodera bakeri* n. sp. [Nematoda, Rhabditoidea, Rhabditidae] eine mit *Calvertius tuberosus* Perm. et Germ. [Coleoptera, Curculionidae, Hylobiinae] vergesellschafte Nematodenart an *Araucaria araucana* [Mol.] Koch) (in German). Zool Anzeiger 173:212–220

Rutherford TA, Webster JM (1987) Distribution of pine wilt disease with respect to temperature in North America, Japan, and Europe. Can J For Res 17:1050–1059

Ryss A, Vieira P, Mota M, Kulinich O (2004) Computerized key to the genus *Bursaphelenchus* Fuchs, analysis of species clusters based on morphology, using information of insect vectors and associated plants, with a revision of the genus.(abstract). XXVII Symposium European Society of Nematologists, Rome, 14–18 June 2004

Ryss A, Vieira P, Mota MM, Kulinich O (2005) A synopsis of the genus *Bursaphelenchus* Fuchs, 1937 (Aphelenchida: Parasitaphelenchidae) with keys to species. Nematology 7:393–458

Sakai AK, Allendorf FW, Holt JS, Lodge DM, Molofsky J, With KA, Baughman S, Cabin RJ, Cohen JE, Ellstrand NC, McCauley DE, O'Neil P, Parker IM, Thompson JN, Weller SG (2001) The population biology of invasive species. Annu Rev Ecol Syst 32:305–332

Sakura A, Ishihara T, Kasuya S, Hasegawa S, Kishi Y (1978) Current status in a natural stand of *Pinus parviflora* at Sumi bog in Chiba Experimental Forest, Tokyo University (in Japanese). Trans Mtg Jpn For Soc 89:403–404

Sasaki M, Furukoshi T, Chiba Y, Kawamura K, Okada S (1982a) Breeding for resistance to pine wood nematode—intrespecific hybrids for resistance to pine wood nematode (in Japanese). Annu Rep Kansai Reg Breed Off 21:35–45

Sasaki M, Tashima M, Kawamura K, Okada S, Tsuda T, Kobayashi S, Katayama S, Furukoshi T (1982b) Interspecific crossing for resistance to pine wood nematode—resistance of progenies hybridized of the F_1 hybrids from Japanese black pine and Masson pine (*P. massoniana*) (in Japanese). Annu Rep Kansai Reg Breed Off 18/19:64–67

Sasaki M, Furukoshi T, Kawamura K, Tashima M, Okada S, Tsuda T (1983a) A few pieces of information of pine wood nematode resistance to Japanese black pine (in Japanese). Trans Kansai For Assoc 34:179–183

Sasaki M, Tashima M, Kawamura K, Okada S, Furukoshi T, Tsuda T, Kobayashi S, Katayama S (1983b) Crossing between the F_1 hybrids from Japanese black pine and Masson pine (II) resistance to pine wood nematode (in Japanese). Trans Mtg Jpn For Soc 94:249–250

Sato H, Sakuyama T, Kobayashi M (1987) Transmission of *Bursaphelenchus xylophilus* (Steiner et Buhrer) Nickle (Nematoda, Aphelenchoididae) by *Monochamus saltuarius* (Gebler) (Coleoptera, Cerambycidae) (in Japanese with English abstract). J Jpn For Soc 69: 492–496

Sato M, Islam SQ, Awaya S, Yamasaki T (1999a) Flavanonol glucoside and proanthocyanidins: oviposition stimulants for the cerambycid beetle, *Monochamus alternatus*. J Pestic Sci 24:123–129

Sato M, Islam SQ, Yamasaki T (1999b) Glycosides of a phenylpropanoid and neolignans: oviposition stimulants in pine inner bark for cerambycid beetle, *Monochamus alternatus*. J Pestic Sci 24:397–400

Schierenberg E, Miwa J, von Ehrenstein G (1980) Cell lineages and developmental defects of temperature-sensitive embryonic arrest mutant in *Caenorhabditis elegans*. Dev Biol 76:141–159

Schmidt J, All JN (1978) Chemical attraction of *Neoaplectana carpocapsae* (Nematoda: Steinernematidae) to insect larvae. Environ Entomol 7:605–607

Schmidt J, All JN (1979) Attraction of *Neoaplectana carpocapsae* (Nematoda: Steinernematidae) to common excretory products of insects. Environ Entomol 8:55–61

Schneider SQ, Bowerman B (2003) Cell polarity and the cytoskeleton in the *Caenorhabditis elegans* zygote. Annu Rev Genet 37:21–249

Schoeder WJ, Beavers JB (1987) Movement of entomopathogenic nematodess of the family Heterorhabditidae and Steinernematidae in soil. J Nematol 19:257–259

Schoener TW (1971) Theory of feeding strategies. Annu Rev Ecol Syst 2:369–404

Scholl EH, Thorne JL, McCarter JP, Bird DM (2003) Horizontally transferred genes in plant-parasitic nematodes: a high-throughput genomic approach. Genome Biol 4:R39

Schönfeld U, Braasch H, Burgermeister W (2006) *Bursaphlenchus* spp. (Nematoda: Parasitaphelenchidae) in wood chips from sawmills in Brandenburg and description of *Bursaphelenchus willibaldi* sp. n. Russ J Nematol 14:119–126

Seki N, Muta T, Oda T, Iwaki D, Kuma K, Miyata T, Iwanaga S (1994) Horseshoe-crab (1,3)-beta-D-glucan-sensitive coagulation factor-G—a serine-protease zymogen heterodimer with similarities to beta-glucan-binding proteins. J Biol Chem 269:1370–1374

Senda T, Sato H (1981) Adult occurrence time of *Monochamus alternatus* in Ichinoseki, Iwate Prefecture. Trans Ann Mtg Jpn For Soc Tohoku Br 33:144–145 (in Japanese)

Seydoux G, Strome S (1999) Launching the germline in *Caenorhabditis elegans*: regulation of gene expression in early germ cells. Development 126:3275–3283

Shaheen F, Winter REK, Bolla RI (1984) Phytotoxin production in *Bursaphelenchus xylophilus*-infected *Pinus sylvestris*. J Nematol 16:57–61

Shain L (1967) Resistance of sapwood in stems of loblolly pine to infection by Fomes annosus. Phytopathology 57:1034–1045

Sharon E, Spiegel Y, Salomon R, Curtis RH (2002) Characterization of *Meloidogyne javanica* surface coat with antibodies and their effect on nematode behaviour. Parasitology 125 (Pt 2):177–185

Shibata E (1981) Seasonal fluctuation and spatial pattern of the adult population of the Japanese pine sawyer, *Monochamus alternatus* Hope (Coleoptera: Cerambycidae), in young pine forests. Appl Entomol Zool 16:306–309

Shibata E (1984) Spatial distribution pattern of the Japanese pine sawyer, *Monochamus alternatus* Hope (Coleoptera: Cerambycidae), on dead pine trees. Appl Entomol Zool 19:361–366

Shibata E (1985) Seasonal fluctuation of the pine wood nematode, *Bursaphelenchus xylophilus* (Steiner et Buhrer) Nickle (Nematoda: Aphelenchoididae), transmitted to pine by the Japanese pine sawyer, *Monochamus alternatus* Hope (Coleoptera: Cerambycidae). Appl Entomol Zool 20:241–245

Shibata E (1986) Dispersal movement of the adult Japanese pine sawyer, *Monochamus alternatus* Hope (Coleoptera: Cerambycidae) in a young pine forest. Appl Entomol Zool 21:184–186

Shibata E, Kawasaki K, Takeda T (1986) Dispersal movement of the adult Japanese pine sawyer, *Monochamus alternatus* Hope (Coleoptera: Cerambycidae) in a young pine forest. Appl Entomol Zool 21:184–186

Shigesada N, Kawasaki K (1997) Biological invasions: theory and practice. Oxford Univ Press, Oxfrod

Shigesada N, Kawasaki K, Takeda Y (1995) Modeling stratified diffusion in biological invasions. Am Nat 146:229–251

Shigo AL (1967) Successions of organisms in discoloration and decay of wood. In: Romberger JA, Mikola P (eds) International review of forestry research, vol 2. Academic Press, New York, pp 237–299

Shimazu M (1993) Control of *Monochamus alternatus* using a pathogen, *Beauveria bassiana* cultured on wheat-bran pellets (in Japanese). Shinrin-boeki (For Pests) 42:232–236

Shimazu M (1994) Potential of the cerambycid-parasitic type of *Beauveria brongniartii* (Deuteromycotina: Hyphomycetes) for microbial control of *Monochamus alternatus* Hope (Coleoptera: Cerambycidae). Appl Entomol Zool 29:127–130

Shimazu M (2004a) Effects of temperature on growth of *Beauveria bassiana* F-263, a strain highly virulent to the Japanese pine sawyer, *Monochamus alternatus*, especially its tolerance to high temperatures. Appl Entomol Zool 39:469–475

Shimazu M (2004b) A novel technique to inoculate conidia of entomopathogenic fungi and its application for investigation of susceptibility of the Japanese pine sawyer, *Monochamus alternatus*, to *Beauveria bassiana*. Appl Entomol Zool 39:485–490

Shimazu M, Katagiri K (1981) Pathogens of the pine sawyer, *Monochamus alternatus* Hope, and possible utilization of them in a control program. Proc 17th IUFRO World Congr Div 2:291–295

Shimazu M, Kushida T (1980) Microbial control of *Monochamus alternatus*—treatment of pathogens on the infested pine trees (in Japanese). Trans Ann Mtg Kanto Br Jpn For Soc 32:93–94

Shimazu M, Kushida T (1983) Virulences of the various isolates of entomogenous fungi to *Monochamus alternatus* Hope (in Japanese). Trans Ann Mtg Kanto Br Jpn For Soc 35: 165–166

Shimazu M, Sato H (2003) Effects of larval age on mortality of *Monochamus alternatus* Hope (Coleoptera: Cerambycidae) after application of nonwoven fabric strips with *Beauveria bassiana*. Appl Entomol Zool 38:1–5

Shimazu M, Kushida T, Katagiri K (1982) Microbial control of *Monochamus alternatus*—spray of pathogens onto the infested pine trees just before adult emergence (in Japanese). Trans Mtg Jpn For Soc 93:399–400

Shimazu M, Kushida T, Katagiri K (1983) Microbial control of *Monochamus alternatus*—spraying of pathogens during maturation feeding (in Japanese). Trans Mtg Jpn For Soc 94:485–486

Shimazu M, Kushida T, Tsuchiya D, Mitsuhashi W (1992) Microbial control of *Monochamus alternatus* Hope (Coleoptera: Cerambycidae) by implanting wheat-bran pellets with *Beauveria bassiana* in infested tree trunks. J Jpn For Soc 74:325–330

Shimazu M, Tsuchiya D, Sato H, Kushida T (1995) Microbial control of *Monochamus alternatus* Hope (Coleoptera: Cerambycidae) by application of nonwoven fabric strips with *Beauveria bassiana* (Deuteromycotina: Hyphomycetes) on infested tree trunks. Appl Entomol Zool 30:207–213

Shimazu M, Maehara N, Sato H (2002a) Density dynamics of the entomopathogenic fungus, *Beauveria bassiana* Vuillemin (Deuteromycotina: Hyphomycetes) introduced into forest soil, and its influence on the other soil microorganisms. Appl Entomol Zool 37:263–269

Shimazu M, Sato H, Maehara N (2002b) Density of the entomopathogenic fungus, *Beauveria bassiana* Vuillemin (Deuteromycotina: Hyphomycetes) in forest air and soil. Appl Entomol Zool 37:19–26

Shiraishi S (1996) Construction of a new breeding system using DNA techniques. In: 2nd forest genetic seminar proceedings—forward new clonal forestry in 21th century (in Japanese with English abstract), pp 21–25

Shiraishi S (1997) New tree breeding system using DNA—progeny test introduced DNA parental analysis-(in Japanese). For Technol 668:24–25

Shoda-Kagaya E (2007) Genetic differentiation of the pine wilt disease vector, pine sawyer *Monochamus alternatus* over a mountain range—revealed from microsatellite DNA markers. Bull Entomol Res 97:167–174

Shoji T, Jinno Y (1985) Seasonal time of inoculation of *Bursaphelenchus xylophilus* associated with disease development in pine trees in the Tohoku region (in Japanese). Trans Mtg Jpn For Soc 96:461–462

Siegfried BD (1987) In-flight responses of the pales weevil *Hylobius pales* (Coleoptera: Curculionidae) to monoterpene constituents of southern pine gum turpentine. Fla Entomol 70:97–102

Singson A (2001) Every sperm is sacred: Fertilization in *Caenorhabditis elegans*. Dev Biol 230:101–109

Skarmoutsos G, Braasch H, Michalopoulou H (1998) *Bursaphelenchus hellenicus* sp. n. (Nematoda: Aphelenchoididae) from Greek pine wood. Nematologica 44:623–629

Smant G, Stokkermans JP, Yan Y, de Boer JM, Baum TJ, Wang X, Hussey RS, Gommers FJ, Henrissat B, Davis EL, Helder J, Schots A, Bakker J (1998) Endogenous cellulases in animals: isolation of beta-1,4-endoglucanase genes from two species of plant-parasitic cyst nematodes. Proc Natl Acad Sci USA 95:4906–4911

Sniezko RA (2006) Resistance breeding against nonnative pathogens in forest trees—current successes in North America. Can J Plant Pathol 28:270–279

Sokal RR, Rohlf FJ (1995) Biometry: the principles and practice of statistics in biological research, 3d edn. Freeman, New York, p 887

Song PQ, Wei LY, Zuang MY, Zhang KQ (2000) Study on antivirus, antimicrobial, and nematicidal activities of nucleotide N9705. Virol Sin Special Issue:180–183

Song S, Zhang L, Huang H, Cui X (1991) Preliminary study of biology of *Monochamus alternatus* Hope. For Sci Technol 6:9–13

Song SH, Zhang LQ, Chen MR, Wei XQ, Wang BZ, Yang L, Xie C, Chen JW (1998) Study on the control of pine wood nematode by releasing *Scleroderma guani*. Guangdong For Sci Technol 14(3):42–48

Soper RS, Olson RE (1963) Survey of biota associated with *Monochamus* (Coleoptera: Cerambycidae) in Maine. Can Entomol 95:83–95

Sousa E, Bravo MA, Pires J, Naves PM, Penas AC, Bonifácio L, Mota M (2001) *Bursaphelenchus xylophilus* (Nematoda: Aphelenchoididae) associated with *Monochamus galloprovincialis* (Coleoptera: Cerambycidae) in Portugal. Nematology 3:89–91

Sousa E, Naves P, Bonifácio L, Bravo MA, Penas AC, Pires J, Serrão M (2002) Preliminary survey for insects associated with *Bursaphelenchus xylophilus* in Portugal. Bull OEPP/EPPO Bull 32:499–502

Sperry J, Tyree MT (1988) Mechanism of water stress-induced xylem embolism. Plant Physiol 88:581–587

Spiegel Y, McClure MA (1995) The surface coat of plant parasitic nematodes: chemical composition origin and biological role, a review. J Nematol 27:127–134

Sritunyalucksana K, Lee SY, Soderhall K (2002) A beta-1,3-glucan binding protein from the black tiger shrimp, *Penaeus monodon*. Dev Comp Immunol 26:237–245

Sriwati R, Takemoto S, Futai K (2006) Seasonal changes in the nematode fauna in pine trees killed by the pinewood nematode, *Bursaphelenchus xylophilus*. Jpn J Nematol 36: 87–100

Sriwati R, Takemoto S, Futai K (2007) Cohabitation of the pine wood nematode, *Bursaphelenchus xylophilus*, and fungal species in pine trees inoculated with *B. xylophilus*. Nematology 9:77–86

Sriwati R, Kanzaki N, Phan LK, Futai K (2008) *Bursaphelenchus eproctatus* n. sp. (Nematoda: Parasitaphelenchidae) isolated from dead Japanese black pine, *Pinus thunbergii* Pars. Nematology 10:1–7

Stamps WT, Linit MJ (1998a) Neutral storage lipid and exit behavior of *Bursaphelenchus xylophilus* fourth-stage dispersal juveniles from their beetle vectors. J Nematol 30: 255–261

Stamps WT, Linit MJ (1998b) Chemotactic response of propagative and dispersal forms of the pinewood nematode *Bursaphelenchus xylophilus* to beetle and pine derived compounds. Fundam Appl Nematol 21:243–250

Stamps WT, Linit MJ (2001) Interaction of intrinsic and extrinsic chemical cues in the behavior of *Bursaphelenchus xylophilus* (Aphelenchida: Aphelenchoididae) in relation to its beetle vectors. Nematology 3:295–301

Steiner G (1932) Some nemic parasites and associates of the mountain pine beetle (*Dendroctonus monticola*). J Agric Res 45:437–444

Steiner G (1935) Opuscula miscellanea nematologica, II. Proc Helminthol Soc Washington 2:104–110

Steiner G, Buhrer EM (1934) *Aphelenchoides xylophilus* n. sp., a nematode associated with blue-stain and other fungi in timber. J Agric Res 48:949–951

Suarez AV, Holway DA, Case TJ (2001) Patterns of spread in biological invasions dominated by long-distance jump dispersal: Insight from Argentine ants. Proc Natl Acad Sci USA 98:1095–1100

Suga T, Ohta S, Munesada K, Ide N, Kurokawa M, Shimizu M, Ohta E (1993) Endogenous pine wood nematicidal substances in pines, *Pinus massoniana*, *P. strobus* and *P. palustris*. Phytochemistry 33:1395–1401

Sulston JE, Horvitz HR (1977) Post-embryonic cell lineages of the nematode *Caenorhabditis elegans*. Dev Biol 56:110–156

Sulston JE, Schierenberg E, White JG, Thomson JN (1983) The embryonic cell lineage of the nematode *Caenorhabditis elegans*. Dev Biol 100:64–119

Sumimoto M, Shiraga M, Kondo T (1975) Ethane in pine needles preventing the feeding of the beetle, *Monochamus alternatus*. J Insect Physiol 21:713–722

Sun JH (1997) Studies on inhibitory effects of soil isolated fungi cultures to growth and propagation of pine wood nematode. Acta Sci Nat Univ Nakaiiensis 30(3):82–87

Sun Y (1982) Pine wood nematode detected in Zhongshanling, Nanjing (in Chinese). Jiangsu For Technol 7:47

Sutherland JR, Ring FM, Seed JE (1991) Canadian conifers as hosts of the pinewood nematode (*Bursaphelenchus xylophilus*): results of seedling inoculations. Scand J For Res 6:209–216

Suzuki A, Ohno S (2006) The PAR-aPKC system: lessons in polarity. J Cell Sci 119:979–987

Suzuki K (1984) General effect of water stress on the development of pine wilting disease caused by *Bursaphelenchus xylophilus* (in Japanese with English abstract). Bull For Forest Prod Res Inst 325:97–126

Suzuki K (1992) Mechanism of wilting in pine wilt disease (in Japanese). Shinrin-boeki (For Pests) 41:59–64

Suzuki K (2002) Pine wilt disease—a threat to pine forest in Europe. Dendrobiology 48:71–74

Suzuki K, Kiyohara T (1978) Influence of water stress on development of pine wilting disease caused by *Bursaphelenchus* lignicolus. Eur J For Path 8:97–107

Suzuki S, Makihara H, Fujioka H (1991) Effect of releasing pine bark beetle contaminated with *Beauveria bassiana* on silkworm (in Japanese). Tohoku Sanshi Kenkyu Hokoku 16: 13–14

Tabara H, Grishok A, Mello CC (1998) RNAi in *C. elegans*: soaking in the genome sequence. Science 282:430–431

Tabuse Y, Izumi Y, Piano F, Kemphues KJ, Miwa J, Ohno S (1998) Atypical protein kinase c cooperates with PAR-3 to establish embryonic polarity in *Caenorhabditis elegans*. Development 125:3607–3614

Taguchi F, Shimizu R, Nakajima R, Toyoda K, Shiraish T, Ichinose Y (2003) Differential effects of flagellins from *Pseudomonas syringae* pv. tabaci, tomato and glycinea on plant defense response. Plant Physiol Biochem 41:165–174

Takahasi F (1977) Generation carryover of a fraction of population members as an animal adaptation to unstable environmental conditions. Res Popul Ecol 18:235–242

Takasu F, Yamamoto N, Kawasaki K, Togashi K, Kishi Y, Shigesada N (2000) Modeling the expansion of an introduced tree disease. Biol Invasions 2:141–150

Takemoto S (2007) Virulence evolution of the pinewood nematode, *Bursaphelenchus xylophilus*. In: Program and abstracts of international symposium on pine wilt disease in Asia, Laboratory of Environmental Mycoscience, Kyoto University, Kyoto, p 20

Takemoto S, Futai K (2006) Avirulent strain of the pinewood nematode, *Bursaphelenchus xylophilus* in Japan and evolution of its virulence. In: Proceedings of XXVIII symposium of the European Society of Nematologists, Blagoevgrad, Bulgaria, June 2006

Takemoto S, Futai K (2007) Polymorphism of Japanese isolates of the pinewood nematode, *Bursaphelenchus xylophilus* (Aphelenchida: Aphelenchoididae), at heat-shock protein 70A locus and the field detection of polymorphic populations. Appl Entomol Zool 42:247–253

Takemoto S, Kanzaki N, Futai K (2005) PCR–RFLP image analysis—a practical method for estimating isolate-specific allele frequency in a population consisting of two different strains of the pinewood nematode, *Bursaphelenchus xylophilus* (Aphelenchida: Aphelencoididae). Appl Entomol Zool 40:529–535

Taketsune A (1982) *Dastarcus helophoroides*, a natural enemy of *Monochamus alternatus* (in Japanese). Shinrin-boeki (For Pests) 31:228–230

Taketsune A (1983) Studies on the population dynamics of the pine sawyer, *Monochamus alternatus* Hope (Coleoptera: Cerambycidae) and its control by some pathogens (in Japanese with English abstract). Bull Hiroshima For Exp Stn 18:39–62

Takeuchi Y, Futai K (2007a) Asymptomatic carrier trees in pine stands naturally infected with *Bursaphelenchus xylophilus*. Nematology 9:243–250

Takeuchi Y, Futai K (2007b) Avirulent isolate of the pinewood nematode (C14–5), *Bursaphelenchus xylophilus*, survives seven months in asymptomatic host seedlings. For Path 37:289–291

Takeuchi Y, Kanzaki N, Futai K (2005) A nested PCR-based method for detecting the pinewood nematode, *Bursaphelenchus xylophilus*, from pine wood. Nematology 7:775–782

Takeuchi Y, Kanzaki N, Futai K (2006a) How different is induced host resistance against the pine wood nematode, *Bursaphelenchus xylophilus*, by two avirulent microbes? Nematology 8:435–442

Takeuchi Y, Kanzaki N, Futai K (2006b) Volatile compounds in pine stands suffering from pine wilt disease; qualitative and quantitative evaluation. Nematology 8:867–877

Takizawa Y (1979) Ecology of the pine sawyer in Tohoku District (VIII). Periods of pupal and adult stages in the pupal cell under natural temperature conditions in Morioka City (in Japanese). Trans Ann Mtg Tohoku Br Jpn For Soc 31:156–157

Takizawa Y (1982) Biology of *Monochamus alternatus* in Tohoku (XIII) Oviposition in autumn and mortality of eggs (in Japanese). Trans Ann Mtg Tohoku Br Jpn For Soc 34:120–121

Tamaru Y, Doi RH (2001) Pectate lyase A, an enzymatic subunit of the *Clostridium cellulovoran*s cellulosome. Proc Natl Acad Sci USA 98:4125–4129

Tamura H (1973) Recent investigation on the nematode-trapping fungi (in Japanese). Jpn J Nematol 3:9–18

Tamura H (1980) Some nematode-trapping fungi in the pupal chambers of *Monochamus alternatus* (in Japanese). Shinrin-boeki (For Pests) 29:39–42

Tamura H (1983a) Infection of pinewood nematode, *Bursaphelenchus xylophilus*, via fused roots (in Japanese). Trans Ann Mtg Jpn Soc Appl Entomol Zool 27:163

Tamura H (1983b) Pathogenicity of aseptic *Bursaphelenchus xylophilus* and associated bacteria to pine seedlings. Jpn J Nematol 13:1–5

Tamura H (1986) Occurrence of dispersal 3rd-stage larvae of *Bursaphelenchus xylophilus* in logs (in Japanese). Trans Ann Mtg Kansai Br Jpn For Soc 37:201–203

Tamura H, Dropkin V (1984) Resistance of pine trees to pine wilt caused by the nematode, *Bursaphelenchus xylophilus*. J Jpn For Soc 66:306–312

Tamura H, Mamiya Y (1975) Reproduction of *Bursaphelenchus lignicolus* on alfalfa callus tissues. Nematologica 21:449–454

Tamura H, Mamiya Y (1976) Reproduction of *Bursaphelenchus lignicolus* on pine callus tissues (in Japanese). Trans Mtg Jpn For Soc 87:229–230

Tamura H, Mamiya Y (1979) Reproduction of *Bursaphelenchus lignicolus* on pine callus tissues. Nematologica 25:149–151

Tamura H, Mineo K, Yamada T (1987) Blockage of water conduction in *Pinus thunbergii* inoculated with *Bursaphelenchus xylophilus*. Jpn J Nematol 17:23–30

Tamura H, Yamada T, Mineo K (1988) Host responses and nematode distribution in *Pinus strobus* and *P. densiflora* infected with the pine wood nematode, *Bursaphelenchus xylophilus*. Ann Phytopathol Soc Jpn 54:327–331

Tan JJ, Feng ZX (2004) Population dynamics of pine wood nematode and its accompanying bacterium in the host (in Chinese with English abstract). Scientia Silvae Sinicae 40:110–114

Tanaka H (2003) Possibility of infection to pine wilt disease though root system (in Japanese). Kyushu J For Res 56:216–217

Tanaka K (1973) Interspecific variation of genus *Pinus* in resistance to the pinewood nematode—in association with natural infection of *Pinus pinaster* (in Japanese). Shinrin-boeki (For Pests) 22:254–258

Tanaka K (1974) Hypha-feeding behavior of the pine wood nematode (in Japanese). Trans Mtg Jpn For Soc 85:247–249

Tanaka K, Matsumura M (2000) Development of virulence to resistant rice varieties in the brown planthopper, *Nilaparvata lugens* (Homoptera: Delphacidae), immigrating into Japan. Appl Entomol Zool 35:529–533

Tanaka N, Che FS, Watanabe N, Fujiwara S, Takayama S, Isogai A (2003) Flagellin from an Incompatible strain of *Acidovorax avenae* mediates H_2O_2 generation accompanying hypersensitive cell death and expression of PAL, Cht-1, and PBZ1, but not of Lox in rice. Mol Plant Microbe Interact 16:422–428

Tang JG, Zhang CF, Yan AJ, Li ZZ, Sun QJ (1999) A study on pesticide effects and duration of contacted breaking release microcapsules with different wall materials and thickness. J Nanjing For Univ 23:69–72

Taoda H (1996) Standards of management and conversion of species of pine forests (in Japanese). Shinrin-boeki (For Pests) 45:14–19

Tarés S, Abad P, Bruguier N, De Guiran G (1992a) Identification and evidence for relationships among geographical isolates of *Bursaphelenchus* spp. (pinewood nematode) using homologous DNA probes. Heredity 68:157–164

Tarés S, Lemontey J-M, De Guiran G, Abad P (1992b) Use of species-specific satellite DNA from *Bursaphelenchus xylophilus* as a diagnostic probe. Phytopathology 84:294–298

Tarés S, Lemontey JM, De Guiran G, Abad P (1993) Cloning and characterization of a highly conserved satellite DNA sequence specific for the phytoparasitic nematode *Bursaphelenchus xylophilus*. Gene 129:269–273

Tarés S, Lemontey JM, De Guiran G, Abad P (1994) Use of species-specific satellite DNA from *Bursaphelenchus xylophilus* as a diagnostic probe. Phytopathology 84:294–298

Tarjan AC, Baeze-Aragon C (1982) An analysis of the genus *Bursaphelenchus* Fuchs, 1937. Nematologica 12:121–144

Tautz D (1989) Hypervariability of simple sequence as a general source for polymorphic DNA markers. Nucleic Acids Res 17:6463–6471

Tautz D, Renz M (1984) Simple sequences are ubiquitous repetitive components of eukaryotic genomes. Nucleic Acids Res 12:4127–4138

Terada M (1986) Roots fusion and contact as a channel of nematode transmission from pine-wilt-infected trees to surrounding healthy ones (in Japanese). Annual Reports of Studies on Pine Wilt Control, pp 53–93

Terashita T (1975) Relations between water content of wood, *Bursaphelenchus lignicolus* density in wood and the numbers of the nematode carried by beetles emerged from diseased trees (in Japanese). Trans Ann Mtg Kansai Br Jpn For Soc 26:279–281

The *Caenorhabditis elegans* sequencing consortium (1998) Genome sequence of the nematode *C. elegans*: a platform for investigating biology. Science 282:2012–2018

Thorne G (1935) Nemic parasites and associates of the mountain pine beetle (*Dendroctonus monticolae*) in Utah. J Agric Res 51:131–144

Thorne G (1961) Principles of nematology. McGraw-Hill, New York, pp 48–49

Thorn RG, Barron GL (1984) Carnivorous mushrooms. Science 224:76–78

Thong CHS, Webster JM (1983) Nematode parasites and associates of *Dendroctonus* spp. and *Trypodendron lineatum* (Coleoptera: Scolytidae), with a description of *Bursaphelenchus varicauda* n. sp. J Nematol 15:312–318

Thu PQ (2003) Status of a Pine wilt nematode in Vietnam. N Z J For Sci 33:336–342

Thurston GS, Kaya HK, Gaugler R (1994) Characterizing the enhanced susceptibility of milky disease-infected scarabaeid grubs to entomopathogenic nematodes. Biol Control 4:67–73

Tijsterman M, May RC, Simmer F, Okihara KL, Plasterk RHA (2004) Genes required for systemic RNAi interference in *Caenorhabditis elegans*. Curr Biol 14:111–116

Timmons L, Fire A (1998) Specific interference by ingested dsRNA. Nature 395:854

Tobase M (2003) Bring back pine forests using super pine—nursing and production of pines resistance to pine wood nematode (in Japanese). For Tree Breed 203:3–5

Toda T (1996) Breeding for resistance to pine wood nematode (in Japanese). Shinrin-boeki (For Pests) 45:2–7

Toda T (1997) Breeding of pine trees resistant to the pine wilt (in Japanese). In: Tamura T (ed) The pine wilt disease. The review of history and recent researches. Zenkoku Shinrin Byochu-jugai Bojo Kyokai (National Forest Petsts Control Association), Tokyo, pp 168–274

Toda T (1999) Studies on the improvement of resistance to pine wood nematode (in Japanese). For Breed 192:1–4

Toda T (2004) Studies on the breeding for resistance to the pine wilt disease in *Pinus densiflora* and *P. thunbergii* (in Japanese with English abstract). Bull For Tree Breed Cent 20:83–217

Toda T (2006) Overseas tree breeding, genetic resources (5) Will the breeding of resistance to pine wilt disease fix in China? The actions and results of resistance breeding center for pine wood nematode of Anhui Province (in Japanese). Trop For 65:74–81

Toda T, Kurinobu S (2002) Realized genetic gains observed in progeny tolerance of selected red pine (*Pinus densiflora*) and black pine (*P. thunbergii*) to pine wilt disease. Silvae Genet 51:42–44

Toda T, Terada K (2001) Forest tree breeding project (3)—breeding project for resistance tp pine wood nematode (in Japanese). For Tree Breed 198:39–43

Toda T, Tajima M, Nishimura K, Takeuchi H (1993) Breeding projection on resistance to the pine-wood nematode in Kyushu district—progressive study after select resistance clone (in Japanese). Bull For Tree Breed Inst 11:37–88

Toda T, Chigira O, Miyata M (1997) A case of resistance improvement to pine wood nematode in the hybrid from Japanese black pine and Japanese red pine (in Japanese). Trans Kyushu For Assoc 50:41–42

Toda T, Kurinobu S, Sasaki M (2002) Growth and survival rate at the seventh year after selecting the progenies of pine trees for resistance to pine wood nematodes by inoculation test (in Japanese with English abstract). J Jpn For Soc 84:188–192

Togashi K (1980) Preliminary report on the hibernation of the Japanese pine sawyer, *Monochamus alternatus* Hope, in Ishikawa Prefecture (in Japanese with English abstract). Bull Ishikawa-ken For Exp Stn 10:39–50

Togashi K (1985) Transmission curves of *Bursaphelenchus xylophilus* (Nematoda: Aphelenchoididae) from its vector, *Monochamus alternatus* (Coleoptera: Cerambycidae), to pine trees with reference to population performance. Appl Entomol Zool 20:246–251

Togashi K (1986) Effects of the initial density and natural enemies on the survival rate of the Japanese pine sawyer, *Monochamus alternatus* Hope (Coleoptera: Cerambycidae), in pine logs. Appl Entomol Zool 21:244–251

Togashi K (1988) Population density of *Monochamus alternatus* adults (Coleoptera: Cerambycidae) and incidence of pine wilt disease caused by *Bursaphelenchus xylophilus* (Nematode: Aphelenchoididae). Res Popul Ecol 30:177–192

Togashi K (1989a) Development of *Monochamus alternatus* Hope (Coleoptera: Cerambycidae) in relation to oviposition time (in Japanese with English abstract). Jpn J Appl Entomol Zool 33:1–8

Togashi K (1989b) Development of *Monochamus alternatus* Hope (Coleoptera: Cerambycidae) in *Pinus thunbergii* trees weakened at different times (in Japanese with English abstract). J Jpn For Soc 71:383–386

Togashi K (1989c) Studies on population dynamics of *Monochamus alternatus* Hope (Coleoptera: Cerambycidae) and spread of pine wilt disease caused by *Bursaphelenchus xylophilus*

(Nematoda: Aphelenchoididae) (in Japanese with English abstract). Bull Ishikawa For Exp Stn 20:1–142

Togashi K (1989d) Temporal pattern of the occurrence of weakened *Pinus thunbergii* trees and causes for mortality. J Jpn For Soc 71:323–328

Togashi K (1989e) Population density of *Monochamus alternatus* adults (Coleoptera: Cerambycidae) and incidence of pine wilt disease caused by *Bursaphelenchus xylophilus* (Nematoda: Aphelenchoididae). Res Popul Ecol 30:177–192

Togashi K (1989f) Variation in external symptom development of pine wilt disease in field grown *Pinus thunbergii*. J Jpn For Soc 71:442–448

Togashi K (1989g) Factors affecting the number of *Bursaphelenchus xylophilus* (Nematoda: Aphelenchoididae) carried by newly emerged adults of *Monochamus alternatus* (Coleoptera: Cerambycidae). Appl Entomol Zool 24:379–386

Togashi K (1990a) Life table for *Monochamus alternatus* (Coleoptera: Cerambycidae) within dead trees of *Pinus thunbergii*. Jpn J Entomol 58:217–230

Togashi K (1990b) Change in the activity of adult *Monochamus alternatus* Hope (Coleoptera: Cerambycidae) in relation to age. Appl Entomol Zool 25:153–159

Togashi K (1990c) A field experiment on dispersal of newly emerged adults of *Monochamus alternatus* (Coleoptera: Cerambycidae). Res Popul Ecol 32:1–13

Togashi K (1991a) Different developments of overwintered larvae of *Monochamus alternatus* (Coleoptera: Cerambycidae) under a constant temperature. Jpn J Entomol 59:149–154

Togashi K (1991b) Spatial pattern of pine wilt disease caused by *Bursaphelenchus xylophilus* (Nematoda: Aphelenchoididae) within a *Pinus thunbergii* stand. Res Popul Ecol 33: 245–256

Togashi K (1991c) Larval diapause termination of *Monochamus alternatus* Hope (Coleoptera: Cerambycidae) under natural conditions. Appl Entomol Zool 26:381–386

Togashi K (1995) Interacting effects of temperature and photoperiod on diapause in larvae of *Monochamus alternatus* (Coleoptera: Cerambycidae). Jpn J Entomol 63:43–252

Togashi K (1997) Lifetime fecundity and body size of *Monochamus alternatus* (Coleoptera: Cerambycidae) at a constant temperature. Jpn J Entomol 65:458–470

Togashi K (2001) Insect vector-nematode relationship and virulence of nematode against host plants. In: Alfaro RI, Day KR, Salom S, Nair KSS, Evans H, Liebhold A, Lieutier F, Wagner M, Futai K, Suzuki K (eds) Protection of world forests from insect pests: advances in research. IUFRO World Series, vol 11, IUFRO Secretariat, Vienna, pp149–155

Togashi K (2002) Life history of Japanese pine sawyer, *Monochamus alternatus*, and characteristics of larval food resources (in Japanese). Jpn J Ecol 52:69–74

Togashi K (2004) A new method for loading *Bursaphelenchus xylophilus* (Nematoda: Aphelenchoididae) on adult *Monochamus alternatus* (Coleoptera: Cerambycidae). J Econ Entomol 97:941–945

Togashi K (2006a) Life of the Japanese pine sawyer, *Monochamus alternatus* Hope (in Japanese). In: Shibata E, Togashi K (eds) The fascinating lives of insects residing in tree trunks: an introduction to tree-boring insects. Tokai University Press, Hadano, pp 83–106

Togashi K (2006b) Life history of tree boring insects in relation to physiological status of host trees. In: Shibata E, Togashi K (eds) The fascinating lives of insects residing in tree trunks: an introduction to tree-boring insects (in Japanese). Tokai University Press, Hadano, pp 259–267

Togashi K, Arakawa Y (2003) Horizontal transmission of *Bursaphelenchus xylophilus* between sexes of *Monochamus alternatus*. J Nematol 35:7–16

Togashi K, Magira H (1981) Age-specific survival rate and fecundity of the adult Japanese pine sawyer, *Monochamus alternatus* Hope (Coleoptera: Cerambycidae), at different emergence times. Appl Entomol Zool 16:351–361

Togashi K, Matsunaga K (2003) Between-isolate difference in dispersal ability of *Bursaphelenchus xylophilus* and vulnerability to inhibition by *Pinus densiflora*. Nematology 5:559–564

Togashi K, Sekizuka H (1982) Influence of the pine wood nematode, *Bursaphelenchus lignicolus* (Nematoda: Aphelenchoididae), on longevity of its vector, *Monochamus alternatus* (Coleoptera: Cerambycidae). Appl Entomol Zool 17:160–165

Togashi K, Nakamura K, Takahashi, F (1992) An index of susceptibility of pine stands to pine wilt disease. Appl Entomol Zool 27:341–347

Togashi K, Matsunaga K, Arakawa Y, Miyamoto N (2003) The random dispersal of *Bursaphelenchus xylophilus* in pine twigs (in Japanese). Trans Jpn For Soc 114:753

Togashi K, Chung YJ, Shibata E (2004) Spread of an introduced tree pest organism—the pinewood nematode. In: Hong S-K, Lee JA, Ihm B-S, Farina A, Son Y, Kim E-S, Choe JC (eds) Ecological issues in a changing world-status, response and strategy. Kluwer, Dordrecht, pp 173–188

Tokushige Y, Kiyohara T (1969) *Bursaphelenchus* sp. in the wood of dead pine trees. J Jpn For Soc 51:193–195

Tomminen J, Halik S, Bergdahl DR (1991) Incubation temperature and time effects of life stages of *Bursaphelenchus xylophilus* in wood chips. J Nematol 23:477–484

Triantaphyllou AC (1971) Genetics and cytology. In: Zuckerman BM, Mai WF, Rohde RA (eds) Plant parasitic nematodes volume II. Academic press, New York, pp 1–34

Tsuda T, Kobayashi S, Katayama S, Ida S (1978) Studies for resistant breeding. Kansai For Tree Breed Inst Annu Rep 14:48–61

Tyree MT, Sperry JS (1989) Characterization and propagation of acoustic emission signals in woody plants towards an improved acoustic emission counter. Plant Cell Environ 12: 371–382

Tyree MT, Zimmermann MH (2002) Xylem structure and the ascent of sap. 2nd edition. Springer, Berlin, Heidelberg, New York. p 304

Ueda T, Oku H, Tomita K, Sato K, Shiraishi T (1984) Isolation, identification, and bioassay of toxic compounds from pine tree naturally infected by pine wood nematode. Ann Phytopathol Soc Jpn 50:166–175

Uehara T, Kushida A, Momota Y (2001) PCR-based cloning of two beta-1,4-endoglucanases from the root-lesion nematode *Pratylenchus penetrans*. Nematology 3:335–341 (337)

Urano T (2003) Preliminary release experiments in laboratory and outdoor cages of *Dastarcus helophoroides* (Fairmaire) (Coleoptera: Bothrideridae) for biological control of *Monochamus alternatus* Hope (Coleoptera: Cerambycidae). Bull For Forest Prod Res Inst 2:255–262

Urano T (2004) Experimental release of a parasitoid, *Dastarcus helophoroides* (Coleoptera: Bothrideridae), on *Monochamus alternatus* (Coleoptera: Cerambycidae) infesting *Pinus densiflora* in the field. Bull For Forest Prod Res Inst 3:205–211

Urwin PE, Lilley CJ, Atkinson HJ (2002) Ingestion of double-stranded RNA by preparasitic juvenile cyst nematodes leads to RNA interference. Mol Plant Microbe Interact 15:747–752

Vago C (1963) A new type of insect virus. J Insect Pathol 5:275–276

van Tol RWHM, Schepman AC (1999) Influence of host and plant roots on the migration of *Heterorhabditis* sp. (NEW) in peat soil. In: Gwynn RL, Smits PH, Griffin C, Ehlers RU, Boemare N, Masson JP (eds) Application and Persistence of Entomopathogenic Nematodes. Office for Official Publications of the EU, Luxembourg, pp 117–121

Vega FE, Lacey LA, Reid AP, Herard F, Pilarska D, Danova E, Tomov R, Kaya HK (2000) Infectivity of a Bulgarian and an American strain of *Steinernema carpocapsae* against codling moth. Biocontrol 45:337–343

Veit RR, Lewis MA (1996) Dispersal, population growth, and the Allee effect: dynamics of the house finch invasion of eastern North America. Am Nat 148:255–274

Victorsson J, Wikars LO (1996) Sound production and cannibalism in larvae of the pine-sawyer beetle *Monochamus sutor* L. Entomologisk Tidskrift 117:29–33

Vieira P, Mota M, Eisenback JD (2004) PWN-CD: A taxonomic database for the pinewood nematode *Bursaphelenchus xylophilus*, and other *Bursaphelenchus* species, pp 165. In: Mota M, Vieira P (eds) The pinewood nematode, *Bursaphelenchus xylophilus*. Nematology monographs & perspectives, vol 1. E.J. Brill, Leiden

Vieira P, Mota M, Eisenback JD (2006) Pine wood nematode taxonomic database. 2nd edn. Mactode Publications, Blacksburg (CD-ROM)

Vieira P, Burgermeister W, Mota M, Metge K, Silva G (2007) Lack of genetic variation of *Bursaphelenchus xylophilus* in Portugal revealed by RAPD–PCR analyses. J Nematol 39:118–126

Vincent, B, Altemeyer V, Roux-Morabito G, Lieutier F (2008) Occurrence of *B. mucronatus* in France and its association with *Monochamus galloprovincialis*. In: Mota M, Vieira P (eds) Pine wilt disease: a worldwide threat to forest ecosystems. Springer, The Netherlands, p 36

Voronov DA, Panchin YV (1998) Cell lineage in marine nematode *Enoplus brevis*. Development 125:143–150

Wada S, Miyamoto K (1997) Effect of larval stage on the susceptibility of Silkworm, *Bombyx mori*, to *Beauveria bassiana* isolated from cerambycid beetle, *Monochamus alternatus* (in Japanese). Trans Ann Mtg Kanto Br Jpn Soc Ser Sci 48:2

Wada S, Miyamoto K (1999) Effects of *Beauveria bassiana* isolated from *Monochamus alternatus* on the silkworm (in Japanese). Trans Ann Mtg Jpn Ser Soc 69:79

Walia KK, Negi S, Bajaj HK, Klia DC (2003) Two new species of *Bursaphelenchus* Fuchs, 1937 (Nematoda: Aphelenchoididae) from pine wood and insect frass from India. Ind J Nematol 33:1–5

Walton JD (1994) Deconstructing the cell wall. Plant Physiol 104:1113–1118

Wang J, Yu B, Lin M (2005) Description of a new species (Nematoda: Aphelenchoididae) isolated from wood packing material. Acta Zootaxonomica Sin 30:314–319

Wang L, Xu F, Jiang L, Zhang P, Yang Z (2004) Pathogens of the pine sawyer, *Monochamus alternatus*, in China. in The pinewood nematode, *Bursaphelenchus xylophilus*. In: Mota M, Vieira P (eds) Nematology monographs and perspectives, vol 1, E.J. Brill, Leiden, pp 283–289

Wang M, Cao FX, Long JX, Teng T (2006) Advance in research of masson pine resistance diversity to pine wilt disease. J Central South For Univ 26(3):128–132

Wang MM, Ye JR, Wang YH (2006) Aplication of attractant for controlling pine wilt disease and interrelated techniques. J Nanjing For Univ (Nat Sci Edn) 30:129–131

Wang MX (2004) Study progress of identification techniques of molecular biology on pine wood nematode, *Bursaphelenchus xylophilus*. Hunan For Sci Technol 31:1–3

Wang SB, Liu YP, Fan MZ, Miao XX, Zhao XQ, Li ZZ (2005) Field attraction effects of different trapping methods of *Monochamus alternatus*. Chin J Appl Ecol 16:505–508

Wang WD, Ishibashi N (1999) Infection of the entomopathogenic nematode *Steinernema carpocapsae*, as affected by the presence of *Steinernema glaseri*. J Nematol 31:207–211

Wang WH (2004) Axenic culture of *Bursaphelenchus xylophilus* (in Chinese with English abstract). Masteral Thesis, Nanjing For Univ, China, p 60

Wang Y, Yamada T, Sakaue D, Suzuki K (2005) Influence of fungi on multiplication and distribution of the pine wood nematode, *Bursaphelenchus xylophilus*, in axenized *Pinus thunbergii* cuttings. Nematology 7:809–817

Wang YY, Li HY, Shu CR, Guo ZH (2001) A study on difference of Ph between healthy and pine wilt diseased woods of several pine species. Acta Phytopathol Sin 31:342–348

Ward S, Carrel JS (1979) Fertilization and sperm competition in the nematode *Caenorhabditis elegans*. Dev Biol 73:304–321

Warren JE, Linit MJ (1993) Effect of *Monochamus carolinensis* on the life history of the pinewood nematode, *Bursaphelenchus xylophilus*. J Nematol 25:703–709

Warren JE, Edwards OR, Linit MJ (1995) Influence of bluestain fungi on laboratory rearing of pinewood nematode infested beetles. Fundam Appl Nematol 18:95–98

Watanabe A, Isoda K, Higashihara T, Ozawa H (2007) Gene expression of Japanese black pine inoculated with pine wood nematode (in Japanese). Trans Ann Mtg Jpn For Soc 118:O19

Watanabe A, Isoda K, Kondo T, Ozawa H (2006a) EST analysis of Japanese black pine (in Japanese). Trans Jpn For Soc 117:L07

Watanabe A, Iwaizumi MG, Ubukata M, Kondo T, Lian C, Hogetsu T (2006b) Isolation of microsatellite markers from *Pinus densiflora* Sieb. et Zucc. using a dual PCR technique. Mol Ecol Notes 6:80–82

Watanabe A, Isoda K, Higashihara T, Ozawa H (2007) Gene expression of Japanese black pine inoculated with pine wood nematode (in Japanese). Trans Jpn For Soc 118:O19

Webster JM, Anderson RV, Baillie DL, Beckenbach K, Curran J, Rutherford TA (1990) DNA probes for differentiating isolates of the pinewood nematode species complex. Rev Nématol 13:255–263

Wen YH, Feng ZX, Xu HH, Chen L (2001) Screening for nemaicidal activity of some Chinese plant extracts against plant parasitic nematodes, *Bursaphelenchus xylophilus*, *Meloidogyne arenaria* and *Hirschmanniella oryzae*. J Huazhong Agric Univ 20:235–238

Westermark U, Hardell HL, Iversen T (1986) The content of protein and pectin in the lignified middle lamella primary wall from spruce fibers. Holzforschung 40:65–68

White JG, Southgate E, Thomson JN, Brenner S (1986) The structure of the nervous system of the nematode *Caenorhabditis elegans*. Philos Trans R Soc Lond B Biol Sci 314:1–340

Wiegner O, Schierenberg E (1999) Regulative development in a nematode embryo: a hierarchy of cell fate transfirnations. Dev Boil 215:1–12

Williams DJ (1980) First record of pinewood nematode transmission by cerambycid adults to red pine. Coop Plant Pest Rep 33:627

Wingfield MJ (1983) Transmission of pine wood nematode to cut timber and girdled trees. Plant Dis 67:35–37

Wingfield MJ (ed) (1987a) Pathogenicity of pine wood nematode. APS Press, The American Phytopathological Society, St. Paul, Minnesota

Wingfield MJ (1987b) Fungi associated with the pine wood nematode, *Bursaphelenchus xylophilus*, and cerambycid beetles in Wisconsin. Mycologia 79:325–328

Wingfield MJ, Blanchette RA (1983) The pine-wood nematode, *Bursaphelenchus xylophilus*, in Minnesota and Wisconsin: insect associates and transmission studies. Can J For Res 13:1068–1076

Wingfield MJ, Blanchette RA, Nichollas TH, Robbins K (1982) Association of pine wood nematode with stressed trees in Minnesota, Iowa, and Wisconsin. Plant Dis 66:934–937

Wingfield MJ, Blanchette RA, Kondo E (1983) Comparison of the pine wood nematode *Bursaphelenchus xylophilus* from pine and balsam fir. Eur J For Path 13:360–372

Wingfield MJ, Bedker PJ, Blanchette RA (1986) Pathogenicity of *Bursaphelenchus xylophilus* on pines in Minnesota and Wisconsin. J Nematol 18:44–49

Winter SMJ, Rajcan I, Shelp BJ (2005) Soybean cyst nematode: Challenges and opportunities. Can J Plant Sci 86:25–32

Wouts WM (1991) *Steinernema* (*Neoaplectana*) and *Heterorhabditis* species. In: Nickle WR (ed) Manual of agricultural nematology. Marcel Dekker, New York, pp 855–897

Xie C, Chen MR (2003) Application of bag-fumigation measure to kill pests on the dead woods caused by the pine nematode disease. J Guangdong For Sci Technol 19:53–54

Xie C, Xu S (1998) Efficiency of multistage marker-assisted selection in the improvement of multiple quantitative traits. Heredity 80:489–498

Xie LQ (2003) Pine wood nematode-carrying bacteria and their dynamics in host and their effects on the host plant. Nanjing, China, PhD Thesis, Nanjing For Univ, China, pp 1–71

Xie LQ, Ju YW, Zhao BG (2004) Dynamics of populations of nematode and bacteria in the process of pine wilt disease (in Chinese with English abstract). Scientia Silvae Sinicae 40:124–129

Xie LQ, Ju YW, Yang ZD, Zhao BG (2005) Dynamics of densities of bacteria and nematode in the branches of *Pinus thunbergii* inoculated with *Bursaphelenchus xylophilus* (in Chinese with English abstract). J Zhejiang For Univ 22:310–314

Xu FY, Xi K, Yang BJ (1991) Effective of hot-water and high temperature treatment for control of the pine wood nematode and the vector of pine wilt disease died wood. Scientia Silvae Sinicae 27:179–185

Xu FY, Yang BJ, Ge MH (1994a) Studies on the emergence, replenishing feeding methods of adults of *Monochamus alternatus* in the area of Nanjing and its control. For Res 7:215–219

Xu FY, Xi K, Xu G, Zhou YS, Xu WL, Xu KQ, Ge MH (1994b) Study on the resistances of various year classes of *Pinus massoniana* to pine wood nematode (PWN), *Bursaphelenchus xylophilus*. J Nanjing For Univ 18(3):27–33

Xu FY, Ge MH, Zhao ZD, Zhu KG (1998) Studies on the relationship of amino acid content of different masson pine provenances and their resistance to pine wood nematode (PWN). For Res 11:313–318

Xu FY, Ge MH, Zhang P, Zhao Z, Sun Z (1999) Studies on provenance of masson pine resistance to pine wood nematode (PWN) and resistance mechanisms. In: Futai K, Togashi K, Ikeda T (eds) Sustainability of pine forests in relation to pine wilt and decline. Proceedings of the international symposium, Tokyo, 27–28 October 1998, Shokado, Kyoto, Japan, pp 213–216

Xu FY, Ge MH, Zhang P, Zhao ZD, Sun Z (2000) Studies on resistance mechanisms of masson pine provenance resistance to pine wood nematode (PWN). J Nanjing For Univ 24:85–88

Xu FY, Ge MH, Xu KQ, Zhang P, Xie CX (2004) Studies on the techniques of integrated management to PWN in Jiangsu. In: Proceedings of the international symposium on ecology and management of pine wilt disease. Seoul, 7–9 October 2004, pp 63–75

Xu FY, Xu KQ, Xie CX, Z Pei (2008) Studies on the *Scleroderma guani* Xiao et Wu to control the pine wood nematodes. In: Mota M, Vieira P (eds) Pine wilt disease: a worldwide threat to forest ecosystem. Springer, The Netherlands, pp 51–52

Xu KQ, Xu FY (2002) The techniques of *Scleroderma guani* Xiao et Wu to control pine sawyer beetles. J Nanjing For Univ (Nat Sci Edn) 26:48–53

Yamada T (1987) Lipid peroxidation during the development of pine wilt disease. Ann Phytopathol Soc Jpn 53:523–530

Yamada T (2006) Biochemical responses in pines infected with *Bursaphelenchus xylophilus* (in Japanese with English abstract). J Jpn For Soc 88:370–382

Yamada T, Ito S (1993a) Histological observations on the response of pine species, *Pinus strobes* and *P. taeda*, resistant to *Bursaphelenchus xylophilus* infection. Ann Phytopathol Soc Jpn 59:659–665

Yamada T, Ito S (1993b) Chemical defense responses of wilt-resistant pine species, *Pinus strobus* and *P. taeda*, against *Bursaphelenchus xylophilus* infection. Ann Phytopathol Soc Jpn 59:666–672

Yamada T, Mineo K, Suzuki K (1984) Isozyme patterns of peroxidase and polyphenol oxidase in Japanese black pine inoculated with the pine wood nematode (in Japanese). Trans Mtg Jpn For Soc 95:469–470

Yamamoto N, Takasu F, Kawasaki K, Togashi K, Kishi Y, Shigesada N (2000) Local dynamics and global spread of pine wilt disease (in Japanese). Jpn J Ecol 50:269–276

Yamanaka S (1993) Field control of the Japanese pine sawyer, *Monochamus alternatus* (Coleoptera: Cerambycidae) larvae by *Steinernema carpocapsae* (Nematoda: Rhabditida). Jpn J Nematol 23:71–78

Yamane A (1974a) Changes in head capsule width and body weight of *Monochamus alternatus* larvae (in Japanese). Trans Mtg Jpn For Soc 85:234–235

Yamane A (1974b) Pupal period of *Monochamus alternatus* and changes in morphology and body weight after eclosion (in Japanese). Trans Mtg Jpn For Soc 85:239–240

Yamane A, Akimoto T (1974) Observation of feeding behavior of *Monochamus alternatus* adults (in Japanese). Trans Mtg Jpn For Soc 85:246–247

Yamasaki T, Hata K, Okamoto H (1980) Luring of Japanese pine sawyer *Monochamus alternatus* Hope by paraquat-treated pine trees (I) (in Japanese). J Jpn For Soc 62:99–102

Yamate K, Okubo T (1977) Relationship between the number of inoculation spots and the mortarity (in Japanese). Trans Kyushu For Assoc 30:63–64

Yan AJ, Zhang CF, Liu FG, Tang JG, Xia CS, Liang H (1999) A study on the processing technique and initial use of contacted-breaking release microcapsules of pesticides. J Nanjing For Univ 23:65–68

Yan Y, Smant G, Stokkermans J, Qin L, Helder J, Baum T, Schots A, Davis E (1998) Genomic organization of four beta-1,4-endoglucanase genes in plant-parasitic cyst nematodes and its evolutionary implications. Gene 220:61–70

Yang BJ (1995) The pine wood nematode in China. In: Proceedings of the international symposium on pine wilt disease caused by pine wood nematode. Beijing, 31 October–5 November 1995, pp 207–209

Yang BJ (2002) Advance in research of pathogenetic mechanism of pine wood nematode (in Chinese with English abstract). For Pest Dis (1):27–31

Yang BJ (2004) The history, dispersal and potential threat of pine wood nematode in China. In: Mota M, Vieira P (eds) The pinewood nematode, *Bursaphelenchus xylophilus*. Proceeding of an international workshop, University of Évora, Portugal, 20–22 August 2001. Brill, Leiden-Boston, pp 21–24

Yang B, Wang Q (1989) Distribution of the pinewood nematode in China and susceptibility of some Chinese and exotic pines to the nematode. Can J For Res 19:1527–1530

Yang BJ, Pan HY, Tang J, Wang YY, Wang LF (2003) Pine wilt disease. Chinese Forestry Press, Beijing

Yang X, Huang JS, He XY, Kang WT, Li WX, Tang CS (2005) Experiment of controlling *Monochamus alternatus* larva with *Sxleroderma guani* and its carried fungus in door. J Fujian For Sci Tech 3(32):94–99

Yang ZQ (2004) Advance in bio-control researches of the important forest insect pests with natural enemies in China. Chin J Biol Control 20:221–227

Yano S (1913) Investigation on pine death in Nagasaki prefecture (in Japanese). Sanrin-Kouhou 4:1–14

Ye W, Giblin-Davis RM, Braasch H, Morris K, Thomas WK (2007) Phylogenetic relationships among *Bursaphelenchus* species (Nematoda: Parasitaphelenchidae) inferred from nuclear ribosomal and mitochondrial DNA sequence data. Mol Phylogen Evol 43:1185–1197

Ye YH, Yu HB, Lin XP (2005) Current situation and control strategy of pine wood nematode disease in Guangdong Province. Guangdong For Sci Tech 21:73–75

Yi CK, Byun BH, Park JD, Yang SI, Chang KH (1989) First finding of the pine wood nematode, *Bursaphelenchus xylophilus* (Steiner et Buhrer) Nickle and its insect vector in Korea. Res Rep For Res Inst 38:141–149

Yin K, Fang Y, Tarjan AC (1988) A key to species in the genus *Bursaphelenchus* with a description of *Bursaphelenchus hunanensis* sp. n. (Nematoda: Aphelenchoididae) found in pine wood in Hunan Province, China. Proc Helminthol Soc Washington 55:1–11

Yin YH, Guo DS, Zhao BG, Du XH, Zhao ZT, Li RG (2007) Primary study on the toxins secreted by *Pseudomonas fluorescens* GcM5–1A carried by pine wood nematodes (in Chinese with English abstract). Chin Bull Bot 24:147–153

Yokoi N (1989) Observation on the mating behavior of the yellow spotted longicorn beetle, *Psacothea hilaris* Pascoe (Coleoptera: Cerambycidae) (in Japanese with English abstract). Jpn J Appl Entomol Zool 33:175–179

Yoshida N (2005) The necessary control ratio to prevent pine wilt disease (in Japanese). Shinrin-boeki (For Pests) 54:2–6

Yoshida N (2006) A strategy for controlling pine wilt disease and its application on-site (in Japanese with English abstract). J Jpn For Soc 88:422–428

Yoshikawa K (1987) A study of the subcortical insect community in pine trees II. Vertical distribution. Appl Entomol Zool 22:195–206

Yoshikawa K, Takeda H, Soné K, Shibata E (1986) A study of the subcortical insect community in pine trees I. Oviposition and emergence periods of each species. Appl Entomol Zool 21:258–268

Yoshimura A, Kawasaki K, Takasu F, Togashi K, Futai K, Shigesada N (1999) Modeling the spread of pine wilt disease caused by nematodes with pine sawyers as vector. Ecology 80:1691–1702

You JX, Song Z, He WL, Wang YY, Song YS (1994) Study on distinguishing between pine wilt diseased wood and sound wood. Scientia Silvae Sinicae 30:145–150

Yui M (1980) Birds prey on *Monochamus alternatus* (in Japanese). Shinrin-boeki (For Pests) 29:34–36

Yui M, Suzuki S, Aoyama I (1985) Cases of utilization of logs for nidification by woodpeckers (in Japanese). Trans Ann Mtg Tohoku Br Jpn For Soc 37:202–204

Yui M, Suzuki S, Nakamura M (1993) Woodpeckers as a predator of *Monochamus alternatus* and their conservation (in Japanese). Shinrin-boeki (For Pests) 42:105–109

Zang X, Linit MJ (1998) Comparison of oviposition and longevity of *Monochamus alternatus* and *M. carolinensis* (Coleoptera: Cerambycidae) under laboratory conditions. J Nematol 27:36–41

Zevos S, Johnson SC, Webster JM (1991) Effect of temperature and inoculum size on reproduction and development of *Heterorhabditis heliothidis* and *Steinernema glaseri* (Nematoda: Rhabditoidea) in *Galleria melonella*. Can J Zool 69:1261–1264

Zhang H, Kanzaki H, Kawazu K (1997) Benzoic acid accumulation in the *Pinus thunbergii* callus inoculated with the pine wood nematode, *Bursaphelenchus xylophilus*. Z Naturforsch 52:329–332

Zhang JP, Zhao BG, Xie LQ (2004) Isolation of fungi and bacteria from the dead pine trees caused by *Bursaphelenchus xylophilus* (in Chinese with English abstract). J Nanjing For Univ (Nat Sci Edn) 28:87–89

Zhang L, Song S (1991) Study on control of *Monochamus alternatus* with attractant and *Scleroderma guani*. For Res 4:285–290

Zhang L, Liu J, Wu H (2000) The screening virulent strain of *Beauveria bassiana* to *Monochamus alternatus*. J Nanjing For Univ 24:33–37

Zhang R, Cho HY, Kim HS, Ma YG, Osaki T, Kawabata S, Soderhall K, Lee BL (2003) Characterization and properties of a 1,3-beta-D-glucan pattern recognition protein of *Tenebrio molitor* larvae that is specifically degraded by serine protease during prophenoloxidase activation. J Biol Chem 278:42072–42079

Zhang XF (2005) Strategies to prevent pine wilt disease in Huangshan City (in Chinese). For Pests Dis (1):39–41

Zhang X, Stamps WT, Linit MJ (1995) A nondestructive method of determining *Bursaphelenchus xylophilus* infestation of *Monochamus* spp. vectors. J Nematol 27:36–41

Zhang YN, Yang ZQ (2006) Studies on the natural enemies and biocontrol of *Monochamus alternatus* Hope (Coleopteran: Cerambycidae). Plant Prot 32:9–14

Zhang ZC, Sun JH, Huang BX, Zhang F, OuYang H (2006) Study of the pine wilt disease occurrence and its range expansion pattern based on GIS. J Zhejiang Univ 32:551–556

Zhao BG (1996) Nematicidal activity of aloperine against pine wood nematodes. Scientia Silvae Sinicae 32:243–247

Zhao BG (1999) Nematicidal activity of quinolizidine alkaloids and the functional group pairs in their molecular structure. J Chem Ecol 25:2205–2214

Zhao B, Lin F (2005) Mutualistic symbiosis between *Bursaphelenchus xylophilus* and bacteria of the genus *Pseudomonas*. For Path 35:39–345

Zhao BG, Wu RQ, Li XP (1998) Field tests of controlling the pine wilt disease with the alkaloid, aloperine. Scientia Silvae Sinicae 34:113–117

Zhao BG, Guo DS, Gao R (2000a) Observation of the body site of pine wood nematode where bacteria are carried with SEM and TEM (in Chinese with English abstract). J Nanjing For Univ 24:69–71

Zhao BG, Gao R, Ju YW, Guo DS, Guo J (2000b) Effects of antibiotics on pine wilt disease. J Nanjing For Univ 24:75–77

Zhao BG, Wang HL, Han SF and Han ZM (2003) Distribution and pathogenicity of bacteria species carried by *Bursaphelenchus xylophilus* in China. Nematology 5:899–906

Zhao BG, Liang B, Zhao LG, Xu M (2005) Inference of pine wood nematode on production of phytotoxins of an accompanying pathogenic bacterial strain (in Chinese with English abstract). J Beijing For Univ 27:71–75

Zhao BG, Liu Y, Lin F (2007) Effects of bacteria associated with pine wood nematode (*Bursaphelenchus xylophilus*) on development and egg production of the nematode. J Phytopathol 155:26–30

Zhao JN, Jiang P, Wu CS, Sun SL, Jian LY, Lin CC (2000) Studies on *Monochamus alternatus* attractant and attractability. For Res 13:262–267

Zheng Y, Fukuda K, Suzuki K (1993) Electrolytes leakage from Japanese black pine (*Pinus thunbergii*) tissue inoculated with pinewood nematode (*Bursaphelenchus xylophilus*) (in Japanese). Trans Jpn For Soc 104:639–640

Zhou K, Li X, Li D, Yu S, Liao J (2007) *Bursaphelenchus uncispicularis* n. sp. (Nematoda: Parasitaphelenchidae) from *Pinus yunnanensis* in China. Nematology 9:237–242

Zhou N, Yao SZ, Hu DF, Song SX, Zhai JL (2005) Advances in artificial propagation and applied study of *Scleroderma* guani. Arid Zone Res 22:569–575

Zhou ZH, Sakaue D, Wu BG, Hogetsu T (2007) Genetic structure of populations of the pinewood nematode *Bursaphelenchus xylophilus*, the pathogen of pine wilt disease, between and within pine forests. Phytopathology 97:304–310

Index

Genus Species Index

n

o

p

r